# Lecture Notes in Computer Science 12918

Debin Gao · Qi Li · Xiaohong Guan ·
Xiaofeng Liao (Eds.)

# Information and Communications Security

23rd International Conference, ICICS 2021
Chongqing, China, November 19–21, 2021
Proceedings, Part I

 Springer

*Editors*
Debin Gao ⓘ
Singapore Management University
Singapore, Singapore

Qi Li ⓘ
Tsinghua University
Beijing, China

Xiaohong Guan
Xi'an Jiaotong University
Xi'an, China

Xiaofeng Liao ⓘ
Chongqing University
Chongqing, China

ISSN 0302-9743   ISSN 1611-3349  (electronic)
Lecture Notes in Computer Science
ISBN 978-3-030-86889-5   ISBN 978-3-030-86890-1  (eBook)
https://doi.org/10.1007/978-3-030-86890-1

LNCS Sublibrary: SL4 – Security and Cryptology

This Springer imprint is published by the registered company Springer Nature Switzerland AG
The registered company address is: Gewerbestrasse 11, 6330 Cham, Switzerland

# Preface

This volume contains papers that were selected for presentation and publication at the 23rd International Conference on Information and Communications Security (ICICS 2021), which was jointly organized by Chongqing University, Xi'an Jiaotong University, and Peking University in China during November 19–21, 2021. ICICS is one of the mainstream security conferences with the longest history. It started in 1997 and aims at bringing together leading researchers and practitioners from both academia and industry to discuss and exchange their experiences, lessons learned, and insights related to computer and communication security.

This year's Program Committee (PC) consisted of 141 members with diverse backgrounds and broad research interests. A total of 202 valid paper submissions were received. The review process was double blind, and the papers were evaluated on the basis of their significance, novelty, and technical quality. Most papers were reviewed by four or more PC members. The PC meeting was held online with intensive discussion over more than two weeks. Finally, 49 papers were selected for presentation at the conference giving an acceptance rate of 24%.

A "Best Paper Selection Committee" with five PC members of diverse backgrounds from around the world was formed, which selected the two best papers after a lengthy discussion. The paper "Rethinking Adversarial Examples Exploiting Frequency-Based Analysis" authored by Sicong Han, Chenhao Lin, Chao Shen, and Qian Wang received the Best Paper Award, while the paper "CyberRel: Joint Entity and Relation Extraction for Cybersecurity Concepts" authored by Yongyan Guo, Zhengyu Liu, Cheng Huang, Jiayong Liu, Wangyuan Jing, Ziwang Wang, and Yanghao Wang received the Best Student Paper Award. Both awards were generously sponsored by Springer.

ICICS 2021 was honored to offer two outstanding keynote talks: "Engineering Trustworthy Data-Centric Software: Intelligent Software Engineering and Beyond" by Tao Xie and "Securing Smart Cars – Opportunities and Challenges" by Long Lu. Our deepest gratitude to Tao and Long for sharing their insights during the conference.

For the success of ICICS 2021, we would like to first thank the authors of all submissions and the PC members for their great effort in selecting the papers. We also thank all the external reviewers for assisting the reviewing process. For the conference organization, we would like to thank the ICICS Steering Committee, the general chairs, Xiaohong Guan and Xiaofeng Liao, the publicity chairs, Qingni Shen, Qiang Tang, and Yang Zhang, and the publication chair, Dongmei Liu. Special thanks to Tao Xiang for the local arrangements. Finally, we thank everyone else, speakers, session chairs, and volunteer helpers for their contributions to the program of ICICS 2021.

Last but not least, we wish to extend a huge thank you to healthcare frontliners and our colleagues in the research of vaccine and immunization in fighting COVID-19. ICICS 2021 could not have become one of the first mainstream security conferences returning to an in-person setting without their enormous contribution.

November 2021

Debin Gao
Qi Li

# Organization

## Steering Committee

| | |
|---|---|
| Robert Deng | Singapore Management University, Singapore |
| Dieter Gollmann | Hamburg University of Technology, Germany |
| Javier Lopez | University of Malaga, Spain |
| Qingni Shen | Peking University, China |
| Zhen Xu | Institute of Information Engineering, CAS, China |
| Jianying Zhou | Singapore University of Technology and Design, Singapore |

## General Chairs

| | |
|---|---|
| Xiaohong Guan | Xi'an Jiaotong University, China |
| Xiaofeng Liao | Chongqing University, China |

## Program Committee Chairs

| | |
|---|---|
| Debin Gao | Singapore Management University, Singapore |
| Qi Li | Tsinghua University, China |

## Program Committee

| | |
|---|---|
| Chuadhry M. Ahmed | University of Strathclyde, UK |
| Cristina Alcaraz | University of Malaga, Spain |
| Man Ho Au | The University of Hong Kong, Hong Kong, China |
| Zhongjie Ba | Zhejiang University, China |
| Joonsang Baek | University of Wollongong, Australia |
| Guangdong Bai | The University of Queensland, Australia |
| Jia-Ju Bai | Tsinghua University, China |
| Diogo Barradas | Universidade de Lisboa, Portugal |
| Yinzhi Cao | Johns Hopkins University, USA |
| Guangke Chen | ShanghaiTech University, China |
| Rongmao Chen | National University of Defense Technology, China |
| Songqing Chen | George Mason University, USA |
| Ting Chen | University of Electronic Science and Technology of China, China |
| Xiaofeng Chen | Xidian University, China |
| Xun Chen | Samsung Research America, USA |
| Yaohui Chen | Facebook, USA |
| Sherman S. M. Chow | The Chinese University of Hong Kong, Hong Kong, China |

| | |
|---|---|
| Kangjie Lu | University of Minnesota, USA |
| Bo Luo | The University of Kansas, USA |
| Xiapu Luo | The Hong Kong Polytechnic University, Hong Kong, China |
| Haoyu Ma | Xidian University, China |
| Christian Mainka | Ruhr University Bochum, Germany |
| Daisuke Mashima | Advanced Digital Sciences Center, Singapore |
| Jake Massimo | Amazon Web Services, USA |
| Weizhi Meng | Technical Universtiy of Denmark, Denmark |
| Jiang Ming | UTA, USA |
| Chris Mitchell | Royal Holloway, University of London, UK |
| Yuhong Nan | Purdue University, USA |
| Jianbing Ni | Queen's University, Canada |
| Jianting Ning | Singapore Management University, Singapore |
| Liang Niu | New York University, USA |
| Satoshi Obana | Hosei University, Japan |
| Rolf Oppliger | eSECURITY Technologies, Switzerland |
| Roberto Di Pietro | Hamad Bin Khalifa University, Qatar |
| Joachim Posegga | University of Passau, Germany |
| Giovanni Russello | The University of Auckland, New Zealand |
| Nitesh Saxena | Texas A&M University, USA |
| Shawn Shan | University of Chicago, USA |
| Vishal Sharma | Queen's University Belfast, UK |
| Qingni Shen | Peking University, China |
| Wenbo Shen | Zhejiang University, China |
| Purui Su | CAS, China |
| Hung-Min Sun | National Tsing Hua University, Taiwan, China |
| Kun Sun | George Mason University, USA |
| Willy Susilo | University of Wollongong, Australia |
| Qiang Tang | Luxembourg Institute of Science and Technology, Luxemburg |
| Yuzhe Tang | Syracuse University, USA |
| Luca Viganò | King's College London, UK |
| Binghui Wang | Duke University, USA |
| Cong Wang | City University of Hong Kong, Hong Kong, China |
| Ding Wang | Nankai University, China |
| Gang Wang | University of Illinois at Urbana-Champaign, USA |
| Haining Wang | Virginia Tech, USA |
| Haoyu Wang | Beijing University of Posts and Telecommunications, China |
| Lei Wang | Shanghai Jiao Tong University, China |
| Lingyu Wang | Concordia University, Canada |
| Shuai Wang | The Hong Kong University of Science and Technology, Hong Kong, China |
| Ting Wang | East China Normal University, China |

| | |
|---|---|
| Xiuhua Wang | Huazhong University of Science and Technology, China |
| Zhe Wang | ICT, China |
| Jinpeng Wei | University of North Carolina at Charlotte, USA |
| Weiping Wen | Peking University, China |
| Daoyuan Wu | The Chinese University of Hong Kong, Hong Kong, China |
| Zhe Xia | Wuhan University of Technology, China |
| Xiaofei Xie | Nanyang Technological University, Singapore |
| Dongpeng Xu | University of New Hampshire, USA |
| Jia Xu | NUS-Singtel Cyber Security R&D Lab, Singapore |
| Jun Xu | Stevens Institute of Technology, USA |
| Minhui Xue | The University of Adelaide, Australia |
| Toshihiro Yamauchi | Okayama University, Japan |
| Feng Yan | University of Nevada, Reno, USA |
| Qiben Yan | Michigan State University, USA |
| Guomin Yang | University of Wollongong, Australia |
| Zheng Yang | Southwest University, China |
| Roland Yap | National University of Singapore, Singapore |
| Xun Yi | RMIT University, Australia |
| Qilei Yin | Tsinghua University, China |
| Meng Yu | Roosevelt University, USA |
| Yu Yu | Shanghai Jiao Tong University, China |
| Xingliang Yuan | Monash University, Australia |
| Chuan Yue | Colorado School of Mines, USA |
| Tsz Hon Yuen | The University of Hong Kong, Hong Kong, China |
| Chao Zhang | Tsinghua University, China |
| Fan Zhang | Zhejiang University, China |
| Fengwei Zhang | SUSTech, China |
| Jialong Zhang | ByteDance, China |
| Jiang Zhang | State Key Laboratory of Cryptology, China |
| Kehuan Zhang | The Chinese University of Hong Kong, Hong Kong, China |
| Yang Zhang | CISPA Helmholtz Center for Information Security, Germany |
| Yinqian Zhang | Southern University of Science and Technology, China |
| Lei Zhao | Computer School of Wuhan University, China |
| Qingchuan Zhao | Ohio State University, USA |
| Tianwei Zhang | Nanyang Technological University, Singapore |
| Yuan Zhang | Fudan University, China |
| Yongjun Zhao | Nanyang Technological University, Singapore |
| Yunlei Zhao | Fudan University, China |
| Yajin Zhou | Zhejiang University, China |
| Yongbin Zhou | Chinese Academy of Sciences, China |
| Shuofei Zhu | Pennsylvania State University, USA |

# Additional Reviewers

Isaac Agudo
Md Rabbi Alam
Cristina Alcaraz
Ahsan Ali
Saed Alsayigh
Enkeleda Bardhi
Christof Beierle
Christian Berger
Alessandro Brighente
Cailing Cai
Giovanni Calore
Xinle Cao
Kwan Yin Chan
Jinrong Chen
Long Chen
Min Chen
Tianyang Chen
Tommy Chin
Murilo Coutinho
Andrei Cozma
Handong Cui
Vasiliki Diamantopoulou
Qiying Dong
Minxin Du
Orr Dunkelman
Alexandros Fakis
Pengbin Feng
Ankit Gangwal
Yiwen Gao
Nicholas Genise
Junqing Gong
Qingyuan Gong
Kamil D. Gur
Yonglin Hao
Ke He
Xu He
Jiaqi Hong
Xinyue Hu
Yupu Hu
Mengdie Huang
Huiwen Jia
Xiangkun Jia
Ziming Jiang

Georgios Karopoulos
Maria Karyda
Andrei Kelarev
Minjune Kim
Felix Klement
Vasileios Kouliaridis
Gulshan Kumar
Jianchang Lai
Qiqi Lai
Chhagan Lal
Gregor Leander
Bo Li
Huizhong Li
Shaofeng Li
Wanpeng Li
Yannan Li
Zheng Li
Ziyuan Liang
Kyungchan Lim
Chaoge Liu
Gang Liu
Songsong Liu
Xiaoning Liu
Xueqiao Liu
Yichen Liu
Yiyong Liu
Yuejun Liu
Yunpeng Liu
Zengrui Liu
Eleonora Losiouk
Xin Lou
Junwei Luo
Lan Luo
Xiaolong Ma
Zhou Ma
Ahmed Tanvir Mahdad
Fei Meng
Vladislav Mladenov
William H. Y. Mui
Lucien K. L. Ng
Shimin Pan
Dimitris Papamartzivanos
Bryan Pearson

Henrich C. Pöhls
Hunter Price
Xianrui Qin
Yue Qin
Tingting Rao
Pengcheng Ren
Yujie Ren
Ruben Rios
Shalini Saini
Md Sajidul Islam Sajid
Stewart Santanoe
Shiqi Shen
Siyu Shen
Menghan Sun
Shuo Sun
Azadeh Tabiban
Fei Tang
Jiaxun Steven Tang
Utku Tefek
Guangwei Tian
Guohua Tian
Zhihua Tian
Yosuke Todo
Zisis Tsiatsikas
Payton Walker
Hongbing Wang
Jiafan Wang
Jianfeng Wang
Kailong Wang
Lihchung Wang
Lu Wang
Shu Wang
Ti Wang
Ting Wang
Wenhao Wang

Xinda Wang
Xinying Wang
Yunling Wang
Rui Wen
Mingli Wu
Yi Xie
Guorui Xu
Jing Xu
Shengmin Xu
Bolin Yang
Fan Yang
Hanmei Yang
Shishuai Yang
Wenjie Yang
Xu Yang
Zhichao Yang
Amirhesam Yazdi
Quanqi Ye
Jun Yi
Xiao Yi
Qilei Yin
Pinghai Yuan
Syed Zawad
Zhe Zhao
Zhiyu Zhao
Ziming Zhao
Chennan Zhang
Yuexin Zhang
Yubo Zheng
Ce Zhou
Jin Zhou
Rahman Ziaur
Max Zinkus
Yang Zou
Yunkai Zou

# Sponsors

**Gold Sponsor**

**Silver Sponsors**

# Keynotes

# Engineering Trustworthy Data-Centric Software: Intelligent Software Engineering and Beyond

Tao Xie

Peking University

**Abstract.** As an example of exploiting the synergy between AI and software engineering, the field of intelligent software engineering has emerged with various advances in recent years. Such field broadly addresses issues on intelligent [software engineering] and [intelligence software] engineering. The former, intelligent [software engineering], focuses on instilling intelligence in approaches developed to address various software engineering tasks to accomplish high effectiveness and efficiency. The latter, [intelligence software] engineering, focuses on addressing various software engineering tasks for intelligence software, e.g., AI software. However, engineering trustworthy data-centric software (which AI software components are part of) requires research contributions from compiler, programming languages, formal verification, security, and software engineering besides systems and hardware. This talk will discuss recent research and future directions in the field of intelligent software engineering along with the broad scope of engineering trustworthy data-centric software.

# Securing Smart Cars – Opportunities and Challenges

Long Lu

NIO

**Abstract.** As cars become more intelligent and connected, the security of on-car systems, software, and data has caught heavy attention from academia, industry, and regulators. This talk will discuss the key technical aspects of smart car security, including low-level system security, secure and robust autonomous driving, V2X security, data security, etc., highlighting the research and technical opportunities and challenges.

# Contents – Part I

## Software Security

## Internet Security

**Data-Driven Cybersecurity**

# Contents – Part II

## Security Analysis

## Post-quantum Cryptography

## Applied Cryptography

# Blockchain and Federated Learning

# The Golden Snitch: A Byzantine Fault Tolerant Protocol with Activity

Huimei Liao[1,2,3], Haixia Xu[1,2,3]([✉]), and Peili Li[1,2,3]

[1] State Key Laboratory of Information Security,
Institute of Information Engineering, Beijing, China
{liaohuimei,xuhaixia,lipeili}@iie.ac.cn
[2] Data Assurance and Communication Security Research Center, Beijing, China
[3] School of Cyber Security, University of Chinese Academy of Science, Beijing, China

**Abstract.** The increasing popularity of blockchain-based cryptocurrencies has revitalized the search for efficient Byzantine fault-tolerant (BFT) protocols. Many existing BFT protocols can achieve good performance in fault-free cases but suffer severe performance degradation when faults occur. This is also a problem with DiemBFT. To mitigate performance attacks in DiemBFT, we present an improved BFT protocol with optimal liveness called the Golden Snitch. The core idea is to introduce unbiased randomness in leader selection and improve the voting mechanism to protect honest leaders from being dragged down by the previous leader. The performance of the Golden Snitch is evaluated through experiments, turning out it outperforms DiemBFT in the presence of faults.

**Keywords:** Blockchain · Consensus · Byzantine fault tolerance · Randomness · Certificate

## 1 Introduction

With the advent of Bitcoin [21], cryptocurrencies have been growing in popularity. The cryptocurrency protocol aims to reach a consensus on a distributed real-time public ledger attacked by potential adversaries.

**BFT Consensus.** The existing consensus solutions are classified into two broad categories: Nakamoto consensus and BFT consensus. BFT consensus is commonly used in the *permissioned* blockchain and has the good potential for significant improvements in performance as opposed to Nakamoto consensus, especially with regard to transaction confirmation time, for transactions in BFT consensus are finalized when enough votes are gathered several times. Moreover, a novel blockchain design approach hybridizes PoW with BFT in various ways [1,7,17,22]. Such a hybrid design can accomplish higher performance and scalability in comparison common Nakamoto consensus.

© Springer Nature Switzerland AG 2021
D. Gao et al. (Eds.): ICICS 2021, LNCS 12918, pp. 3–21, 2021.
https://doi.org/10.1007/978-3-030-86890-1_1

The study of consensus in the face of Byzantine failures [8] originated from the Byzantine General problem [19]. Pease et al. [23] first came up with a synchronous solution, which was then optimized by Dolev et al. [12] later, having communication complexity as $O(n^3)$. In 1999, Castro et al. [10] developed PBFT, an efficient leader-based Byzantine agreement protocol, whose stable leader required $O(n^2)$ communication and view-change incurred $O(n^3)$ communication. PBFT has been deployed in multiple systems [6], followed by a series of improved protocols, such as Zyzzyva presented by Kotla et al. [18].

BFT protocols were initially conceived as being deployed in a small-sized system. A renewed focus on these protocols by their applications to the blockchain would pose a challenge involving large-scale communications.

Many methods have been employed to reduce the cost of reaching a consensus in BFT protocols [11,18,25]. More recently, Yin et al. proposed HotStuff [26] by changing the mesh communication network in PBFT to the star-like communication network. It standardizes each phase to simplify the process of view-change and results in the reduction of communication complexity. Besides, it pioneers the chaining paradigm by adopting a brilliant but straightforward commit rule. This idea has made great progress on BFT protocols recently and is adopted in the Diem Blockchain [3,4] by Facebook.

In 2018, Facebook presented a consensus protocol DiemBFT for the Diem Blockchain. DiemBFT is an instance of HotStuff, where the round duration is about three times the network latency. DiemBFT instantiated the pacemaker module of HotStuff through a timeout mechanism and a leader selection mechanism. Paralleling with Dfinity [15] and Algorand [14], DiemBFT injected randomness into the leader selection mechanism by invoking the VRF.

In April 2020, Facebook released an updated version of DiemBFT, DiemBFT 2.0. In this version, validators send their votes to the leader of the next round to accelerate the process of committing. It is a trick to further couple the vote with the view-change, reducing the communication overhead in a round and therefore causing a reduction in round duration - perhaps down to just twice the network latency as soon as the leader is stable. Besides, DiemBFT 2.0 has had an acute analysis of temporary forks in possible scenarios, where the safety is still maintained as a consensus will be reached finally due to the fact that the fork will be dismissed after several rounds. Moreover, it has been noticed that a leader rotates among nodes in the new version, suggesting performance improvement at the cost of randomness in leader selection. While predictable leaders are likely to increase the system risk of forking, compromising the throughput.

**Randomness.** Introducing randomness in leader selection is proved to be a vital component of BFT protocols [5,14,16]. The most commonly used algorithm to generate random numbers in the existing BFT based blockchains is to invoke a verifiable random function (VRF) [20]. There exists a general problem that the adversary can cheat in the leader selection by selectively discarding blocks, compromising the unbiasedness of random numbers. To mitigate this problem, a recent protocol HydRand [24] was proposed, which incorporates the publicly verifiable secret sharing (PVSS) scheme to generate unbiased random numbers.

In Table 1, we summarize the main properties of the proposed work and related works on BFT protocols.

**Table 1.** Comparison with related works.

|  | Network | Comm. complexity | Randomness | Bias-assistance |
|---|---|---|---|---|
| Algorand [14] | Partial-sync | $O(n)$ | VRF | Biased |
| Ouroboros [16] | Sync | $O(n^3)$ | PVSS | Unbiased |
| RBFT [2] | Async | $O(n^3)$ | – | – |
| Hydrand [24] | Sync | $O(n^2)$ | PVSS | Unbiased |
| DiemBFT 2.0 [3] | Partial-sync | Sync $O(n)$ | – | - - |
| the Golden Snitch | Partial-sync | Sync $O(n)$ | PVSS | Unbiased |

**Design Challenge.** Sending votes to the next leader causes two implicit combinations of messages. First, the leader collects the votes for the previous round as well as proposes its proposal in the current round. Second, replicas send their votes and the latest confirmation certificates known to them. Nevertheless, several potential vulnerabilities must be considered as follows:

(a) *"Waste Attack"*. In the chaining paradigm, it is intuitive to allow replicas to change their minds after the first vote while a consensus has not been reached. Therefore, as analyzed in DiemBFT 2.0, the leader can fork the chain at the previous two blocks, causing some uncommitted (but maybe certified) blocks to be abandoned. The forking does causes a waste of blocks, especially for those generated by honest leaders. Furthermore, if the leader in the future round can be predicated, adversaries can attack through the corruption of leaders in every other round to hinder the agreement process by deliberately forking. This could compromise the throughput.

(b) *"Being Dragged Down"*. Operations such as proposal and confirmation of a single block are launched by separate leaders. The leaders of two consecutive rounds can be read as collaborators. If malicious leaders are allowed, the worst case is that the previous leader is dragged down and the latter one.

We will elaborate on more details later in Sect. 3.

**Our Contribution.** Focusing on the potential aforesaid concerns, our proposed consensus protocol, the Golden Snitch[1], made several improvements on Diem-BFT 2.0. Overall, the core idea is to optimize the performance of the protocol by introducing unbiased randomness and resist potential malicious behaviors. Specifically, the main contributions of this paper can be summarised as follows:

---

[1] The Golden Snitch was originally created by writer J.K. Rowling in "Harry Potter". The Golden Snitch, often simply called the Snitch, is the third and smallest ball used in Quidditch. It appears randomly on the court and moves very fast.

(1) We analyzed DiemBFT and found that the adversary can impede the agreement progress through continuous forks, as permitted by the rules.
(2) Given the inevitability of forking in the protocol that embraces the chaining paradigm, we introduce unbiased randomness in leader selection. Similar to Hydrand, we consider utilizing the PVSS scheme to provide continuous unbiased random numbers with fault tolerance in the proposed leader selection mechanism without increasing communication overhead.
(3) We elaborate on an enhancement called "veto certificate" to balance the relationship between the two consecutive leaders to keep each leader isolated and independent of others to a certain extent in order to differentiate between their responsibilities in terms of cooperation.
(4) We report the experimental performance evaluation results in a comparison between the Snitch and DiemBFT, showing that the Snitch outperforms DiemBFT in the presence of faults.

**Paper Structure.** The remainder of this paper is organized as follows. We present the preliminary knowledge in Sect. 2 and then analyze the limitation of DiemBFT and overview the Snitch in Sect. 3. The Snitch is outlined in detail in Sect. 4 and experiment results are presented in Sect. 5. Finally, we discuss the scalability of the Snitch further and conclude the paper in Sect. 6. Noted that the correctness properties, i.e., safety and liveness, of the Snitch are rigorously proved in Appendix A as space is limited.

## 2   Preliminaries

We propose the Snitch in a *permissioned* setting. Assumed a system that is equipped with at least $3f + 1$ nodes $\{P_1, \ldots, P_n\}$ to tolerate $f$ faults. Faults can behave arbitrarily and coordinate to take down the system. Nevertheless, they cannot break cryptographic techniques, showing up as hash functions are resistant to collision and signatures can not be forged. The Snitch proceeds in rounds. In each round, there is a single designated leader (in fact, we typically identify the leader with the round-number $r$ as $l_r$). Furthermore, we adopt the partially synchronous model of Dwork et al. [13], in which there is a known delay bound of message transmission after an unknown moment, called the global stable time (GST). Hence, the Snitch will provide safety all the time while assure liveness when the system becomes synchronous. Specifically, we use the instance of $(f + 1, n)$ Scrape's PVSS [9] as a common building block to generate randomness. It allows a leader to share its secret value among $n$ participants in the system and then ensures that the $f$ faulty nodes cannot collude to recover the secret without requiring the collaboration of a correct node.

Next, in conjunction with two different types of sub-blockchain in Fig. 1(a) and Fig. 1(b), we present several relevant terminologies in existing chain-based BFT protocols [3, 25, 26].

**Block:** The basic data structure is a block $B_r =< r, D_r, CC_k, H(B_r) >$. Transactions are batched into blocks with some predefined ordering as $D_r$. Generally, we use $parent(B_r)$ to refer to the parent of $B_r$ and specify it as $parent(B_r) = B_k (k < r)$. In fact, $B_r$ is chained to its parent block $B_k$ via $CC_k$. $H(B_r)$ is the unique hash digest of $B_r$.

**Vote:** Replica $P_i$ sends a vote $V_i(B_r) =< H(B_r), HCC_i >$ while receiving $B_r$. $H(B_r)$ is the digest of $B_r$ and $HCC_i$ is the confirmation certificate in the highest round the replica maintains, serving as the New-View message in PBFT. Replicas also track the round of its latest vote as $r_v$ (i.e., round $r + 2$ in Fig. 1).

**Confirmation Certificate:** A confirmation certificate (CC) is a set of signed votes for a block by a quorum of replicas, from $n - f = 2f + 1$ (out of $n = 3f + 1$) distinct replicas. A block $B_r$ is certified if there exists $CC_r$ for it. Replicas maintain the locked round $r_l$ that is defined as the round number of the second-previous certified block. As presented in Fig. 1(a), replicas that contribute to the generation of $CC_{r+2}$ remember round $r + 1$ as the locked round.

**Timeout Message:** In the setup, the system initializes a maximum delay $\Delta$ to ensure the duration of each round does not exceed a specific time. Each replica maintains a local timer. The timer for round $r$ is denoted by $Timer_r$. If replicas have not received any message from $l_r$ up to $Timer_r$ expired, they would broadcast their signatures on $r$ to each other.

**Timeout Certificate:** A timeout certificate (TC) is a set of timeout messages. The generation of $TC_r$ implies that most replicas give up on round $r$ and move to round $r + 1$ (i.e., round $r + 2$ in Fig. 1(b)).

**Vote Rules:** To validate a proposal, replicas check its timeliness first according to the following two constraints:

* *Vote in strictly increasing rounds.* Replicas vote for block $B_r$, only if $r > r_v$.
* *Be consistent with the locked round.* Replicas vote for block $B_r$, only if the round of $parent(B_r)$ is no less than $r_l$ .

**Commit Rule:** Take view-change into consideration, three consecutive polls on a block are necessary for committing it. Accordingly, the commit rule in the chaining paradigm can be interpreted as: a block is committed as soon as it has been followed by three consecutive certified blocks.

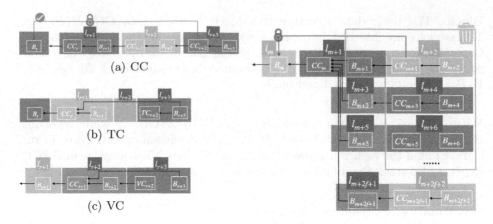

**Fig. 1.** Three manifestations of the chain.

**Fig. 2.** An example of the waste attack. (Color figure online)

## 3  Protocol Overview

This section shows our improvements on DiemBFT 2.0. These amelioration measures could have a real impact on liveness on the premise of ensuring safety.

**Leader Selection.** As discussed earlier, randomness is given greater importance and forking makes it more necessary. According to Fig. 2, if an adversary can corrupt $f$ nodes selectively with the knowledge of the future leader, DiemBFT will be progression-free for $2f$ rounds in the worst-case scenario.

The adversary attacks by corrupting the leader every other round. Without generality loss, assuming that leaders in rounds $m + 3, m + 5, \ldots, m + 2f + 1$ are corrupted. In Fig. 2, $CC_m$ and $CC_{m+1}$ are generated by the honest leaders $l_{m+1}$ and $l_{m+2}$, for certifying blocks $B_m$ and $B_{m+1}$. At this point, at least $2f + 1$ replicas who contributed to the formation of $CC_{m+1}$ remember $m$ as their locked round. Then the corrupted leader $l_{m+3}$ forks the chain at block $B_m$, and causes $B_{m+1}$ and $B_{m+2}$ to be abandoned. Replicas vote it because there is no violation of the vote rules. Owing that $B_{m+3}$ is the highest certified block with the generation of $CC_{m+3}$, the honest leader $l_{m+4}$ forms $CC_{m+3}$ and extends the tail of $B_{m+3}$ with $B_{m+4}$. Similarly, the corrupted leaders $l_{m+2a+1}$ $(2 \leqslant a \leqslant f)$ still fork the chain at block $B_m$ (the gray area in Fig. 2). Finally, the blocks (marked in red box of Fig. 2) generated by honest leaders during $2f$ rounds after round $m$ will be wasted. And no block will be committed. What's worse is that this attack is difficult to detect, because they behave normally.

So it is dreadful if leaders in the future rounds are predicted, facilitating attacks on the targeted leaders. Randomness has a pivotal role in Diem.

The Snitch uses PVSS as an underlying primitive to generate a consistent and unbiased random number in each round for leader selection. Each node maintains a roundup of the possible leader that would change regarding the discovery of malicious nodes. The set of nodes with leader candidacy is represented by $L_r$ in round $r$. It is stipulated that the leader should reveal the previously committed secret value and attach the next commitment simultaneously as a preparation for the subsequent selection. A corrupted leader may decide not to reveal its secret in time. Therefore, replicas enter the reconstruction phase of PVSS where they broadcast their decrypted secret shares and the corresponding correctness proofs. Upon the receipt of $f + 1$ secret shares, the secret can then be recovered to decide on the next leader $l_{r+1}$.

Penalties are meted out to faults as they are being excluded from the eligible set of leaders in future rounds. It is also be adapted to facilitate that temporarily failed nodes could rejoin $f + 1$ rounds after publishing their fresh commitments.

**Veto Certificate.** Owing to the separation of the proposer and confirmer of a block, the malicious leader can publish an invalid block to drag the next leader down. If replicas simply fail to send votes for the invalid block proposed by leader $l_r$, it will be impossible for the leader $l_{r+1}$ to collect enough votes. Therefore $l_{r+1}$ will be forced to lead a timeout in round $r + 1$. This wastes an opportunity to propose and deprives the innocent leader of candidacy.

It will be unfair for a leader to be punished for cooperating with a malicious leader. To address it, we add a field *Attitude* in vote. *Attitude* is a Boolean value filled with *YES* or *NO* on behalf of the opinion of the replica. To reject the proposal, replicas send negative votes instead of being silenced. Likewise, enough negative votes can form a veto certificate $VC_r$, identifying $B_r$ as an *invalid block* (i.e., $B_{r+2}$ in Fig. 1(c)) whose dataset would be passed over. In this way, replicas advance to the next round without waiting for the timer to expire, accelerating the protocol process. And balance is established between independence and cooperation for leaders in two consecutive rounds.

## 4   The Golden Snitch Protocol

This section gives a complete and detailed description of the Snitch protocol.

On the whole, the Snitch proceeds in a pipeline of rounds, aiming to commit blocks in sequence. There is a designated leader in each round to proposing a block. Replicas send their potential positive votes or negative votes to the next leader and enter the next round. As soon as a certificate is formed, the leader of the next round publishes its proposed block that comes with the certificate it forms. To ensure that the system can continue progress, if a quorum of replicas suspects that the current leader is faulty, then a recovery occurs. Replicas obtain the next leader through the proposed leader selection mechanism so that the next leader will not be affected by the current leader.

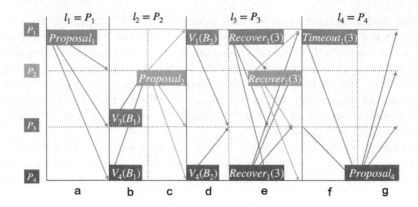

**Fig. 3.** The overview of the consensus process in the Snitch.

To gain more insight, the protocol is specified as operations triggered by messages or timer. Specifically, the system is initiated as shown in Algorithm 1. The algorithms for replicas are defined as Algorithm 2 and Algorithm 3, describing how replicas perform in a round led by an honest leader or in a timeout round respectively. The algorithm for the leader is presented in Algorithm 4. To simplify the description, Table 2 lists the notations used later.

In the Snitch, it is assumed that the round segmentation rests on a leader's tenure. To make this same point a little more visually, Fig. 3 shows four rounds

**Table 2.** Notations used in Algorithms.

| Notation | Meaning |
|---|---|
| $A$ | All replicas |
| $m$ | All transmitted messages, e.g. block, vote, certificate |
| $round(m)$ | The round in which message $m$ is generated |
| $share_i(r)$ | $P_i$'s decrypted share of the secret value committed by $l_r$ previously |
| $MP$ | The malicious proofs of corrupted leaders |
| $F_R()$ | Random number generation function(used once in setup) |
| $s$ | Secret value |
| $H()$ | Cryptographic hash function |
| $Dist()$ | Secret share generation |
| $Dec()$ | Secret share decryption |
| $Rec()$ | Secret reconstruction |
| $VerShare()$ | Secret share verification |
| $VerSig()$ | Signature verification |
| $VerData()$ | Dataset verification |

of an example execution of the Snitch, where $l_r = P_r$ and $l_3$ is faulty. In round 1, $l_1$ proposes $Proposal_1$ that includes $B_1$ and its revealed secret $s_1$. Replicas then send votes for $B_1$ to $l_2$. Next, $l_2$ collects votes to generate the certificate ($CC_1$ or $VC_1$) for $B_1$ and proposes $Proposal_2$. Replicas vote for $B_2$ and move to round 3. While at round 3, $l_3$ experiences a temporary disruption. Round 3 times out and a reconstruction occurs. Replicas broadcast the decrypted secret shares to recover the secret $l_3$ committed before and advance to round 4. Therefore, $l_4$ collects timeout messages to generate $TC_3$ for round 3 and extends the chain.

## 4.1 Setup

For the setup, it is assumed that each participant $P_i$ will be part of the set of initial potential leaders $L_0$. They mark a fixed genesis block as $B_0$ and exchange their public keys $pk_i$ with commitments $Com(s_i)$ to initial selected secrets.

It should be pointed out that they agree on the random beacon $R_0$ that becomes public knowledge only after the set of commitments was defined. $R_0$ is used to select the leader of round 1 that can be obtained via PoW, the method in [16] and so on. After the first round, replicas enter rounds upon the receipt of a certificate for the previous round.

## 4.2 Replicas Vote in an Honest Round

**Advance to the Next Round.** While receiving a proposal from $l_r$, replicas first validate its integrity and validity. The replica accepts the proposal provided that it is constructed properly as $\langle Propose, B_r, MP, CC_{r-1}, s_l, Com(s'_l) \rangle_{\sigma_l}$. Then replicas stop $Timer_r$ and advance to round $r + 1$.

**Process the Certificates Included in the Proposal.** The local states of the replicas are updated according to the received certificates as follows:

– Remove faulty nodes from $L_r$ according to the VC or TC in $MP$, if $MP \neq \bot$.
– Update $r_l$ it maintains and commit block by the commit rule if a CC occurs.

---

**Algorithm 1:** Setup - for the system:

---

1  $r \leftarrow 0$, $L_0 \leftarrow A$, $B_0 \leftarrow GenesisBlock$, $(pk_i, sk_i) \leftarrow KeyGen(1^\lambda)$
2  **for** $P_i \in A$ **do**
3  $\quad$ **selects** a secret $s_i$,
4  $\quad$ **broadcasts** $(pk_i, Com(s_i))$, where $Com(s_i) \leftarrow Dist(s_i)$
5  **end**
6  $R_0 \leftarrow F_R(1^\lambda)$

---

---

**Algorithm 2:** Vote - for the replica $P_i$:

1   **Initialization:**
2   $l \leftarrow l_r$                             { the current leader is the leader of round $r$}
3   $SS_{s_l} \leftarrow \perp$                      { set of decrypted secret shares for $s_l$}
4   $L_{r-1}$                          { set of potential leaders for round $r$ }
5   $R_{r-1}$                         { the random beacon in round $r-1$}

6   **on event** $\langle Propose, B_r, MP, C_{r-1}, s_l, Com(s'_l)\rangle_{\sigma_l}$,
     where $B_r = \langle r, D_r, CC_k, H(B_r)\rangle_{\sigma_l}$
7   **if** $VerSig(pk_l, \sigma_l) = 1 \wedge r > r_v$ **then**
8      **stop** $Timer_r$    // update local states
9      **if** $MP \perp$ **then**
10       | $L_r \leftarrow L_{r-1}$
11      **else**
12       | $L_r \leftarrow L_{r-1} - \{l_{k+1}, l_{k+2}, \ldots, l_{r-1}\}$
13      **end**
14      **if** $k > round(HCC_i)$ **then**
15        $HCC_i \leftarrow CC_k$
16        **if** $round(parent(B_k)) > r_l$ **then**
17          $r_l \leftarrow round(parent(B_k))$
18          **if** $round(parent(B_k)) = k-1 \wedge round(parent(B_{k-1})) = k-2$ **then**
19           | **commit** $B_{k-2}$
20
21
22      **end**
23      $R_r = H(s_l \parallel R_{r-1})$,    $l_{r+1} = l_{R_r mod |L_r|}$    // vote for the block
24      **if** $round(parent(B_r)) \geqslant r_l$ **then**
25        **if** $VerData(B_r) = 1$ **then**
26         | $V_i(B_r) = \langle Vote, H(B_r), YES, HCC_i\rangle_{\sigma_i}$
27        **else**
28         | $V_i(B_r) = \langle Vote, H(B_r), NO, HCC_i\rangle_{\sigma_i}$
29        **end**
30        send $V_i(B_r)$ to $l_{r+1}$, $r_v = r$ and **start** $Timer_{r+1}$
31      **end**
32 **end**

---

**Send Votes to the Leader in the Next Round.** To make a vote for block $B_r$, it is necessary for replicas to obtain the leader $l_{r+1}$ first. They select the leader $l_{r+1}$ from $L_r$, depending on the random beacon $R_r$ that is computed from $R_{r-1}$ and the secret $s_l$ revealed in proposal:

$$R_r = H(s_l \parallel R_{r-1}) \tag{1}$$

$$l_{r+1} = l_{R_r mod |L_r|} \tag{2}$$

Then replicas check its validity, mainly from the following several aspects:

– Check whether the proposal is in a timely manner according to vote rules.

- Check whether the digest of $B_r$ is computed correctly.
- Check the validity of $D_r$.

For a valid proposal, the replica constructs a positive or negative vote as $\langle Vote, H(B_r), YES/NO, HCC_i \rangle_{\sigma_i}$ and sends to the leader $l_{r+1}$ (Fig. 3, b, d).

As described before, the vote message also contains the highest confirmation certificate $HCC_i$ the replica maintains, providing branch choices for leader $l_{r+1}$ to extend. As the close of the polling, replicas update $r_v$ and start $Timer_{r+1}$.

---

**Algorithm 3:** Recover - for the replica $P_i$:

---

1 **Initialization:**
2 $\quad l \leftarrow l_r$                                  { the current leader is the leader of round $r$}

3 **if** $Timer_r$ *expired* **then**
4 $\quad\quad$ $share_i(r) \leftarrow Dec(sk_i, Com(s_l))$ // local timeout
5 $\quad\quad$ **send** $Recover_i(r) = \langle Recover, r-1, share_i(r) \rangle_{\sigma_i}$ to A
6 **end**
7
8 $\quad$ **on event** $Recover_j(r)$
9 **if** $VerShare\left(share_j(r), Com(s_l)\right) = 1$ **then**
10 $\quad\quad$ $SS_{s_l} \leftarrow SS_{s_l} \cup \sigma_j$
11 **end**
12 **if** $|SS_{s_l}| = f + 1$ **then**
13 $\quad\quad$ $s_l \leftarrow Rec(SS_{s_l})$ // recover the secret
14 $\quad\quad$ $l_{r+1} = l_{R_r \bmod |L_r|}$, where $R_r = H(s_l \parallel R_{r-1})$
15 $\quad\quad$ **send** $Timeout_i(r) = \langle Timeout, r, HCC_i \rangle_{\sigma_i}$ to $l_{r+1}$,
16 $\quad\quad$ $r_v = r$ and **start** $Timer_{r+1}$
17 **end**

---

### 4.3 Replicas Recover in a Timeout Round

The current leader is malicious or a benign crash are some reasons for a timeout. Replicas can skip the timeout round and keep the proceedings remain unaffected.

**Local Timeout Triggers the Recovery.** The replica $P_i$ moves into the recovery phase when $Timer_r$ expired without receiving the proposal from $l_r$. To this end, $P_i$ uses $sk_i$ to compute its decrypted share $share_i(r)$ from the commitment that $l_r$ committed before and broadcasts $Recover_i(r)$ to others (Fig. 3, e).

**Recover the Secret and Inform the New Leader.** After receiving more than $f + 1$ decrypted shares for $s_{l_r}$, replicas recover it consistently. Afterwards, replicas send $Timeout_i(r)$ to $l_{r+1}$ (Fig. 3, f), who is obtained from the recovered secret. Similarly, replicas update $r_v$ and start $Timer_{r+1}$, moving to round $r + 1$.

---

**Algorithm 4:** Propose - for the leader $l_r$ in round $r$:

---

1 **Initialization:**
2 $l \leftarrow l_r$                                                    { the current leader is the leader of round $r$}
3 $CC_{r-1} \leftarrow \perp$                                          {set of positive votes for round $r-1$}
4 $VC_{r-1} \leftarrow \perp$                                          {set of negative votes for round $r-1$}
5 $TC_{r-1} \leftarrow \perp$                                          {set of timeout messages for round $r-1$}
6 $MP \leftarrow \perp$                                                {set of TC or VC }

7 **select** a new secret $s'_l$ and **computes** $Com(s'_l) \leftarrow Dist(s'_l)$

8 **on event** $V_i(B_{r-1}) = \langle Vote, H(B_{r-1}), YES, HCC_i \rangle_{\sigma_i}$
9     $CC_{r-1} \leftarrow CC_{r-1} \cup \sigma_i$   // collect positive votes
10    **if** $|CC_{r-1}| = 2f + 1$ **then**
11        send $\langle Propose, B_r, CC_{r-1}, s_l, Com(s'_l) \rangle_{\sigma_l}$ to $A$,
          where $B_r = \langle r, D_r, CC_{r-1}, H(B_r) \rangle_{\sigma_l}$
12        $s_l \leftarrow s'_l$
13 **end**
14
15 **on event** $V_i(B_{r-1}) = \langle Vote, H(B_{r-1}), NO, HCC_i \rangle_{\sigma_i}$
16    $VC_{r-1} \leftarrow VC_{r-1} \cup \sigma_i$    // collect negative votes
17    **if** $|VC_{r-1}| = 2f + 1$ **then**
18        $CC_k \leftarrow \left( \underset{V_i(B_{r-1}) \in VC_{r-1}}{argmax} \{V_i(B_{r-1}).round(HCC_i)\} \right).HCC_i$
19        $k \leftarrow round(CC_k)$
20        **if** $k < r - 2$ **then**
21        $\quad$ | $\quad$ $MP \leftarrow \{C_{k+1}, C_{k+2}, \ldots, C_{r-2}, VC_{r-1}\}$
22        **end**
23        send $\langle Propose, B_r, MP, VC_{r-1}, s_l, Com(s'_l) \rangle_{\sigma_l}$ to $A$,
          where $B_r = \langle r, D_r, CC_k, H(B_r) \rangle_{\sigma_l}$
24        $s_l \leftarrow s'_l$
25 **end**
26
27 **on event** $Timeout_i(r) = \langle Timeout, r, HCC_i \rangle_{\sigma_i}$
28    $TC_{r-1} \leftarrow TC_{r-1} \cup \sigma_i$   // collect timeout messages
29    **if** $|TC_{r-1}| = 2f + 1$ **then**
30        $CC_k \leftarrow \left( \underset{Timeout_i(r) \in TC_{r-1}}{argmax} \{round(Timeout_i(r).HCC_i)\} \right).HCC_i$
31        $k \leftarrow round(CC_k)$
32        **if** $k < r - 2$ **then**
33        $\quad$ | $\quad$ $MP \leftarrow \{C_{k+1}, C_{k+2}, \ldots, C_{r-2}, TC_{r-1}\}$
34        **end**
35        send $\langle Propose, B_r, MP, TC_{r-1}, s_l, Com(s'_l) \rangle_{\sigma_l}$ to $A$,
          where $B_r = \langle r, D_r, CC_k, H(B_r) \rangle_{\sigma_l}$
36        $s_l \leftarrow s'_l$
37 **end**

---

### 4.4   A Leader Proposes Proposal

In particular, $l_1$ is only responsible for proposing $B_1$ on the basis of $B_0$ (Fig. 3, a), given the fact that $B_0$ is agreed upon as part of the setup for the protocol. Later, $l_r(r > 1)$ is required to finish the following tasks (Fig. 3, c, g):

**Commit to a New Secret.** As a leader of round $r$, $l_r$ selects a new secret and computes its commitment while revealing its old secret. The commitment will be in public, containing $n$ secret shares and the corresponding correctness proof.

**Generate the Certificate.** The leader $l_r$ also collects positive votes, negative votes or timeout messages for round $r - 1$ and then forms the corresponding certificate $C_{r-1}$ as $CC_{r-1}$, $VC_{r-1}$, or $TC_{r-1}$.

**Propose its Proposal.** There are different designs of the proposal depending on the certificate for the previous round as follows:

$CC_{r-1}$: With the formation of $CC_{r-1}$, the leader $l_r$ extends the tail of $B_{r-1}$ with a new proposal as embedding $CC_{r-1}$ in $B_r$. The proposal is constructed as $\langle Propose, B_r, CC_{r-1}, s_l, Com(s'_l)\rangle_{\sigma_l}$, where $B_r = \langle r, D_r, CC_{r-1}, H(B_r)\rangle_{\sigma_l}$.

$VC_{r-1}$ **or** $TC_{r-1}$: In case that the leader behaves abnormally in the previous round, the leader $l_r$ chooses the highest confirmation certificate from votes it collected, denoted by $CC_k$. Then $l_r$ packages transactions into $B_r$ based on the block that $CC_k$ certified. Concerning the situation where leaders corrupted in several consecutive rounds, $l_r$ piggybacks the certificates (VC or TC) for rounds from $k + 1$ to $r - 1$ for the sake of round continuity. If no such intermediate round exists, then this set only contains a single value. Hence, the proposal will be constructed and signed as $\langle Propose, B_r, MP, VC_{r-1}/TC_{r-1}, s_l, Com(s'_l)\rangle_{\sigma_l}$, where $B_r = \langle r, D_r, CC_k, H(B_r)\rangle_{\sigma_l}$ and $MP = \{C_{k+1}, \ldots, C_{r-1}\}$.

## 5   Performance

This section evaluates the performance of the Snitch from two perspectives: throughput and latency. Throughput refers to the number of transactions that can be processed by replicas per second while latency is the time duration between the sending of a request and the completion of the request at clients. n order to make the experimental results more obvious,a series of comparative experiments are conducted to compare DiemBFT 2.0(short for Diem) and the Snitch in different parameters and conditions.

**Experimental Setup.** All our experiments are conducted over Aliyun ECS where replicas were executed on a hfc7.8xlarge instance with 32 vCPUs supported by Intel Xeon Platinum 8369HB processors, 64GiB RAM, and Ubuntu 18.04 as OS. Our implementation is an adaptation of the open-source implementation of Diem [3]. We modify the Diem code in primarily the consensus module. Specifically, the implementation of PVSS uses the P256R1 elliptic curve.

**Baseline.** We compare with Diem, for the reason that the Snitch shares the same code base as Diem enabling a fair comparison and Diem is an acknowledged implementation of HotStuff. Concretely, Diem is a payment system and includes many components that are not the focus of our evaluations. The purpose of our experiments is to compare the performance of two consensus algorithms. Thus, to have a fair comparison, our implementation is distinguished from Diem by the consensus module.

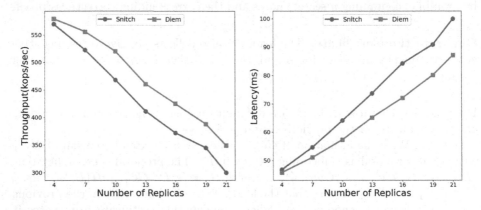

**Fig. 4.** Throughput and latency as the network size varies in fault-free cases

**Implementation Details.** We deploy our private blockchain with the number of replicas assigned to be 4, 7, 10, 13, 16, 19, 21 respectively. And we run clients, which is a separate process different from those for replicas, to inject transactions into the system. Specifically, We assign the number of clients to be 320 for fixing the number of transactions to be about 30000. We set the duration of each experiment as one minute and obtain experimental data every ten seconds. All results were the average value of at least five independent experiments.

### 5.1  Fault-Free Cases

In this part, we first evaluate the performance of Diem and the Snitch in fault-free cases, where the difference between of two protocols is that in the Snitch, the leader needs to distribute its secret in each round. Figure 4 presents the latency and throughput achieved by the different protocols as a function of the number of replicas. Generally, the throughput decreases and the latency increases for both protocols along with the network size increases. This is because the system is bottlenecked by a leader communicating with all other replicas. We notice that two protocols behave almost the same in the network with four replicas. At more replicas, the throughput of the Snitch tends to be slightly worse than Diem. This is mainly due to the cost of distributing secret.

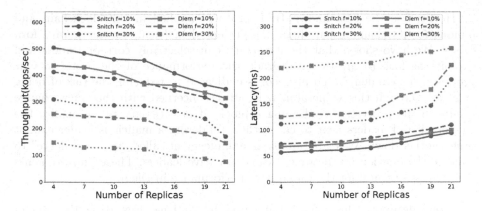

**Fig. 5.** Throughput and latency as the network size varies in normal cases

## 5.2   Normal Cases

Our next experiment evaluates the faulty scalability of protocols by observing the performance change as the percentage of faults increases. We stimulate the faulty replicas by not responding or proposing invalid proposals arbitrarily. We did not implement the "waste attack" discussed in Sect. 2, which will only further hurt the performance of Diem. Since partially synchronous protocols tolerate one-third faults, we conduct several experiments where the percentage of faults varies from 10% to 30%. Figure 5 reports the results. As expected, the Snitch and Diem have the best performance for the 10% faults. As the ratio of faulty replicas raises, the performance of both protocols is degraded while Diem degrades more sharply. Therefore, for each set of faults, the Snitch always achieves higher throughput and lower latency than Diem. When the proportion of faults is up to 30%, the performance of the Snitch scales much better than Diem. Overall, the higher the ratio of faults in the network and the more significant the performance superiority of the Snitch.

## 6   Discussion and Conclusion

As discussed in Sect. 1, BFT protocols were initially conceived as being deployed in a small-sized system, with a static group of participants. And we proposed the Snitch in a *permissioned* setting above. If necessary, the Snitch can become more scalable in two ways. On the one hand, the Snitch can be designed to facilitate open and dynamic participation via reconfiguration [1,3,6]. This demonstrates the desire shift from *permissioned* system to *permissionless* system. On the other hand, we consider a large-sized system. As analyzed in [24], a system equipped with 100 replicas can realize the PVSS. Hence, if there are more than 100 replicas in the system, some replicas are selected by PoW or PoS to execute the Snitch [17,22]. It makes sure that the protocol is always executed in a system of acceptable size, regardless of the total number of replicas.

HotStuff is the best known BFT protocol applied in the blockchain and famous for its chaining paradigm. As presented in Diem, the temporary fork exists. We have shown that the adversary can selectively corrupt leaders and exploit the fork to drastically degrade the performance of the protocol. Our proposed the Golden Snitch protocol is dedicated to reducing the risk of being attacked through this vulnerability. Besides, concerned with that view-change occurs in the process of confirming a block, it is necessary to rationalize the cooperation of leaders next to each other, rendering a malicious leader cannot drag another down. We elaborate on an enhancement called "veto certificate" to balance the relationship between two consecutive leaders. These improvements will be instrumental for the progressive performance in Diem.

**Acknowledgment.** This work is supported by National Key R&D Program of China (2017YFB0802500), Beijing Municipal Science and Technology Project (No Z191100007119007) and Shandong province major science and technology innovation project (2019JZZY020129).

## A    Analysis of Correctness

The correctness of BFT protocols is usually defined by two properties: safety and liveness. This section surveys that the Snitch can guarantee safety and liveness under some reasonable assumptions mentioned previously.

### A.1    Safety

The safety of the Snitch is proved to be provided regardless of the network status.

**Definition 1 (safety).** *The protocol provides safety if it satisfies agreement and validity simultaneously.*

**Lemma 1 (validity).** *Any block containing invalid data can not be confirmed.*

*Proof.* Because a valid CC can be formed only with $n - f = 2f + 1$ votes for it, there must be a correct replica who voted it. In other words, Malicious replicas cannot generate a certificate without the cooperation of a correct node. As a correct replica, it is impossible to send a positive vote for a block with invalid data. Trivially ensures that only valid blocks can be confirmed.

**Lemma 2 (agreement).** *In BFT model, for round $r$, the replica $P_i$ maintains a block $B_r$ with its confirmation certificate $CC_r$. If there exits any other replica $P_j$ that maintains a block $B'_r$ with its confirmation certificate $CC'_r$ in the same round, it must be $B_r = B'_r$ and $CC_r = CC'_r$.*

*Proof.* We prove Lemma 2 by contradiction. It is assumed that there exists a round $r$, where two conflicting blocks $B_r$ and $B'_r$ are both confirmed, each by a correct replica. As defined before, $2f+1$ positive votes would be required to form a CC. Hence, $CC_r$ and $CC'_r$ need $2(2f + 1) = n + f + 1$ votes simultaneously.

It implies that at least $f + 1$ replicas vote twice with the original setting of $n$. It then goes against the assumption that at most $f$ malicious replicas exist. Consequently, there is at most one valid block in a round.

**Theorem 1 (safety).** *Two conflicting blocks can not be committed according to the commit rule.*

*Proof.* It is assumed that there are three certified blocks chained in consecutive rounds, as shown in the example in Fig. 1(a):

$$B_r \leftarrow CC_r \leftarrow B_{r+1} \leftarrow CC_{r+1} \leftarrow B_{r+2} \leftarrow CC_{r+2}.$$

Here, $n - f$ votes were cast for $CC_2$ to commit $B_r$, out of which at least $f + 1$ were from correct replicas. These correct replicas that contributed to the generation of $CC_2$ remember $r$ as their locked round $r_l$, and would not vote for block $B$ if it is not a descendant of block $B_r$, according to the vote rules. Hence, if there exist another three certified blocks chained in consecutive rounds as:

$$B'_r \leftarrow CC'_r \leftarrow B'_{r+1} \leftarrow CC'_{r+1} \leftarrow B'_{r+2} \leftarrow CC'_{r+2}.$$

Then $B'_r$ must be a descendant of $B_r$.

## A.2    Liveness

Liveness can be provided after GST. Next, we set the timer for $\Delta$ to denote the maximum round duration and use $\delta$ to denote the network delay.

**Definition 2 (Liveness).** *Whenever the network becomes synchronous, the algorithm provides liveness as to commits will be produced in a timely manner.*

**Lemma 3 (Round Sync).** *For two consecutive rounds, round $r - 1$ is led by a correct leader and round $r$ is a timeout round. If a correct replica first switches to a new maximum round $r + 1$ at time $t$, then others will move to round $r + 1$ at time $t + 2\delta$.*

*Proof.* It is assumed that leader $l_{r-1}$ published $B_{r-1}$ at time $t'$. Replicas can be divided into three parts and discussed as follows:

– Replica A receives $B_{r-1}$ at time $t'$:A moves to round $r$ and and starts the timer at time $t'$. Then the timer expired at time $t' + \Delta$.
– Replica B receives $B_{r-1}$ at time $t' + \delta$: Similarly, B sends starts its timer at time $t' + \delta$ and then broadcasts a timeout message at time $t' + \delta + \Delta$.
– Replica C receives $B_{r-1}$ at time $t' + \delta$: C starts its timer at time $t' + \delta$. While replica C received more than $f + 1$ timeout messages leading to its broadcast for timeout message before the timer expired.

In consequence, replicas broadcast respective timeout messages before time $t' + \delta + \Delta$. Then replicas complete the recovery within $\delta$. Finally, we came to the conclusion that replicas would be synchronized to the same round within $2\delta$.

**Lemma 4.** *In the situation of all leaders in three consecutive rounds, $l_r, \ldots, l_{r+3}$ are correct, $B_r$ can be committed within $4 * 2\delta$ after it is proposed.*

*Proof.* According to the commit rule, three consecutive confirmation certificates are required to commit a block. For these rounds that generate certificates, each round duration is at most $2\delta$. Taking round synchronization into consideration, no more than $3 * 2\delta + 2\delta$ is required to commit a block.

**Lemma 5.** *Malicious nodes cannot impede progress infinitely.*

*Proof.* The recover threshold is $f + 1$, ensuring that malicious nodes can not reveal others' secrets without at least one correct replica. It is inevitable that a malicious leader can block the process temporarily by means of no response or publishing an invalid block. While these behaviors cannot affect the next-round leader at all. Thanks to the punishment, in the worst case, the process will be made after $f + 1$ rounds.

**Theorem 2 (Liveness).** *The request issued by a correct client eventually completes.*

*Proof.* By Lemma 3, it can be concluded that all replicas enter the same round in time. Besides, owing to Lemma 4 combined with Lemma 5, transaction finality is guaranteed when a succession of three consecutive leaders behave correctly one after another. Hence, correct clients will receive replies to their requests eventually. It means liveness is guaranteed.

# References

1. Abraham, I., Malkhi, D., Nayak, K., Ren, L., Spiegelman, A.: Solida: a blockchain protocol based on reconfigurable byzantine consensus. arXiv preprint arXiv:1612.02916 (2016)
2. Aublin, P.L., Mokhtar, S.B., Quéma, V.: Rbft: redundant byzantine fault tolerance. In: 2013 IEEE 33rd International Conference on Distributed Computing Systems, pp. 297–306. IEEE (2013)
3. Bano, S., et al.: State machine replication in the libra blockchain (2020). https://developers.libra.org/docs/state-machine-replication-paper, Accessed 19 Dec 2020
4. Baudet, M., et al.: State machine replication in the libra blockchain. The Libra Assn., Technical Report (2019)
5. Bentov, I., Pass, R., Shi, E.: Snow white: provably secure proofs of stake. IACR Cryptol. ePrint Arch. **2016**, 919 (2016)
6. Bessani, A., Sousa, J., Alchieri, E.E.: State machine replication for the masses with bft-smart. In: 2014 44th Annual IEEE/IFIP International Conference on Dependable Systems and Networks, pp. 355–362. IEEE (2014)
7. Buterin, V., Griffith, V.: Casper the friendly finality gadget. arXiv preprint arXiv:1710.09437 (2017)
8. Canetti, R., Rabin, T.: Fast asynchronous byzantine agreement with optimal resilience. In: Proceedings of the twenty-fifth annual ACM symposium on Theory of computing, pp. 42–51 (1993)

9. Cascudo, I., David, B.: SCRAPE: scalable randomness attested by public entities. In: Gollmann, D., Miyaji, A., Kikuchi, H. (eds.) ACNS 2017. LNCS, vol. 10355, pp. 537–556. Springer, Cham (2017). https://doi.org/10.1007/978-3-319-61204-1_27

10. Castro, M., Liskov, B., et al.: Practical byzantine fault tolerance. In: OSDI, vol. 99, pp. 173–186 (1999)

11. Cowling, J., Myers, D., Liskov, B., Rodrigues, R., Shrira, L.: Hq replication: A hybrid quorum protocol for byzantine fault tolerance. In: Proceedings of the 7th Symposium on Operating Systems Design and Implementation, pp. 177–190 (2006)

12. Dolev, D., Strong, H.R.: Polynomial algorithms for multiple processor agreement. In: Proceedings of the Fourteenth Annual ACM Symposium on Theory of Computing, pp. 401–407 (1982)

13. Dwork, C., Lynch, N., Stockmeyer, L.: Consensus in the presence of partial synchrony. J. ACM (JACM) **35**(2), 288–323 (1988)

14. Gilad, Y., Hemo, R., Micali, S., Vlachos, G., Zeldovich, N.: Algorand: scaling byzantine agreements for cryptocurrencies. In: Proceedings of the 26th Symposium on Operating Systems Principles, pp. 51–68 (2017)

15. Hanke, T., Movahedi, M., Williams, D.: Dfinity technology overview series, consensus system. arXiv preprint arXiv:1805.04548 (2018)

16. Kiayias, A., Russell, A., David, B., Oliynykov, R.: Ouroboros: a provably secure proof-of-stake blockchain protocol. In: Katz, J., Shacham, H. (eds.) CRYPTO 2017. LNCS, vol. 10401, pp. 357–388. Springer, Cham (2017). https://doi.org/10.1007/978-3-319-63688-7_12

17. Kogias, E.K., Jovanovic, P., Gailly, N., Khoffi, I., Gasser, L., Ford, B.: Enhancing bitcoin security and performance with strong consistency via collective signing. In: 25th {usenix} Security Symposium ({usenix} Security 16), pp. 279–296 (2016)

18. Kotla, R., Alvisi, L., Dahlin, M., Clement, A., Wong, E.: Zyzzyva: speculative byzantine fault tolerance. In: Proceedings of Twenty-First ACM SIGOPS Symposium on Operating Systems Principles, pp. 45–58 (2007)

19. Lamport, L., Shostak, R., Pease, M.: The byzantine generals problem. In: Concurrency: The Works of Leslie Lamport, pp. 203–226 (2019)

20. Micali, S., Rabin, M., Vadhan, S.: Verifiable random functions. In: 40th Annual Symposium on Foundations of Computer cience (cat. No. 99CB37039), pp. 120–130. IEEE (1999)

21. Nakamoto, S.: Bitcoin: a peer-to-peer electronic cash system (2008). https://bitcoin.org/bitcoin.pdf

22. Pass, R., Shi, E.: Hybrid consensus: efficient consensus in the permissionless model. In: 31st International Symposium on Distributed Computing (DISC 2017). Schloss Dagstuhl-Leibniz-Zentrum fuer Informatik (2017)

23. Pease, M., Shostak, R., Lamport, L.: Reaching agreement in the presence of faults. J. ACM (JACM) **27**(2), 228–234 (1980)

24. Schindler, P., Judmayer, A., Stifter, N., Weippl, E.R.: Hydrand: efficient continuous distributed randomness. In: 2020 IEEE Symposium on Security and Privacy, SP 2020, San Francisco, CA, USA, 18–21 May 2020 pp. 73–89. IEEE (2020). https://doi.org/10.1109/SP40000.2020.00003

25. Veronese, G.S., Correia, M., Bessani, A.N., Lung, L.C., Verissimo, P.: Efficient byzantine fault-tolerance. IEEE Trans. Comput. **62**(1), 16–30 (2011)

26. Yin, M., Malkhi, D., Reiter, M.K., Gueta, G.G., Abraham, I.: Hotstuff: bft consensus with linearity and responsiveness. In: Proceedings of the 2019 ACM Symposium on Principles of Distributed Computing, pp. 347–356 (2019)

# Rectifying Administrated ERC20 Tokens

Nikolay Ivanov$^{(\boxtimes)}$, Hanqing Guo, and Qiben Yan

SEIT Lab, Michigan State University, East Lansing, MI 48824, USA
{ivanovn1,guohanqi,qyan}@msu.edu

**Abstract.** ERC20 token is the most popular type of Ethereum smart contract. The daily transaction volume of these tokens exceeds 100 billion dollars, which agitates the popular notions of "decentralized banking" and "tokenized economy". Yet, it is a common misconception to assume that the decentralization of blockchain entails the decentralization of smart contracts deployed on this blockchain. In practice, the developers of smart contracts implement *administrating patterns*, such as censoring certain users, creating or destroying balances on demand, destroying smart contracts, or injecting arbitrary code. These routines, which are designed to tightly control the operation of these smart contracts, turn an ERC20 token into an *administrated token*—the type of Ethereum smart contract that we scrutinize in this research.

We discover that many smart contracts are administrated, which means that their owners solely possess an omnipotent power over these contracts. Moreover, the owners of these tokens carry lesser social and legal responsibilities compared to the traditional centralized actors that those tokens intend to disrupt. This entails two major problems: a) the owners of the tokens have the ability to quickly steal all the funds and disappear from the market; and b) if the private key of the owner's account is stolen, all the assets might immediately turn into the property of the attacker. Therefore, the administrated ERC20 tokens are not only dissimilar to the traditional centralized asset management tools, such as banks, but they are also more vulnerable to adversarial actions by their owners or attackers. We develop a pattern recognition framework based on 9 syntactic features characterizing administrated ERC20 tokens, which we use to analyze existing smart contracts deployed on Ethereum Mainnet. Our analysis of 84,062 unique Ethereum smart contracts reveals that nearly 58% of them are administrated ERC20 tokens, which accounts for almost 90% of all ERC20 tokens deployed on Ethereum.

To protect users from the frivolousness of unregulated token owners without depriving the ability of these owners to properly manage their tokens, we introduce *SafelyAdministrated*—a library that enforces a responsible ownership and management of ERC20 tokens. The library introduces three mechanisms: *deferred maintenance, board of trustees* and *safe pause*. We implement and test *SafelyAdministrated* in the form of Solidity abstract contract, which is ready to be used by the next generation of safely administrated ERC20 tokens.

**Keywords:** Ethereum · Blockchain · Smart contracts · Security

© Springer Nature Switzerland AG 2021
D. Gao et al. (Eds.): ICICS 2021, LNCS 12918, pp. 22–37, 2021.
https://doi.org/10.1007/978-3-030-86890-1_2

# 1    Introduction

Millions of Ethereum smart contracts operate hundreds of billions of dollars worth of assets. ERC20 fungible token is the most popular type of smart contract in Ethereum, often compared to decentralized bank account. Ethereum has two type of accounts: externally owned accounts (EOAs) and smart contracts. An EOA has an associated private key and can deploy smart contracts, but cannot execute custom code. On the other hand, a smart contract can execute custom code, but it does not have any associated private key for determining its owner. The deploying EOA of the contract does not automatically own this smart contract, unless this functionality is manually implemented by the contract developer. Moreover, any functionality related to ownership, role-based access, or other special permissions must be manually implemented by the developer; otherwise, the contract becomes orphaned at the moment it is deployed.

Many smart contracts use routines from the OpenZeppelin Contracts [3] library for implementing ownership and role-based access in the smart contracts. A recent analysis by Zhou et al. [17] shows that at least 2.1 million Ethereum smart contracts, out of 5.8 million total, use the `onlyOwner` modifier from the OpenZeppelin Contracts library, which allows only a certain user (i.e., owner) to call the functions of the smart contract implemented with this modifier. Figure 1 shows a Venn diagram of the relationships between different subsets of Ethereum smart contracts from the perspective of this research. Specifically, we subdivide all smart contracts into two major categories: *administrated contracts*, and *effectively ungoverned smart contracts*, particularly emphasizing that not all contracts that have an owner are necessarily administrated, as the ownership may be purely symbolic sometimes or only allows harmless operations. The administrated smart contracts are characterized by two major properties: a) there is at least one Ethereum account whose owner possesses a unique privileged status; b) the privileged status allows the user to perform actions that may affect other users of the smart contract. These two properties constitute the difference between the administrated and ownable smart contracts: the ownable smart contract must only meet the first property; however, there are smart contracts that have an owner, but this owner has no power to disrupt the operation of the smart contract.[1] We further refer to non-administrated smart contracts as *effectively ungoverned*, the set that includes the ownable non-administrated contracts, and many of them are ERC20 tokens.[2] In this work, however, we zero in on the administrated ERC20 tokens, and our goal is to introduce a novel subset of these tokens—*safely administrated ERC20 tokens*.

The obvious popularity of owned smart contracts and ERC20 tokens leads us to the following research question: *how many unique administrated ERC20 tokens*

---

[1] The smart contracts deployed at `0xdf4df8ee1bd1c9f01e60ee15e4c2f7643b690699` and `0x5dc60c4d5e75d22588fa17ffeb90a63e535efce0` are two (out of many) examples of ownable non-administrated contracts.

[2] A typical example of an effectively ungoverned token is the popular ChainLink Token deployed at `0x514910771AF9Ca656af840dff83E8264EcF986CA`.

*are deployed on Ethereum?* To answer this question, we develop an extractor of 9 syntactic features characterizing administrated ERC20 tokens. We then gather 1,173,271 open source smart contracts written in Solidity programming language, and by removing the duplicates, we reduce the dataset to 84,062 unique, independent, and identically distributed (i.i.d.) smart contracts. We further select 385 random contracts for manual labeling in order to choose the most accurate classifier among several candidates. Finally, we use the 9 features and the chosen classifier to determine the approximate percentage of administrated ERC20 contracts deployed on the Ethereum Mainnet blockchain. Our evaluation shows that nearly 58% of all the smart contracts and almost 90% of all ERC20 tokens are administrated ERC20 tokens. To the best of our knowledge, *we are the first to conduct the Ethereum-wide evaluation of administrated ERC20 tokens and quantify their ubiquity.*

To mitigate the potential adverse effects of administrated ERC20 tokens in a low-regulated economic environment, we propose *SafelyAdministered*—a Solidity library that allows developers of ERC20 tokens to implement most common administrated patterns in a safe and responsible way, thereby increasing the trust towards their products without sacrificing the need to retain control over certain operations (e.g., upgrade).

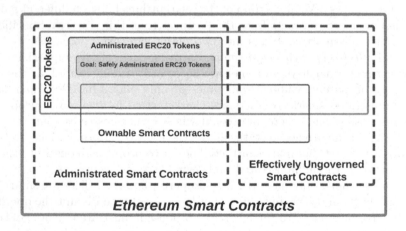

**Fig. 1.** Venn diagram of different types of Ethereum smart contracts.

In summary, we make the following contributions:

- We analyze the class of administrated ERC20 tokens and show that these contracts are more owner-controlled and less safe than the services they try to disrupt, such as banks and centralized online payment systems.
- We develop a binary classifier for identification of administrated ERC20 tokens, and conduct extensive data analysis, which reveals that nearly 9 out of 10 ERC20 tokens on Ethereum are administrated, and thereby unsafe to engage with even under the assumption of trust towards their owners.

– We design and implement *SafelyAdministrated*—a Solidity abstract class that
safeguards users of administrated ERC20 tokens from adversarial attacks or
frivolous behavior of the tokens' owners.

## 2    Background

**Smart Contracts and EVM.** A smart contract is a program deployed on
a blockchain and executed by the blockchain's virtual machine (VM). A smart
contract consists of a set of functions that can be called through blockchain
transactions. Most smart contracts are written in a high-level special-purpose
programming language, such as Solidity or Vyper, and compiled into the bytecode for deployment and execution on a blockchain VM. The Ethereum Virtual
Machine (EVM) is the blockchain VM for executing Ethereum smart contracts.

**Externally Owned Account.** Ethereum blockchain has two types of accounts:
smart contract account and Externally Owned Account (EOA). Both EOAs and
smart contract accounts can be referenced by their 160-bit public addresses.
EOAs can be used to call the functions of smart contracts via signed transactions.

**Solidity.** Solidity is the most popular programming language for EVM smart
contract development, which syntax is similar to JavaScript and C++. The
source code of a smart contract written in Solidity needs to be compiled into
bytecode before being deployed on EVM. All smart contracts analyzed in this
study are written in Solidity.

**ERC20 Tokens.** ERC20 is the most popular standard for implementing fungible tokens[3] in Ethereum smart contracts. Some of the most traded alternative
cryptocurrencies (altcoins) are ERC20-compatible smart contracts deployed on
Ethereum Mainnet, such as ChainLink and BinanceCoin. The ERC20 standard
defines an interface with 6 mandatory functions, 2 mandatory events, and 3
optional properties that a smart contract should implement in order to become
an ERC20 token to interact with ERC20-compliant clients.[4]

**OpenZeppelin Contracts.** *OpenZeppelin Contracts* is a library of smart contracts that have been extensively tested for adherence to best security practices.
These smart contracts are considered to be the de-facto standardized implementations of popular smart contract code patterns [4]. The OpenZeppelin project
provides a rich code base for ERC20 token developers [2]. Most ERC20 tokens, as
well as the administrated patterns in these tokens, are implemented by inheriting
routines from the OpenZeppelin Contracts library.

---

[3] Each fungible token has the same value and does not possess any special characteristics compared with other tokens of the same type.
[4] https://eips.ethereum.org/EIPS/eip-20.

```
1 function kill() public onlyAdmin {
2     selfdestruct(payable(msg.sender));
3 }
```

**Fig. 2.** A snippet of an administrated self-destruction pattern in the contract deployed at 0xbF3d14995D4A4A719A3B9101DE60baa47De60F39.

# 3   Administrated ERC20 Patterns

In this section, we elaborate upon five general re-centralization patterns that we observe in Ethereum smart contracts.[5]

## 3.1   Self-destruction

EVM opcode SELFDESTRUCT[6] allows to remove a smart contract from the blockchain. To provide further incentive for owners to remove unused contracts, the address supplied as an argument of SELFDESTRUCT call receives the entire Ether cryptocurrency balance of the smart contract. Solidity uses the built-in function selfdestruct() to initiate the removal of the smart contract—if this functionality is implemented, the administrator (or an attacker impersonating the administrator) can trigger it at any moment, effectively destroying all users' assets with a single transaction. Figure 2 shows a real-world example of such a pattern.

## 3.2   Deprecation

With the exception of self-destruction, the source code of an Ethereum smart contract is immutable, which impedes the ability for developers to deliver new features or fix existing bugs. To address this limitation, some developers of smart contracts implement a bypass scheme, in which a contract can be declared as *deprecated* by the owner, resulting in the redirection of the users' transactions towards functions of a new contract. The danger of this scheme stems from the fact that it grants the owner of the contract an ability to replace the code of some critical functions with arbitrary ones. Figure 3 shows a real-world example of the deprecation pattern.

---

[5] The discovery of these patterns has been largely facilitated by a manual examination of approximately 3,800 source codes of smart contracts in the course of our previous research.

[6] This opcode is formerly known as SUICIDE. In this context, the word "remove" means that the contract is no longer available for transactions; however the entire transaction history of the contract is still retained by the blockchain.

```
1 // deprecate current contract in favour of a new one
2 function deprecate(address _upgradedAddress) public
      onlyOwner {
3   deprecated = true;
4   upgradedAddress = _upgradedAddress;
5   Deprecate(_upgradedAddress);
6 }
```

**Fig. 3.** A snippet of an administrated deprecation pattern in the TetherUSD smart contract deployed at 0xdAC17F958D2ee523a2206206994597C13D831ec7, which allows the owner to effectively inject the code of the contract with an arbitrary one.

```
1 function setFee(address to) public onlyOwner{
2   fee = to;
3 }
```

**Fig. 4.** A snippet of a change-of-address pattern in the smart contract deployed at 0x350BDC46d931712d83ef989725Ba4904C487F360. The exploitation of such pattern has been demonstrated in previous research.

### 3.3 Change of Address

Another administration strategy is the ability for the owner of a smart contract to change certain critical addresses, such as recipients of fees or accounts associated with certain roles. As shown in our previous study [12], a replacement of a public address in a smart contract can lead to an acquisition of the funds by the owner of the contract. Figure 4 demonstrates such an address changing pattern.

### 3.4 Change of Parameters

Another administration pattern is characterized by the change of certain parameters by the owner, which may affect the ability by a user of the contract to perform certain operations. For example, if the owner is allowed to arbitrarily change the amount of withdrawal fees, this parameter might be set to a very large value (e.g., 99%), effectively preventing withdrawal of funds by the user. Another example of this pattern is shown in Fig. 5, where the owner of the contract exercises an unbounded power to manage administrators of the smart contract.

```
1 function setAdmin(address newAdmin, bool activate)
      onlyOwner public {
2   admins[newAdmin] = activate;
3 }
```

**Fig. 5.** A snippet of a change-parameter pattern in the smart contract deployed at 0x18c210013ea6cbe99b2dacdc9cfcb6e07458f0ca.

```
1 function mint(address account, uint amount) public
     onlyOwner {
2   _mint(account, amount);
3 }
4 function burn(address account, uint amount) public
     onlyOwner {
5   _burn(account, amount);
6 }
```

**Fig. 6.** A snippet of a minting and burning patterns in the smart contract deployed at 0x82bfdd53dd95efa2c3e92543f28d46c566bf4b8a.

### 3.5  Minting and Burning

An increase of a token supply of an ERC20 contract is called *token minting*, and the reduction of supply of tokens is called *burning*. Since the entire supply of tokens is partitioned between owners in a way that there are no balances belonging to nobody, minting a token means to increase someone's balance, and burning a token means to reduce someone's balance. Although most tokens are minted or burned as a result of a certain event, such as token creation, token swap, crowdsale, or exchange into Ether balance, some contracts allow privileged users to arbitrarily mint or burn tokens, which is a dangerous action that even highly centralized commercial banks normally cannot do. Figure 6 demonstrates an example of the minting and burning pattern implemented in a deployed Ethereum smart contract.

## 4  Administrated Tokens in the Wild

In this section, we use a pattern recognition method to search for administrated ERC20 tokens in the Ethereum Mainnet network, as shown in Fig. 7. We start the process with preprocessing all the input samples by removing comments and extracting source codes from multi-part JSON files.[7] Then we randomly select 385 samples from 84,062 unique source code files and manually assign (label) them into two classes: a) administrated ERC20 tokens, and b) others. After that, we extract 385 9-dimensional feature vectors corresponding to the labeled samples, with the assumption that all the samples are identical and independently distributed (i.i.d). Then we use 385 labeled samples and the corresponding feature vectors to evaluate the performance of 9 different classifiers using the K-fold method (with $k = 5$). Next, we choose the best performing classifier, i.e., the one that demonstrated the higher accuracy during the evaluation stage (i.e., SVC). After that, we extract 84,062 feature vectors corresponding to the entire data set. Next, we train the SVC classifier with the 385 labelled samples. Due to the

---

[7] The smart contracts that include several files are represented as JSON arrays in our dataset. Preprocessing these arrays also includes an additional step of replacing the escaped characters, such as newlines and quotes, with their original ASCII codes.

i.i.d. assumption, we can now classify all the samples using the trained SVC model. Finally, we gather the output and analyze the results.

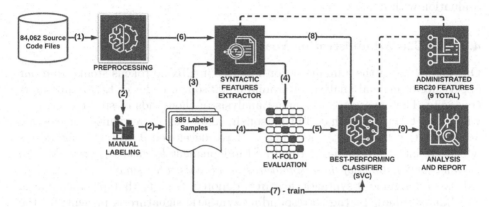

**Fig. 7. General worflow of the analysis of administrated ERC20 tokens.** The workflow includes 9 major steps. (**1**): Pre-process input samples to remove comments and parse multi-part JSON files. (**2**): Pick 385 samples from 84,062 unique source code files and manually assign them into two classes: a) administrated ERC20 tokens, and b) others. (**3**): Extract 385 feature vectors corresponding to the labeled samples. (**4**): Use 385 labeled samples and the corresponding feature vectors to evaluate the performance of 9 different classifiers using the K-fold methods (with $k = 5$). (**5**): Choose the best performing classifier on the 385 labeled samples with the given 9 features. (**6**): Extract 84,062 feature vectors corresponding to the entire data set. (**7**): Train the classifier with the 385 labelled samples. (**8**): Classify all the samples using the trained classifier. (**9**): Analyze and report the results.

## 4.1 Data Set

First, we gather 1,173,271 open-source smart contracts from Etherscan,[8] and by removing duplicates (using `fdupes`[9]), reduce the size of the database to 84,062 distinct smart contracts. Then, we remove all comments from the data points (i.e., source code files), and select 385 random contracts for manual labelling using the following formula:

$$n = \frac{N}{1 + N \cdot (1 - c)^2}. \tag{1}$$

Equation 1 is the Slovin's formula [6], which statistically determines a required representative sample size for a given data size and desired confidence level. $N$ is the original population of smart contracts, i.e., $N = 84,062$, and $n$ is the sample size that we choose to represent the population. $c$ is the confidence

---

[8] https://etherscan.io/.
[9] https://github.com/adrianlopezroche/fdupes.

level that represents the certainty that the sample size represents the population. We set the confidence level as 95% (precisely, 94.915%), leading to sample size $n = 385$, which can be split into two partitions of 77 and 308 samples for k-fold evaluation with $k = 5$.

## 4.2 ERC20 Administration Features

Our knowledge of the administration features in ERC20 tokens stems from our experience of manual analysis of around 3,800 source codes of Ethereum smart contracts. The experience of manual analysis of thousands of smart contracts, which has taken more than 140 person/h, allows us to recognize all existing administration patterns. As a result, we have developed 9 syntactic signatures which are intuitively well-separated and independent *because we have observed various combinations of these signatures in administrated smart contracts.* This led us to designing 9 syntactic features, denoted $f_1 \ldots f_9$ that produce one of two binary values: 1—the corresponding syntactic signature is present; 0—the signature is absent. Below is the brief description of the syntactic signatures that the 9 features correspond to.

$f_1$: **ERC20 Interface Implementation.** The goal of this research is to identify administrated ERC20 tokens. In order to separate ERC20 tokens from other types of smart contracts, feature $f_1$ extractor detects the simultaneous presence of syntactic identifiers corresponding to the eight mandatory items of the ERC20 interface, as described in the EIP-20 standard.

$f_2$: **Administrated Self-destruction Signature.** If the owner of a smart contract implements a self-destruction procedure, they may remove the contract from the Ethereum ecosystem with a single transaction, simultaneously acquiring all the Ether balance of the contract. Feature $f_2$ detects such a signature, both in old versions of Solidity and the modern ones (the exact procedure differs for different versions of the language).

$f_3$: **Pausable Functionality Signature.** The owner of a smart contract can inhibit any operations with the contract at their will for indefinite period of time. Although pausing a smart contract does not allow to directly acquire Ether or token balances, it may have dire consequences if the owner's private key is stolen by an attacker or lost while the token is paused. Feature $f_3$ is intended to identify signatures of such pausable tokens.

$f_4$: **Contract Deprecation Signature.** Since Ethereum smart contracts are non-modifiable, the only means of upgrading the contract is to deprecate the existing contract and refer the users to the new one using inter-contract calls (ICCs). Unfortunately, this procedure allows the owner of the smart contract to effectively introduce any arbitrary code. Feature $f_4$ extracts the signatures of contract deprecation functionality, which is one of the most dangerous patterns in administrated ERC20 tokens.

$f_5$: **Minting and Burning Signatures.** The ability for a privileged user to arbitrary create and remove tokens, known as minting and burning respectively, is a major concern associated with administrated ERC20 tokens. Feature $f_5$ represents the signature of a minting and/or burning in the smart contract, which execution can only be triggered by a privileged user (administrator).

$f_6$: **Role-Restricted Transfers and Withdrawals.** Another signature of an administrated ERC20 token is the ability for a privileged user to perform arbitrary token or Ether cryptocurrency transfers and withdrawals of the funds that do not belong to these users. Feature $f_6$ corresponds to the syntactic signature related to such transfer and withdrawal functionality under a privileged access.

$f_7$: **Function-Disabling Modifiers.** Some function modifiers do not directly check for the identity of privileged users; instead, they use the parameters previously changed by an administrator to decide whether the function needs to be executed. Feature $f_7$ is related to such modifiers that are capable of disabling the execution of a function based on a parameter adjustable by the contract's privileged user.

$f_8$: **Direct Checks of a Sender Address.** Although modifiers are popular means of granting privileged access to certain functions of a smart contract, some administrated contracts use direct checks of the `msg.sender` or `msg.origin` values. Feature $f_8$ targets the direct (i.e., bypassing Solidity modifiers) transaction identity checks, which predominantly make sense within the administrated smart contracts context.

$f_9$: **Freezing, Halting, or Killing Methods.** A list of some specific frequently occurring function names, such as "freeze", "halt", and "kill" empirically strongly correlate with the administrated property of ERC20 tokens. Feature $f_9$ detects the presence of such frequently used functions that almost always indicate an administration pattern.

### 4.3 Classifier Evaluation and Model Selection

We use 385 manually labeled samples to evaluate the performance of 9 popular classifiers using the K-fold method with $k = 5$. Table 1 summarises the classification models used for evaluation and the accuracy of each of these models using the K-fold evaluation method with 385 labeled samples. The evaluation demonstrates that 8 out of 9 classifiers stay within the $95\% \ldots 97\%$ accuracy range, except for the Gaussian Naive Bayes classifier, which performance is slightly above $61\%$.

**Table 1.** Tested classifiers.

| Model | Parameters | Accuracy |
|---|---|---|
| Support Vector Classifier (SVC) | scikit-learn default | 96.6233% |
| Decision Tree | $max.depth = 9$ | 96.3636% |
| K-Nearest Neighbors (K-NN) | $k = 1$ | 95.5844% |
| Random Forest | scikit-learn default | 96.3636% |
| Gaussian Naive Bayes | scikit-learn default | 61.0389% |
| Linear Discriminant Analysis (LDA) | $n\_components = 1$ | 96.3636% |
| Gradient Boosting | scikit-learn default | 96.3636% |
| Adaptive Boosting (AdaBoost) | scikit-learn default | 95.0649% |
| Multi-Layer Perc. Classifier (MLPC) | $alpha = 1, max\_iter = 1000$ | 96.6233% |

### 4.4  Implementation and Evaluation of the Analysis Workflow

We implement the extractors of all the 9 syntactic features using Python 3.8.5 and **re** regular expressions library. We implement the K-fold evaluation and dataset analysis using Python 3.8.5 with `sckit-learn` 0.24.1 and `numpy` 1.20.0 libraries. We randomly selected 385 smart contracts from the i.i.d. set of 84,062 and manually labeled them by human comprehension of the semantics of each of the smart contracts, which took approximately 40 person/h of total effort.

### 4.5  Results

Out of 84,062 evaluated smart contracts, 54,626 have been identified as ERC20 tokens, which is around 64.6%. As many as 39,034 contracts have been classified as administrated ERC20 tokens (by counting the occurrences of $f_1 = 1$), which is 57.96% of all the evaluated smart contracts, and 89.76% of all ERC20 tokens. Subsequently, only about 10% of all ERC20 tokens are non-administrated, i.e., exhibit full decentralization and permissionless design, while the vast majority of the tokens are tightly controlled by their owners and other privileged users, effectively overriding the decentralization capability of the hosting blockchain. Figure 8 shows the summary of the results of our analysis.

## 5  SafelyAdministrated Library

Existing administrated ERC20 tokens are generally unsafe because they are loosely regulated and their functionality often hinges upon a single account's private key, which can be abused by its owner or stolen by an adversary. To mitigate such an unsafe arrangement without denouncing the idea of administration or boycotting the administrated tokens, we propose a novel solution for making these smart contracts safe. As shown in Sect. 4, most ERC20 tokens are administrated, and therefore potentially unsafe. However, due to their ubiquity, it would

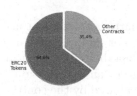 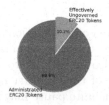 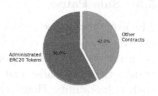

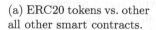

(a) ERC20 tokens vs. other all other smart contracts.

(b) Administrated ERC20 tokens vs. effectively ungoverned ERC20 tokens.

(c) Administrated ERC20 tokens vs. other types of smart contracts.

**Fig. 8.** Results of processing of 84,602 unique source codes of Ethereum smart contracts using the SVC classifier and the 9 developed syntactic features.

be naive to urge users to boycott 9 out of 10 of currently deployed ERC20 tokens. In this work, we propose a feasible "evolutionary" fix to the existing problem. Specifically, we realize that administrated patterns can be used by token owners without jeopardizing the safety of the contract and requiring trust from the users. For that, the current primitive administrated routines can be re-implemented to incorporate three novel concepts: *deferred maintenance, board of trustees*, and *safe pausing*. The details of these three approaches are explained below.

### 5.1   Deferred Maintenance

The owners of existing administrated ERC20 tokens have the ability to call the managerial functions without any announcement. In order to prevent unannounced actions, *SafelyAdministrated* library implements a mechanism of *deferred maintenance*, which allows to announce the maintenance action to the users and enact it only after a certain delay. For example, if the contract is about to be upgraded, the users of the contract may be notified and decide whether they agree on the upgrade or not. If the users disagree with the upgrade, they may safely quit (i.e., sell or transfer their tokens) before the action takes into effect.

### 5.2   Contract Board of Trustees

In most administrated smart contracts, the privileged user (administrator) has a sole power to perform critical actions upon the smart contract, which incurs the need of trust from the users of the contract. Moreover, if the private key of the smart contract's administrator is stolen, the attacker *becomes* the administrator of the contract. Essentially, the safety of the contract often hinges on a single private key belonging to a single person, which is the major concern about the administrated smart contracts. The *contract board of trustees* allows to split the administrative power among multiple private keys possessed by different parties, such that the maintenance actions are only possible through a voting consensus with a pre-determined threshold.

## 5.3  Safe Pause

The ability to pause the execution of transactions in a smart contract is not necessarily a whimsical action of the contract administrator. For example, this may be a necessary action upon discovery of a zero-day vulnerability—by pausing transactions, the administrator of the contract may prevent an exploitation of such vulnerability. However, indefinite pause may also be abused by the contract administrator, or it can be triggered by an adversary who stole a private key of the administrator's account. To prevent the adverse effects of the pause functionality, in this work we introduce a *safe pause* routine, which allows to freeze all transactions in the smart contract with a forced un-freeze after a certain deadline. Moreover, once the contract is un-frozen, it cannot be frozen again for some time. This way, any of the trustees of the contract can enact an emergency pause, but no one is able to keep the contract paused indefinitely.

## 5.4  Implementation

We implement *SafelyAdministrated* as an abstract Solidity class, which includes 6 functions, 3 modifiers, and 5 events, summarized in Table 2. We implemented a testing ERC20 token that inherits the *SafelyAdministrated* contract, compiled it using Solc 0.8.1, and thoroughly tested its functionality to confirm that *SafelyAdministrated* allows an ERC20 token to be administrated in a safe manner.

## 5.5  Limitation

One limitation of *SafelyAdministrated* is that the trustee whose vote attains the voting threshold effectively pays fees for the execution of the maintenance

**Table 2.** Inheritable interfaces of *SafelyAdministrated* abstract class.

| Inheritable interface | Type | Description |
|---|---|---|
| actionCleared | function | Check if a given action can be performed |
| safelyPaused | function | Check if contract is paused |
| safelyUnpaused | function | Check if contract is unpaused |
| safelyPause | function | Safely pause the smart contract |
| safelyUnpause | function | Safely un-pause the smart contract |
| whenSafelyPaused | modifier | Check if contract is paused |
| whenSafelyUnpaused | modifier | Check if contract is un-paused |
| trusteeVote | function | Cast trustee vote for an action |
| SafelyPaused | event | A trustee paused the contract |
| SafelyUnpaused | event | A trustee un-paused the contract |
| TrusteeVoted | event | A trustee voted for an action |
| ActionCleared | event | Next vote will activate the action |
| ActionActivated | event | A trustee vote activated a cleared action |
| trusteeAction[0...9] | modifier | Modifiers for nine functions subject to approval |

transaction, while other trustees pay only for execution of recording of their vote. Although we assume that this unfairness is unlikely to be important in most cases, we leave the implementation of fee reimbursement for future work.

## 6   Related Work

Currently, the major concern about the safety of smart contracts comes from security vulnerabilities in them. Researchers have proposed automated tools for detecting known smart contract vulnerabilities. Some notable security scanners for Ethereum include Oyente [13], Mythril [1], and Vandal [5]. Tsankov et al. [16] propose Securify, a tool that analyzes the bytecode of Ethereum smart contracts to detect patterns associated with known security vulnerabilities. Torres et al. [15] present a taxonomy of smart contract *honeypots*, which are deceptive smart contracts targeting users who attempt to exploit known vulnerabilities of smart contracts. Recently, Chen et al. propose TokenScope [7], an automated tool, which detects the discrepancies between syntax and semantics in the functions of ERC20 tokens. In this work, we reach beyond the security vulnerabilities and explore a generally overlooked safety issue in smart contracts, i.e., administrated patterns that allow owners of ERC20 tokens (or adversaries who steal the owner's account private key) to cause a mass damage to the token owners.

The influence of private actors on blockchain resources has been a subject of concern for many years. Raman et al. [14] conduct a case study of decentralized web applications and identify a prevalence of *re-centralization* of such apps. Griffin et al. [11] discover that TetherUSD ERC20 token has been used for manipulating the price of cryptocurrencies. In this work, we expand the discussion about the re-centralization and private manipulation of the services that are intended to be centralized to embrace the realm of ERC20 tokens.

The public trust towards administrated ERC20 tokens may be indicative of a well-studied irrational or semi-rational human behavior. In our previous research [12], we explore social engineering attacks in Ethereum smart contracts by demonstrating how visual cognitive bias and confirmation bias lead a user into engaging with a malicious smart contract. Fenu et al. [9] demonstrate the irrational behavior exhibited by many people when engaging with high-risk smart contracts involved in initial coin offerings (ICOs). In this work, we scrutinize a new facet of semi-rational human behavior: the false assumption that most smart contracts are decentralized, permissionless, and ungoverned *just because* they are deployed on a blockchain that holds these properties.

Previous studies proposed smart contract-level multi-signature voting schemes. ÆGIS [10] implements a voting-based mechanism, in which trusted experts vote for a security patch. Unfortunately, the voting mechanism in ÆGIS has been design for different context and cannot be applied, even with modifications, to the trustee-based contract maintenance scenarios. Christodoulou [8] introduces a decentralized voting scheme similar to the Board of Trustees used in this work. However, all the above solutions are domain-specific, and cannot be directly used for general cases, as we see it in the *SafelyAdministrated* library.

# 7  Conclusion

Unlike banks and other financial institutions, smart contracts are weakly regulated or unregulated at all. Simultaneously, an ERC20 token is often owned by a single account, the security of which hinges on a single private key. At the same time, we observe that market capitalization of some tokens, such as USDT and BNB, reaches billions of dollars, which means that if the administrator's private key is stolen or abused, all the funds from all users in the contract might be stolen immediately. ERC20 fungible tokens have been a hope for the next-generation tokenized economy. However, in this research we demonstrate that approximately 9 out of 10 ERC20 tokens are administrated assets that are generally less secure than traditional financial institutions and accounts. Instead of stigmatizing the widespread administration of the tokens, we deliver a solution for the honest token owners to achieve their goals in a way that is safe for both them and the users—through implementing the novel contract ownership mechanism, which effectively prevents a single point of security failure and enforces prior notice of maintenance. At the time of writing, there is no affiliation or sponsorship, current or arranged, between the authors of this work and any banks, online payment systems, and smart contract developers mentioned or implied in this research.

**Acknowledgements.** We would like to thank Dr. Arun Ross and other anonymous reviewers for providing valuable feedback on our work.

# References

1. Mythril. https://github.com/ConsenSys/mythril. Accessed 06 Jan 2020
2. OpenZeppelin ERC-20 Token Implementations. https://github.com/OpenZepp elin/openzeppelin-contracts/tree/master/contracts/token/ERC20. Accessed 12 Jan 2020
3. Openzeppelin contracts (2021). https://github.com/OpenZeppelin/openzeppelin-contracts
4. Antonopoulos, A.M., Wood, G.: Mastering Ethereum: Building Smart Contracts and DApps. O'Reilly Media (2018)
5. Brent, L., et al.: Vandal: a scalable security analysis framework for smart contracts. arXiv preprint arXiv:1809.03981 (2018)
6. Burt, J.E., Barber, G.M., Rigby, D.L.: Elementary Statistics for Geographers. Guilford Press (2009)
7. Chen, T., et al.: TokenScope: automatically detecting inconsistent behaviors of cryptocurrency tokens in Ethereum. In: Proceedings of the 2019 ACM SIGSAC Conference on Computer and Communications Security, pp. 1503–1520 (2019)
8. Christodoulou, P., Christodoulou, K.: A decentralized voting mechanism: engaging ERC-20 token holders in decision-making. In: 2020 7th International Conference on Software Defined Systems (SDS), pp. 160–164. IEEE (2020)
9. Fenu, G., Marchesi, L., Marchesi, M., Tonelli, R.: The ICO phenomenon and its relationships with Ethereum smart contract environment. In: 2018 International Workshop on Blockchain Oriented Software Engineering (IWBOSE), pp. 26–32. IEEE (2018)

10. Ferreira Torres, C., Baden, M., Norvill, R., Fiz Pontiveros, B.B., Jonker, H., Mauw, S.: ÆGIS: shielding vulnerable smart contracts against attacks. In: Proceedings of the 15th ACM Asia Conference on Computer and Communications Security (2020)
11. Griffin, J.M., Shams, A.: Is Bitcoin really untethered? J. Finan. **75**(4), 1913–1964 (2020)
12. Ivanov, N., Lou, J., Chen, T., Li, J., Yan, Q.: Targeting the weakest link: social engineering attacks in Ethereum smart contracts. In: Proceedings of the 2021 ACM Asia Conference on Computer and Communications Security, pp. 787–801 (2021)
13. Luu, L., Chu, D.H., Olickel, H., Saxena, P., Hobor, A.: Making smart contracts smarter. In: Proceedings of the CCS, pp. 254–269 (2016)
14. Raman, A., Joglekar, S., Cristofaro, E.D., Sastry, N., Tyson, G.: Challenges in the decentralised web: the Mastodon case. In: Proceedings of the Internet Measurement Conference, pp. 217–229 (2019)
15. Torres, C.F., Steichen, M., et al.: The art of the scam: demystifying honeypots in Ethereum smart contracts. In: 28th USENIX Security Symposium, USENIX Security 2019, pp. 1591–1607 (2019)
16. Tsankov, P., Dan, A., Drachsler-Cohen, D., Gervais, A., Buenzli, F., Vechev, M.: Securify: practical security analysis of smart contracts. In: Proceedings of the CCS (2018)
17. Zhou, S., et al.: An ever-evolving game: evaluation of real-world attacks and defenses in Ethereum ecosystem. In: 29th USENIX Security Symposium, USENIX Security 2020, pp. 2793–2810 (2020)

# *Moat*: Model Agnostic Defense against Targeted Poisoning Attacks in Federated Learning

Arpan Manna, Harsh Kasyap$^{(\boxtimes)}$, and Somanath Tripathy

Department of Computer Science and Engineering,
Indian Institute of Technology Patna, Patna, India
{arpan_1911cs05,harsh_1921cs01,som}@iitp.ac.in

**Abstract.** Federated learning has migrated data-driven learning to a model-centric approach. As the server does not have access to the data, the health of the data poses a concern. The malicious participation injects malevolent gradient updates to make the model maleficent. They do not impose an overall ill-behavior. Instead, they target a few classes or patterns to misbehave. Label Flipping and Backdoor attacks belong to targeted poisoning attacks performing adversarial manipulation for targeted misclassification. The state-of-the-art defenses based on statistical similarity or autoencoder credit scores suffer from the number of attackers or ingenious injection of backdoor noise. This paper proposes a universal model-agnostic defense technique (*Moat*) to mitigate different poisoning attacks in Federated Learning. It uses interpretation techniques to measure the marginal contribution of individual features. The aggregation of interpreted values for important features against a baseline input detects the presence of an adversary. The proposed solution scales in terms of attackers and is also robust against adversarial noise in either homogeneous or heterogeneous distribution. The most appealing about *Moat* is that it achieves model convergence even in the presence of 90% attackers. We ran experiments for different combinations of settings, models, and datasets, to verify our claim. The proposed technique is compared with the existing state-of-the-art algorithms and justified that *Moat* outperforms them.

**Keywords:** Federated learning · Label flipping attack · Backdoor attack · Hybrid attack · Model interpretation · Shapley value

## 1 Introduction

Federated Learning (FL) has emerged with growing shares in prominent applications, protecting users' security and privacy [11]. It brings the model to the data existing on the edge devices, felicitating in-place training. It is a promising alternative to centralized training, which has been vulnerable to privacy

A. Manna, H. Kasyap and S. Tripathy—Equal contribution.

D. Gao et al. (Eds.): ICICS 2021, LNCS 12918, pp. 38–55, 2021.
https://doi.org/10.1007/978-3-030-86890-1_3

breaches and data abuse. FL has evolved as a privacy-preserving paradigm for the participants. However, it induces a risky design trade-off where the aggregation has come under the scanner, citing that passive data providers may act as active adversaries [7]. The mere presence of malicious participants will gradually impede the goal of subtle performance.

The adversaries send erroneous model updates injecting malicious and ingenious training strategies, which causes a greater catastrophe. These attacks are more insidious as they maintain the overall accuracy and achieve desired results on attacker-chosen samples. Label Flipping and Backdoor attacks are the targeted poisoning attacks, which are deftly crafted by the attacker and mostly remain untraceable from the existing defense measures. The other poisoning attacks are Additive Noise and Gradient Ascent attacks. We analyzed these attacks in various FL settings by varying numbers of attackers, samples, and triggers through a series of case studies.

Numerous defense strategies have been proposed to mitigate such attacks. The existing approaches rely on similarity-based techniques like euclidean distance and cosine similarity or detecting outlier using mean, median-based algorithms [2,5]. They work only in the presence of a majority of benign participants and are designed as attack-specific. Adversarial defense techniques have been proposed where the server prepares auxiliary data to detect an adversary's presence [18]. This process bears huge computational costs and is impractical to deploy. Credit score-based approach [8] using autoencoder has been effective in detecting Label Flipping attacks and outperforms traditional defense-based approaches.

This paper discusses a novel defense mechanism using model explanation techniques to detect adversaries' presence besides their number and strategy. This analysis does not require any prior knowledge or auxiliary synthesis. It uses Shapley algorithms, game-theory-based mathematical formulations that give the features' marginal contributions based on their importance. It requires some random samples to learn the reference to interpret the model. Further, operating over the attribution values on baseline data helps us to spot the heterogeneity.

The following are the major contributions.

- We propose a model agnostic mitigation strategy to defend against targeted poisoning attacks in Federated Learning. It uses the attribution-based Shapley algorithms to measure the marginal contribution for individual features. Using an additive feature importance strategy, we could successfully figure out the presence of adversaries.
- A hybrid attack is designed by colluding the attackers with the intentions of Label Flipping and Backdoor capabilities.
- The empirical evaluation has been extensively carried over the MNIST and Fashion MNIST datasets under different attack settings and evaluation metrics. It has been tested under both the IID (Independent and Identically Distributed) and non-IID distributions. The results provide a shred of evidence to the proposed conjecture.

– This paper evaluates the existing defense strategies against poisoning attacks. It demonstrates their limitation against ingenious backdoor attacks and working at the cost of honest participants in label flipping attacks.

The remainder of this paper is organized as follows. Section 2 discusses the background and related works. Section 3 briefs Federated Learning and discusses the capabilities of an adversary. Section 4 proposes a model-agnostic defense strategy against poisoning attacks in Federated Learning. Section 5 describes the simulation setup, experiment, and results. Section 6 briefs comparison with the existing state-of-the-art algorithms. Section 7 concludes this work.

## 2   Related Work

Cao et al. [3] designed a distributed poisoning attacks colluding multiple attackers. The authors proposed SNIPER's defense mechanism by constructing a graph based on euclidean distance between the local model updates. Further, they aggregate the models present in the largest clique of the graph. Authors claim to restrict the attack success rate by 2% even in the presence of one-third of attackers. The statistical similarity-based defenses, including Mean-Around-Median, Trimmed Mean [17], Krum and MultiKrum [2] select m out of n similar models and declare the remaining as malicious based on similarity measures. They claim to prevent flipping attacks up to 50% byzantine workers. Fung et al. [5] proposed a novel defense algorithm FoolsGold to evaluate the vulnerabilities of Sybil-based poisoning attacks. They claim for no bounds against the number of attackers present. Li et al. [8] proposed an anomaly detection algorithm based on the credit score to generate low dimensional surrogates, which requires additional computation overheads of pre-training for some initial rounds.

Gu et al. [6] highlighted the risks of outdoor training or pre-trained models using adversarially trained backdoor model aiming to achieve the attacker's intention to predict any input as the attacker-chosen label in the presence of some trigger. Bagdasaryan et al. [1] discussed how attackers could use model replacement to introduce malicious functionality by encoding the backdoor. Xie et al. [16] discussed the vulnerabilities arising due to data heterogeneity and proposed a Distributed Backdoor Attack (DBA) as a novel threat assessment framework in a federated environment. Salem et al. [12] proposed a dynamic backdoor generated as a function of inputs, which can be more effective than semantic or pixel pattern backdoor.

Wang et al. [14] emphasized the requirement of model interpretability, citing the regulatory and legal perspective to avoid unethical cases like discrimination. Their method promises to balance it with data privacy to interpret Federated learning models. Wang et al. [15] measured the contributions of multiple parties in Federated Learning by calculating the grouped feature importance using Shapley values. Takeishi and Kawahara [13] discussed the challenges in implementing Shapley values for anomaly detection.

Based on the survey of the defense strategies to mitigate against the targeted poisoning attacks in Federated learning, we observe a trade-off with the number of attackers present, the encoded backdoor trigger, data heterogeneity, pre-computational overheads, and the lack of robustness.

# 3 Federated Learning and Threat Model

## 3.1 Federated Learning

Federated learning is a distributed training protocol across clients in a multi-round fashion coordinated through a trusted server. Participating clients share their model updates or gradients to the server while preserving their training data. The server has a preliminary untrained global model optionally defined with local epochs, batch size, learning rate, and other parameters. At every iteration t, the server selects a random sample of clients $(S_t)$ and sends them a copy of the global model$(w_t)$. Client k on receiving the global model trains on its private data by running the optimization algorithm and outputs a new local model $w_{t+1}^k$ as $\leftarrow w_t^k - \eta \nabla l(w_t^k; b)$ where $\eta$ is the local learning rate of clients and b is the local batch size. Then, each client shares its gradient updates with the server. Consequently, the server aggregates the received local gradients and updates the new global model $w_{t+1}^k$ to the random set of clients selected for the next round. This process iterates until a predefined accuracy or performance metric is reached. A typically used aggregation algorithm (FedAvg [10]) is the weighted averaging of the shared gradients. Federated learning securely aggregates the individual client updates. However, we assume that these types of obfuscations are not used, and the central server is able to observe any individual client's model update at each iteration.

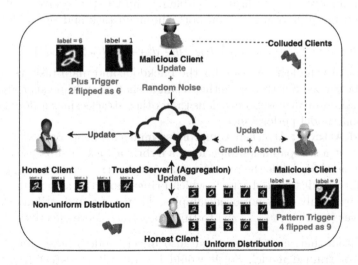

**Fig. 1.** Threat model

## 3.2   Threat Model

We consider an adversary that can evade the holistic training cycle and induce insidious behavior by crafting different poisoning attacks. The attacker aims to poison the global model for high error rates discriminately to victimize a selected class. It is deftly crafted for targeted misclassification, and becomes substantially worse. The malicious clients craft the poisoning attacks to miscalibrate the prediction accuracy. They exploit the fact of provable opacity of data to any trusted entity and gradually impede the model. Figure 1 illustrates our threat model considering label-flipping, backdoor-ed induced noise, and different data distributions. The potential poisoning attacks are summarized below.

1. **Label Flipping Attack:** In this form of attack, the attacker poisons a specific class § by inverting its label to another chosen label $\tau$. For example, every '4' is predicted as '9' in the handwritten digit recognition model. The attacker may pick a rare class from a skewed distribution and flip its label. It goes undetected in most cases due to little effect on the overall accuracy. An attacker can also target multiple classes to flip to a single or a subset of chosen targets.
2. **Backdoor Attack:** In a backdoor attack, the attacker alters training data with certain features ($\psi$) to a target label ($\tau$). The resultant model predicts the wrong label $\tau$ in the presence of these features. The attacker can also embed some trigger to induce malicious behavior. For example, plus trigger is embedded to an image of class '2' for getting misclassified as '6' as shown in Fig. 1. These patterns $\psi$ may vary in shape, size, and position. Attackers generate adversarial samples $d_p^k$ by inserting trigger ($\psi$) using some backdoor function $b_f$, $d_p^k = [b_f(d^k, \psi) = \tau]$ such that $d_c^k \cup d_p^k = d^k$ and $d_c^k \cap d_p^k = \phi$. Further, it solves an optimization problem to update the received global model by effectively performing well on both the main task and the backdoor task as

$$w_{t+1}^k \leftarrow w_t^k - \eta_p \nabla l(w_t^k; b \in d_p^k) - \eta \nabla l(w_t^k; b \in d_c^k)$$

   $\eta_p$ and $\eta$ are the learning rate for the backdoor and main task, respectively. It can be carried out through either a single-shot or multiple-shot attack [16]. The backdoor attack is more vicious than other attacks since it does not affect the global model performance.
3. **Hybrid Attack :** Attackers compromise multiple clients or adversaries with different attack capabilities to perform a hybrid attack. A subset of clients can perform label flip while the remaining can induce trigger to attempt backdoor. It is a form of distributed attack, which brings more threats and degrades the system with only a few poisoned samples. They can also distribute the attack objectives ($\psi$ in $\psi_1, \psi_2, ..., \psi_k$) to turn the attack more effective.

Apart from that, an adversary can also inject random noise into its locally held data or trained model. So, it would be difficult to detect fraud in one-class non-IID distribution. We have considered non-uniform or skewed distribution using a two-class setting and Dirichlet distribution with different hyperparameters for varying heterogeneity.

# 4    *Moat*: The Proposed Defense Technique

Here, we present our proposed defense approach against targeted poisoning attacks named *Moat* (**Mo**del **A**gnostic Defense against **T**argeted Poisoning Attack).

## 4.1    Overview

*Moat* exploits the model's interpretability and is inspired by partial dependence plots, individual conditional expectation, and accumulated feature importance. The partial dependence marginalizes the model for the selected feature $x_S$ to other features $x_S^C$ present in the input. However, the partial dependence obscures the heterogeneity based on the interaction between the features. The intuition is to calculate conditional effects to uncover the heterogeneity. It makes us realize that the induced trigger or the flipped label's accumulated feature would impact the model $f(.)$ to detect the attacker-chosen label. $f(x)$ is the output to be approximated on instance $x$ with $d$ features, and $g$ is an explanatory model for calculating the additive feature attributions. It can be expressed as a linear function of binary variables in Eq. 1.

$$g\left(z'\right) = \phi_0 + \sum_{j=1}^{d} \phi_j z_j' \tag{1}$$

It uses simplified features $x'$ as input using a mapping function such that, $x = h_x(x')$. The explanation for the prediction over $x'$ is based on the local methods, which try to ensure $g\left(z'\right) \approx f\left(h_x\left(z'\right)\right)$ where $z' \approx x'$, $z' \in \{0,1\}^d$, and $\phi_j \in R$. $\phi_j$ is the corresponding attribution to each feature.

*Moat* uses the Shapley algorithm to calculate individual feature attributions $(\phi_j)$ for interpreting the predictions. It assigns an importance score to each feature value with all possible coalitions for a particular prediction. Equation 2 explains the calculation using a value function where $S$ is the selected set of features, $x$ is the data vector selected for interpretation, and $p$ be the number of features considered.

$$\phi_j(\text{val}) = \sum_{S \subseteq [x_1,\ldots,x_p]\backslash(x_j)} \frac{|S|!(p-|S|-1)!}{p!} \left(\text{val}\left(S \cup \{x_1\}\right) - \text{val}(S)\right) \tag{2}$$

$val_x(S)$ is the prediction for feature values in set S that are marginalized over features not included in the set S. It is demonstrated in the Eq. 3 and is calculated through multiple integration's.

$$\text{val}_x(S) = \int \hat{f}\left(x_1,\ldots,x_p\right) dP_{x\notin S} - E_X(\hat{f}(X)) \tag{3}$$

Before discussing the proposed algorithm (*Moat*), we list our observations that conceptualize our conjecture for defense against targeted poisoning attacks.

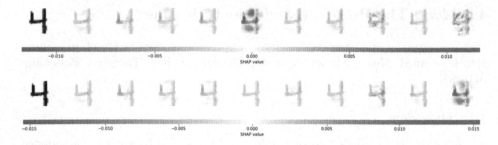

**Fig. 2.** Feature attributions of '4' for 10 labels in a benign ($f$) and malign model ($f'$)

We start with a label flipping attack and perform preliminary experiments to demonstrate how it is spotting heterogeneity among labels. Let us consider that an honest participant trains model $f$, while the adversary trains $f'$ with '4' flipped as '9' in the handwritten digit recognition dataset. The server calculates interpretable approximations for the predictions against an image of '4'. Attribution effect $\phi_j$ for every feature j in the image is calculated using Eq. 2 by operating over the received gradients at the server. The additive attribution of the features is calculated for model interpretation using Eq. 1 against each label. Figure 2 demonstrates the graphical representation of attributions for every labels (0, 1, 2 ..., 9) in both $f$ (top image) and $f'$ (bottom image), where red color depicts positive influence, and the blue negates it. It is pertinent to observe that red spectrum of label '4' in $f$ shifts to '9' in $f'$. After analyzing the attribution tensors, $g\left(z'\right)$, two outliers could be easily spotted in the malicious model $f'$ over the sum of $\phi_j$.

### 4.2   Algorithm

As described in Algorithm 1, it takes client uploaded gradients ($w_t$), some reference data (R), and baseline data (B) as input. The server decides an external input as threshold $\xi$. It is defined for varying the strictness to identify the adversary based on the application's sensitivity. Interpreting the model and post-analyzing, we expect it to converge for the next iteration. It will also provide the probable attacked labels ($l_m$), set of malign ($m_a$) and benign models ($m_b$).

For each client $k$, the server interprets their uploaded model $w_t^k$ with some reference inputs (R). It only requires black-box access to the model. The reference data may be some auxiliary data held by the server for validation or some neutral data. The interpretation $r_k$ learns all the changes in prediction to spot the heterogeneity. Further, it is fed with a baseline input data B for calculating the additive feature attributions against all the labels. The baseline input can be real data owned by the server. However, we suggest using a zero vector input, which acts as a fair baseline and is equitably distributed against all the inputs. The returned output is the attribution value for every feature against each label $(a_1, a_2 \ldots a_l)$, with all the possible coalitions having an influential impact on the

---

**Algorithm 1.** Model-agnostic Defense: Moat

---

**Require:** Client's gradients $(w_t^k)$, Threshold $\xi$
**Ensure:** Global model $(w_{t+1})$, Set of possible attackers $(m_a)$ and attacked labels $(l_m)$

1: $R \leftarrow$ reference data {Auxiliary / Neutral Data}
2: $B \leftarrow$ baseline data {Zero / Mean Vector}
3: $m_b \leftarrow \phi$ {Set of benign models}
4: $m_a \leftarrow \phi$ {Set of attackers}
5: $l_m \leftarrow \phi$ {Set of possible attacked labels}
6: $S_t \leftarrow$ {k} {Set of clients in the round t}
7: **for each client** $k \in S_t$ **do**
8:     $r_k \leftarrow InterpretModel(w_t^k, R)$
9:     $a_1, a_2 \ldots a_l \leftarrow r_k(B)$          //calculation of feature attribution of labels 1,2...,l
10:     **for j = 1 to l do**
11:         $AF_k^j \leftarrow \sum_{i=1}^f \phi_i | \phi_i \in a_j$          //f is the number of features
12:     **end for**
13:     $z \leftarrow Z - score(AF_k)$
14:     **for j = 1 to l do**
15:         **if** $z_j > \xi$ **then**
16:             $l_m \leftarrow l_m \cup$ j
17:         **end if**
18:     **end for**
19:     **if** $|l_m| > 0$ **then**
20:         $m_a \leftarrow m_a \cup w_t^k$
21:     **else**
22:         $m_b \leftarrow m_b \cup w_t^k$
23:     **end if**
24: **end for**
25: **if** $|m_b| == 0$ **then**
26:     $w_{t+1} \leftarrow w_t$
27: **else**
28:     $w_{t+1} \leftarrow \sum_{k \in m_b} \frac{n_k}{n} w_k^k$          //Aggregate using FedAvg
29: **end if**
30: **return** $w_{t+1}$, $m_a$ and $l_m$

---

prediction. We suggest adding the individual label attributions $a_i$, either positive and negative, or their combinations. It results in an array $AF_k$ with value for all the labels.

Further, we perform a Z-score numerical evaluation with varying threshold $\xi$ to detect the presence of an adversary. It describes the relationship of a value with the mean of a group of values and is measured in terms of standard deviations($\sigma$) from the mean $\mu$. For every instance $x$, it is calculated as $(x - \mu)/\sigma$. We have calculated Z-score for label-wise additive attributions $AF_k$. A label with Z-score exceeding a threshold is identified as a malicious label $(l_m)$. Clients with at least one possible malicious label will be treated as malicious. The server computes aggregation of the benign clients $(m_b)$. The algorithm is expected to converge for the next update over all the benign models as long as the convergence of FedAvg is guaranteed. In the case of no benign model, it returns the previous global model $w_t$ to the newly selected set of clients.

After extensive analysis, it is inferred that different combinations of reference (R) and baseline data (B) have different detection capabilities. Possible choices of R and B can be real data, zero-vector data, or single input averaged over specific samples. The attribution effects can be operated with different cases as mentioned in the Appendix A with a detailed analysis for all combinations of references and baselines.

## 5   Experiment and Result Analysis

*Moat* is evaluated with varying numbers of attackers, poisoned samples, heterogeneous distribution, and different attack strategies. The experiments have been run for federated learning architecture with $n = 50$ clients. In each round of training, the server randomly selects $n_k = 30$ clients. Each of the client runs mini-batch stochastic gradient descent locally with a batch size $b = 32$ and a learning rate $\eta = 0.01$ for 2 local epochs. The complete training process has been carried for 50 iterations.

The proposed defense is evaluated against the widely used MNIST and Fashion-MNIST datasets. MNIST[1] is a dataset of handwritten grayscale digits. Fashion-MNIST[2] is a dataset of grayscale fashion products. Both are equally distributed among 10 classes with 7000 images each. Data has been distributed equally, 1200 samples per client, around 120 samples of each category. A convolution Neural Network (CNN) is used for local training at the client device. CNN has two convolution layers with 32 and 64 kernels of the size of 3X3, followed by a max-pooling layer and two fully connected layers with 9216 and 128 neurons, respectively. ReLU activation is used in each layer with a dropout of 0.25.

These experiments are run for testing label-flipping and backdoor attacks against the proposed defense mechanism. For the label flipping attack, we have flipped images of label '6' to label '2' in MNIST, and images of label '8' to label '3' in F-MNIST. Similarly, in the backdoor attack, attackers inject pixel-pattern triggers as shown in Fig. 3 by changing pixels at appropriate locations. Each attacker poisons 500 samples and alters the corresponding class label to target label '2'. The rest of the local data are kept unchanged. We have only considered multiple-shot distributed attack (Attack A-M) following [1] as the default backdoor setting so that the backdoor contribution of attackers does not get weakened after global aggregation.

In each case, the number of compromised workers $n_m$ varies from 10% to 90%. The global model accuracy is tested against 8000 test samples uniformly distributed across all classes, and backdoor attack success rates are tested against 1500 trojan-ed samples. Label flipping attack is evaluated with Attack Success Rate ($l_{asr}$) and Targeted Misclassification Rate ($t_{mcr}$). We evaluated Backdoor

---

[1] http://yann.lecun.com/exdb/mnist/.
[2] https://github.com/zalandoresearch/fashion-mnist.

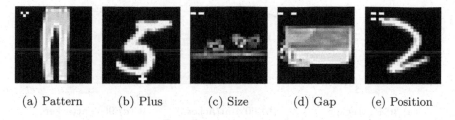

(a) Pattern      (b) Plus      (c) Size      (d) Gap      (e) Position

**Fig. 3.** Variations of backdoor triggers

Attack Success Rate ($b_{asr}$) to calibrate the success of the attacker's intention using an embedded trigger. Attackers aim to achieve high $l_{asr}$, $t_{mcr}$, and $b_{asr}$ while maintaining the global accuracy.

We have used Deep SHAP [9] as the model-agnostic approximation method for the computation of SHAP (SHapley Additive exPlanations) values. The input for the SHAP is the model and some background images. Throughout the experiments, we have used 100 samples uniformly distributed over the classes as the background images (R) for learning average attributions $r_k$ of the $k^{th}$ client models' prediction. Increasing the number of background images (R), results in learning better attribution, and helps in easy detection with fewer rounds. However, to simulate a worst case scenario and to reduce the impact on learning, we simulate R with less than 10% of individual client contribution. A zero vector image of $28 \times 28$ is used as the baseline image B. Attribution results have been discussed and plotted against the threats mentioned above for visualization in Appendix A.

### 5.1   Results

**Defense Against Backdoor Attack.** We have implemented *Moat* defense against distributed pixel-pattern backdoor attack on MNIST and F-MNIST dataset. Figure 4 plots the result for backdoor attack success rates ($b_{asr}$) and corresponding defense for 30%, 60%, and 90% attackers. The external threshold ($\xi$) has been set to 1.8 for MNIST. It is observed from Fig. 4a that global model accuracy (main-acc) does not get impacted by backdoor injection in either of attack or defense. $b_{asr}$ increases gradually from 0.2 to 0.8 in 50 rounds. It is restricted to 0.1 using *Moat*, which clearly suggests that no malicious contribution has been taken while aggregation. Figure 4b and 4c illustrate the defense results with $n_m = 60\%$ and $n_m = 90\%$, respectively. Similar observation have been plotted in Fig. 5 on F-MNIST, keeping threshold as 2. *Moat* has also been tested against varying trigger size (Fig. 3c), gap (Fig. 3d) and position (Fig. 3e) as suggested in [16]. Number of attackers have been set to 20% and $\xi$ as 2. Figure 6 demonstrates the behaviour of *Moat* against all of these trigger factors on F-MNIST. It is prominent that *Moat* successfully defends backdoor attacks and retains $b_{asr}$ to 9–11% for varying size and gap. With little complex trigger (Fig. 3e), $b_{asr}$ is around 15%, which is bit higher compared to other trigger cases as shown in Fig. 6c.

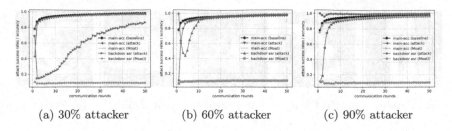

(a) 30% attacker        (b) 60% attacker        (c) 90% attacker

**Fig. 4.** *Moat* defense against backdoor attack on MNIST

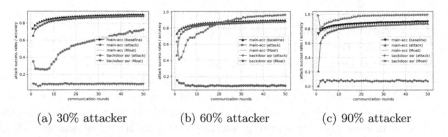

(a) 30% attacker        (b) 60% attacker        (c) 90% attacker

**Fig. 5.** *Moat* defense against backdoor attack on F-MNIST

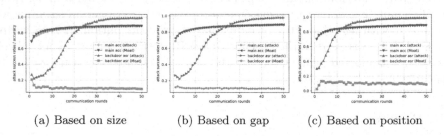

(a) Based on size        (b) Based on gap        (c) Based on position

**Fig. 6.** *Moat* defense against backdoor attack on F-MNIST over variations in trigger

**Defense Against Label Flipping Attack:** Figure 7 illustrates the result for label flipping attacks and defenses on MNIST. We have plotted the results for 30%, 60%, and 90% attackers. The baseline in all the plots depicts the nature of performance in the ideal conditions with no attacker present. With $n_m = 30\%$ and 36% flipped samples, global accuracy decreases around $4-5\%$ as compared to baseline. However, *Moat* has successfully managed to recover to baseline accuracy. After 40 rounds, it overlaps with the baseline. $l_{asr}$ and $t_{mcr}$ increases around 20% and 30% respectively. It is successfully defended by *Moat* by retaining $l_{asr}$ below 10% and $t_{mcr}$ below 20%, respectively. Figure 7d, 7e, and 7f show the results with $n_m = 60\%$ and 72% poisonous samples. Figure 7g shows that accuracy is largely impacted by the attack with $n_m = 60\%$. However, *Moat* recovers the accuracy to baseline, when run for more rounds. $l_{asr}$ and $t_{mcr}$ are restricted at 0.1 and 0.2 respectively as shown in Fig. 7h and 7i. *Moat* forces to continue the training with the remaining 10% benign clients, although the convergence is delayed as compared to other attack cases. It converges in 15–20 iterations while reducing the impact by a margin in initial iterations only. We have also analyzed distributed attack with malicious intention of flipping multiple labels. *Moat* can also defend against such attacks and converges with good accuracy as illustrated in the Fig. 13 of the Appendix B.

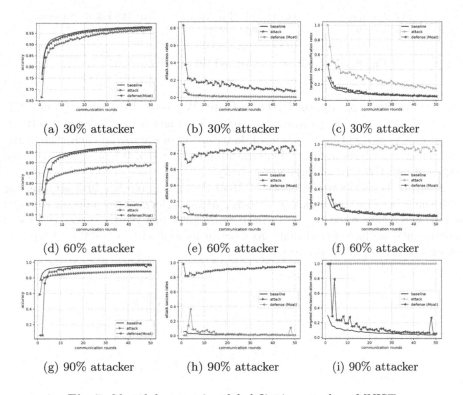

**Fig. 7.** *Moat* defense against label flipping attack on MNIST

**Defense Against Hybrid Attack:** Considering that the attacker's intention is unknown before training, we formulated a hybrid attack with $n_m = 50\%$ malicious clients. 25% of the attacker is actively doing label flipping, and the rest 25% are inducing 'plus' triggers to generate backdoor samples. *Moat* has been analyzed against this attack and proves to be robust against it. It is run for 50 iterations to check the fast convergence on F-MNIST as illustrated in Fig. 8. Figure 8a illustrates the improved main accuracy of 4–5% against the attacked accuracy plot. It achieves almost 40–45% $l_{asr}$ in 50 iterations as illustrated in Fig. 8b. *Moat* brings down $l_{asr}$ to 10% after certain rounds and keeps $l_{asr}$ stable. $b_{asr}$ is also constant around 10% compared to 90% in attack, as shown in Fig. 8c.

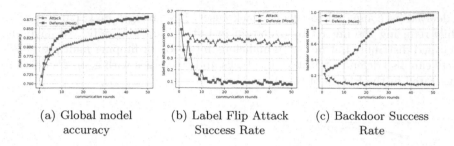

(a) Global model
accuracy

(b) Label Flip Attack
Success Rate

(c) Backdoor Success
Rate

**Fig. 8.** *Moat* defense against hybrid attack on F-MNIST

**Impact of Heterogeneous Distribution:** We have also performed non-IID distribution of the data to simulate a real-life situation in a federated setting. We supply each client an unbalanced distribution of data from each class using a Dirichlet distribution with a hyperparameter $\alpha = 0.5$, $0.7$, and $0.9$. $\alpha$ is the degree of non-IID for varying heterogeneity.

Figure 9 shows the performance of *Moat* on varying $\alpha$ with 30% attacker for both the label flip and backdoor attacks. *Moat* performs very well on $\alpha = 0.9$ and $\alpha = 0.5$ but when $\alpha = 0.7$ the defense performs poorly. After extensive experiments, we have found that *Moat* performs well with $\alpha = 0.9$ on backdoor attack but not on label flip attack.

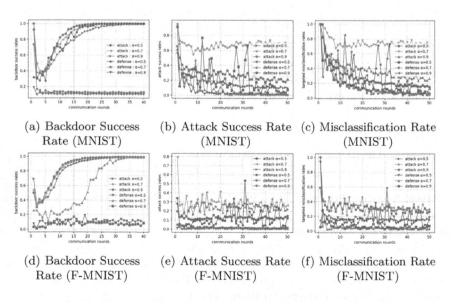

(a) Backdoor Success
Rate (MNIST)

(b) Attack Success Rate
(MNIST)

(c) Misclassification Rate
(MNIST)

(d) Backdoor Success
Rate (F-MNIST)

(e) Attack Success Rate
(F-MNIST)

(f) Misclassification Rate
(F-MNIST)

**Fig. 9.** *Moat* Defense over non-IID data with varying degree

We have considered a more strict set of non-IID data distribution as suggested in [10]. First, the training data is sorted by class labels. It is split into chunks of

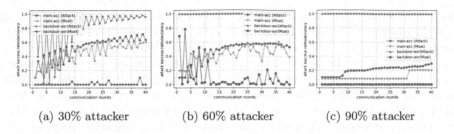

(a) 30% attacker                (b) 60% attacker                (c) 90% attacker

**Fig. 10.** *Moat* Defense against backdoor attack in non-IID setting on F-MNIST

300, and each participant is assigned 2 different chunks from 2 different classes. Figure 10 shows the performance of *Moat* over this non-uniform distribution. It is clear that the backdoor attack success rate ($b_{asr}$) fluctuates rapidly in initial rounds and takes many rounds to converge. However, $b_{asr}$ is stable at 0.0 except for a few rounds 14, 21, and 37. It is pertinent to observe that in a non-IID setting, main-accuracy is somehow impacted by *Moat*. With $n_m = 60\%$, $b_{asr}$ is constant at 100%. After applying *Moat*, it keeps fluctuating between 0 and 0.8 at the initial rounds and gets stable after round 10, with some spikes observed afterward. However, $b_{asr}$ is way below the attack-ed backdoor success rate, which validates the effectiveness of *Moat*.

## 6   Discussion

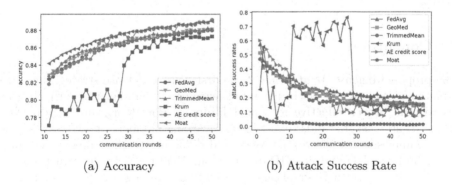

(a) Accuracy                          (b) Attack Success Rate

**Fig. 11.** Comparison with existing defenses

We have evaluated some of the existing defense strategies for targeted poisoning attacks, namely Krum [2], Autoencoder (AE) credit score [8], GeoMed, and Trimmed Mean [17]. We ran experiments on F-MNIST with $n = 30$ clients, and $n_m$ is set to 20% with number of attackers $m = 6$. For GeoMed, we have taken the layerwise median of the uploaded client gradients. For Trimmed Mean, top $n - 2\beta$ values closest to median are chosen with $\beta = 13$. For implementing Krum, we have set the multi-Krum parameter ($f = n - m - 2$) as 22 for meeting the assumption $2m + 2 < n$. For the AE credit score-based approach, an autoencoder is trained at the server with client updates accumulated till 20 rounds. The autoencoder input is a vector of 4096 chosen randomly from the first fully connected layer of the CNN model described above. It is trained for 20 local epochs with a batch size $b = 32$, $\eta = 0.0001$ and a dropout rate of 0.2. The pre-trained autoencoder is used for detecting malicious behavior. A client with a credit score exceeding a threshold is considered an attacker. The threshold is set to mean anomaly score following [8].

All of the above algorithms are run for 50 communication rounds with the attacker actively performing label flip with the similar settings discussed above. Figure 11a and 11b illustrates the result of comparison in terms of global model accuracy and attack success rates. Similarity-based approaches could achieve accuracy around 86–88%, while AE looks to outperform them by achieving 90% accuracy. *Moat* achieves 86–88% accuracy in merely 15–20 iterations and surpasses 90% with 50 iterations. These are designed attack-specific and are not robust against different targeted poisoning attacks. GeoMed, Krum, and Trimmed Mean are only robust up to $\lceil \frac{n}{2} \rceil - 1$ attackers. They require prior knowledge of malicious participation as well as white-box access to the client gradients. They suffer badly against a non-IID setting. Authors in [4] claim their defense strategy is secure against existing attacks and strong adaptive attacks. However, they suffer with a root dataset having less than 100 examples and proves to be robust only against 60% attackers. AE credit score-based approach also requires white-box access to client gradients and is robust only up to $\lceil \frac{n}{3} \rceil$ attackers. *Moat* scales with the number of attackers and are robust against $n - 1$ attackers.

## 7   Conclusion

The proposed defense *Moat* is one of the first generic defense strategies against targeted poisoning attacks in federated learning. It has been extensively run, tested, and proven effective for Label Flipping, Backdoor, and Hybrid (Distributed) attacks. It stands robust against varying numbers of adversaries, poisoned samples, architectures, datasets, and different attack strategies. It works under both the IID and non-IID distribution. *Moat* restricts the attack success rate to 5–10%, which is significantly lower than existing defense strategies. It converges with the set of benign clients even in the presence of a majority of compromised workers (90%).

**Acknowledgement.** We acknowledge the Ministry of Human Resource Development, Government of India, for providing fellowship to complete this work.

# Appendix

## A    SHAP Analysis

Figure 12a shows the SHAP attribution results for all 10 labels $a_0$ to $a_9$, of 'pattern' triggered backdoor model with target label '2'. Figure 12b illustrates the attributions of a honest model. It is pertinent to observe the additive increment of the blue spectrum of $a_2$ for the malign model. Figure 12c illustrates the SHAP attribution results in a label flipped attack-ed model for class '6' flipped to '2', where $a_6$ produces more red spectrum as compared to the benign model. It allows the server to detect the presence of backdoor without any knowledge of embedded trigger since the additive attributions of attacked label in malicious model and benign model differs for the same reference and baseline data.

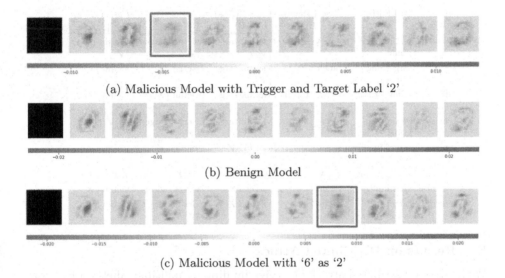

(a) Malicious Model with Trigger and Target Label '2'

(b) Benign Model

(c) Malicious Model with '6' as '2'

**Fig. 12.** SHAP attributions over a zero-vector baseline

**Different Combinations of References and Baseline:** We have analyzed different combinations of reference data (R) for learning feature attributions and the baseline data (B). R can be a subset of real data (real), a single sample with a mean of a set of real data (mean), and zero-vector data (neutral). We have analyzed this with an instance of label flip attack where '4' is flipped to '9' and backdoor attack where trojan is inserted in '4' and the class label changed to '9'. A image of '4' (victim_img), '9' (target_img) and a zero-vector(z_img) image is used as B in label flip attack. For backdoor attack, a benign sample of '4' (g_img), a triggered sample of '4' (b_img) and a zero-vector(z_img) image is used as baseline. After getting label-wise attributions $(a_1, a_2, \ldots, a_l)$, we can perform outlier detection by calculating as, Case 1: total influence ($\sum_{i=1}^{f} \phi_i \in a_j$), Case 2: absolute influences ($\sum_{i=1}^{f} |\phi_i| \in a_j$), Case 3: positive influences ($\sum_{i=1}^{f} \{\phi_i \in a_j | \phi_i > 0\}$), and Case 4: negative influences ($\sum_{i=1}^{f} \{\phi_i \in a_j | \phi_i < 0\}$).

Table 1 tabulates the detection capabilities, whether *Moat* can detect the victim, target, or both the label for label flipping attack. It can be observed that when R is real with B set to z_image, it detects both the victim and flipped label in all 4 cases. The absolute sum of the attribution influences across all R can detect either victim, attacked, or both labels. In backdoor attack, we check whether the altered label ('9') of backdoor-ed input can be detected or not, as listed in the Table below.

**Table 1.** Detection capabilities for various combinations of R and B against Label Flip (LF) and Backdoor (B) attacks

| R | B (LF \| B) | Case 1 (Sum) (LF \| B) | Case 2 (Abs) (LF \| B) | Case 3 (Pos) (LF \| B) | Case 4 (Neg) (LF \| B) |
|---|---|---|---|---|---|
| real | victim_img \| g_img | target \| ✓ | victim, target \| ✓ | victim, target \| ✓ | victim, target \| ✓ |
| real | target_img \| b_img | target \| ✓ | victim, target \| ✓ | victim \| ✓ | victim \| ✓ |
| real | z_img \| z_img | victim, target \| ✓ | victim, target \| ✗ | victim \| ✗ | victim, target \| ✗ |
| mean | victim_img \| g_img | target \| ✓ | victim, target \| ✓ | victim, target \| ✓ | − \| ✓ |
| mean | target_img \| b_img | target \| ✓ | victim \| ✓ | victim \| ✓ | victim \| ✓ |
| mean | z_img \| z_img | − \| ✗ | victim \| ✗ | − \| ✗ | victim \| ✗ |
| neutral | victim_img | target \| ✓ | target \| ✗ | target \| ✓ | victim \| ✗ |
| neutral | target_img \| b_img | target \| ✓ | victim \| ✗ | victim, target \| ✗ | victim \| ✗ |

## B     Results on Distributed Attack

We analysed distributed attack objective by flipping multiple labels on Fashion-MNIST dataset. Label '8' is flipped to '3', '6' to '2' and '9' to '1'. The results are plotted in Fig. 13a, 13b and 13c and show good convergence for Moat.

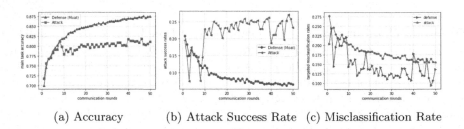

(a) Accuracy          (b) Attack Success Rate     (c) Misclassification Rate

**Fig. 13.** Distributed attack on Fashion-MNIST dataset

# References

1. Bagdasaryan, E., Veit, A., Hua, Y., Estrin, D., Shmatikov, V.: How to backdoor federated learning. In: International Conference on Artificial Intelligence and Statistics, pp. 2938–2948. PMLR (2020)
2. Blanchard, P., Guerraoui, R., Stainer, J., et al.: Machine learning with adversaries: Byzantine tolerant gradient descent. In: Advances in Neural Information Processing Systems, pp. 119–129 (2017)
3. Cao, D., Chang, S., Lin, Z., Liu, G., Sun, D.: Understanding distributed poisoning attack in federated learning. In: 2019 IEEE 25th International Conference on Parallel and Distributed Systems (ICPADS), pp. 233–239. IEEE (2019)
4. Cao, X., Fang, M., Liu, J., Gong, N.Z.: Fltrust: Byzantine-robust federated learning via trust bootstrapping. arXiv preprint arXiv:2012.13995 (2020)
5. Fung, C., Yoon, C.J., Beschastnikh, I.: Mitigating sybils in federated learning poisoning. arXiv preprint arXiv:1808.04866 (2018)
6. Gu, T., Dolan-Gavitt, B., Garg, S.: Badnets: Identifying vulnerabilities in the machine learning model supply chain. arXiv preprint arXiv:1708.06733 (2017)
7. Kasyap, H., Tripathy, S.: Privacy-preserving decentralized learning framework for healthcare system. ACM Trans. Multimed. Comput. Commun. Appl. **17**(2s), 1–24 (2021)
8. Li, S., Cheng, Y., Liu, Y., Wang, W., Chen, T.: Abnormal client behavior detection in federated learning. arXiv preprint arXiv:1910.09933 (2019)
9. Lundberg, S.M., Lee, S.I.: A unified approach to interpreting model predictions. In: Advances in Neural Information Processing Systems, pp. 4765–4774 (2017)
10. McMahan, B., Moore, E., Ramage, D., Hampson, S., Arcas, B.A.: Communication-efficient learning of deep networks from decentralized data. In: Artificial Intelligence and Statistics, pp. 1273–1282. PMLR (2017)
11. McMahan, H.B., Moore, E., Ramage, D., y Arcas, B.A.: Federated learning of deep networks using model averaging. CoRR abs/1602.05629 (2016). http://arxiv.org/abs/1602.05629
12. Salem, A., Wen, R., Backes, M., Ma, S., Zhang, Y.: Dynamic backdoor attacks against machine learning models. arXiv preprint arXiv:2003.03675 (2020)
13. Takeishi, N., Kawahara, Y.: On anomaly interpretation via shapley values. arXiv preprint arXiv:2004.04464 (2020)
14. Wang, G.: Interpret federated learning with shapley values. arXiv preprint arXiv:1905.04519 (2019)
15. Wang, G., Dang, C.X., Zhou, Z.: Measure contribution of participants in federated learning. In: 2019 IEEE International Conference on Big Data (Big Data), pp. 2597–2604. IEEE (2019)
16. Xie, C., Huang, K., Chen, P.Y., Li, B.: Dba: distributed backdoor attacks against federated learning. In: International Conference on Learning Representations (2019)
17. Yin, D., Chen, Y., Kannan, R., Bartlett, P.: Byzantine-robust distributed learning: Towards optimal statistical rates. In: International Conference on Machine Learning, pp. 5650–5659. PMLR (2018)
18. Zhao, Y., Chen, J., Zhang, J., Wu, D., Teng, J., Yu, S.: Pdgan: A novel poisoning defense method in federated learning using generative adversarial network. In: International Conference on Algorithms and Architectures for Parallel Processing, pp. 595–609. Springer (2019)

# Malware Analysis and Detection

# Certified Malware in South Korea: A Localized Study of Breaches of Trust in Code-Signing PKI Ecosystem

Bumjun Kwon[1], Sanghyun Hong[2], Yuseok Jeon[3], and Doowon Kim[4(✉)]

[1] The Affiliated Institute of ETRI, Daejeon, South Korea
[2] Oregon State University, Corvallis, USA
[3] Ulsan National Institute of Science and Technology (UNIST), Ulsan, South Korea
[4] University of Tennessee, Knoxville, USA
doowon@utk.edu

**Abstract.** Code-signing PKI ecosystems are vulnerable to abusers. Kim et al. reported such abuse cases, *e.g.,* malware authors misused the stolen private keys of the reputable code-signing certificates to sign their malicious programs. This *certified* malware exploits the chain of the trust established in the ecosystem and helps an adversary readily bypass security mechanisms such as anti-virus engines. Prior work analyzed the large corpus of certificates collected from the wild to characterize the security problems. However, this practice was typically performed in a global perspective and often left the issues that could happen at a local level behind. Our work revisits the investigations conducted by previous studies with a local perspective. In particular, we focus on code-signing certificates issued to South Korean companies. South Korea employs the code-signing PKI ecosystem with its own regional adaptations; thus, it is a perfect candidate to make a comparison. To begin with, we build a data collection pipeline and collect 455 certificates issued for South Korean companies and are potentially misused. We analyze those certificates based on three dimensions: (i) abusers, (ii) issuers, and (iii) the life-cycle of the certificate. We first identify that the strong regulation of a government can affect the market share of CAs. We also observe that several problems in certificate revocation: (i) the certificates had issued by local companies that closed the code-signing business still exist, (ii) only 6.8% of the abused certificates are revoked, and (iii) eight certificates are not revoked properly. All of those could lead to extending the validity of certified malware in the wild. Moreover, we show that the number of abuse cases is high in South Korea, even though it has a small population. Our study implies that Korean security practitioners require immediate attention to code-signing PKI abuse cases to safeguard the entire ecosystem.

## 1 Introduction

The establishment of trust in software distributed over the Internet is challenging due to the nature of software distribution: unknown sources and a high chance

© Springer Nature Switzerland AG 2021
D. Gao et al. (Eds.): ICICS 2021, LNCS 12918, pp. 59–77, 2021.
https://doi.org/10.1007/978-3-030-86890-1_4

of tampering during distribution. To overcome these challenges and to guarantee the authenticity and integrity of software, Code-signing PKI is designed and now becomes a de-facto standard in the software ecosystem. Similar to other PKIs such as the Web's PKI, the code-signing PKI also requires Certificate Authorities (CAs) to attest that a certificate belongs to a legitimate software publisher. The CAs issue code-signing certificates for publishers, after the vetting process. Software publishers sign their software with their issued certificates to warrant the authenticity and the integrity of the software. In turn, clients can establish trust in the signed software by verifying the digital code-signing signature. They can know not only the identity of the publisher but also that the software has not been altered during the distribution.

A security rule of thumb for a user is to only execute or install software that contains valid signatures from reputable software publishers with whom she can establish trust. However, anecdotal evidence has shown that the security rule cannot be guaranteed since software properly signed by legitimate publishers can be severe malware [10,24,31]. For example, the Stuxnet worm included device drivers that had been properly signed with the private keys stolen from two Taiwanese semiconductor companies, located in close proximity [10]. The fact is that the malicious usage of these stolen private keys helps remain undetected for a longer period than the other malware [10]. Furthermore, the abuse of code-signing is also prevalent among Potentially Unwanted Programs (PUPs) [5,16,17,32].

This observation has sparked an interest in the real-world breaches of trust in the code-signing ecosystem. In particular, Kim *et al.* [13,14] conducted a large-scale analysis of code-signing abuse cases in the Windows code-signing PKI ecosystem. However, these studies were mostly conducted from a global perspective; hence, they often left the breaches that would happen in sub-populations overlooked. Local software publishers may mainly target local customers; so in this case, the local publishers should have regional adaptations in their code-signing ecosystem, considering the environmental factors of their countries or regions—*e.g.,* because of law[1]. For instance, regulations may state the qualification of a CA or force how the PKI should be operating. Thus, the characteristics of abuse cases can be different from the previous studies. Moreover, the analysis tools in prior work focus on emphasizing the most prevalent findings in the collected datasets.

In this paper, we tackle the prior emphasis on the global perspective and make a first step towards understanding the breach cases in the sub-populations. Specifically, we ask: *What characteristics can we find from an analysis of a specific country?* To answer this question, we give an eye to the Windows code-signing PKI in South Korea. South Korea is known to have its unique PKI ecosystem, developed alongside the digital signature act (DSA), which was established in 1999 [6,15]. DSA states that only a signature is valid if it is endorsed by an accredited CA. South Korean users are also known to be exposed to various "security software" necessary for web activities where identity verification is

---

[1] PKI in Asia – Case Study and Recommendations: https://fidoalliance.org/wp-content/uploads/FIDO-UAF-and-PKI-in-Asia-White-Paper.pdf.

required *e.g.,* banking, e-commerce [28,29]. The electronic financial transaction act, which became active in 2007, has fostered such an environment. Therefore, we may expect to see unique characteristics reflected in the code-signing PKI ecosystem as well. Nevertheless, little is known about the regional differences; the same applies to the Korean code-signing PKI abuse.

We design a system, which extracts code-signing certificates and identifies Korean certificates that are likely compromised. We utilize information from the certificate and the scanning reports of the binary for identification. We examine the characteristics of Korean signed malicious samples and compromised certificates. Specifically, we investigate how prevalent code-signing abuse is, who are the abusers, who issue the certificates, and whether the compromised certificates are adequately revoked or not.

We found code-signing abuse is prevalent in Korea for its population. The number of signed malicious samples accounted for 1.8% of the total samples, whereas the population is nearly 1% among the global internet users. We also find the unique distribution of the CAs. Thawte dominates the population and a local CA Yessign is observed. Yessign is out of the code-signing business and that could be a potential problem in revocation. Such characteristics might be due to the web environment in Korea, cultivated by its regional PKI laws. Besides, we observe revocation is not done properly in Korea as well. Only 6.8% of the certificates are revoked, and eight certificates have set revocation dates ineffectively. It endangers users of the signed malicious binaries.

**Contributions.** In summary, we make the following contributions:

- We design a system that collects the malicious programs and compromised certificates from South Korea. We identified 455 certificates that are issued for South Korean companies and are potentially misused by malware authors.
- We highlight the abuse cases in the code-signing ecosystem in South Korea. Using the observations in the previous studies as our baseline, we report the commonalities and differences in our findings.
- Using those differences, we analyze and identify the distinct characteristics of Korean compromised certificates that are fostered by the regional laws.

## 2   Background and Motivation

In this section, we briefly overview the code-signing PKI; especially, the code-signing process, the distinct characteristic of the code-singing PKI that is mainly different from the Web's PKI, and revocation that can cause extra security threats. We then highlight our motivation why we need to study the unique characteristics of the Korean code-signing PKI ecosystem.

### 2.1   Overview of the Code-Signing PKI

Code-signing is a security technology that utilizes the digital signature mechanism. It helps authenticate the publisher of a software program and guarantees the software's integrity after signing. It requires creating a digital signature using the publisher's private key (i.e., signing), and then embed the digital signature into the software. In turn, for clients, when verifying the signed software, they need the public key associated with the publisher's private key to verify the signature.

The code-signing also relies on Public Key Infrastructure (PKI), called the Code-signing PKI. As the nature of the Internet, clients cannot trust any public key transferred over the Internet that claims to be legitimate. It is because public keys do not have any information about the ownership. To mend this problem, third-parties, called Certificate Authorities (CAs), attest that a public key belongs to a particular owner (in this case, a software publisher or developer) who possesses the associated private key. We call this endorsed key a certificate. As long as we trust the CAs, we trust all certificates issued by the CAs except for revoked certificates. This chain of trust starts from the end entity (i.e., publisher) to the root certificate pre-installed in client-side systems such as operating systems or web browsers.

## 2.2  Code-Signing Process

Like the Web's PKI (e.g., TLS), a software publisher first applies for code-signing certificates to CAs with the applicant's public key. After verifying the publisher's identity, the CA issues a code-signing certificate based on the X.509 v3 certificate standard [8]. The software publisher uses its private key associated with the issued certificate to sign its software. Specifically, in the signing process, the hash value of the software is first computed, and then, the hash value is digitally signed with the publisher's private key. Finally, the digital signature and the chain of the certificates are bundled with the original software. This whole process is illustrated in Fig. 5. In turn, the client has to verify the signature with the public key embedded in the certificate when encountering the signed software. The verification process allows clients to recognize any modifications of the program when verifying the signed software.

**Trust Timestamping.** The distinct difference between the Web's PKI and the code-signing PKI is *trust timestamping*. The trust timestamping guarantees when a binary file is signed, and if a binary is signed before the certificate's expiration date, the validity extends after the certificate expires, which is different from the Web's PKI where the validity of a domain is no longer ensured after the certificate expires.

As illustrated in Fig. 5, when a binary file is signed, the hash value of the binary file is sent to a Time Stamping Authority (TSA), and the TSA issues a trusted timestamp. The TSA signs the timestamp and the hash value with its certificate. This so-called trust timestamp is sent back to the publisher. Then the software publisher embeds the trust-timestamp signature and the TSA's certificate into the signed software.

## 2.3  Revocation

Another important role for CAs besides issuing certificates is to revoke the compromised certificates that they have issued. There are various reasons for CAs to revoke their issued certificates; (1) when the private key associated with a certificate is stolen and used to sign malware samples [13], (2) when a weak cryptographic key is used to generate a certificate [33], (3) when CAs are hacked and compromised, and then issue certificates for adversaries [24], and (4) when a certificate is issued under the name of a shell company or through impersonations, etc. [13].

There are two primitive ways for CAs to disseminate the revocation status information; (1) Certificate Revocation List (CRL) and (2) Online Certificate Status Protocol (OCSP). In CRL, clients need to download the revocation lists periodically and to check if the certificate is on the lists. If the certificate's serial number is on the list, clients can consider the certificate is revoked and no longer valid. OCSP is the successor to CRL, and it allows clients to query a CA for the revocation status of a certain certificate rather than downloading a bulk of the serial numbers using CRLs. Both CRLs and OCSP responses are signed with CAs' certificates to guarantee their integrity.

**Erroneous Revocation Data Setting.** When revoking certificates, CAs must set the effective revocation date (c.f., Sect. 2.3). Kim et al. [14] have examined the security problems of the current code-signing revocations. If CAs erroneously set an effective revocation date, all signed programs (including malware) signed before the effective revocation date can remain valid even though the certificate is revoked. It is due to the trust timestamp mechanism.

## 2.4   Motivation

**Code-Signing Abuse.** Recent measurement studies   [5,13,14,16–18] have reported that adversaries have attempted to compromise the code-signing PKI for their malicious purpose; their main purposes are 1) to efficiently distribute their malware and 2) to lure clients into installing their malware. Attackers can make a bold move of stealing the private keys of benign software companies and use the keys to sign their malware, which makes a much powerful attack. The signed malware now looks like a legitimate product from a benign software company, which misleads clients to believe the signed malware is safe to execute. Furthermore, adversaries incorporate shell companies and use this fake company information to get issued code-signing certificates legally and legitimately from the code-signing CAs.

**Motivation for a Regional Study.** Previous measurement studies have been conducted from a global perspective considering software publishers and CA as global entities. However, this global perspective analysis can lead to misunderstanding or neglecting local characteristics because it mainly focuses on global-scale cases. In other words, the code-signing abuse cases may vary depending on the locality of the attackers and their targets. Thus, to enhance the security of the code-signing PKI ecosystem, we need to understand 1) the local characteristics of the code-signing abuse cases and 2) adversaries who compromise local software publishers targeting local victims. Moreover, in terms of data collection, the previous methods and results may often be biased to the majority population and a limited number of countries. Thus, a regional target attack campaign with a small number of malware samples could have been neglected or overlooked.

**South Korean Web Environment.** We focus on South Korea for this study. South Korea has a unique environment fostered by the regulations. Two acts played as the dominant factor. The digital signature act (DSA) established in 1999 has restricted the "valid" form of digital signatures [6,15]. It only concedes signatures issued from "accredited CA"s to be legitimate. There are six accredited CAs, including KFTC, KICA, Koscom, KECA, KTD, and Initech [4]. Among them, KFTC once served as a code-signing CA under the name of "yessign" (https://www.yessign.or.kr/). KICA and KECA act as a distributor of the global code-signing CA. KICA (https://www.kicassl.com/) is a relay of Comodo; KECA (https://cert.crosscert.com/) offers Digicert and Thawte products. Next, the electronic financial transaction act, which became active in 2007, is known as the main cause of the notorious Korean web environment. Due to this act, Korean users have been forced to install various "security software" such as keylogger detection for web activities [28,29]. The flood of these mandatory "legitimate" software, which are digitally signed, may have introduced side effects that incapacitate the defense mechanism of code-signing. For instance, a survey was conducted on Korean adware victims [1], which reported that only 2.8% consciously clicked "allow install" the adware. Moreover, anecdotes [2,3] show that South Korean software companies have become an attractive target for adversaries. Specifically, many South Korean software companies were stolen the private keys of their code-signing certificates, and the stolen private keys were misused to sign malware. Therefore, we believe South Korea is an attractive candidate for studying the local characteristics of the code-signing PKI, and understanding such characteristics may help improve the security of the entire code-signing PKI ecosystem.

## 3   Data Collection

To better understand the landscape of code-signing abuse in South Korea, we first need to capture signed malware and PUPs in the wild and extract code-signing certificates. From the code-signing certificates, we need to obtain information such as publisher names (common names), locality addresses, issue dates, expiration dates, issuers (CAs), and more. However, due to the nature of software distribution, it is significantly challenging to collect all signed malicious samples and their code-signing certificates in the wild. Whereas in the Web's PKI, a comprehensive list of TLS certificates can be readily collected by scanning the entire IPv4 addresses with a network scanner (e.g., ZMap [9]). This is because signed malware samples can be distributed through a pre-installed updater/installer tool; or others can be distributed from external storage or directly from websites. To overcome these challenges, we present a new collection pipeline for Korean code-signing certificates that are likely compromised, as illustrated in Fig. 1.

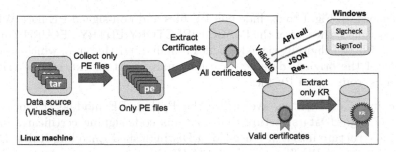

**Fig. 1. The overview of our Korean compromised certificates collection pipeline.** (1) Malicious files are collected from VirusShare, (2) filter out non-PE files, (3) extract code signing certificates from PE files, (4) validate PE files and certificates using the Windows SigCheck and SignTool, and (5) extract only Korean certificates.

### 3.1 Data Source

We utilize *VirusShare* (http://virusshare.com), the large corpus of malware, to collect signed malware and to extract Korean compromised code signing certificates from the corpus. We also utilize VirusTotal (http://virustotal.com) to label the collected signed malware samples.

**VirusShare and VirusTotal.** We collect malicious binaries from VirusShare that is one of the most extensive sets of malware samples available to the public. Since the data sets are freely downloadable, many security research works have utilized them. The malicious samples consist of not only Windows Portable Executable (PE) files, but also HTML files including malicious JavaScript code. We sample 57 tar files (out of 312 tar files) from VirusShare. Each tar file contains either 131,072 or 65,536 malware samples. We collect a total of 5,934,399 malicious files.

To classify the malicious samples, we use VirusTotal. VirusTotal is a Web service where users can freely upload executable samples (including malware and benign samples) and analyze the samples to classify with up to 63 different Anti-Virus (AV) engines. The service provides a report containing the number of AV engines that detect the samples as malicious and the corresponding labels. In our work, we utilize that information to classify the collected samples (c.f., Sect. 3.3).

### 3.2 System Overview

In this section, we describe our new system. As presented in Fig. 1, the new system is a pipeline for identifying digitally signed PE files and extracting compromised Korean code-signing certificates.

**Identifying PE Files.** We first filter out non-PE files from the total of 5,934,399 malicious files (out of 57 VirusShare tar files) since the files include not only PE files, but also JavaScript code. The 5,934,399 samples are fed into our system shown in Fig. 1 to exclude non-PE files and non-signed PE files. When a PE file is signed, the size of the IMAGE_DIRECTORY_ENTRY_SECURITY field is non-zero. 525,071 digitally signed PE files remain after this step, which accounts for 8.8% of the original data set. We now move next to extract code-signing certificates from signed PE files.

**Extracting Certificates.** We utilize the Python PE module[2] to locate the PKCS #7 SignedData structure that contains code-signing certificates and to dump the structure into a file encoded in Distinguished Encoding Rules (DER). Of 525,071 signed PE files, only 495,124 PKCS #7 files are extracted due to parsing errors. Then, we extract all code-signing certificates (including TSA certificates and code-signing intermediate certificates) from the DER-encoded PKCS #7 files, and then we filter out non-leaf code signing certificates using the keyword of "CodeSigning" in the *extendedKeyUsage* extension field and the "Basic Constraints" field.

**Valid Korean Certificates.** The last part is where we obtain the set of Korean certificates that are valid. We can specify the certificates belonging to a particular country by looking at the country code in the leaf certificate's subject field. If the country code of a leaf certificate is "KR," we know that the certificate is issued for a Korean publisher. Using this concise but effective method, we identify 844 certificates issued to Korean identities and 8,815 malicious PE files signed with those certificates. The number of PE files accounts for 1.8% of the initial data set.

Now we explain the verification process. Only valid certificates remain after this step. We first verify the digital signatures and code-signing certificates embedded in PE files using both SignTool[3] and SigCheck[4] tools in the Windows Sever 2016. SignTool returns error code with a message for the scanned certificate. Table 4 enumerates the error code returned by SignTool and the associated messages. We consider the three messages of *"Successfully Verified,"* *"0x800B0101,"* and *"0x800B010C"* valid since the two error code, *"0x800B0101"* and *"0x800B010C"* are returned only when PE samples have been properly signed. Specifically, *0x800B0101* returns when a PE file has not been trust-timestamped and its certificate expires, and when a certificate is revoked, *0x800B 010C* returns. Detailed information is described in Table 4. In the end, we have 783 valid Korean certificates and 8,093 PE files signed with the valid Korean certificates as described in Table 5(left). In other words, 94.2% of signed samples have a proper PKCS#7 structure.

### 3.3   Binary Labeling

Samples from VirusShare may contain false positives. Here, we describe the line of efforts we made to reduce the false positives. First, we re-scan the malicious

---

[2] https://github.com/erocarrera/pefile.

[3] https://docs.microsoft.com/en-us/windows/desktop/seccrypto/signtool.

[4] https://docs.microsoft.com/en-us/sysinternals/downloads/sigcheck.

samples using VirusTotal. It is known that AV engines' labels may change over time as more evidence is gathered. Thus, some samples may be re-labeled as benign. We observe 234 PE samples among the malicious samples signed with Korean code-signing certificates are no longer malicious after the re-scan.

Next, we set a threshold to filter out samples with less confidence. For each signed PE sample, we define $c_{mal}$ as the number of AV engines in VirusTotal that label the sample as malware. We consider a signed PE file as malware when $c_{mal} \geq 10$. For example, $c_{mal} \geq 10$ means that about 15% of AV engines (out of more than 60 AV engines in VirusTotal) detect the samples as malware. This approach is presented in prior works [13,19]. After this step, we now have 455 valid Korean certificates used to sign malicious samples detected by more than 10 AV engines in VirusTotal.

As a final step, we utilize a malware labeling tool, called AVClass [30] to label our malicious samples and classify them into malware and Potential Unwanted Program (PUP).

# 4   Code-Signing PKI Abuse in Korea

Table 5(right) summarizes the breakdown of PE malicious files, signed PE malicious files, Korean compromised certificates, and Korean malicious PE files signed with the Korean certificates. With this data, we investigate the characteristics of code-signing PKI and the abuse within Korea. Here, we try to answer the following research questions.

1. *Q1: How prevalent is code-signing abuse in Korea?*
2. *Q2. Who abuses the code-signing in Korea?*
3. *Q3: Who issued the certificate?*
4. *Q4: Are the certificates issued with safe cryptographic guarantees?*
5. *Q5: How long do the abusive certificates survive in Korea?*

The final goal for these questions is to ask the main research question we raised in the introduction: *Q: What characteristics can we find from an analysis of a specific country?*

## 4.1   Abusers

We answer a couple of questions *Q1. How prevalent is code-signing abuse in Korea?* and *Q2. Who abuses the code-signing in Korea?* in this section. We initiate with simple statistics to answer the first question. For the second question, we investigate the problem from two different angles 1) the malicious sample family based on their labels and 2) the publisher's information stated on the certificate.

**Prevalence.** As presented in Table 5(right), signed malicious binaries with Korean certificates are 844  in numbers. It accounts for 1.8% of the data set, which is a global collection. The Korean internet population is about 5 million,

**Table 1. Top 10 Malware/PUP label breakdown.** SINGLETON is labeled when AVCLass is unable to find a family name for a malware sample such as generics. On average, a Korean certificate is used to sign 2.5 different family of malicious samples. Bold malware families are considered as trojan or severe threats.

| Family label | PE | Certificate |
|---|---|---|
| **Kraddare** | 2,850 (41.70%) | 198 (43.52%) |
| **Onescan** | 798 (11.68%) | 50 (10.99%) |
| SINGLETON* | 418 (6.12%) | 191 (41.98%) |
| Sidetab | 298 (4.23%) | 5 (1.10%) |
| Hotclip | 177 (2.59%) | 6 (1.32%) |
| Openshopper | 169 (2.47%) | 7 (1.54%) |
| **Delf** | 158 (2.31%) | 32 (7.03%) |
| Viruscure | 243 (3.23%) | 36 (6.79%) |
| Adkor | 135 (1.98%) | 63 (13.85%) |
| Hebogo | 121 (1.77%) | 7 (1.54%) |
| **Total** | 6,835 (100%) | 1,123 (246.81%) |

**Table 2. Top 10 common name, issuer, and region breakdown.** N/A in region means that neither province nor locality name information are specified.

| Common Name | PE | Cert. | Issuer | Cert. | Region | Cert. |
|---|---|---|---|---|---|---|
| cloudweb Inc | 1,040 (15.22%) | 3 (0.66%) | Thawte | 336 (73.85%) | Seoul | 267 (58.68%) |
| nbiz Ltd. | 702 (10.27%) | 5 (1.10%) | VeriSign | 74 (16.26%) | Busan | 63 (13.85%) |
| UCF | 489 (7.15%) | 4 (0.88%) | YesSign | 19 (4.18%) | Gyeonggi-do | 63 (13.85%) |
| NKsolution Corp. | 358 (5.24%) | 5 (1.10%) | eBiz Networks | 10 (2.20%) | N/A | 20 (4.40%) |
| Akorea | 306 (4.48%) | 5 (1.10%) | Symantec | 7 (1.54%) | Incheon | 11 (2.42%) |
| SearchLink Co., Ltd. | 263 (3.85%) | 3 (0.66%) | GlobalSign | 6 (1.32%) | Gyeongsangbuk-do | 7 (1.54%) |
| TGSM Inc. | 194 (2.84%) | 5 (1.10%) | COMODO | 3 (0.66%) | Daegu | 7 (1.54%) |
| JE communication | 166 (2.43%) | 4 (0.88%) | | | Gyeongsangnam-do | 6 (1.32%) |
| 씨큐미디어 | 161 (2.36%) | 2 (0.44%) | | | Jeollanam-do | 2 (0.44%) |
| OPEN.co., ltd | 158 (2.31%) | 4 (0.88%) | | | Ulsan | 2 (0.44%) |
| **Total** | 6,835 (100%) | 455 (100%) | **Total** | 455 (100%) | **Total** | 455 (100%) |

which occupies about 1% of all internet users worldwide[5]. Compared to its population, code-signing abuse is quite prevalent in Korea. It may imply that Koreans tend to be vulnerable to code-signing abuse and attackers are exploiting it. Such a tendency might have been formed due to its web usability environment, as mentioned in  Subsect. 2.4.

**Malicious Sample Family.** To better understand what kind of malware family used Korean code-signing certificates, we utilize the VirusTotal reports of our collected malicious samples and AVClass [30] to label the samples. We identify 278 different malicious sample families, and we break down the top ten malware

---

[5] Internet world stats: https://www.internetworldstats.com/stats3.htm.

and PUP labels as described in Table 1. The SINGLETON label indicates that AVClass is unable to classify malicious samples.

About 41% malicious samples signed with Korean certificates is kraddare. This family is considered PUP/PUA, which redirects to unwanted homepages without user action, changes the browser settings, shows unwanted advertisements using pup-ups [11]. Microsoft Defender Antivirus [22] classifies the malicious sample as a "severe" threat and removes the sample as encountered. The following malware family is onescan. The family is considered as a "severe" threat by Windows Defender Antivirus, and called "fakeAV [23]." The malware pretends to scan victims' computers, and reports to them that their computers have been infected by any malware, and then asks them to pay for cleaning up the reported malware. However, the victims' computers are not infected by any malware, and nothing is actually done by the malware, but victims pay for it. Delf [21] is a trojan that redirects Web traffic, downloads malicious programs, etc. On average, a Korean certificate is misused to sign 2.5 different families of malware/PUP samples. It would imply either 1) a couple of malware groups share a code-signing certificate to sign their malware or 2) a malware group produces a couple of malware families. However, we have little evidence to specify which.

**Publisher.** In Windows, when executing/installing a signed PE file, a client is prompted a request that shows the publisher name of the PE file by the system. Only after the client accepts the request, the signed PE files will be executed. Details about the publishers' information are available when clients look at the certificates since certificates include publishers' information such as the company/individual name, physical address (country, province, street address, and zip code), etc.

We start the investigation from the publisher's name stated in the Common Name (OID: 2.5.4.3) field. The common name is a required field in Subject of the X.509 v3 standard. It is used to identify the legal name of a publisher. The Legal names can be specified in the field only when verified by CAs using notarized documents or legal documents from attorneys. Unlike TLS, where the common name should have a domain name to be verified, in the code-signing PKI, the common name is usually an organization's name such as Google Inc. and Microsoft Corporation. We observe 330 common names in 455 Korean certificates; on average, a company has 1.4 different code-signing certificates to sign malicious samples. The top 10 publishers are enumerated in Table 2. cloudweb Inc has the largest signed malicious samples in our data set. The publisher had three different certificates to sign 1,040 malicious samples.

Furthermore, we could find some reputable Korean companies within the certificates misused to sign malicious samples. We believe that their private keys associated with the certificates were likely stolen and used to sign malicious samples. For example, the certificate of a Korean software company that develops not only software tools but also an AV product was misused to sign malware, called "plugx." The malware is a kind of Remote Access Trojan (RAT). Fortunately, the certificate was explicitly revoked, and the malware is no longer valid. Moreover, an English education company located in the Gangnam district released a program with a Trojan downloader malware payload. The malware was distributed at a legitimate website. Since it is a reputable and legitimate company, we believe that the development infrastructures were compromised, and the payload was injected into the legitimate program.

Next, we take a look at where these publishers are located. We use the Province filed (OID: 2.5.4.8) to locate the regions of the publishers. According to the minimum requirements [7], the Province field is required to be specified when the Locality Name field (OID: 2.5.4.7) is absent. However, we observe that 20 certificates issued by YesSign do not include any information in both the Province field and the Locality Name field; YesSign does not obey the requirement. Specifically, YesSign specifies their CA name on the Organization Name filed (OID: 2.5.4.10) rather than the publisher's organization name. Most malicious publishers (58.7%) are located in Seoul as depicted in Table 2. We also manually investigate certificates located in a small, rural, agricultural area where IT companies are less likely to exist. We observe that two certificates located in a small agricultural area are issued to non-existing IT companies. The same name of the IT companies exists, but they are located in Seoul, not the small rural area. Moreover, the two certificates were issued on the same day, and the certificates were misused to sign the same malicious sample families; onescan, kraddare, and jaik. Therefore, we believe that the two publishers are related to each other, even though they use different publisher names. This goes along with our findings from analyzing the malware families.

### 4.2 Issuer

In this section, we answer the questions: *Q2: Who issued the certificate?* and *Q3: Are the certificates issued with safe cryptographic guarantees?.*

**Certificate Authority (CA).** CAs issue code-signing certificates to software publishers (e.g., software developers). In the certificates, CAs specify their information such as the country, address, name of the issuer CAs. Similar to the Subject field, the issuer information is located in the Issuer field.

We observe only seven CAs, and certificates issued by Thawte are the majority (73.8%), which contradicts the finding [13] that VeriSign dominates the code signing certificate market share. We believe it is because Thawte is distributed by one of the accredited CAs in Korea, as we described in Subsect. 2.4. Also, Thawte allows publisher names with Korean alphabets[6], which may have boosted the market share. In addition, we find "YesSign[7]" in our data set, a CA which is hardly observed in prior works [13,14]. YesSign is one of the largest Korean CAs, and is operated by Korea Telecommunications and Clearings Institute (KFTC). The CA no longer issues code-signing certificates, but it still provides the OCSP and TSA service. However, as they stopped the business, there is a chance the revocation checking services may shut down in the future, which may make users vulnerable.

---

[6] Provided by crosscert: https://www.crosscert.com/symantec/02_1_04.jsp.
[7] https://www.yessign.or.kr.

**Table 3.** Signature and public key algorithm breakdown.

| Signature algorithm | Count | Public key algorithm | Count |
|---|---|---|---|
| MD5 With RSA | 6 (1.31%) | RSA | 455 (100%) |
| SHA1 With RSA | 413 (90.77%) | DSA | 0 (0%) |
| SHA256 With RSA | 36 (7.91%) | ECDSA | 0 (0%) |
| Total | 455 (100%) | Total | 455 (100%) |

**Cryptography Algorithm.** It is important to use strong cryptography algorithms for the certificates. Certificates with a weak algorithm may be utilized for collision attacks. It is critical in code-signing as an attacker could perform collision attacks on time-stamped binary samples with weak algorithms. MD5 and SHA1 are weak hash algorithms, vulnerable to collision attacks. We have observed a severe security threat where Flame malware exploited an unknown chosen prefix collision attack against the MD5 hash algorithm [31]. Google and CWI Amsterdam demonstrated that two different files could have the same SHA1 hash [12]. Although the SHA1 collision attack against certificates is not yet reported, it could be exploited to create fake certificates in the near future. Therefore, Microsoft deprecates MD5 and SHA1 hash algorithms in 2013 and 2015, respectively [20,26]. Still, CAs should be aware of this fact and move on to SHA256.

We examine what cryptography algorithms are used for signature and public keys in Korean certificates. As depicted in Table 3, all certificates in our data set use RSA for public key generation. For the signature algorithm, the majority (around 91%) use SHA1. We can also see the use of MD5 in a few certificates (6, 1.31%). It implies that weak algorithms are still prevalent in Korea, which has the potential to lead to serious security problems.

### 4.3  Certificate Life-Cycle

A life cycle of a certificate starts from its issue date and ends at its expiration date. In case it is compromised, a revocation is conducted to invalidate the certificate. However, we know that some signed binaries may survive even after their expiration and revocation due to the trusted timestamp. To answer the next research question *Q4: How long do the abusive certificates survive in Korea?*, we start the examination from the validity period of the Korean certificates. Then we check how prevalent trust timestamping is among the signed malicious binaries. In the end, we investigate if the revocation is performed effectively for those certificates, invalidating all the signed malware.

**Validity Period.** Each certificate has two fields, notBefore and notAfter for validity period; notBefore is an issue date and notAfter is an expiration date. In other words, a certificate is only valid between notBefore and notAfter, inclusive. As shown in Fig. 2, most certificates (69.43%) were issued between 2009 and 2012. It does not indicate that the signed malware was collected between the

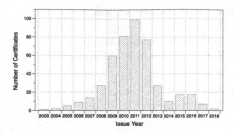

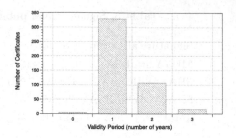

**Fig. 2. Issue year.** Around 70% of certificates in our data set were issued between 2009 and 2012.

**Fig. 3. Validity year.** The majority is one-year-valid certificates since CAs usually issue one-year-valid certificates.

periods because the signed samples are still being valid even seven or eight years have passed due to the trust timestamping.

Figure 3 shows that most certificates (70.57%) are only valid for one year as expected since CAs typically issue one-year-valid code signing certificates. However, interestingly, the validity period of four certificates is less than one year. Two certificates issued by Thawte are valid only for three and nine months; one certificate done by VeriSign is valid only for 11 months, and a certificate issued by YesSign is valid for four months. Unlike the TLS certificates, because the code signing PKI has the trust timestamping, signed binary samples can be valid even after their certificate expiration date as long as the samples are trust timestamped. Therefore, expiration dates do not count as much as the Web's PKI. We do not observe that certificates are valid for more than three years.

**Trust Timestamp (Signing Date).** The distinct difference between the Web's PKI and the code signing PKI is the trust timestamping mechanism (c.f., Sect. 2.2). We measure how many Korean malicious samples are trust-timestamped. Of 8,815 Korean malicious signed PE samples, we observe that 6,190 samples (70.2%) are trust-timestamped. Only when we consider the valid malicious samples ($c_{mal} \geq 10$), 4,625 samples (67.7%, out of 6,835) contain the signing date (trusted timestamps). It means that most malicious samples use trust timestamping to extend their validation period beyond their certificate's expiration date. We also examine when the malicious samples are signed; we utilize issue dates and expiration dates. More than 50% of malicious samples are signed about 200 days before their expiration dates, as shown in Fig. 4. It indicates that most malicious samples are consistently signed with compromised certificates during the validity periods.

**Revocation Status.** All certificates we have identified in the paper are misused to sign malicious binary samples. Therefore, they should be revoked. We check whether or not the certificates in our data set are revoked using CRLs. We observed three security threats that let signed malware alive. First, only 31 (6.81%) of 455 Korean certificates are explicitly revoked. It implies that mal-

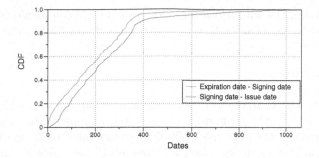

**Fig. 4.** The difference in days between signing dates and issue dates, and between expiration dates and signing dates.

ware signed with not-revoked certificates may remain valid even after the certificate's expiration date, due to the trust timestamping mechanism (c.f., Sect. 2.2).

Second, we encountered CRLs that are unreachable. Two reasons interfered with accessing and fetching the CRLs; (1) the CRL domain was taken by a domain re-seller, and (2) the CRL file was moved/removed, returning a 404 error. Those findings are in line with prior work [14]. Clients exposed to these certificates become vulnerable as they cannot check the revocation status of these certificates.

Lastly, several certificates were not effectively revoked. Signed malware can continue to be valid, although its certificate is revoked if the revocation date is set erroneously (c.f., Sect. 2.3). We measure if Korean signed malicious samples are still valid as CAs erroneously set the revocation dates after the samples' signing dates. We find that 321 malicious samples are still valid, and eight Korean certificates are used to sign the samples. The average difference between the signing date and the revocation date is 6,013.01 h (250.51 days); the shortest difference is 11.04 h (0.46 days), and the longest one is 25,389.82 h (1,057.91 days, 2.9 years).

Although the certificates are mostly issued to be valid for a year, several signed malware remain a threat for an extended period due to time-stamping and the clumsy set of revocation dates.

## 5    Related Work

Compared to the Web's PKI, little research has been conducted on the code signing PKI. The first attempt [27] was done in 2010 by *F-Secure*. In the attempt, they introduced the ways of abusing Microsoft Authenticode [25]. However, the work was presentation slides focusing on introducing new threat models rather than a research paper. In 2015, Kotzias et al. [17] examined 356,000 digitally signed samples collected between 2006 and 2015. They observed that most of the collected signed samples were Potentially Unwanted Programs (PUP), while signed malware was relatively uncommon in their corpus. Kim et al. presented new threat models that highlight the breaches of trust in the code signing PKI. Kim et al. also identified the security problems of the revocation mechanisms currently deployed in the wild. However, those studies are conducted from a global perspective while we measure Korean compromised certificates' characteristics.

## 6    Conclusion

We investigate the characteristics of code-signing abuse in South Korea. We design a system that extracts abusive Korean code-signing certificates with a simple but effective method. A couple of findings were related to its unique web environment fostered by regulations. South Korea has its own government-accredited CAs, and these CAs affect the certificate landscape. We observe Thawte, re-selled by one of the accredited CAs, dominating the population. Another accredited CA even acted as a code-signing CA. However, the CA is no longer in business, which is a potential threat as they might stop the revocation service. We observed that code-signing abuse is quite prevalent in Korea, and it might be due to the exposure of mandatory installation for using the web. Besides, we also found a common vulnerability reported in prior works. Only 6.8% certificates have been revoked, and eight certificates of them have erroneous effective revocation dates, which extends the validity of signed malicious samples.

**Acknowledgements.** We thank the anonymous referees for their constructive feedback. This research was supported by Basic Science Research Program through the National Research Foundation of Korea (NRF) funded by the Ministry of Education (2021R1F 1A1049822). Any opinions, findings, and conclusions or recommendations expressed in this material are those of the authors and do not necessarily reflect the views of the sponsor.

## A    Appendix

**Table 4.** SignTool error code & message.

| Validation | Error code | Message |
| --- | --- | --- |
| Valid | N/A | Successfully verified |
| Invalid | 0x800B0101 | Expired certificates |
| | 0x800B010C | Revoked |
| | 0x800B010A | Not a trusted root CA |
| | 0x80096010 | Signature does not match the file |
| | Terminated in a root cert | Not trusted by the trust provider |
| | No signature found | No signature found |

**Table 5. Breakdowns.** Error code of Korean malicious PE files (left), PE files and certificates (right).

| Validation | Error code | KR malware |
|---|---|---|
| Valid | Successfully Verified | 5,714 |
| | 0x800B0101 | 558 |
| | 0x800B010C | 1,821 |
| | Valid total number | 8,093 |
| Invalid | 0x800B010A | 405 |
| | 0x80096010 | 94 |
| | Terminated in a root cert. | 24 |
| | No signature found | 199 |
| | Invalid total number | 722 |
| Total number | | 8,815 |

| Type | PE & Cert. | Number |
|---|---|---|
| Total | All malicious sample | 5,934,399 |
| | PE file | 3,240,176 |
| | Signed PE file | 525,071 |
| | PKCS #7 | 495,124 |
| Korean | Malicious signed PE | 8,815 |
| | Malicious cert. | 844 |
| | Valid malicious signed PE | 8,093 |
| | Valid malicious cert. | 783 |
| | Valid malicious signed PE ($c_{mal} \geq 10$) | 6,835 |
| | Valid malicious cert. ($c_{mal} \geq 10$) | 455 |

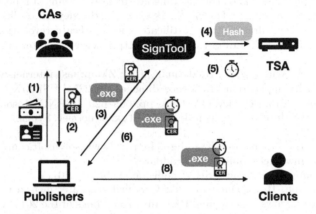

**Fig. 5. Code-signing process.** (1) A publisher applies for a code-signing certificate to a code-signing CA with her/his identifications such as government-issued photo IDs, (2) After vetting, the CA issues a code-signing certificate to the publisher, (3) Using the `SignTool` (a signing tool provided by Microsoft), the software publisher signs a binary sample with the certificate, (4) when a TimeStamp Authority (TSA) is specified for timestamping (c.f., Sect. 2.2), the signing tool sends the hash value of the binary sample to the TSA server, (5) The TSA server issues the timestamp and signs the timestamp with the TSA's private key, and send them back to the signing tool, (6) The signing tool finally embeds the code-signing and the TSA certificate chain, the digital signature, and the timestamp into the binary sample, and (7) Finally, the publisher distributes the signed binary sample in the wild.

# References

1. What should i do with the annoying ads? (in Korean) https://www.donga.com/news/Economy/article/all/20140914/66399483/1. Accessed 03 Sept 2020
2. N. Korea fakes 'code signing' to spread spyware. KBS world radio. http://world.kbs.co.kr/service/news_view.htm?lang=e&Seq_Code=119375. Accessed 30 Aug 2020

3. To bypass code-signing checks, malware gang steals lots of certificates. ars technica. https://arstechnica.com/information-technology/2016/03/to-bypass-code-signing-checks-malware-gang-steals-lots-of-certificates/. Accessed 30 Aug 2020
4. Adobe. Electronic Signature Laws and Regulations - South Korea (2020). https://helpx.adobe.com/sign/using/legality-south-korea.html
5. Alrawi, O., Mohaisen, A.: Chains of distrust: towards understanding certificates used for signing malicious applications. In: WWW 2016, Republic and Canton of Geneva, Switzerland (2016)
6. Chai, S.-W., Min, K.-S., Lee, J.-H.: A study of issues about accredited certification methods in Korea. Int. J. Secur. Appl. **9**(3), 77–84 (2015)
7. Code Signing Working Group. Minimum requirements for the issuance and management of publicly-trusted code signing certificates. Technical report (2016)
8. Cooper, D., Santesson, S., Farrell, S., Boeyen, S., Housley, R., Polk, W.: Internet X.509 public key infrastructure certificate and certificate revocation list (CRL) profile. RFC 5280. RFC Editor (May 2008). http://www.rfc-editor.org/rfc/rfc5280.txt
9. Durumeric, Z., Wustrow, E., Halderman, J.A.: ZMap: fast internet-wide scanning and its security applications. In: Proceedings of the 22Nd USENIX Conference on Security, SEC 2013, Berkeley, CA, USA, pp. 605–620. USENIX Association (2013)
10. Falliere, N., O'Murchu, L., Chien, E.: W32.Stuxnet dossier. Symantec Whitepaper (February 2011)
11. Geater, J.: How to remove Kraddare. https://www.solvusoft.com/en/malware/potentially-unwanted-application/kraddare/
12. Google: Announcing the first SHA1 collision (February 2017)
13. Kim, D., Kwon, B. J., Dumitras, T.: Certified malware: measuring breaches of trust in the windows code-signing PKI. In: Proceedings of the 2017 ACM SIGSAC Conference on Computer and Communications Security, CCS 2017 (2017)
14. Kim, D., Kwon, B.J., Kozák, K., Gates, C., Dumitras, T.: The broken shield: measuring revocation effectiveness in the windows code-signing PKI. In: 27th USENIX Security Symposium, USENIX Security 2018. USENIX Association (2018)
15. KLRI: Digital Signature Act, 2017. https://elaw.klri.re.kr/eng_service/lawView.do?hseq=42625&lang=ENG
16. Kotzias, P., Bilge, L., Caballero, J.: Measuring PUP prevalence and pup distribution through pay-per-install services. In: Proceedings of the USENIX Security Symposium (2016)
17. Kotzias, P., Matic, S., Rivera, R., Caballero, J.: Certified PUP: abuse in authenticode code signing. In Proceedings of the 22nd ACM SIGSAC Conference on Computer and Communications Security, CCS 2015. ACM, New York (2015)
18. Kozák, K., Kwon, B.J., Kim, D., Gates, C., Dumitraş, T.: Issued for abuse: measuring the underground trade in code signing certificate. In: 17th Annual Workshop on the Economics of Information Security (WEIS) (2018)
19. Kwon, B.J., Srinivas, V., Deshpande, A., Dumitras, T.: Catching worms, trojan horses and pups: unsupervised detection of silent delivery campaigns. In: 24th Annual Network and Distributed System Security Symposium, NDSS 2017 (2017)
20. Microsoft: Microsoft security advisory: update for deprecation of MD5 hashing algorithm for Microsoft root certificate program, 13 August 2013
21. Microsoft: Trojan:win32/delf. https://www.microsoft.com/en-us/wdsi/threats/malware-encyclopedia-description?Name=Trojan:Win32/Delf
22. Microsoft: Trojan:win32/kraddare. https://www.microsoft.com/en-us/wdsi/threats/malware-encyclopedia-description?Name=Trojan:Win32/Kraddare

23. Microsoft: Win32/onescan. https://www.microsoft.com/en-us/wdsi/threats/malware-encyclopedia-description?name=win32%2Fonescan
24. Microsoft: Erroneous VeriSign-issued Digital Certificates Pose Spoofing Hazard (2001)
25. Microsoft: Windows Authenticode portable executable signature format (March 2008). http://download.microsoft.com/download/9/c/5/9c5b2167-8017-4bae-9fde-d599bac8184a/Authenticode_PE.docx
26. Morowczynski, M.: SHA-1 deprecation and changing the root CA's hash algorithm (2018)
27. Niemela, J.: It's Signed, therefore it's Clean, right? (2010)
28. NLIC: Electronic Financial Transaction Act, 2017. http://www.law.go.kr/eng/engLsSc.do?menuId=1&query=electronic+financial+transactions+act&x=0&y=0#liBgcolor0
29. Park, H.M.: The web accessibility crisis of the Korea's electronic government: fatal consequences of the digital signature law and public key certificate. In: 2012 45th Hawaii International Conference on System Sciences, pp. 2319–2328. IEEE (2012)
30. Sebastián, M., Rivera, R., Kotzias, P., Caballero, J.: AVCLASS: a tool for massive malware labeling. In: Monrose, F., Dacier, M., Blanc, G., Garcia-Alfaro, J. (eds.) RAID 2016. LNCS, vol. 9854, pp. 230–253. Springer, Cham (2016). https://doi.org/10.1007/978-3-319-45719-2_11
31. Swiat: Flame malware collision attack explained (June 2012)
32. Wood, M.: Want my autograph? The use and abuse of digital signatures by malware. In: Virus Bulletin Conference, September 2010, pp. 1–8 (September 2010)
33. Yilek, S., Rescorla, E., Shacham, H., Enright, B., Savage, S.: When private keys are public: results from the 2008 Debian OpenSSL vulnerability. In: Proceedings of the 9th ACM SIGCOMM Conference on Internet Measurement, IMC 2009. ACM (2009)

# GAN-Based Adversarial Patch for Malware C2 Traffic to Bypass DL Detector

Junnan Wang[1,2], Qixu Liu[1,2], Chaoge Liu[1,2(✉)], and Jie Yin[1]

[1] Institute of Information Engineering, Chinese Academy of Sciences,
Beijing 100093, China
liuchaoge@iie.ac.cn
[2] School of Cyber Security, University of Chinese Academy of Sciences,
Beijing 100049, China

**Abstract.** The constantly evolving malware brings great challenges to network security defense. Fortunately, deep learning (DL)-based system achieved good performance in the malware command and control (C2) traffic detection field due to its excellent representation capabilities. However, DL models have been shown to be vulnerable to evasion attacks, that is, DL models can easily be misled by adding subtle perturbations to the original samples. In this paper, we propose a GAN-based evasion method, which can help malware C2 traffic bypass the DL detector. Our main contributions contain: (1) directly generate adversarial traffic that can implement malicious functions by inserting additional adversarial patches in the original flow; (2) adaptively imitating victim's normal traffic by training GAN in victim environment, and introducing transfer learning to reduce the additional victim resource usage caused by GAN training. Results show that the adversarial patch generated by GAN can prevent malware C2 traffic from being detected with 51.4% success rate. The higher time efficiency and smaller malware impact make our method more suitable for real attacks.

**Keywords:** Malware C2 traffic · Evasion attacks · GAN · Transfer learning

## 1 Introduction

Malware allows attackers to remotely control computers to perform criminal activities using Command and Control (C2) channels, which has posed great

This work is supported by the Youth Innovation Promotion Association CAS (No. 2019163), the National Natural Science Foundation of China (No. 61902396), the Strategic Priority Research Program of Chinese Academy of Sciences (No. XDC02040100), the Key Laboratory of Network Assessment Technology at Chinese Academy of Sciences and Beijing Key Laboratory of Network security and Protection Technology.

D. Gao et al. (Eds.): ICICS 2021, LNCS 12918, pp. 78–96, 2021.
https://doi.org/10.1007/978-3-030-86890-1_5

challenges to network security. Fortunately, it can be mitigated by detecting C2 channels on the network.

Among the rich malware C2 traffic detection methods, the deep learning (DL)-based detection method has been widely used and researched because it is an end-to-end solution that can automatically learn feature representations from raw traffic data [13,15,16,24,25]. In this paper, we mainly focus on the DL-based malware C2 detection model taking raw malware C2 traffic data as input, which is a state-of-the-art detection method [19].

After the success of DL in the field of malicious traffic identification, their robustness and security issues have become the subject of much discussion by security researchers. In 2014, Szegedy et al. [23] first discovered that due to their linear nature, well-performing DL models are vulnerable to adversarial examples, which are intentionally crafted by adding tiny perturbations to mislead the DL model. After that, how to use adversarial machine learning (AML) ideas to construct adversarial malicious traffic to bypass detection also received attention.

Different from AML in the image recognition field, the construction of adversarial malware traffic has many unique constraints and challenges:

1. Ensure that the generated adversarial traffic can retain the original malicious functions, and the basic network protocol format will not be destroyed.
2. Directly generate adversarial traffic without the help of other attachments, rather than generating adversarial features that are just intermediate results of evasion attacks.
3. How to make adversarial traffic adaptively imitate the normal traffic of individual victims, so as to ensure that it can be applied to a variety of terminals. While those imitations that are limited to specific normal application traffic will fail when the application is rarely used on some victims.

These three challenges are progressive. Challenge-1 represents effectiveness, challenge-2 means usability, and challenges-3 is a practical requirement that proposed based on real attack scenarios.

Unfortunately, none of the existing work can solve the above problems at the same time. [8] and [10] directly treat traffic samples as image samples, even cannot meet challenge-1. [14] and [7] can only generate adversarial features violate challenge-2. What counts is, most of the current work does not consider challenge-3, which is the most realistic requirement in the malware traffic evasion field.

In light of the challenges, we present an adaptive evasion attack on DL-based detectors in practical settings. Specifically, we propose a GAN-based method that can directly generate sample-independent adversarial patches (*adv_patch*es ). Malware can directly send a packet encapsulating the *adv_patch* in C2 communication to bypass the DL-based detector, without other attachments' help or complex source code modification. And the C2 flow that encapsulates *adv_patch* is called adversarial flow, which can directly bypass the DL detector. Therefore, our method can solve challenge-1 and challenge-2 mentioned above.

In order to adaptively simulate a specific victim's traffic, there are two solutions. One is to collect large-scale normal traffics on the victim and send them back to train GAN, but that is unrealistic because it will increase the exposure risk of the C2 channel. The other is to train the GAN model on the bot, which will increase the exposure risk of malware on the victim. We choose the latter to solve challenge-3. At the same time, in order to reduce the extra resource utilization caused by GAN training on the victim, we introduced transfer learning (TL) technology to further improve the similarity between malware C2 traffic and victim normal traffic at a small cost. Results show that our method can not only achieve a success rate of 51.4%, but also has a good performance in time efficiency and a minor negative impact on malware.

Our major contributions are elaborated as follows:

1. We propose a GAN-based black-box malware C2 traffic evasion method to bypass the DL detector. Under the premise of functionality preserving and network protocol compliance, we can directly obtain adversarial traffic by inserting an additional *adv_patch* packet, without other attachments or complex source code modifications.
2. Our method enables adversarial traffic to adaptively imitate host-side normal traffic, that is, dynamically adjust adversarial traffic according to the traffic characteristics of different victim terminals, which is more practical and concealed. We also introduce TL to alleviate the additional system resource occupation caused by GAN training on the victim.
3. We design a real-life experiment to evaluate the proposed method, and proved its practicability and efficiency from the perspectives of evasion performance, time performance, and impact on malware.

As far as we know, this is the first work on adaptive evasion method, that considers and comprehensively evaluates the negative impact of the evasion method on malware.

The rest of the paper is organized as follows: We start by providing backgrounds and related works in Sect. 2. Section 3 introduces the overview of our evasion method. Section 4 elaborate experimental setting up. Experimental results and findings are shown in Sect. 5. Finally, we conclude in Sect. 7.

## 2    Background and Related Work

### 2.1    Background–Malware Traffic Detection

With the development of machine learning technology, DL technology has been widely used in the malware C2 traffic detection field. On the one hand, DL-based methods can automatically learn deep abstract feature representations, thereby solving the dilemma of manual feature engineering. On the other hand, compared with the traditional ML methods, DL-based methods also have a considerably higher capacity to learn complex patterns, so they can deal with large-scale encryption and unknown malicious traffic detection well.

According to the different model inputs, DL-based classifiers can be divided into statistic feature-based and raw data-based. [18] and [19] have proved that DL-based model, using raw flow representations as input, can outperform other detectors, while without requiring any prior knowledge.

In this work, we particularly focus on the vulnerability of DL-based malware C2 traffic detector, which taking raw byte stream flow data as input.

[25] proposed a stacked autoencoder (SAE) based network protocol identification method using raw traffic data, and achieved high accuracy.

[24] proposed an end-to-end malware traffic classification method with 2D-CNN taking the first 784 bytes of flow. Lotfollahi et al. [16] combines SAE and 1D-CNN, and takes the first 1500 bytes of IP header and payload data as input.

Byte Segment Neural Network (BSNN) [13] and Flow Sequence Network (FS-Net) [15] are both RNN-based traffic classification methods. The difference is that BSNN takes raw payload as input, while FS-Net's input is raw flow.

In summary, the current DL model for malware traffic detection often takes the first few bytes of the raw byte stream as input, then learns the abstract representation through multi-layer neural networks, and the final prediction is calculated by the softmax layer.

## 2.2 Related Work–Malware Traffic Evasion

While the malware traffic detection method is constantly improving, attackers are also exploring evasion techniques to avoid detection. Evasion and detection technologies are innovating in the tit-for-tat game, trying to be able to overwhelm the opponent.

In order to bypass blacklist-based detection, attackers introduced dynamic resolution technologies such as DGA and Fast-Flux to replace the hard-coding method. Introducing techniques such as encryption and data encoding to cover up the payload, so the payload-based detection is invalidation. To bypass the detector based on statistical characteristics, the attacker introduces technologies such as protocol tunnels and online-social networks (OSN) to construct covert channels and overwhelms malicious traffic in mass normal traffic.

In recent years, with the widespread application of DL in the field of malicious traffic detection, many researchers have also tried to use the inherent security vulnerabilities of DL to bypass DL-based detectors. We divide these tasks into two categories according to the adversarial output.

**Feature-space attack** refers to a type of attack method that can only generate adversarial feature vectors. However, the mapping process from traffic samples to traffic characteristics is irreversible and non-differentiable. So, it is difficult to reversely infer traffic samples, even if the adversarial feature vector is known. In other words, this attack method is just theoretical proof that DL-detector is vulnerable to evasion attacks, and cannot be directly used for malicious delivery. This attack method can only be used as theoretical proof that the detection system is vulnerable to attack.

Clements [8] and Ibitoye [10] used classic AML algorithms (FGSM [9], BIM [11], PGD [17], C&W [5], JSMA [21] etc.) to evaluate the robustness of DL-based network intrusion detection system (NIDS) against adversarial attacks in a white box scenario. They directly convert the traffic samples into gray images and perturb the 'pixel' indiscriminately. No consideration is given to the fine structure of traffic samples and the constraints of maintaining malicious functions.

Lin et al. [14] proposed a black-box evasion attack method-IDSGAN, which uses GAN to generate adversarial statistical features of malware traffic. Although IDSGAN can ensure the effectiveness of the intrusion by changing only non-functional features, it does not consider the dependence between statistical features. FENCE [7] solves this problem by combining gradient-based methods and mathematical constraints to maintain consistency in a family of dependencies.

**Traffic-space attack** refers to attack methods that can generate adversarial traffic samples. Unlike feature-space attacks, traffic-space attack methods can be powerful weapons for attackers to bypass malware traffic detectors.

Novo [20] used the classic AML algorithm FGSM [9] to perturb the encrypted C&C malware traffic characteristics and achieved a white-box adversarial attack against the detector. It requires additional traffic proxy or complex source code modification to obtain the final adversarial traffic. And white-box attacks require a full understanding of the detector, which is difficult to attain in real life.

Rigaki et al. [22] proposed a method that uses GAN to generate statistical features similar to Facebook traffic, thus adjust the behavior of the malware C2 traffic to avoid detection. FlowGAN [12] is no longer limited to Facebook traffic, can dynamically morph traffic features as any other "normal" network flow to bypass censorship. However, in these two works, GAN can only output adversarial features. If the attacker wants to obtain adversarial traffic based on these adversarial features, he needs to make complex modifications to the malware source code, which will cause delays to the malware's communication channel.

In Attack-GAN [6], the generator is viewed as an agent in RL, which can craft adversarial traffic conditioned to the security domain constraints to ensure attaining the attack functionality. But Attack-GAN needs to constantly access IDS to obtain prediction results, which is unrealistic in real-attack.

Unlike the works we reviewed in this section, in this paper we focus specifically on how to directly generate adaptive adversarial traffic without the help of any other additional components. Only by adaptively imitating victim traffic, can the adversarial traffic seemed to be normal-like in bots with different characteristics. Moreover, while most related work assesses the performance of the evasion attack on malware traffic detectors, they do not consider the impact of the proposed methods on malware, nor do they consider the practicality of the method. We properly solve these problems by performing a real-life experiment in this work.

## 3   Method

In this section, we use some technical terms to represent various roles in a malware C2 traffic evasion attack. *Malware* means the code used to achieve C2, *master* means the computer of the attacker, *victim* means malware-infected hosts. The *adversary* tries to control the victim by malware, while the *defender* tries to protect the victim through a DL-based malware C2 traffic detector.

### 3.1   Thread Model

**Adversary's Goal.** From the perspective of the CIA (confidentiality, integrity, and availability), attackers try to reduce the availability of detectors by camouflaging malware C2 flow.

**Adversary's Knowledge.** The attacker knows that the target network may be protected by a flow-level detector based on DL. However, the attacker does not need to master any prior knowledge about the detector, such as the architecture, parameters, or training data.

**Adversary's Capability.** The attacker has full control of the C2 server and partial control of the victims, so he can update victims to change their communication behaviors as he wants.

### 3.2   Framework

Our framework is inspired by [4], that attackers can mislead the classifier by placing a gradient-based sample-independent *adv_patch* in a specific area. *Adv_patch* is effective because it can calculate the most effective perturbation to the DL model by using gradient backpropagation according to the gradient passed by the discriminator. When inputting the detector, *adv_patch* can dominate the feature learning of the detector, thereby misleading the detector

The idea of *adv_patch*es suits malware C2 traffic evasion well. On the one hand, through this method, we can directly operate on the traffic samples and output traffic samples with actual attack functions.

On the other hand, traffic samples have more complex network protocol constraints than images, and there is a need to keep malicious functionality in the perturbed sample. That makes many AML algorithms designed for images unavailable. While our method can better meet the constraints of functionality preserving and network protocol.

Specifically, our method includes two modules, a GAN-based generation module and a TL-based transfer module.

To better illustrate our method, we propose two terms. We define **universal benign communication (UBC)** traffic as benign communication traffic that has multiple types benign communication traffic and can cover a variety of benign communication behavior characteristics, while **host benign communication (HBC)** traffic only includes benign communication traffic from a specific host. HBC is more specific and targeted, while UBC is more versatile and generalized.

In the generation module, we use GAN to imitate the normal traffic to generate *adv_patch*. By inserting it into the original flow, we can obtain adversarial malware C2 traffic that can mislead the DL-based malware C2 traffic detector.

In the transfer module, we retrain the GAN model in the victim environment to adaptively simulate the victim's normal flow. TL is used in this module because it can realize the transfer from imitating UBC traffic tasks to adaptively imitating specific HBC traffic tasks with a smaller data scale requirement and system resource cost.

The workflow of our method is shown in Fig. 1. It can be divided into three stages: the pre-training stage in the master environment, the fine-tuning stage in the victim environment, and the practical stage.

The pre-training stage in the master environment refers to the pre-training of GAN performed by the attacker before the weapon is delivered. In a fully controllable master environment, the attacker can construct a training dataset by capturing the original malware C2 traffic and the UBC flow. After the pre-training stage is completed, the attacker compresses and packs the GAN model together with the malware, and delivers them to the victim.

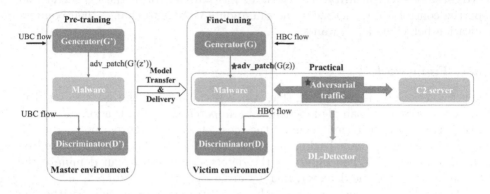

**Fig. 1.** System framework

Next is the fine-tuning stage in the victim environment. The malware will call the packet sniffer module to build up the HBC traffic profile, which is used for fine-tuning the GAN model so that the specific characteristics of the victim's normal traffic can be more accurately embedded in output *adv_patch*.

Finally, in the practical stage, the fine-tuned GAN model can be used to camouflage malware C2 traffic. Specifically, the malware will first access the generator to obtain the *adv_patch* before communicating with the C2 server, and send out the packet encapsulating the *adv_patch* after the TCP three-way handshake, followed by other original malicious packets.

### 3.3   Generation Module – WGAN

As the core of the method, we choose GAN as the generation module. Generative Adversarial Networks (GAN), are a class of DL-based generative model. The GAN model architecture involves two sub-models: a generator (G) that is trained to generate new examples, and a discriminator (D) that tries to classify examples as either real or fake. The final goal is to make the data obtained by the generator becoming more similar to the real data.

In the context of malware C2 traffic evasion, the generator is responsible for learning the characteristics of the normal communication traffic and generating fixed-length *adv_patch*es to help malware C2 traffic evading the DL-based detector. While the discriminator plays a similar role to the detector, which is used to determine whether the generated confrontation traffic is sufficiently similar to the normal traffic and pass gradients to the generator for parameter tuning.

Specifically, we use Wasserstein GAN (WGAN) [3]. Instead of JS divergence, WGAN introduces Wasserstein distance (calculate as Eq. 1) to calculate the distance between the generated distribution and the real distribution as the loss function. WGAN can solve many problems of vanilla GAN, such as unstable training and collapse mode, and Wasserstein distance can be used as an indicator of training progress.

We choose WGAN not only because of its excellent learning ability, but also because its adversarial fits well with the confrontation scenarios of malicious traffic detection and evasion attacks.

$$W(p_r, p_g) = \inf_{\gamma \sim \prod(p_r, p_g)} E_{(x,y)\sim\gamma} [\|x - y\|] \tag{1}$$

The loss function of WGAN is:

$$L^D = E_{x\sim p_{data}}[D(x)] - E_{z\sim p_{(z)}}[D(G(z))] \tag{2}$$

$$L^G = E_{z\sim p_{(z)}}[D(G(z))] \tag{3}$$

$$W_D \leftarrow clip\_by\_value(W_D, -0.01, 0.01) \tag{4}$$

In our method, during the training process, the generator will take benign communication flow as input, attempt to generate a fixed-length $adv\_patch$, and return it to the malware. The discriminator takes the new malware C2 flow and the benign communication flow as input, and learns how to distinguish between them. During the application process, the generator will be requested by malware to obtain a new $adv\_patch$.

We adopted the classic model in [3] as our generation module. One small difference is that in order to avoid that the discriminator is too powerful to guide the parameter learning of the generator well, we have removed several convolutional layers in the discriminator to reduce the complexity of the discriminator. At the same time, this can also further reduce the size and parameter number of the GAN model. It is worth mentioning that in order to insert $adv\_patch$ into

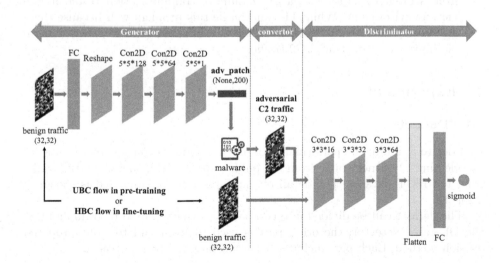

**Fig. 2.** The architecture of the GAN model we used

the original malicious traffic, we built a concatenate layer between the generator and the discriminator to facilitate gradient propagation. The architecture and hyperparameters setting of the GAN model are shown in Fig. 2.

### 3.4   Transfer Module–Transfer Learning

Transfer Learning is a machine learning method that transfers knowledge from the source domain (task A) to the target domain (task B), so that task B can achieve better learning results. Usually applicable to situations where the amount of data in the source field is sufficient, but in the target field is small.

In the context of malware C2 traffic evasion, we regard the pre-training process in a fully controllable master environment as task A. Task A attempts to train the generator to generate a fixed-length payload and insert it into the malicious communication flow, making it difficult for the discriminator to distinguish the newly constructed malicious flow from the UBC flow.

Task B is a fine-tuning process that occurs in the victim environment. In this process, the generator will use the HBC traffic captured in the victim as a template to learn how to construct malware C2 traffic.

The difference between the two tasks is that the traffic distribution of task B is more specific and concentrated. To some extent, the distribution of UBC traffic and HBC traffic is similar, so it is very suitable to use parameter-TL.

Specifically, on the premise of further improving evasion performance, applying TL has the following two advantages:

(1) Reduce the training cost in the victim environment: Parameter-TL can reduce the later training cost, by only training a small part of the parameters. Therefore, we can reduce the victim's perception of the fine-tuning process and avoid being detected due to taking up too many system resources.
(2) Suitable for small datasets: It is unrealistic to train a large neural network from scratch to capture a large amount of communication traffic in the victim environment. While TL can handle this problem well because there are fewer parameters to learn. Besides, we can rely on TL to generate more victim-specific adversarial C2 traffic.

## 4   Experiment

### 4.1   Dataset

In order to evaluate the performance of our method, we constructed a data set by selecting 12 botnet traffic from the public dataset CTU and the UBC traffic from the ISOT dataset. The detail of the dataset we summarized is shown in Table 1.

The dataset can be divided into two parts, one part is used to train and test the DL-based detector, the other part is used to train and test our proposed evasion method. Each part includes both malware and benign traffic.

The dataset for the detector is the dataset used by the defender. In this part, malware traffic includes 9 malware families from the CTU covering a variety of commonly used C2 channels. The benign traffic is captured from the 10 computers in our laboratory environment, which can cover many different types of normal traffic. The reason for this setting is that in order to protect the internal network in a targeted manner, the defender often uses the specific internal normal traffic to train the DL-based detector.

The dataset for WGAN is the dataset used by the attacker. The malicious traffic is 5 malware families selected from CTU, Neris and Virut are also used to train the detector, but the Neris traffic files come from different captures. The benign traffic includes UBC traffic from the ISOT for pre-training, and internal capture id01 for fine-tuning. The reason for this setting is the fact that it is difficult for an attacker to obtain a large amount of internal traffic. Therefore, pre-training can only use public datasets, and in fine-tuning stage, a small volume of traffic samples can be used to adaptively simulate specific HBC traffic.

The original data needs to be preprocessed before inputting into the model. The data preprocessing process mainly includes three steps.

1. Split. The captured pcap file is divided into bidirectional flows according to the five-tuple $<sip, dip, sport, dport, protocol>$. We use the open-source tool pkt2flow [1] to complete this operation.
2. Filter. After the split, we only keep the flow with valid data transmission, and filter out the flow that the TCP connection is not fully established or is closed immediately after establishment.
3. Anonymization. We perform anonymization on traffic data to avoid specific information such as $IP$ and $MAC$ misleading the detection model. Specifically, we replace them all with 0.

**Table 1.** Details of the dataset

|          |         | Malware family     | Flow num. | C2 channel   |
|----------|---------|--------------------|-----------|--------------|
| Detector | Malware | CTU-44-Rbot        | 2745      | IRC          |
|          |         | CTU-47-Menti       | 216       | TCP          |
|          |         | CTU-49-Murlo       | 1986      | TCP          |
|          |         | CTU-42-Neris       | 1583      | HTTP         |
|          |         | CTU-54-Virut       | 3451      | HTTP         |
|          |         | CTU-127-Miuref     | 1286      | HTTP         |
|          |         | CTU-125-Geodo      | 6320      | HTTP         |
|          |         | CTU-141-1-Bunitu   | 6143      | HTTP/HTTPS   |
|          |         | CTU-348-1-HTbot    | 10000     | HTTP/HTTPS   |
|          | Benign  | id01-id10          | 39452     | –            |
| WGAN     | Malware | CTU-50-Neris       | 19282     | HTTP         |
|          |         | CTU-54-Virut       | 3451      | HTTP         |
|          |         | CTU-264-2-Emotet   | 10000     | HTTPS        |
|          |         | CTU-346-1-Dridex   | 8022      | HTTPS        |
|          |         | CTU-327-1-Trickbot | 25924     | HTTPS        |
|          | Benign  | UBC traffic        | 17144     | –            |
|          |         | HBC traffic-id01   | 9706      | –            |

## 4.2 Hyperparameters

### The Length of the New Packet

The output dimension of GAN is fixed, so we need to determine the length of the generated *adv_patch*.

In order to simulate the normal data packet as much as possible, the mode of the payload length of the UBC traffic is selected as the length of the *adv_patch*. This can make the newly added packet look closer to the normal data packet at least in terms of statistical characteristics.

We perform statistics on the captured UBC traffic, and obtain the distribution of its payload length, which is shown in Fig. 3.

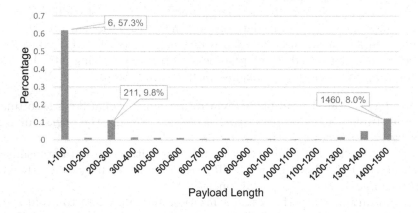

**Fig. 3.** The payload length distribution of universality normal traffic

Through the inspection of the original traffic data, we found that the reason for many packets with 6 bytes payload is that the Ethernet data link layer will automatically pad the frame with 0 to ensure that the minimum length of the frame is 64 bytes. While 1460 is the maximum segment size of TCP transmission. TCP segmentation will be performed when large-size data is transmitted, resulting in a large number of packets with a payload length of 1460.

Based on the above findings, we set the length of the *adv_patch* to 200, which is close to the second most frequent payload length 211, and it is also convenient for quantification and calculation.

### Insertion Position of the New Packet

After getting *adv_patch*, we need to decide when to send it. Through the investigation of the current DL-based detection work, we found that in order to balance the model accuracy and complexity, researchers often intercept part of the traffic data for learning deep representations as the basis for classification. Wang et al. [24] proved that the first few packets, up to the first 20 packets, are sufficient for correct accuracy, even for encrypted traffic.

In this case, to guarantee the impact on the DL-based detector, we decided to add the *adv_patch* packet after the TCP three-way handshake process. That means malware should send out the packet encapsulating the *adv_patch* once the TCP connection is established. In this way, it can be ensured that the carefully crafted *adv_patch* can appear in the visual field of the detector.

### 4.3 Detector

Marín et al. [19] designed a series of experiments to prove that the raw flows & DL-based malicious traffic detection models outperform traditional ML-based models, which use specific hand-crafted features based on domain expert knowledge as input.

RawFlows's input is a tensor of size (n, 1, m), where n is the number of bytes, and m represents the number of packets. They set $n = 100$ and $m = 2$, that is, only the first 100 bytes of the first two packets in a flow are considered.

In our work, we refer to their DL architecture and make certain extensions on it. Our adjustment is mainly reflected in the hyperparameters of the model. On the one hand, considering that our purpose is to detect malware C2 traffic, only sampling the first 2 packets may lose a lot of flow information. Moreover, because the proportion of C2 traffic is relatively small, our data volume does not reach the scale of RawFlows. In order to provide more flow information to the model, we set n = 200 and m = 8.

At the same time, to accommodate the expansion of input, we also need to adjust the model structure accordingly to increase the expressive ability of the model. Specifically, we refer to the DL architecture of raw packets in [19] to reshape our DL-based malware C2 traffic detector. The DL architecture of the target detector in this article is shown in Fig. 4.

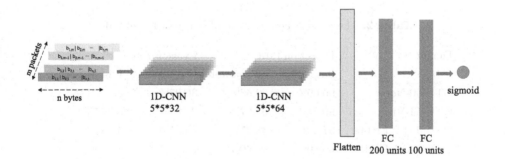

**Fig. 4.** DL architecture for detector

## 5   Results

As mentioned earlier, we designed comprehensively evaluate the effectiveness of our method from the perspectives of evasion performance, time performance, and impact on malware.

Specifically, in the pre-training stage, we will use the traffic of a specific family and the UBC traffic to train the GAN model. Then fix the discriminator and all convolutional layers of the generator, and fine-tune the model using the victim's locally captured benign samples and 1000 malicious samples. Finally, in the practical stage, the generator will directly or be called by malware to generate the *adv_patch*, which is inserted into the original malicious C2 flow to construct the adv-C2 traffic. The newly crafted samples are input to the detector for prediction. So we can evaluate the evasion performance through the change in the detector's recall score.

## 5.1   Evasion Performance

We randomly select 3000 flow samples from each of the 5 families for testing, and get the results in Table 2. We use *DetectionRate* and *EvasionRate* to measure the performance of the detector and our method, respectively. They are calculated according to Eq. 5 and Eq. 6.

$$DetectionRate = Recall = \frac{Number\ of\ detected\ malware\ flow}{Number\ of\ all\ malware\ flow} \quad (5)$$

$$EvasionRate = \frac{Successful\ evasion\ attempts}{All\ evasion\ attempts} \quad (6)$$

The first column in Table 2 is the original detector recall score for 5 malware families, indicating that the detector we use has good accuracy and generalization performance.

From the table, we can see that the proposed method can achieve an evasion rate up to 51.4%, while reducing the recall of the detector to 45.4%, less than 50%, which means that it is difficult for the detector to resist our attack method.

**Table 2.** The evasion performance of our attack method

| Family | Detection rate | | Evasion rate | |
|---|---|---|---|---|
| | init_sample | adv_sample | Pre-training | Fine-tuning |
| CTU-50-Neris | 99.00% | 62.63% | 30.07% | 36.37% |
| CTU-54-Virut | 99.83% | 67.07% | 24.93% | 32.77% |
| CTU-264-2-Emotet | 97.47% | 48.67% | 45.37% | 48.80% |
| CTU-346-1-Dridex | 96.80% | 45.40% | 47.50% | 51.40% |
| CTU-327-1-Trickbot | 99.93% | 63.37% | 31.40% | 36.57% |

By comparing the last two columns in the table, we can find that after the fine-tuning process in the victim environment, the evasion rate generally increases by 7%–21%, which can prove the effectiveness of the TL.

In addition, we have also observed that there are certain differences in the evasion performance of different malware families, from 51.4% in Dridex to 32.77%

in Virut. Through the analysis of the original flow samples, we believe that the bypass rate is mainly affected by the following factors: 1) The detection rate of the detector to the malware. Take Dridex and Neris as examples. The detector has seen the Neris samples during the training phase, so it can easily get the characteristics of Neris and achieve a relatively high recall, while the initial recall of the unseen Dridex family is relatively low. Therefore, a carefully generated *adv_patch* is more likely to mislead the detector, who has not seen Dridex before.

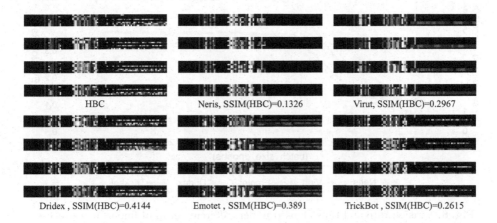

**Fig. 5.** Visualize flow samples of different families

2) The similarity between the malware C2 sample and the target benign sample. We turn the input of the detector into a grayscale image for direct observation, which is shown in Fig. 5. We also calculate the average structural similarity score (SSIM, a common measure of image similarity) between each malware family and HBC to measure how similar each family's traffic is to HBC. Take Dridex and Trickbot as examples, both of them are mostly TLS traffic and have high similarities. But through the visualization of the samples, we found that the Dridex C2 samples are more similar to the target benign C2 samples, so it is easier to fool the detector by adding disturbances on Dridex C2 flow.

## 5.2 Time Performance

In order to evaluate the time performance of our method, we recorded the fine-tuning time elapsed and the evasion rate of 5 CTU malware families under different training hyperparameters (batch_size, epochs).

From Fig. 6, we can find that the time elapsed of batch_size = 128 is roughly 1.4 times that of batch_size = 256, but what needs to be noted is that larger batch_size often means larger memory consumption. The overall fine-tuning time consumption will increase linearly with the increase of epochs number, while the evasion rate is different. The evasion rate at epochs = 100 is significantly higher than that of epochs = 50, but the evasion rate at epochs = 200 is not significantly improved compared to epochs = 100. Therefore, for efficiency considerations, we think it is reasonable to set the training hyperparameters to (batch_size = 128, epochs = 100) or (batch_size = 256, epochs = 100). The attacker can trade-off between shorter training time and lower resource occupancy as needed.

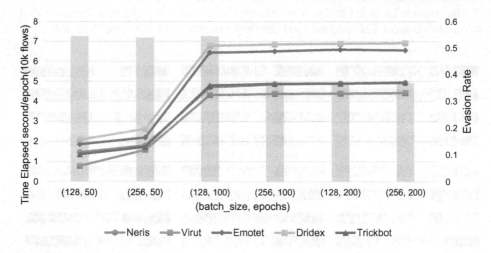

**Fig. 6.** Evasion rate and time elapsed under different setup of fine-tuning process

# 6    Real-Life Experiment

We believe that in order to develop evasion methods that can be applied in real attacks, a comprehensive evaluation from the perspective of practicality must be conducted, rather than just proving the effectiveness.

In this work, we complete the evaluation of practicability by designing a real-life experiment. Specifically, we built a custom malware on the basis of Byob. By requesting GAN in real-time to obtain *adv_patch*es and communicating with the server through C2 channel, we obtain a real-life scenario.

It should be pointed out that the real-life experimental settings are exactly the same as those described in Sect. 4, except that the source of the malware C2 traffic. The effectiveness experiment uses public traffic dataset, while the real-life experiment uses traffic that generated by our custom malware.

## 6.1    Custom Malware

To evaluate our method we used the open-source post-exploitation framework called Byob [2]. Byob was modified to receive the *adv_patch* from the GAN generator and send it after TCP three-way handshake. Byob consists of a client and a C2 server that is written in python. We deploy the C2 server in a Linux virtual machine and the infected victim in a Windows 8 virtual machine respectively.

The communication between client and server is established over HTTP. In order to allow the client to receive the payload generated by the generator, we modify the core module of the client so that every time the client communicates with the server, it will first call the trained generator to obtain the *adv_patch*, and send it once the connection is successfully established.

To continuously obtain malicious communication traffic, we write a script to let Byob client performs the following actions in sequence:

- checks if the server is online.
- sends a heartbeat message with a unique identifier.
- retrieves a command id from server.
- executes the corresponding module.

In this way, we obtained 17,536 Byob C2 flow, which is used as the malware C2 flow dataset for training GAN.

## 6.2   Impact on Malware

We evaluated the practicability of our method from two perspectives: malware C2 channel efficiency impact and resource utilization.

### Malware C2 Channel Efficiency

For malware, the transmission efficiency of the C2 channel is a very important requirement. The significant C2 channel delay caused by evasion methods will reduce the communication efficiency of the victim and the C2 server.

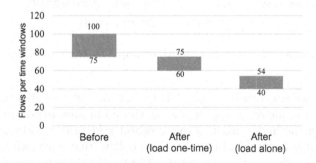

**Fig. 7.** Malware C2 channel efficiency before and after applying our method

Therefore, we evaluated malware C2 channel efficiency before and after applying our method. By calculating the number of C2 flows sent by malware before and after applying our method in a time window, we obtained Fig. 7.

Before applying, we program the malware to communicate with the C2 server every 3 s, so there will be 75–100 flows within a 5 min time window. After applying our method, this number dropped to 40–75. That is to say, whether we choose to access a large number of *adv_patch*es at one time, or request the generator every time before the start of each communication, we can guarantee at least 8 C2 communications per minute, which makes it a feasible channel for a C2.

**Resource Utilization**

Another major impact of the evasion method on malware is that it will increase resource usage in the victim environment. To measure the resource utilization of our method, we recorded the CPU and memory usage of GAN training and inferring, as shown in Fig. 8. Figure 8 shows that a large amount of resource occupancy is mainly caused by GAN training, and the resource usage of the GAN generation process (malware calls GAN to generate adv-patch) is relatively equivalent to malware calling other malicious functions.

As for GAN training, although the fine-tuning stage has obvious optimizations in memory utilization compared to the pre-training stage, it seems that there is no improvement in CPU utilization. That is constrained by the maximum CPU capacity. The CPU utilization in the fine-tuning stage and the pre-training stage is close to 100%, but by comparing the training time of each epoch, we can find that the pre-training is about 14.35 times that of the fine-tuning.

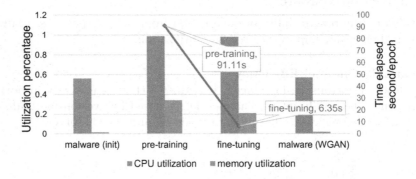

**Fig. 8.** Changes in resource utilization caused by GAN model training

Therefore, we can conclude that TL can reduce the resource utilization in the victim environment, thereby reducing the additional exposure risk caused by our evasion method. Although our method still brings additional resource consumption, it is inevitable. In any case, we believe that our method has certain advantages over other methods in terms of impacts on victim environment.

## 7   Conclusion

In this paper, we focus on how to use the DL model's vulnerability to craft adversarial samples in the field of malware C2 traffic, and propose a GAN-based evasion attack method. Specifically, GAN generates *adv_patch* by simulating the distribution of benign samples, so that malware C2 traffic containing that *adv_patch* can mislead DL-based detectors. Our method is not only able to adaptively simulate the normal traffic of the victim, but also has less negative impact on the malware. These two advantages make our method more suitable for real attack scenarios. The results show that our method can not only achieve a bypass rate of 51.4%, but also has relatively little impact on malware C2 channel and less victim resource usage.

In future work, we plan to explore the influence of hyperparameters such as patch length and embedding position on the evasion rate. At the same time, we will seek ways to further reduce the negative impact of evasion methods on malware, such as model size and resource utilization.

# References

1. https://github.com/caesar0301/pkt2flow
2. https://github.com/malwaredllc/byob
3. Arjovsky, M., Chintala, S., Bottou, L.: Wasserstein GAN (2017)
4. Brown, T.B., Mané, D., Roy, A., Abadi, M., Gilmer, J.: Adversarial patch. CoRR abs/1712.09665 (2017). http://arxiv.org/abs/1712.09665
5. Carlini, N., Wagner, D.: Towards evaluating the robustness of neural networks. In: 2017 IEEE Symposium on Security and Privacy (SP), pp. 39–57. IEEE (2017)
6. Cheng, Q., Zhou, S., Shen, Y., Kong, D., Wu, C.: Packet-level adversarial network traffic crafting using sequence generative adversarial networks (2021)
7. Chernikova, A., Oprea, A.: FENCE: feasible evasion attacks on neural networks in constrained environments (2020)
8. Clements, J., Yang, Y., Sharma, A., Hu, H., Lao, Y.: Rallying adversarial techniques against deep learning for network security (2019)
9. Goodfellow, I.J., Shlens, J., Szegedy, C.: Explaining and harnessing adversarial examples (2015)
10. Ibitoye, O., Shafiq, O., Matrawy, A.: Analyzing adversarial attacks against deep learning for intrusion detection in IoT networks. In: 2019 IEEE Global Communications Conference (GLOBECOM), pp. 1–6 (2019)
11. Kurakin, A., Goodfellow, I., Bengio, S.: Adversarial machine learning at scale. arXiv preprint arXiv:1611.01236 (2016)
12. Li, J., Zhou, L., Li, H., Yan, L., Zhu, H.: Dynamic traffic feature camouflaging via generative adversarial networks. In: 2019 IEEE Conference on Communications and Network Security (CNS), pp. 268–276 (2019)
13. Li, R., Xiao, X., Ni, S., Zheng, H., Xia, S.: Byte segment neural network for network traffic classification. In: 2018 IEEE/ACM 26th International Symposium on Quality of Service (IWQoS), pp. 1–10 (2018)
14. Lin, Z., Shi, Y., Xue, Z.: IDSGAN: generative adversarial networks for attack generation against intrusion detection (2019)
15. Liu, C., He, L., Xiong, G., Cao, Z., Li, Z.: FS-Net: a flow sequence network for encrypted traffic classification. In: IEEE Conference on Computer Communications, IEEE INFOCOM 2019, pp. 1171–1179 (2019)
16. Lotfollahi, M., Siavoshani, M.J., Zade, R.S.H., Saberian, M.: Deep Packet: a novel approach for encrypted traffic classification using deep learning. Soft. Comput. **24**(3), 1999–2012 (2020)
17. Madry, A., Makelov, A., Schmidt, L., Tsipras, D., Vladu, A.: Towards deep learning models resistant to adversarial attacks. arXiv preprint arXiv:1706.06083 (2017)
18. Marín, G., Casas, P., Capdehourat, G.: RawPower: deep learning based anomaly detection from raw network traffic measurements. In: Proceedings of the ACM SIGCOMM 2018 Conference on Posters and Demos, pp. 75–77 (2018)

19. Marín, G., Casas, P., Capdehourat, G.: Deep in the dark - deep learning-based malware traffic detection without expert knowledge. In: 2019 IEEE Security and Privacy Workshops (SPW), pp. 36–42 (2019)
20. Novo, C., Morla, R.: Flow-based detection and proxy-based evasion of encrypted malware c2 traffic. In: Proceedings of the 13th ACM Workshop on Artificial Intelligence and Security, AISec 2020, pp. 83–91. Association for Computing Machinery, New York (2020)
21. Papernot, N., McDaniel, P., Jha, S., Fredrikson, M., Celik, Z.B., Swami, A.: The limitations of deep learning in adversarial settings. In: 2016 IEEE European Symposium on Security and Privacy (EuroS&P), pp. 372–387. IEEE (2016)
22. Rigaki, M., Garcia, S.: Bringing a GAN to a knife-fight: adapting malware communication to avoid detection. In: 2018 IEEE Security and Privacy Workshops (SPW), pp. 70–75 (2018)
23. Szegedy, C., Zaremba, W., Sutskever, I.: Intriguing properties of neural networks. arXiv preprint arXiv:1312.6199 (2013)
24. Wang, W., Zhu, M., Zeng, X., Ye, X., Sheng, Y.: Malware traffic classification using convolutional neural network for representation learning. In: 2017 International Conference on Information Networking (ICOIN), pp. 712–717 (2017)
25. Wang, Z.: The applications of deep learning on traffic identification. BlackHat USA **24**(11), 1–10 (2015)

# Analyzing the Security of OTP 2FA in the Face of Malicious Terminals

Ahmed Tanvir Mahdad[1(✉)], Mohammed Jubur[2], and Nitesh Saxena[1]

[1] Texas A&M University, College Station, TX, USA
{mahdad,nsaxena}@tamu.edu
[2] University of Alabama at Birmingham, Birmingham, AL, USA
mjabour@uab.edu

**Abstract.** One Time Password (OTP) is the most prevalent 2FA method among users and service providers worldwide. It is imperative to assess this 2FA scheme's security from multiple perspectives, considering its ubiquitous presence in the user's day-to-day activities. In this work, we assess the security of seven commercially deployed *OTP-2FA* schemes against malware in the terminal attack model without compromising any 2FA device or authentication services. To implement this attack scenario, we develop a combination of attack modules that will capture password and OTP in different ways during the user's login attempt. At the same time, it would originate a fresh concurrent hidden session from within the terminal or remotely to get possession to the user account without compromising the service or network or any external device. We examine implemented attack against seven different popular public services, which mostly use two variants of *OTP-2FA* and observed that almost all of them are vulnerable to this attack. Here, the threat model is practical as the attack components can be installed in the user's terminal without any root/administrator privilege. Moreover, the attack modules require a small number of resources to run. The whole procedure would run from the background that makes the attack very hidden in nature and attain low detectability after examining against prominent anti-malware programs that indicate a real-world threat. Our findings after the analysis of the *OTP-2FA* schemes indicate that an adversary who can install malware on the user's terminal can defeat almost all popular and widely used *OTP-2FA* schemes, which are vital security components of online accounts and secure financial transactions. The result also points out that the *OTP-2FA* scheme does not add extra security on top of the password in the presence of the malicious program in the terminal.

## 1 Introduction

Password-only authentication is the most widely used and deployed authentication method. Adversaries can steal the password using some well-known method (e.g., phishing [13], dictionary attack [2], man-in-the-middle attack [10]), thus the security of this authentication method always raises questions. Two-Factor Authentication (2FA) is introduced to provide an extra layer of security over the password-only authentication system to address the risk of password leakage. 2FA requires another factor of authentication (*Something the user has* or *Something the user is*) besides a password (*Something the user knows*).

© Springer Nature Switzerland AG 2021
D. Gao et al. (Eds.): ICICS 2021, LNCS 12918, pp. 97–115, 2021.
https://doi.org/10.1007/978-3-030-86890-1_6

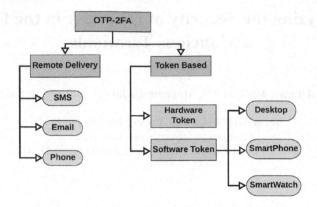

**Fig. 1.** Overview of OTP variant covered in this work

The most primitive and popular form of the 2FA method is One-Time Pin (OTP). Here, a temporary PIN is generated on or delivered to a user registered device. The user has to provide this PIN as proof of his possession of the registered device. The OTP can be generated on a remote server and delivered to the user device (e.g., SMS or email-based OTP) or generated on the user device itself (e.g., hardware or software token). We denote this particular 2FA scheme as *OTP-2FA* throughout the paper.

As a 2FA method, *OTP-2FA* is popular among both users and service providers. In a study conducted by *Duo* in 2017 [4], in the United States, 86% of 2FA users use email/SMS OTP, and 19% of them use hardware tokens, and 52% of them use software tokens. *OTP-2FA* is popular among the users and the service providers for various reasons. It requires a simple deployment and does not require internet connectivity on a 2FA device in most cases (e.g., SMS, software token, hardware token).

The primary motivation of deploying 2FA schemes is to elevate a digital entity's security in the face of malicious attacks. Adversaries can launch an attack on authentication using the user terminal or network components. The malware starts most of the attacks from the user terminal where it resides. There is an increased risk of malware infection in shared terminals (e.g., computers in airports and public libraries). These computing and internet facilities are prevalent among people living in under-developed countries and under the poverty line. The voluminous number [32] of people habituated to use public computers increases the chance of being affected by malware.

As *OTP-2FA* is one of the most deployed 2FA schemes among service providers (e.g., financial/banking services, email, social networks), security analysis of this 2FA method always draws the attention of the researchers. They have studied deployment challenges, usability [16,43,44], and security [21,45] and reveal many issues. The most common vulnerabilities are wireless attack, SIM swap attack, and attack by mobile malware [21]. They also did vulnerability analysis of alternative OTP deployment (e.g., encap OTP [30], gridcode OTP [26], 2GR protocol of OPTAP-scheme [22]) and banking system OTP vulnerabilities [45]. Some of the researches also emphasize using an alternative method of *OTP-2FA* that will address common usability and security issues.

We classify studied *OTP-2FA* schemes in this work into two primary categories. One is *Remote Delivery*, where the OTP is generated in remote service and communicated with the user through a trusted channel (e.g., SMS, email). Here, the user has to prove possession of an entity (e.g., phone number in case of SMS, a specific account in case of email) as second-factor authentication (*Something the user has*). Another is *Token-based* OTP, where it is generated in a local device. It can be a software token where OTP is generated in an app that resides in a 2FA device (e.g., smartphone or smartwatch) or a hardware token where it is generated in a token like portable hardware devices (shown in Fig. 7a). Figure 1 illustrates the summary of our covered *OTP-2FA* categories.

**Our Work: Vulnerability analysis of popular *OTP-2FA* deployment in the face of malware in the terminal:** In this paper, we do a security analysis of 2FA methods that uses *OTP-2FA* as the second factor of authentication. We implement malware components to generate malware-infected terminal scenarios and evaluate how much protection the *OTP-2FA* schemes provide in the face of these attacks. To accomplish that, we implement two different variants of attack. One of them is launched from the user terminal, and the other is from a remote attacker's machine. Both of the attack variants can make concurrent attacks to defeat the *OTP-2FA* scheme. Our observations indicate that these attacks can overcome almost all the *OTP-2FA* schemes we have studied.

**Our Contributions:** Our contributions to this work are two-fold:

1. *Implementation of carefully crafted attack to test the vulnerability of OTP-2FA schemes*: We design two variants of the attack on *OTP-2FA* aided by malware residing in the terminal. One of them is *Internal Attack*, which can launch attacks within the user's terminal with the keylogger, headless browser, and browser extensions as attack components. Another attack is *Remote Attack*, where malware components residing in the browser send captured information to the remote attacker. The attack components used in this technical study are designed to evade the common anti-virus and anti-malware programs that represent a real-world attack scenario.
2. *Vulnerability analysis of OTP-2FA deployed in commonly used service providers.*: We analyze *OTP-2FA* scheme deployed in different popular service providers that the users use for different sensitive purposes (e.g., e-commerce, email, online storage service, social networks). Our analysis covers two primary types of *OTP-2FA* schemes, *Remote Delivery* and *Token Based* OTP. Our observation indicates that they are all vulnerable to malware in terminal attack where the attacker does not have to compromise the 2FA device or service itself.

**Attack Demonstrations and Paper Outline:** We provide our attack implementation demonstrations at https://sites.google.com/view/otp-attack-demo/home.

## 2   Background

### 2.1   One Time Pin Based 2FA

One Time PIN (OTP) is one of the most widely used 2FA methods. The main idea is to generate a temporary PIN as proof of the user's possession of a device (e.g., smartphone, smartphone) or entity (e.g., phone number, email account). OTP can be generated on a remote service and communicated with the user or generated on the 2FA device itself.

For generating OTP, generally, two types of algorithms are used , Time-Based One Time Password (TOTP) [28] and HMAC-based One-time Password (HOTP) [27,34]. The generation of HOTP depends on a secret key and a "counter". This "counter' is known only to the second-factor device and the server. The "counter" in the token increments itself each time after a new OTP is generated. The server used it to check the validity of OTP. In TOTP, instead of "counter", the incremental factor uses a timestamp, which would be incremented every 30 or 60 s. The algorithms also take other parameters that are used to determine the length of the generated OTP.

We have classified the studied *OTP-2FA* scheme into two main categories that are illustrated in Fig. 1. In *Remote Delivery* method, the OTP is generated in an external server and communicated with the user by a delivery channel. The most common delivery channel is the mobile phone network used by SMS (Short Messaging Service) based OTP. The OTP receiving endpoint is a phone which is subject to mobile network connectivity. It does not depend on internet connectivity, leading the SMS-OTP to widespread acceptance, especially in remote areas. We show snapshots of the user interface from the user terminal and phone for SMS-OTP in Appendix Fig. 7e and 7f. Email and a phone call can also be other mediums of communication in that case.

The second category is *Token Based OTP-2FA* , which is generated locally in the device. As this variant does not need any delivery channel, it is safe from an attack like man-in-the-middle (MitM). Based on the generating device, it has two variants, hardware token and software token. An example of the hardware token is RSA SecureID [31], where the user has to carry a small device that can generate the OTP anytime on demand. A snapshot of the RSA SecureID device is shown in Appendix Fig. 7a. In software token, OTP is generated by an OTP generator software that can be deployed in computers (i.e., desktop, laptop), smartphones, or smartwatches. We show the user interface snapshot of smartphone-based software token in Appendix Fig. 7c and 7d. *Token based* OTP also has a desktop-based variant with a "Copy-to-Clipboard" feature that allows users to copy and paste the generated OTP without typing it on the terminal. This feature increases both usability (user does not have to type it) and security (cannot be captured by keylogger-based malware). A snapshot of the desktop-based software token is shown in Appendix Fig. 7b.

## 2.2 Malware on Terminal

Malware on the terminal is a common threat for users nowadays. Adversaries infect the terminal by installing malicious software to steal user data and control sensitive information and programs. They can target terminals that the users are habituated to use (e.g., personal computer, smartphone). We focus on desktop-based terminals for this work. In the operating system market share, about 30.57% of users use Windows, and approximately 7.65% use other desktop-based operating systems [37]. The rest of the users use smartphone-based operating systems (e.g., Android, iOS). So, 38.22% of the internet traffic comes from desktop-based operating systems that indicate that many users still use desktop-based terminals to do their daily online activities.

The threat introduced by malware on the terminal is prevalent in the shared terminals, where more than one person uses it. People tend to use public computing resources (e.g., computer terminals, internet connection) in universities, offices, airports, and many other public places, and the number is significantly large [32]. On

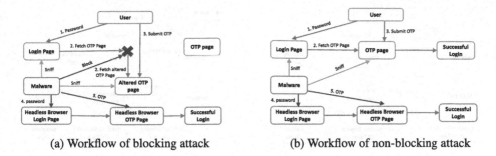

(a) Workflow of blocking attack                      (b) Workflow of non-blocking attack

**Fig. 2.** Workflow of blocking and non-blocking variant of *Internal Attack*

the other hand, personal computers have also become one of the prominent targets of malware creators. They can easily be infected by malware when connected to unsafe networks (i.e., public network in the airport, coffee shop) and devices (e.g., USB flash drive, CD).

## 3 Our Attack: Overview and Design

### 3.1 Attack Overview

We implement two attack variants, *Internal Attack* and *Remote Attack*. We design *Internal Attack* in two ways, blocking and non-blocking attacks. The attack's primary focus is to learn the password and OTP and request service concurrently using a headless browser session. In blocking attack, after recording the password, the user's request will be blocked by a malicious browser extension and redirects her to an altered page at the same time. The altered page is the exact replication of the OTP page where the user is supposed to type the OTP. The keylogger can capture the typed OTP here and can launch a concurrent attack. After the successful attack, the user will be redirected to the login page again. A high-level overview of *Internal Attack* is shown in Fig. 2.

Another variant, *Remote Attack* is more hidden as most of its attack components are not residing in the user's terminal. Here, the keylogger is implemented on the user's side to capture passwords like *Internal Attack*. Moreover, it can send the credential through a secure channel to the remote attacker. The attack component on the user's end can also take a screenshot at any time. Some desktop-based software tokens (e.g., Authy [40]) used the "copy-to-clipboard" feature where the user does not have to type the OTP in the terminal. In that case, the captured screenshot has been sent to the remote attacker via a secure channel where the attacker can extract the OTP by Optical Character Reader (OCR). OCR is implemented on the attacker's end as part of attack implementation. We have portrayed a high-level overview of *Remote Attack* in Fig. 3.

### 3.2 Attack Assumptions

– The attacker has the capability of installing malicious attack components in any desktop-based terminals running on Windows 10 or earlier. The installation of attack components does not need any administrator/root privilege for being installed.

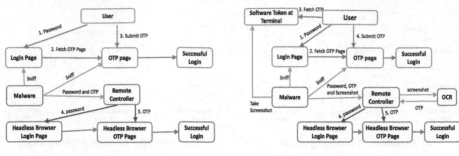

(a) OTP capture by keylogger workflow    (b) OTP capture by screenshot workflow

**Fig. 3.** OTP capture and screenshot capture workflow of *Remote Attack*

- The user is habituated to use *OTP-2FA* as the 2FA method and will not use any other 2FA variant or will not remember her browser's session.

### 3.3    Attack Implementations Vs. Other Known Attack

The attack implementation is significantly different from other known forms of attacks. *Session Hijacking* attack rides on the user's existing session to attack authentication schemes, where our implemented attack creates another independent session (in both Internal and Remote attacks). As a separate session has been created, the attacker has more control (i.e., not dependent on expiry of user session and user's activity) than session hijacking attack. As our study's focus is malware residing in the terminal, external attacks like active phishing attacks are out of the scope of this comparison.

### 3.4    Attack Components

We implement different attack components for both variants of our attack. Keylogger is used in both *Internal Attack* and *Remote Attack* to capture password and OTP typed in keyboard. Along with the keylogger, the malware can eavesdrop on mouse events. For example, after a specific web page has been opened, the keylogger program can record the keystroke unless it records an "ENTER" key or right mouse click (submit button or placing the cursor to password field) and save it as username. Similarly, it can record and save the password. The saved pattern can be used for future attack initialization. For *Internal Attack*, an automated hidden browser has been used to launch the concurrent attack. Also, it uses a browser extension to monitor, block and redirect any user requests.

*Remote Attack* variant needs implementation in both user terminal and remote attacker's side. Along with the keylogger, this variant has a component that is capable of taking a screenshot. Implementation on the attacker's side includes an OCR to analyze received screenshots and extract OTP.

## 3.5    Internal Attack

All the *OTP-2FA* variants require the user to type OTP on the terminal to complete the challenge (except the "copy-to-clipboard" feature of Authy, which we discuss in the next subsection). As generation algorithms and service implementations are different, different schemes show different characteristics. Generated OTP from different service implementations can have a different validity period or reusability.

We observed that for some deployed *OTP-2FA* allows the same OTP to be reused in concurrent log in (e.g., Google). Others prohibit reusing the same OTP in more than one session. We use *non-blocking* variant of attack to the service that allows reusing the same OTP in multiple sessions. For other services, we use *Blocking* variant of attack. We illustrate the block diagram of the *Internal Attack* in detail in Appendix Fig. 6b. The step-by-step workflow is listed down here:

1. Keylogger component scans for a predefined pattern (e.g., username or part of already known password) which works as an indicator that the user is trying to log in. It then sends a command to other attack modules to launch a concurrent attack.
2. For *Blocking* attack, the browser extension will block the user's request and redirect her to an altered OTP entry page. *Non-blocking* attack will let the request reach the service, and the service would redirect the user to the OTP entry page.
3. In OTP capture state, the keylogger records the next **N** digit (N= Length of OTP).
4. When the Nth digit has been recorded, it would save the OTP in a local file and instantly launches a headless browser session. With captured password and OTP, the headless browser session would complete authentication.
5. For *Blocking* attack, the user would be redirected to the login page after some time.

## 3.6    Remote Attack

In Remote Attack, user credentials and OTP captured from the user terminal are sent to the remote attacker's machine using a secure channel. One of the advantages of *Remote Attack* is that it can overcome the keylogger's limitations to capture the OTP, which uses the "copy-to-clipboard" feature of Desktop-based software tokens. The step-by-step procedure is discussed here, and the diagram is shown in Appendix Fig. 6a.

1. Keylogger looks for a predefined pattern like *Internal Attack*. It then sends a command to the remote machine to launch the attack.
2. After typing the password, when the user presses the "Enter" key or clicks on the "Submit" button, the keylogger will send the captured password to the remote attacker. The automated browser session on the attacker's machine launches an attack and waits for OTP to capture.
3. OTP can be captured in two ways. In the case of the "copy-to-clipboard" feature on desktop-based software-token, the attack component would capture a screenshot when it finds the software-token app is in the foreground (opened in the screen and not minimized) and send it to the remote attacker's machine. The OCR that resides on the remote attacker's machine would extract the OTP from the screenshot and transfer it to the automated browser session to complete the attack.

# 4   Implementation

## 4.1   Attack Components of *Internal Attack*

**Keylogger and Controller:** We develop the "Keylogger and Controller" module using Python 3.7.5 using some standard python keylogger libraries. The detection submodule is used to match and detect any pre-configured pattern. The capture submodule can capture any patterns on demand, which can be password or OTP. It can also save captured patterns on local machines for future use. After matching any pre-configured pattern, the connector submodule can send the command to launch other modules.

**Automated Headless Browser Session:** This module's primary purpose is to launch an automated browser session in the background to launch a concurrent attack. We use *Selenium Webdriver* [35] and *PhantomJS* [29] to implement this module. The automated browser session can request and log in to any web-based online service using captured login credentials and OTP. It also can take any screenshot of a web page or download an HTML source of a page on demand in a hidden manner. We develop different versions of this module that can work with different services.

**Chrome Extension for Request Control:** This attack module is developed as a chrome extension that can monitor, block, and redirect the user's request to a malicious web page. It accomplishes its task in such a covert way that the user would not get any error message in the browser. This component continuously monitors the URL in the address bar to match a predefined pattern (i.e., part of the URL of a service provider's authentication page). It takes action (e.g., block or redirect) after a match is found. We develop the extension as a standalone Chrome extension. The malicious code can be ported inside any benign extension without the user's knowledge.

## 4.2   Attack Components of *Remote Attack*

**Keylogger and Controller:** The *Keylogger and Controller* component of *Remote Attack* can also encrypt the captured password and OTP and send them to a remote attacker's machine through a secure channel along with the key pattern detection and recording.

**Screenshot Capture Module:** This component resides on the user's terminal and can capture a screenshot of a specific window. We develop this module using python 3.7.5 using some standard libraries of python. It can capture screenshots without any administrator permission or user notification. It can monitor active programs (that are opened in the foreground or background) and take a screenshot of a specific program's window when it comes to the foreground. This attack component can also send the captured image to the remote attacker via a secure channel.

**Remote Attack Controller:** This module resides in the attacker's machine. It waits for the command to launch a concurrent attack from the "Keylogger and Controller" component in the user's terminal. When it receives the command, it activates itself and collects the user's terminal information. It can decrypt the encrypted password and

OTP. This component is also equipped with a specially designed OCR (Optical Character Reader) that can extract OTP from the user's terminal screenshot. This module is developed in python 3.7.5 using some standard libraries.

**Remote Automated Browser Session:** *Remote Attack Controller* is another automated browser session that is implemented on the remote attacker's side. After getting commands from attack components from the user's terminal, it activates itself to run a concurrent attack. We implement it using the *Selenium Web Driver* framework of Java.

**Table 1.** Evaluation with *OTP-2FA* deployments of popular service providers

| Service name | Remote delivery | Token based |
|---|---|---|
| Microsoft Outlook | ✓ | ✓ |
| Facebook | ✓ | ✓ |
| Duo | N/A | ✓ |
| Google | ✓ | ✓ |
| LastPass | N/A | ✓ |
| Amazon | ✓ | ✓ |
| Twitter | ✓ | N/A |

✓ - Attack Successful, ✗- Not Successful

(a) Snapshot of Google UI after a successful attack

(b) Snapshot of Outlook UI after a successful attack

**Fig. 4.** Snapshot after login in to some popular service after successful attack

## 5    Evaluation

### 5.1    Evaluation of Commercially Deployed *OTP-2FA* Schemes in the Face of the Attack

We examine the *OTP-2FA* scheme of seven well-known services, Microsoft Outlook [1], Facebook [18], Duo [17], Google [19], LastPass [23], Amazon [3], and Twitter [41], against our implemented attack. Almost all of them use both *Remote Delivery* and *Token Based OTP-2FA* schemes. Some of the *Remote Delivery OTP-2FA* schemes allow the same OTP to be used multiple times (e.g., Google), which might open attack opportunities. *Token Based OTP-2FA* schemes normally allow a limited amount of time (e.g., 30 s) as OTP validity. These schemes also can be reusable or not-reusable, depending on service implementation. We check that almost all of the schemes mentioned above are vulnerable to our attack. We present a summary in Table 1. We also show snapshots of the successful attack on Google and Outlook in Fig. 4a and 4b.

### 5.2    Detectability from Terminal and 2FA Device

**Internal Attack:** The attack design is straightforward in case of a non-blocking attack on *OTP-2FA Remote Delivery*, where reuse of OTP is allowed for a certain period (e.g., 30 s). After capturing the user's OTP, a headless browser can quickly log in to another independent session using the same OTP. So, no trace of concurrent login would be visible in the user's terminal. From the 2FA device's point of view, multiple SMS would be received. As they would display the same OTP (OTP is reused in this case), it is very unlikely that the user would be suspicious.

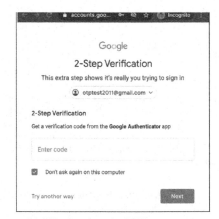

(a) UI of altered OTP entry page    (b) UI of authentic OTP entry page

**Fig. 5.** Comparison between user interface of OTP entry page when altered page has been shown to user as a part of blocking attack of *Internal Attack* variant

The blocking variant of the attack does not send any duplicate OTP on the user's 2FA device, thus also not detectable from the 2FA device's point of view. This attack variant is also useful for the *OTP-2FA* schemes where reuse of the same OTP in different sessions is not allowed. The browser extension would block the user request to the server and redirect the user to a duplicate OTP entry page. It is a similar-looking page and hard to detect unless the user pays close attention to the changing URL in the address bar, which is very unlikely in a real-world scenario. When the user proceeds after entering OTP on that duplicate page, the extension will redirect the user to the original login page, which would convince her to think that her login failed due to some glitch. If the user can detect at this point, she can do nothing much as the attacker has full control of her account in the meantime. We present a similar-looking malicious page and authentic page in Fig. 5a and 5b for comparison.

**Remote Attack:** In the case of the *Remote Attack* variant, it shows minimal activity on the user's terminal. After capturing credentials on the user's terminal, the attack components send them to the remote attacker's machine. Detection scenarios are similar to *Internal Attack* from the user terminal and 2FA devices. The screenshot capturing process also does not show any visible trace.

## 5.3   Detectability from Service

The *Non-blocking* variant of the attack sends only one concurrent request along with the user's request. As the user's request is blocked in the *Blocking* attack, only a single request would be sent. So, this low amount of request and activity, which is similar to the user's benign activity, would not raise a warning to the service provider's end.

**Table 2.** Evaluation with free desktop based antivirus and web based malware scanning tool

| Desktop based scanning | | | | | Web based scanning |
|---|---|---|---|---|---|
| Antivirus name | Quick scan | Full scan | Runtime warning | Version | Web-based antivirus Detection |
| Bitdefender | ✗ | ✗ | ✗ | Free | ✗ |
| Avast | ✗ | ✗ | ✗ | Free | ✗ |
| Avira | ✗ | ✗ | ✗ | Free | ✗ |
| AVG | ✗ | ✗ | ✗ | Free | ✗ |
| Sophos Home | ✗ | ✗ | ✗ | Free | ✗ |
| Kaspersky Security Cloud | ✗ | ✗ | ✗ | Premium (Free Trial) | ✗ |
| ZoneAlarm | ✗ | ✗ | ✗ | Free | ✗ |
| Mcafee total protection (Free Trial) | ✗ | ✗ | ✗ | Premium (Free Trial) | ✗ |

✓ - Detected, ✗- Not Detected

### 5.4 Detectability in the Presence of Anti-Malware Program

Our implemented attack is tested in the presence of some well-known and free anti-malware solutions for the desktop terminal. We also did web-based malware analysis for attack components. We examine attack script against Bitdefender [11], Avast Free Edition [7], Malwarebytes anti-malware [24], Kaspersky Security Cloud [5], Sophos Home [36], Avira [9], AVG [8], ZoneAlarm [14] and Mcafee Total Protection [25]. We also evaluate executable attack files with 68 antivirus engines in VirusTotal [42], where the above-mentioned antivirus engines are also included. We present the detailed and comparative results of this analysis in Table 2.

We focus on two types of detection of anti-malware softwares, which are *Signature-based analysis* and *Behavior-based analysis*. In the signature-based analysis, an antivirus engine matches a predefined signature (i.e., pattern or part of a previously detected malicious program) in a file. To evade signature-based detection, we obfuscate attack script code and use some customized libraries (i.e., a customized version of python standard libraries). We observed almost zero detection with the desktop-based antivirus. Only 2 out of 68 engines in Virustotal [42] detect our executable as malware. We later analyzed to extract the root cause of the detection and found that the detections were false alarms. It detects other benign python executables in the same way.

We develop the attack script in such a way so that it consumes minimal CPU, memory, and network resources. We present the average usage of resources during the active and idle states in Appendix Table 4. From the usage table, it is obvious that resource consumption is very low for *Remote Attack*, which is expected as the attack uses minimal resources on the user's terminal. For *Internal Attack*, peak usage is more, although peak usage is observed when an attack is underway. Peak usage remains for a few seconds and remains idle for the rest of the time. As the script demonstrates a small amount of activity, a runtime scan cannot detect the malware based on *Behavior-based Analysis*.

### 5.5 Detectability During Attack Module Deployment

An attacker can deploy the attack components on the Windows platform without any root/administrator privilege. The "Keylogger and Controller" component and headless browser session implemented on phantomJS requires no installation. The headless browser session developed using the "Selenium web driver" requires only java installation in the user's terminal, which also can be avoided if the attacker uses the portable version of java. A hidden malicious batch script file can copy attack modules in the user's terminal promptly.

**Table 3.** Summary of analysis

| OTP variant | Generation/ Delivery method | Attack variant | Browser dependency | Detectable in 2FA device | Detectable in rerminal | Attack success |
|---|---|---|---|---|---|---|
| Token based | Hardware roken | Internal attack (Blocking) | ✓ | No | Very low | ✓ |
| | | Internal attack (Non-blocking) | ✗ | No | No | ✓ |
| | | Remote attack | ✗ | No | No | ✓ |
| | Software token (2FA device) | Internal attack (Blocking) | ✓ | No | Very low | ✓ |
| | | Internal attack (Non-Blocking) | ✗ | No | No | ✓ |
| | | Remote attack | ✗ | No | No | ✓ |
| | Software token (Terminal) | Internal attack (Blocking) | ✓ | N/A | N/A | ✗ |
| | | Internal attack (Non-blocking) | ✗ | N/A | N/A | ✗ |
| | | Remote attack | ✗ | No | No | ✓ |
| Remote delivery | SMS, Email, Phone | Internal attack (Blocking) | ✓ | No | Very low | ✓ |
| | | Internal attack (Non-blocking) | ✗ | Medium | No | ✓ |
| | | Remote attack | ✗ | No | No | ✓ |

# 6  Discussion and Future Work

## 6.1  Attack Summary

We can observe from Table 3 that the attack is successful for both *OTP-2FA* variants. As we described in previous sections, both of the attack variants are a combination of multiple attack components. The "Keylogger and Controller" component records the user's password and OTP and can launch the attack both from the user's terminal and remote attacker's end. Our designed attack can defeat *OTP-2FA* schemes whether OTP can be reused in multiple sessions or not. The advantage of the implemented attack is that it can start simultaneously and promptly when the targeted user attempts to authenticate. The attack begins as soon as a match with a predefined pattern is found. They wait for OTP to capture and can complete the attack immediately. We have demonstrated that, after capturing OTP, the attack can be done within seconds.

Desktop-based OTP generator with a "Copy-to-Clipboard" feature secures OTP from keyloggers. However, our implemented attack can monitor every active window in the user's terminal, and take a screenshot of them when they become active and can send it to the remote attacker. The attack components in the remote attacker's end can extract the OTP from the screenshot instantly and can send a concurrent request to the service. The attack can be done within seconds, which we have shown in the demonstrations. It is prompt and stealthy, demonstrating the real threat in real-world scenarios.

### 6.2    General Discussion

**Malware Infection Risk of Windows-based Terminal:** Malware infection on Windows-based desktop terminals is prevalent nowadays. According to the AV test, 90.83 million malware and 5.43 million potentially unwanted programs are reported for Windows OS [6]. In the United States, 30% of computers are infected with some form of malware [15]. Moreover, Windows is the most common target of malware writers as it has 357 dangerous security gaps, according to a recent report [46].

**2FA System Vs. Malware in Terminal**: Previous literature suggests that the attacker has to compromise both the terminal and 2FA device to compromise any 2FA system successfully. According to Bonneau et al. [12], 2FA schemes are *resilient-to-internal-observation*, which indicates that they cannot be bypassed by compromising only a single entity (i.e., the terminal). Our work contradicts this prior line of reasoning.

**Feasibility of Installing Keylogger/Browser Extension on User Terminal**: According to [38], the development of keylogger-based malware is on the rise, and 80% of keyloggers are not detectable to anti-malware programs. Our evaluation against standard anti-malware programs in Sect. 5.4 also supports this claim. Furthermore, a recent article reveals that 500 malicious Chrome extensions have been identified with a similar ability to our attack component (i.e., redirecting victims to a malicious website) [39]. Most of them are hidden inside benign and useful user extensions.

### 6.3    Mitigation Strategy

We propose some prevention techniques that the service providers and the users can take to prevent themselves from a similar attack scenario.

– Service providers should block reusing the same OTP in two different sessions. Although that does not prevent the users from all attacks discussed above, this can make the attacker's task difficult and challenging.
– To prevent the concurrent attack, the service should discard the concurrent request if that arrives before another request is not completed.
– The desktop-based operating systems (e.g., Windows) and the software token developers should work together to prevent taking hidden screenshots by unauthorized applications.

### 6.4    Limitations and Future Work

From the attack design perspective, we have some limitations. To design the browser extension component, we only use "Google Chrome" as the browser and "Windows" as the user's terminal operating system. To design a more stealthy attack, we have a plan to implement browser-independent and platform-independent components in the future. We will also work on designing a component that can start the software token in the user's terminal and capture the OTP in a hidden way without the user's assistance.

We cover commercially deployed *OTP-2FA* schemes in this work. We plan to examine more sophisticated and sensitive *OTP-2FA* schemes (i.e., academic and banking *OTP-2FA* schemes) for more innovative threat models in the future.

## 7 Related Work

The authors [45] discussed the vulnerabilities of *OTP-2FA* implementation in some of the bank's online networks in South Korea in this work. Here they intercepted password and OTP, which is similar to our attack design. The attack we demonstrated can make both internal and remote attacks. It can create a hidden browser session from the victim's terminal to instantly initiate the attack, where the authors in this work send OTP credentials to the outside attacker only. Our designed "Internal Attack" can launch itself more promptly compared to them. Not only that, the attack can do similar damage in the absence of the attacker.

In the work [33], *Siadati et al.* discussed a social engineering method to steal SMS-based 2FA, which tricks a victim into sending SMS with authentication code to the attacker's device. Our demonstrated attack also uses social engineering, but it is more hidden compared to the author's approach, and the user does not have to send the SMS anywhere. Instead, the user types the OTP to the terminal, and attack components automatically capture it and initiate an attack.

Some works focus on common *OTP-2FA* vulnerabilities. Examples of similar work are [21] and [20]. In this work [21], authors focus on wireless attacks, SIM swap attacks, or mobile malware attacks. We do not compromise mobile phones or any other 2FA device or network devices in our designed attack. Instead, this attack can start a concurrent attack from the terminal or a remote computer. Similarly, in this work [20], the author focuses on some vulnerabilities like cookie theft, subject hijacking, SiM swap, forged SMS recovery messages, and duplicate OTP generators. Our designed attack is simpler, and it does not need to compromise anything outside of the terminal.

There are many custom *OTP-2FA* deployments, and researchers analyzed those schemes and found vulnerabilities. Examples of such work are [22, 26, 30, 45]. In contrast, we designed an attack on the "malware-in-terminal" attack scenario and evaluated commercial and popular *OTP-2FA* schemes against it.

## 8 Conclusion

There is a common belief that second-factor authentication schemes should be more secure deployments that give users an extra step of security to safeguard them from adversaries. *OTP-2FA* is the most widely used and acknowledged 2FA implementation. We evaluated the security of *OTP-2FA* in the malicious program's presence in the user terminal and tried to assess if it adds any additional protection in that circumstance. We designed and prepared some attack modules and examined them against seven prominent service providers to answer that question. Our findings signify that the concurrent attack can authenticate the attacker from the terminal or a remote PC without compromising any external factor (e.g., service, password, database, 2FA device, network). The attack also accomplishes low detectability for its hidden nature and little activity in the user's terminal for a small amount of time. Our approach is unique as it involves the users during the attack without their notice and achieves success by only compromising the terminal. More work can be done focusing on concurrent attacks on the different second-factor schemes and how to safeguard sensitive accounts and resources from those attacks.

112     A. T. Mahdad et al.

**Acknowledgements.** Authors are thankful to the shepherd Yuhong Nan and the anonymous reviewers for their feedback. This work is funded in part by NSF grants: CNS-1547350, CNS-1526524 and CNS-1714807.

# A    Appendix

## A.1    Tables

**Table 4.** Attack variant resource consumption

| Variant | State | CPU | Memory | Power | Network |
|---|---|---|---|---|---|
| Internal attack | Idle | 0.01% | 27.5 MB | Very low | 0.0 MB |
| | Peak | 20.0% | 108.83 MB | Low | 0.1 MB |
| Remote attack (Password and OTP capture) | Idle | 0.01% | 23.7 MB | Very low | 0.0 MB |
| | Peak | 0.1% | 24.5 MB | Low | 0.1 MB |
| Remote attack (Software token screenshot capture) | Idle | 0.01% | 29.5 MB | Very low | 0.0 MB |
| | Peak | 0.5% | 32.8 MB | Low | 0.8 MB |

## A.2    Other snapshots

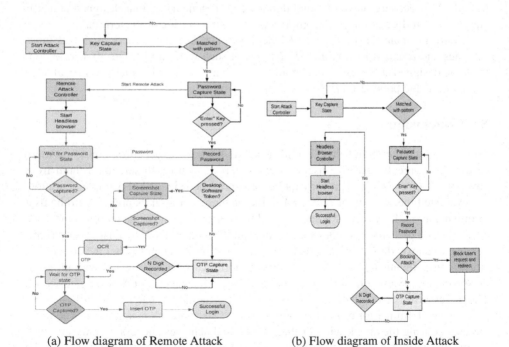

(a) Flow diagram of Remote Attack          (b) Flow diagram of Inside Attack

**Fig. 6.** Flow diagrams of remote and inside attack on OTP

(a) RSA SecureID, that was widely used as hardware token *OTP-2FA*

(b) Snapshot from Authy, which has a desktop based software token variant.

(c) Snapshot of UI from terminal during authentication using software-token variant of *OTP-2FA*

(d) Snapshot of UI from 2FA device during authentication using Software-token variant of *OTP-2FA*

(e) Snapshot of UI from the user terminal during authentication using remote-delivery variant of *OTP-2FA* (SMS)

(f) Snapshot of UI from the 2FA device (Phone) during authentication using remote-delivery variant of *OTP-2FA* (SMS)

**Fig. 7.** Collection of snapshots of *OTP-2FA* prompt UI from user terminal and 2FA device

# References

1. Outlook- free personal email and calender from microsoft (2020). https://outlook.live.com/owa/
2. Adams, C.: Dictionary Attack, pp. 332. Springer, Boston (2011). https://doi.org/10.1007/978-1-4419-5906-5_74
3. Amazon.com Inc: Amazon.com: Online shopping for electronics, apparels, computer, books & dvd and more (2020). https://www.amazon.com
4. Anise, O., Lady, K.: State of the auth: experiences and perceptions of multi-factor authentication. Duo security (2017)
5. AO Kaspersky Lab: Kaspersky security cloud - free (2020). https://www.kaspersky.com/free-cloud-antivirus
6. AV-TEST - The independent IT Security Institute: Malware statistics & trends report (2020). https://www.av-test.org/en/statistics/malware
7. Avast Foundation: Avast free antivirus (2020). https://www.avast.com/en-us
8. Avast Software s.r.o.: Avg free antivirus (2020). https://www.avg.com/en-us/
9. Avira Operations GmbH & Co. KG.: Avira antivirus (2020). https://www.avira.com/
10. Bhushan, B., Sahoo, G., Rai, A.K.: Man-in-the-middle attack in wireless and computer networking – a review. In: 2017 3rd International Conference on Advances in Computing, Communication Automation (ICACCA) (Fall), pp. 1–6 (2017)
11. Bitdefender: Bitdefender - global leader in cybersecurity software (2020). https://www.bitdefender.com/
12. Bonneau, J., Herley, C., Van Oorschot, P.C., Stajano, F.: The quest to replace passwords: a framework for comparative evaluation of web authentication schemes. In: 2012 IEEE Symposium on Security and Privacy, pp. 553–567. IEEE (2012)
13. Chandel, A., Kumar, P., Yadav, D.K.: Phishing attack and its countermeasures. IEEE Electron. Device Lett 7, 569–571 (1999)
14. Check Point Software Technologies Inc: Pc and mobile security software - zonealarm (2020). https://www.zonealarm.com/
15. DataProt: A not-so-common cold: malware statistics in 2021 (2021). https://dataprot.net/statistics/malware-statistics/
16. De Cristofaro, E., Du, H., Freudiger, J., Norcie, G.: A comparative usability study of two-factor authentication. arXiv preprint arXiv:1309.5344 (2013)
17. Duo: Duo two factor authentication and endpoint security (2020). https://duo.com
18. Facebook: Facebook (2020). https://www.facebook.com/
19. Google: Google accounts (2020). https://accounts.google.com
20. Grimes, R.: The many ways to hack 2fa. Netw. Secur. 2019(9), 8–13 (2019)
21. Karia, M.A.R., Patankar, A., Tawde, M.P.: SMS-based one time password vulnerabilities and safeguarding OTP over network. Int. J. Eng. Res. Technol. (IJERT) 3(5), 1339–1343 (2014)
22. Kuo, W.C., Lee, Y.C.: Attack and improvement on the one-time password authentication protocol against theft attacks. In: 2007 International Conference on Machine Learning and Cybernetics, vol. 4, pp. 1918–1922. IEEE (2007)
23. Logmeln Inc.: Lastpass - password manager & vault app (2020). https://www.lastpass.com/
24. MalwareBytes: Malwarebytes cybersecurity for home and business (2020). https://www.malwarebytes.com/
25. McAfee: Mcafee total protection (2020). https://www.mcafee.com/en-us/antivirus/free.html
26. Molloy, I., Li, N.: Attack on the gridcode one-time password. In: Proceedings of the 6th ACM Symposium on Information, Computer and Communications Security, pp. 306–315 (2011)

27. M'Raihi, D., Bellare, M., Hoornaert, F., Naccache, D., Ranen, O.: Hotp: an hmac-based one-time password algorithm. The Internet Society, Network Working Group. RFC4226 (2005)
28. M'Raihi, D., Machani, S., Pei, M., Rydell, J.: Totp: time-based one-time password algorithm. Internet Request for Comments (2011)
29. PhantomJS Contributors: Phantomjs- scriptable headless browser (2010–2018). https://phantomjs.org
30. Raddum, H., Nestås, L.H., Hole, K.J.: Security analysis of mobile phones used as OTP generators. In: Samarati, P., Tunstall, M., Posegga, J., Markantonakis, K., Sauveron, D. (eds.) WISTP 2010. LNCS, vol. 6033, pp. 324–331. Springer, Heidelberg (2010). https://doi.org/10.1007/978-3-642-12368-9_26
31. RSA Security LLC.: RSA securid hardware tokens (2020). https://www.rsa.com/en-us/products/rsa-securid-suite/rsa-securid-access/securid-hardware-tokens/rsa-securid-hardware-tokens
32. Separd, D.: First-ever national study: millions of people rely on library computers for employment, health, and education (2020). https://www.gatesfoundation.org/media-center/press-releases/2010/03/millions-of-people-rely-on-library-computers-for-employment-health-and-education
33. Siadati, H., Nguyen, T., Gupta, P., Jakobsson, M., Memon, N.: Mind your SMSes: mitigating social engineering in second factor authentication. Comput. Secur. **65**, 14–28 (2017)
34. Smith, N.: Hotp vs totp, whats the difference? (2018). https://www.microcosm.com/blog/hotp-totp-what-is-the-difference
35. Software Freedom Conservancy- Selenium Project: Selenium webdriver (2020). https://www.selenium.dev/projects/
36. Sophos Ltd: Sophos home - cybersecurity made simple (2020). https://home.sophos.com/en-us.aspx
37. StatCounter: Operating system market share worldwide- february 2020 (2020). https://gs.statcounter.com/os-market-share
38. TechJury.net: What is a keylogger? [everything you need to know] (2021). https://techjury.net/blog/what-is-a-keylogger/
39. ThreatPost: 500 malicious chrome extensions impact millions of users (2020). https://threatpost.com/500-malicious-chrome-extensions-millions/152918/
40. Twilio Inc.: Authy- two factor authentication (2fa) app & guides (2020). https://authy.com/
41. Twitter Inc: Explore twitter (2020). https://twitter.com/explore
42. VirusTotal: Virustotal (2020). https://www.virustotal.com/gui/home/upload
43. Weir, C.S., Douglas, G., Carruthers, M., Jack, M.: User perceptions of security, convenience and usability for ebanking authentication tokens. Comput. Secur. **28**(1–2), 47–62 (2009)
44. Weir, C.S., Douglas, G., Richardson, T., Jack, M.: Usable security: user preferences for authentication methods in ebanking and the effects of experience. Interact. Comput. **22**(3), 153–164 (2010)
45. Yoo, C., Kang, B.T., Kim, H.K.: Case study of the vulnerability of OTP implemented in internet banking systems of South Korea. Multimedia Tools Appl. **74**(10), 3289–3303 (2015)
46. Ziff Davis, LLC. PCMAG Digital Group: Windows computers were targets of 83% of all malware attacks in q1 2020 (2020). https://www.pcmag.com/news/windows-computers-account-for-83-of-all-malware-attacks-in-q1-2020

# IoT Security

# Disappeared Face: A Physical Adversarial Attack Method on Black-Box Face Detection Models

Chuan Zhou[1,2], Huiyun Jing[3(✉)], Xin He[4], Liming Wang[2], Kai Chen[2], and Duohe Ma[2]

[1] School of Cyber Security, University of Chinese Academy of Sciences, Beijing, China
[2] Institute of Information Engineering, Chinese Academy of Sciences, Beijing, China
{zhouchuan1,wangliming,chenkai7274,maduohe}@iie.ac.cn
[3] China Academy of Information and Communications Technology, Beijing, China
jinghuiyun@caict.ac.cn
[4] National Computer Network Emergency Response Technical Team/Coordination Center of China, Beijing, China
hexin@cert.org.cn

**Abstract.** Face detection is a classical problem in the field of computer vision. It has significant application value in face recognition and face recognition related applications such as face-scan payment, identity authentication, and other areas. The emergence of adversarial algorithms on face detection poses a substantial threat to the security of face recognition. The current adversarial attacks on face detection have the limitations of the need to fully understand the attacked face detection model's structure and parameters. Therefore, these methods' transferability, which can measure the attack's effectiveness across many other models, is not high. Moreover, due to the consideration of commercial confidentiality, commercial face detection models deployed in real-world applications cannot be accessed, so we cannot directly launch white-box adversarial attacks against these models. Aiming at solving the above problems, we propose a Black-Box Physical Attack Method on face detection. Through ensemble learning, we can extract the public weakness of the face detection models. The attack against the public weakness has high transferability across models and makes escaping black-box face detection models possible. Our method realizes the successful escape of both the white-box and black-box face detection models in both the PC terminal and the mobile terminal, including the camera module, mobile payment module, selfie beauty module, and official face detection models.

**Keywords:** Adversarial attack · Face detection · Black-box attack · Real-world attack

## 1 Introduction

Face detection is a classical problem in the field of computer vision. It has significant application value in face recognition related applications such as face-scan

D. Gao et al. (Eds.): ICICS 2021, LNCS 12918, pp. 119–135, 2021.
https://doi.org/10.1007/978-3-030-86890-1_7

payment, identity authentication, image focus, and other areas. What's more, face detection is widely integrated into commercial applications in our daily life. For example, Face Unlock Module of commercial mobile phones, payment software Alipay, and selfie beauty application B612 all need to complete the initial positioning of faces through the face detection module.

Unfortunately, the emergence of adversarial examples seriously threatens the real-world applications based on face recognition. For example, one person can escape face detection models by attaching adversarial patches to its face or headwear, which means that the face detection model cannot detect the face in the form of rectangular boxes. Because face detection is the presequence module of face recognition, this may cause the face recognition related application to crash unexpectedly. For example, criminals might use specifically trained adversarial examples to evade facial recognition systems deployed by the police. To investigate the face detection's vulnerability to improve both the face detection security and the face recognition security, it is urgent to study the adversarial attacks on face detection. There are some adversarial attack methods on face detection nowadays.

According to whether they launch adversarial attacks using digital images or real-world objects like masks and patches, attack methods can be roughly categorized into Digital Adversarial Attacks and Physical Adversarial Attacks.

Digital adversarial attacks [16] are methods that rely on an implicit assumption that attackers can directly feed digital adversarial examples to machine learning algorithms. The adversarial perturbation is directly added to the original digital image, and then the perturbation-added digital image is fed into the classification model [22]. Bose et al. [1] generate adversarial examples by solving constrained optimization problems such that the face detector cannot detect the faces in generated adversarial examples. As a white-box attack, their method relies on full access to the attacked face detection model - Faster R-CNN [12], which is unrealistic in real-world and commercial scenarios.

Physical adversarial attacks focus on generating adversarial examples on real objectives such as patches and masks. For example, Adversarial patches are printed by a laser printer and then attached to the physical-domain target like human cheeks to fool image classifiers instead of directly feeding digital images into machine learning algorithms. Zhou et al. [23] use infrared LEDs attached to headwear to create physical adversarial examples to escape the face detection systems. However, their method of configuring the infrared light source is quite complicated. Kaziakhmedov et al. [8] introduce an easily reproducible way to attack the cascade CNN face detection system - Mtcnn by attaching adversarial patches to the face directly. However, similar to Bose et al.'s attack method [1], none of the above attack methods can achieve the black-box adversarial attack on face detection.

In a word, these methods all require direct access to the model's structure and parameters. So, they pose less threat to the real-world commercial applications that integrate black-box face detection models. It is because that the enterprise will not opensource the application source code and the internal details of the face detection model to protect the core interests.

What is more, the current attacks on face detection generally lack transferability [11]. Their adversarial examples can only mislead specific models that take part in training adversarial examples but not other models without participation. Model changes caused by changes of structure, parameters, or just training datasets can seriously affect attack performance.

To solve the above-mentioned problems in existing methods, we propose a black-box physical adversarial attack method on face detection with high transferability. We focus on attacking **black-box** face detection models in the real world. Our method is inspired by the opinion that some public semantic features [3,13] can reflect the public vulnerabilities of unknown DNN models that focus on the same task [3]. Focus on the task of face detection, the detection confidence is such a kind of semantic feature in essence. It is because the detection confidence naturally has the ability to reflect the face detection effect. In this paper, we use ensemble learning [6] to fuse multiple white-box face detection models' detection confidence to construct the total classification loss function. After we optimize the learnable parameters in the loss function through ADAM [9], we believe that loss value calculated by the parameters-optimized loss function can reflect the public vulnerability of face detection task to some extent. Since generating the adversarial patches by attacking the public vulnerabilities of face detection, our method bypasses the strict limitation of needing detailed internal information of the attacked model and has good transferability. By directly posting the printed adversarial patches to the cheeks, we realize the physical adversarial attack on face detection. It also means that we have breached the whole face recognition module from another aspect.

The demonstration video is available on the Internet[1]. We show our escape effect of attacking one real-time black-box face detection model - Official Yoloface [4] for demonstration.

Our specific contributions are as follows.

1. For the first time, we successfully realize black-box physical adversarial attacks on face detection in the real scenes. Our method realizes the escape of face unlock module and built-in camera software of mobile phones such as Samsung, Xiaomi, etc. Simultaneously, we realize the escape of commercial applications, such as Alipay's face payment and B612 selfie beauty applications. Our method provides a new way to evaluate the security of face recognition.
2. We realize the high-transferability adversarial attack on face detection. The adversarial patches generated by specific face detection models can also successfully attack many other face detection models through our method. For example, the adversarial patches generated by our method can successfully escape the black-box official face detection models such as Pyramidbox, light-DSFD, and Yoloface at the same time.

---

[1] https://drive.google.com/file/d/1atQvE9tPMRhKwZw4j-MGt7xFxEY86bBh.

# 2   Related Works

## 2.1   Adversarial Attacks on Face Recognition

There are many adversarial attack methods on face recognition nowadays. According to the different attack principles, we classify adversarial attacks on face recognition as the following.

Debayan Deb et al. propose an automated adversarial examples synthesis method (Advfaces [5]). It uses GAN to limit perturbations on significant pixel locations (e.g., eyebrows and eyeballs) to improve the invisibility of the generated adversarial examples. They successfully mislead the face recognition model by generating a digital adversarial example. However, it is difficult to migrate this attack to the physical domain because it works by making slight changes to pixels at specific locations in specific images.

Mahmood Shari et al. [14] limit the adversarial disturbance to the area where the glasses are worn and use a 3D printer to produce the custom-made glasses. Later, they come up with a method called AGN [15] (Adversarial Generative Nets) to generate universal eyewear, which dramatically improves the attack's effectiveness and allows the resulting eyewear to resemble the shape and texture of real eyewears.

Xiao et al. propose an adversarial attack method on face recognition called StAdv [18]. This method interpolates and merges the neighborhood information through a spatial transformation attack, making the image smoother. Instead of adding irrelevant adversarial perturbations to the image, the method distorts the image by moving the pixels, which can mislead the face recognition model into giving wrong results.

The above methods directly attack the face recognition model to induce the models to give the wrong classification results. Focusing on the presequence module of face recognition, face detection, we will introduce several algorithms that also pose a severe threat to face recognition.

## 2.2   Adversarial Attacks on Face Detection

Different from directly attacking the face recognition module, adversarial attacks on face detection can also threaten the security of face recognition by attacking its presequence face detection module.

AJ Bose et al. [1] realized the white-box attack against the Faster RCNN face detection model by training a generator. The training process can be regarded as a constrained optimization problem, similar to a C&W attack [2]. As a typical white-box attack method, its white-box escape rate is high. However, the need to fully understand the structure and parameters of the model makes its attack capability very limited and lacks in transferability.

Kaziakhmedov et al. [8] were the first to propose the physical adversarial attack on face detection, and they could escape the camera without being detected. Adversarial patches are adjusted by backward propagation of the gradient value of the loss function. As a gradient-based adversarial algorithm, its core advantage lies in fast attack speed and high efficiency. However, its white-box characteristics make it lack transferability, and its attack success rate is relatively low.

Zhou et al. [23] proposed a face detection escape method based on infrared interference. Face recognition systems can be bypassed or misdirected by using special infrared light that cannot be seen by the naked eyes to interfere with facial feature point regions. This attack method has strong concealment, but the escape success rate is relatively low, and it is very complicated to configure the infrared light source and other steps. The experiments show that the attack algorithm's transferability is also low, and it is hard to use their method to escape other unknown black-box face detection models [23].

## 3   Our Proposed Method

Although DNNs that focus on the same task may have different structures and weights, they may share the public semantic feature [3]. The public semantic feature could reflect the specific task's public attention(public weakness) to some extent. Researchers find that by attacking the public attention(public weakness) of white-box DNNs, they could make DNNs' attention lose their focus and there-fore fail in judgment [3]. Inspired by the above idea and ensemble learning [6], we design an algorithm to find out face detection models' public weakness. Then, we train specific adversarial patches to attack the find-out face detection models' public weakness. When the specific tester posts trained patches on his Living Face, he could escape Real-World Black-Box face detection models in front of a

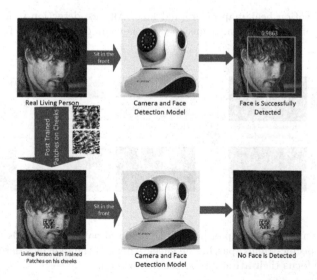

**Fig. 1.** The pipeline of attacking face detection models in the real world. We use a specific tester to post specific-trained patches on his cheeks; then, he sits in front of a camera to attack the black-box face detection model.

camera. The process of attacking black-box face detection models in real scenes using our method is shown in Fig. 1.

When make judgements, human tend to concentrate on certain parts of an object to allocate attention efficiently. In computer vision, the same idea has been applied and becomes an important component in DNNs [3]. Therefore, we show our escape effect from the respect of the attention heat map. The attention heat map shows the DNN's concern area, which is usually essential for its specific mission (e.g. face detection). When no pasting trained patches, Fig. 2 shows that all face detection models can focus attention on the face region. However, when pasting adversarial patches trained by our method, Fig. 3 shows that all four face detection models are no longer able to focus attention on the face region. That is maybe why they cannot find faces anymore. You can also prove this with the average attention heat map of four models in the bottom right corner of Fig. 2 and Fig. 3. (We just average the four attention heat maps to obtain the average attention heat map.)

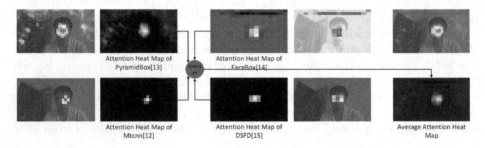

**Fig. 2.** The attention heat map of original face. All face detection models can focus their attention on the face area. It means that they can detect human faces normally.

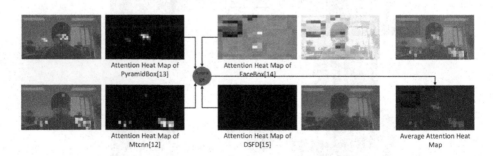

**Fig. 3.** The attention heat map of posting adversarial patches on the Tester's Cheeks. After posting specific-trained patches on cheeks, none of the four face detection models could anymore focus their attention on the face region.

Figure 2 and Fig. 3 essentially illustrate why we can escape the face detection models, and they visually demonstrate the effectiveness of our method from the perspective of heat maps.

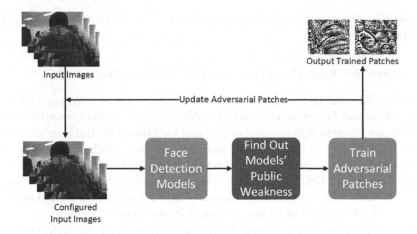

**Fig. 4.** The pipeline of training adversarial patches

The flow chart of training adversarial patches is shown in Fig. 4. Our proposed method mainly includes three parts: **Configure Input Images**, **Search for Face Detection Models' Public Weakness**, and **Update the Adversarial Patches** using the find-out public weakness. We will cover each part in detail in the following subsections.

### 3.1    Configure Input Images

In the "Configure Input Images" part, we take photos to Obtain Input Images. Then we affine the updated adversarial patches (When the first time of training, use the initialized patches) to the projection areas to obtain Configured Input Images. The following describes in detail how to Obtain Input Images and how to Obtain Configured Input Images. For more details of our datasets and training parameter, please refer to Sect. 4.1.

**Obtain Input Images.** Use a laser printer to print out the checkerboard patches and post them on the tester's cheeks. We use a camera to collect face images under different illumination, distance, and angle to enhance the attack's robustness. For anyone who wants to escape black-box face detection models and reach the best escape effect, we should collect his eight face images in different conditions (distance, lights) to train specific patches.

**Obtain Configured Input Images.** We mark the edge of the collected input images' checkerboard patches. Because we need iterative training adversarial patches, if it is the first time, we would affine the **initialized** adversarial patches to the projection areas obtained by identifying the marked edge; otherwise, we would affine the **updated** adversarial patches to the projection areas obtained by identifying the marked edge. At this point, we have obtained the Configured Input Images we need.

## 3.2    Search for Face Detection Models' Public Weakness

In order to find out face detection models' public weakness, we need to select several white-box face detection models. We uniformly collect **unofficial** face detection models with different structures and backbones for training adversarial patches so that we can use the corresponding official models to run the black-box escape test. In this paper, we select **Mtcnn** [20][2], **PyramidBox** [17][3], **Facebox** [21][4], and **DSFD** [10][5]. You can also choose other white-box models, as long as they can perform the relevant calculations described below. (The number of models to choose is also changeable, we choose four here.)

First, we feed all the configured images into our selected face detection models. Due to the uninterpretable nature of artificial intelligence, we cannot obtain entities that can reflect the public weakness of the face detection task. For each face detection model, we calculate its classification loss $L_{clf}$ by formula 1. We assume that the average sum of multiple face detection models' classification losses, $L_{clf_{total}}$, can reflect the public weakness of the face detection task. Our hypothesis is reasonable because the total classification loss of multiple face detection models is an important indicator reflecting the effect of face detection task. The effectiveness of face detection can, of course, reflect the vulnerability of the model; therefore, we can use $L_{clf_{total}}$ to naturally reflect the public weakness of face detection models to some extent.

$$L_{clf_i} = \sum_N \sum_m max(p_m - \gamma_i, 0)^2 \tag{1}$$

$$L_{clf_{total}} = \frac{1}{K} \sum_{i=1}^{K} L_{clf_i} \tag{2}$$

$N$ is the number of face images participating in training adversarial patches. $p_m$ represents the confidence probability that the region $m$ of the face image is judged to contain the detected face by the corresponding face detection model. The trainable parameter $\gamma_i$ could reflect the influence of model $i$ on the public weakness of the face detection task in nature. We will train all $\gamma_i$ to represent the public weakness better in the following.

**Train All $\gamma_i$ to Represent Public Weakness Better.** In order to keep the contribution of each model to face detection public weakness similar, we design the following algorithm to train each $\gamma_i$. We do not hope that the contribution of one model to be so enormous that the other models' contributions would be drowned out. When the contribution of one model is much more outstanding

---

[2] https://github.com/edosedgar/Mtcnnattack/tree/master/Mtcnn.
[3] https://github.com/EricZgw/PyramidBox.
[4] https://github.com/610265158/faceboxes-tensorflow/tree/tf1.
[5] https://github.com/610265158/DSFD-tensorflow.

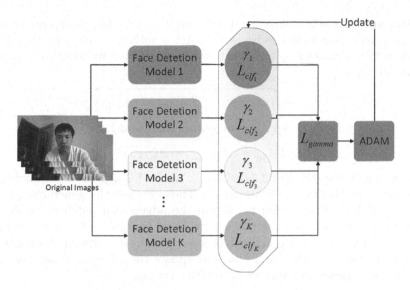

**Fig. 5.** The pipeline of training all $\gamma_i$ in $L_{clf_{total}}$

than that of the others, our algorithm could not extract the public weakness of the face detection task, and it is more likely to extract only one model's weakness (Instead of public weakness).

By optimizing all $\gamma_i$, we kept all the $L_{clf_i}$ at the same order of magnitude. Then $L_{clf_{total}}$ can represent the public weakness better. The specific process is as follows:

Initially, all $\gamma_i$ are set to 0.5. Then, we feed all the original images into selected face detection models to calculate all $L_{clf_i}$ and the $L_{clf_{total}}$. Then, we calculate the $L_{gamma}$ by formula 3. Next, we minimize $L_{gamma}$ through ADAM [9] algorithm and finally obtain all optimized $\gamma_i$ values. The pipeline of training all $\gamma$ is shown in Fig. 5.

$$L_{gamma} = \frac{1}{K} \sum_{i=1}^{K} (L_{clf_i} - L_{clf_{total}})^2 \tag{3}$$

Finally, we can calculate the total classification loss ($L_{clf_{total}}$) through all the optimized $\gamma_i$ values to represent the public weakness of the face detection task better.

### 3.3   Update the Adversarial Patches

At this point, we have finished the training of all optimizable parameters $\gamma_i$ and fixed them in the total classification loss function ($L_{clf_{total}}$). Now, we can use $L_{clf_{total}}$ to represent the public weakness better. The specific process of 'Update the Adversarial Patches' is as follows.

First, we artificially define the Total Variation loss $L_{TV}$ to make the optimization prefer good-looking adversarial patches without sharp color transitions and noise. We calculate $L_{TV}$ from pixel values of the training adversarial patches in position $i, j$.

$$L_{TV} = \sqrt{(p_{i,j} - p_{i+1,j})^2 + (p_{i,j} - p_{i,j+1})^2} \qquad (4)$$

By summing $L_{clf_{total}}$ and $L_{TV}$, we obtain the Total Loss $L_{total}$.

$$L_{total} = \alpha L_{clf_{total}} + \beta L_{TV} \qquad (5)$$

After calculating the Total Loss $L_{total}$, we feed $L_{total}$ as the optimization target into the MI-FGSM [7] algorithm. Then, we use the generated patches from the MI-FGSM algorithm to update the adversarial patches. Finally, we would affine the updated adversarial patches to the projection areas of Input Images and obtain new Configured Images. The new Configured Images are then used to calculate the Total Loss and update the adversarial patches again until reaching the setting number of iterative training.

## 4    Experiments and Result Analysis

In this section, we first introduce the details of our experiment settings. Then, we use living faces to do escape experiments in the real world. Finally, we conduct contrast experiments and ablation experiments to illustrate the effectiveness and feasibility of our method.

### 4.1    Experiment Settings

**Training Datasets.** For anyone who wants to escape face detection models, his 8 face images in different conditions (distance, lights) should be used to train specific adversarial patches. He can then wear the specific-trained patches on his cheeks to attack face detection models in the real world. There are only one tester's eight face images in training sets.

**Testing Datasets.** Unlike digital adversarial attacks that focus on evaluations with well-known digital datasets such as WIDER FACE [19], our physical adversarial attack focuses on launching attacks with a living face. Therefore, we have no test sets, and we use the living face to attack different face detection models in the real world to measure our escape effect. To achieve the best escape effect, the same person who provides the Input Images should be used to attack real-world face detection models.

**Training Parameters.** For all experiments in Chapter 4, the training epoch of adversarial patches is set to 2000, and the training datasets are set to include only 1 tester's 8 face images. $\gamma_i$ of Mtcnn [20], PyramidBox [17], Facebox [21], and DSFD [10] are finally optimized to 0.5, 0.45, 0.6, 0.65, respectively. (Refer to Sect. 3.2 for more details.)

## 4.2    Escape Experiments in the Real World

According to different test content, the real-world escape experiments are finally divided into two parts: escape experiments of face unlock module in mobile phones, and escape experiments of face detection function-related applications in mobile phones.

All the experiments here are set to compare the three cases of posting adversarial patches on cheeks, posting randomly generated patches (subject to a uniform distribution between 0–255), and no posting.

**Escape Experiments of Face Unlock Module in Mobile Phones.** Because the internal details of the face detection models deployed in mobile phones are inaccessible, this part's escape experiments are essentially black-box. We collect mainstream mobile phone brands and finally choose iPhone 11, Samsung Galaxy S10 5G, Xiaomi Redmi K20 Pro. They all support unlocking the phone by face recognition. To control the variables, all experiments use the same lighting conditions, background, and test angle. The tester has registered his face in advance, meaning he could naturally unlock the phone by face recognition under normal circumstances.

The tester respectively wears the adversarial patches, random patches, and nothing. Presses the power button and swipes the screen to trigger the face unlock module. When the mobile phone does not detect a human face, it will prompt "No face detected!" on the screen. When the mobile phone thinks that a face is detected, but the face does not match the registered face, it will give a prompt "Face Does Not Match!" on the screen and refuse to unlock. When the mobile phone thinks that the face is successfully detected and the detected face matches the registered face, the mobile phone will be unlocked successfully. **All videos of attacking the face unlock module are available on the Internet[6].**

We have successfully achieved the escape of both Samsung and Xiaomi phones. It can be seen from the video of Samsung S10 5G and Xiaomi Redmi K20 Pro that when the tester posts trained adversarial patches on his cheeks, the face detection model can not detect the face after a long-time search and gives a prompt of 'No Face Detected!'. When the tester posts randomly generated patches, The phone quickly detects the face and prompts 'Face Mismatch'. The different search times of the above two experiments also prove from another angle that the adversarial patches obtained by our training have indeed achieved escape attacks. When the tester does not post any patches, the mobile terminal is quickly unlocked. These videos demonstrate the excellent escape effect of our adversarial patches. When we conduct the unlock test with IPhone 11, no matter the real situation is "No Face Detected!" or "Face Mismatch!" ', it shows that the "lock" at the top of the screen is not turned on, and the phone is still

---

[6] https://drive.google.com/drive/folders/1LzGVVWl9OMHqXXL5dnh-FrxXsCnG vBJ0.

locked. Therefore, we cannot verify the escape effect of the generated adversarial patches through IPhone 11.

The escape videos above show our excellent escape effect. When the tester posts the specific-trained patches on cheeks, he could successfully escape the face detection model deployed in the mobile phone. Because face detection is the presequence module of face recognition, when face detection breaks down, the entire face-unlock module fails as well.

**Escape Experiments of Face Detection Function-Related Software in Mobile Phones.** To enhance our face detection escape method's credibility and persuasiveness, we also use our method to attack the above mobile phones' **built-in camera** module. At the same time, we conduct escape tests on **Alipay** and selfie beauty software **B612** with the above mobile phones.

**Built-In Camera.** The tester respectively wears the adversarial patches, random patches, and nothing, turns on the built-in camera application, adjusts the camera mode to "portrait mode", switches to the front camera, and faces the camera. When the camera detects a face, it will frame the face with a rectangular box. When the rectangular box cannot select the face or disappears, it means the face detection model cannot detect the face, and we achieve the escape of the built-in camera module. The test results are saved in Fig. 6 as screenshots.

The figures A1~F1, A2~F2, A3~F3 show that only when testers post adversarial patches generated by our method can they successfully escape the face detection module in mobile phones. Posting random patches and No Posting can not escape. A2 shows the automatic Soft-light function fails due to the inability to detect the face, and A2 is significantly colder than B2 and C2. G1~I1, G2~I2, G3~I3 show our excellent escape effect for another tester and demonstrate the generality of our method.

**Alipay.** The tester respectively wears the adversarial patches, random patches, and nothing, opens Alipay in the mobile phone, searches for the official service "Alipay Face-Scan Life", and selects "Experience Face-Scan Payment" from the "Face Brushing Settings" to start the face brushing payment test. The test results are also saved in Fig. 6 as screenshots.

A4~F4 show the attack effect against Alipay's face payment module. A4 and D4 prompt: "No face detected." B4 and E4 prompt, "Please show your whole face." C4 and F4 say, "Please blink." D4~F4 show our excellent escape effect for another tester and demonstrate the generality of our method. Only when the testers post specific-trained adversarial patches(A4 and D4) can they escape the Alipay's face detection module. We successfully escape the commercial black-box Alipay Face Payment module.

**B612.** The tester respectively wears the adversarial patches, random patches, and nothing. In each case, tests are performed in two environments, namely,

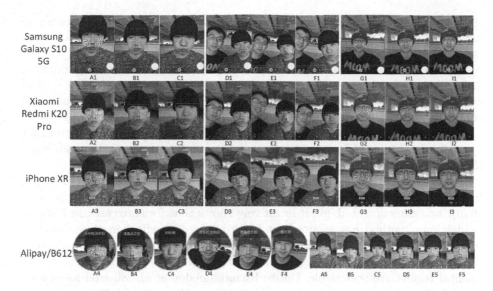

**Fig. 6.** Escape test effect in mobile terminals. Because of the narrow lines of the rectangular boxes, you may need to review this figure carefully.

turning on or turning off beauty function. The test results are also saved in Fig. 6 as screenshots.

A5~F5 show the escape effect of B612. A5, C5, and E5 all turn on beauty function, and B5, D5, F5 not. The selfie beauty app B612 fails to perform its beauty function only when the user wears the adversarial patches because it could not detect the face(A5). We successfully escape from the commercial black-box B612 application.

### 4.3   Contrast Experiments

Nowadays, similar studies include AJ Bose et al. [1], Zhou et al. [23], and Kaziakhmedov et al. [8]. AJ Bose et al. [1] focus on digital attack and is not of comparable value. (We focus on physical attacks.) As for Zhou et al. [23], their method of configuring the infrared light source is much complicated than ours posting patches. Therefore we design relatively fair experiments with Kaziakhmedov et al. [8].

We conduct escape tests for Baidu official face detection model: PyramidBox [7], Tencent official face detection model: light-DSFD [8], and Awesome Open Source recommendation algorithm: Yoloface [9]. These official algorithms such as PyramidBox, lightDSFD, and Yoloface all use the same lighting conditions, the

---

[7] https://github.com/PaddlePaddle/PaddleHub/tree/release/v1.8/demo/mask_detection.

[8] https://github.com/lijiannuist/lightDSFD.

[9] https://github.com/sthanhng/yoloface.

same background, and the same test angle to control the variables. Since we do not use the official face detection model in the training process, the following experiments are essentially black-box escape experiments.

All the contrast experiments are divided into two groups: an experimental group and a control group. The experimental group uses our method. The control group uses the method of Kaziakhmedov et al. [8]. Both the experimental and control group experiments are set to be completed under three distances: close, middle, and far-distance. In each case, the tester shakes his head at a constant speed for 5 s. We use FFmpeg to intercept the video frame by frame and define the escape rate as the percentage of the frames that successfully escape.

Please refer to Fig. 7 to see the results of the contrast experiment. Figure 7 fully demonstrates that our method has much better black-box escape capability than Kaziakhmedov et al. [8]. It intuitively shows that for three attacked official models, our method has an excellent black-box escape effect in close-distance, middle-distance, and far-distance scenarios. The average escape rate at all three distances is also much higher than that of Kaziakhmedov et al. [8]. The experimental results are reasonable. Unlike Kaziakhmedov et al. [8], who focuses on a single face detection model, we extract the public vulnerability of the face detection task by fusing multiple white-box models.

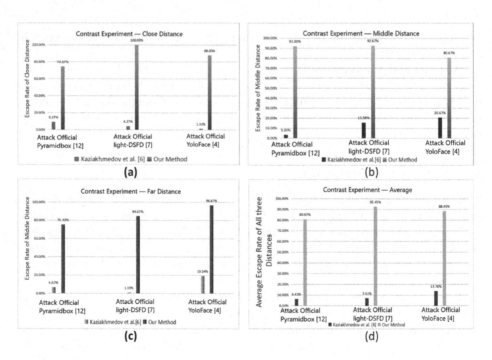

**Fig. 7.** Contrast experiments: escape rate of Close (a), Middle (b), Far (c) Distance and average escape rate of all three distance (d)

## 4.4    Ablation Experiments

To illustrate that our method is indeed better than using the classification loss of only one model, we conduct ablation experiments to investigate the effect of training models' number on the black-box escape effect. Our ablation experiments still focus on conducting escape tests for official PyramidBox, official light-DSFD, and official Yoloface. All the experiments can be divided into four groups: the total classification loss $L_{clf_{total}}$, respectively including 1, 2, 3, and 4 face detection models' classification loss.

Because Kaziakhmedov et al. [8] use only one model(Mtcnn) to train patches, the results of Kaziakhmedov et al. [8] in Fig. 7 are ablation experiments that integrate only 1 model in essence. Since we do not use the official face detection model in the training process, our ablation experiments are essentially black-box escape experiments.

The result of the black-box escape rate of different distance is respectively shown in Fig. 8. Subgraph (a) , (b) and (c) show the escape rates of attacking Official PyramidBox [17], Official light-DSFD [10], and Official Yoloface [4], respectively. Subgraph (d) shows the average escape rate of all three distances for three official models.

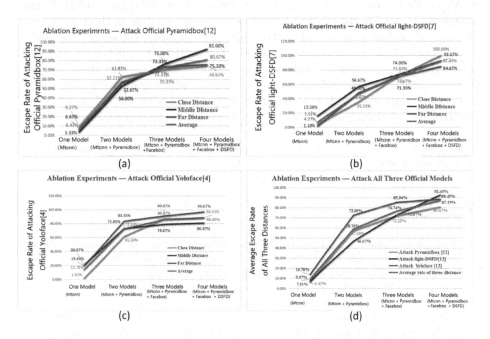

**Fig. 8.** Ablation experiments: escape rate of attacking Official PyramidBox [17] (a), Official light-DSFD [10] (b), and Official Yoloface [4] (c), and average escape rate of three distances for three official models (d)

Subfigures (a), (b), and (c) of Fig. 8 show that we attack different black-box face detection models at different distances and achieve excellent black-box escape effects. Subgraph (d) shows that when compared with the fusion of single model, two models, and three models, the black-box escape rate of fusing four models is increased by about 80%, 30%, and 10%, respectively. (the green line in subfigure (d)).

The above results indicate that the fusion of the four face detection models can achieve the optimal black-box escape effect and prove the effectiveness and reliability of our method.

## 5    Conclusion

We propose a Black-Box Physical Adversarial Attack Method on Face Detection to evaluate face detection security and face recognition security. By calculating the total classification loss function ($L_{clf_{total}}$) and training the parameter $\gamma$ , we successfully extract the face detection models' public vulnerabilities. By executing the adversarial attack on the extracted public weakness, we realize the escape against black-box face detection models with high-transferability. This means that we have broken the face recognition system in another way. In detail, we achieve a high-success-rate black-box escape of commercial applications such as Alipay Face payment, B612, and many official face detection models. Simultaneously, we realize the escape of the face detection module in the mainstream commercial mobile phones and make their Face Unlock, Automatic Soft-light, and other functions break down because the face could not be detected. Through our method, it is possible to evaluate the current commercial face detection and face recognition model's physical-domain security, which helps to further understand the fragility of the face recognition deep neural network and promote the face recognition deep learning model to a safer direction step forward.

**Acknowledgment.** This research was supported by National Research and Development Program of China (No.2019YFB1005203).

## References

1. Bose, A.J., Aarabi, P.: Adversarial attacks on face detectors using neural net based constrained optimization. In: 2018 IEEE 20th International Workshop on Multimedia Signal Processing (MMSP), pp. 1–6. IEEE (2018)
2. Carlini, N., Wagner, D.: Towards evaluating the robustness of neural networks. In: 2017 IEEE Symposium on Security and Privacy (SP) (2017)
3. Chen, S., He, Z., Sun, C., Yang, J., Huang, X.: Universal adversarial attack on attention and the resulting dataset damagenet. IEEE Trans. Pattern Anal. Mach. Intell. (2020)
4. Chen, W., Huang, H., Peng, S., Zhou, C., Zhang, C.: YOLO-face: a real-time face detector. Vis. Comput. **37**(4), 805–813 (2020). https://doi.org/10.1007/s00371-020-01831-7
5. Deb, D., Zhang, J., Jain, A.K.: Advfaces: adversarial face synthesis. In: 2020 IEEE International Joint Conference on Biometrics (IJCB), pp. 1–10. IEEE (2020)
6. Dietterich, T.G., et al.: Ensemble learning. Handb. Brain Theory Neural Netw. **2**, 110–125 (2002)

7. Dong, Y., et al.: Boosting adversarial attacks with momentum. In: Proceedings of the IEEE Conference on Computer Vision and Pattern Recognition, pp. 9185–9193 (2018)

8. Kaziakhmedov, E., Kireev, K., Melnikov, G., Pautov, M., Petiushko, A.: Real-world attack on MTCNN face detection system. arXiv preprint arXiv:1910.06261 (2019)

9. Kingma, D.P., Ba, J.: Adam: a method for stochastic optimization. arXiv preprint arXiv:1412.6980 (2014)

10. Li, J., et al.: DSFD: dual shot face detector. In: Proceedings of the IEEE Conference on Computer Vision and Pattern Recognition, pp. 5060–5069 (2019)

11. Papernot, N., McDaniel, P., Goodfellow, I.: Transferability in machine learning: from phenomena to black-box attacks using adversarial samples. arXiv preprint arXiv:1605.07277 (2016)

12. Ren, S., He, K., Girshick, R., Sun, J.: Faster R-CNN: towards real-time object detection with region proposal networks. IEEE Trans. Pattern Anal. Mach. Intell. **39**(6), 1137–1149 (2016)

13. Selvaraju, R.R., Cogswell, M., Das, A., Vedantam, R., Parikh, D., Batra, D.: Grad-cam: visual explanations from deep networks via gradient-based localization. In: Proceedings of the IEEE International Conference on Computer Vision, pp. 618–626 (2017)

14. Sharif, M., Bhagavatula, S., Bauer, L., Reiter, M.K.: Accessorize to a crime: real and stealthy attacks on state-of-the-art face recognition. In: Proceedings of the 2016 ACM SIGSAC Conference on Computer and Communications Security, pp. 1528–1540 (2016)

15. Sharif, M., Bhagavatula, S., Bauer, L., Rciter, M.K.: Adversarial generative nets: neural network attacks on state-of-the-art face recognition. arXiv preprint arXiv:1801.00349 2(3) (2017)

16. Shen, M., Liao, Z., Zhu, L., Xu, K., Du, X.: Vla: a practical visible light-based attack on face recognition systems in physical world. Proc. ACM Interact. Mobile Wearable Ubiquit. Technol. **3**(3), 1–19 (2019)

17. Tang, X., Du, D.K., He, Z., Liu, J.: Pyramidbox: a context-assisted single shot face detector. In: Proceedings of the European Conference on Computer Vision (ECCV), pp. 797–813 (2018)

18. Xiao, C., Zhu, J.Y., Li, B., He, W., Liu, M., Song, D.: Spatially transformed adversarial examples. arXiv preprint arXiv:1801.02612 (2018)

19. Yang, S., Luo, P., Loy, C.C., Tang, X.: Wider face: a face detection benchmark. In: Proceedings of the IEEE Conference on Computer Vision and Pattern Recognition, pp. 5525–5533 (2016)

20. Zhang, K., Zhang, Z., Li, Z., Qiao, Y.: Joint face detection and alignment using multitask cascaded convolutional networks. IEEE Sig. Process. Lett. **23**(10), 1499–1503 (2016)

21. Zhang, S., Zhu, X., Lei, Z., Shi, H., Wang, X., Li, S.Z.: Faceboxes: a CPU real-time face detector with high accuracy. In: 2017 IEEE International Joint Conference on Biometrics (IJCB), pp. 1–9. IEEE (2017)

22. Zhao, Y., Zhu, H., Liang, R., Shen, Q., Zhang, S., Chen, K.: Seeing isn't believing: towards more robust adversarial attack against real world object detectors. In: Proceedings of the 2019 ACM SIGSAC Conference on Computer and Communications Security, pp. 1989–2004 (2019)

23. Zhou, Z., Tang, D., Wang, X., Han, W., Liu, X., Zhang, K.: Invisible mask: practical attacks on face recognition with infrared. arXiv preprint arXiv:1803.04683 (2018)

# HIAWare: Speculate Handwriting on Mobile Devices with Built-In Sensors

Jing Chen[1], Peidong Jiang[1], Kun He[1(✉)], Cheng Zeng[2], and Ruiying Du[1]

[1] Wuhan University, Wuhan, Hubei 430072, China
{chenjing,jiangpd,hekun,duraying}@whu.edu.cn
[2] Wuhan University of Technology, Wuhan, Hubei 430070, China

**Abstract.** A variety of sensors are built into intelligent mobile devices. However, these sensors can be used as side channels for inferring information. Researchers have shown that some touchscreen information, such as PIN and unlock pattern, can be speculated by background applications with motion sensors. Those attacks mainly focus on the restricted-area input interface (e.g., virtual keyboard). To date, the privacy risk in the unrestricted-area input interface does not receive sufficient attention.

In this paper, we investigate such privacy risk and design an unrestricted-area information speculation framework, called Handwritten Information Awareness (HIAWare). HIAWare exploits the sensors' signals that are affected by handwriting actions to speculate the handwritten characters. To alleviate the impact of different handwriting habits, we utilize the generality patterns of characters. Furthermore, to mitigate the impact of holding posture in handwriting, we propose a user-independent posture-aware approach. As a result, HIAWare can attack any victim without obtaining the victim's information in advance. The experiments show that the speculation accuracy of HIAWare is close to 90.0%, demonstrating the viability of HIAWare.

**Keywords:** Motion sensors · Side channel · Privacy leaks

## 1 Introduction

In recent years, intelligent mobile devices have been equipped a variety of sensors to assist in navigation, gaming, health monitoring and more. However, these sensors can also be used as side channels for inferring information, yet until now the mobile operating systems, Android for example, have no restrictions to sensors by applications.

By exploiting those sensors, touchscreen information can be speculated, such as passwords typed by users on the virtual keyboard [10,21,23]. By far, most researches only focus on the *restricted-area* input interface, in which touchscreen information is entered at a specified position/area on the touchscreen, such as the virtual keyboard [4,11] and the pattern lock screen [2,24]. Mehrnezhad et al. [12] have shown that primitive operation actions (e.g., click and scroll in gesture control) on *unrestricted-area* screen can be recognized with sensors.

© Springer Nature Switzerland AG 2021
D. Gao et al. (Eds.): ICICS 2021, LNCS 12918, pp. 136–152, 2021.
https://doi.org/10.1007/978-3-030-86890-1_8

Nevertheless, the speculation for the more complex and meaningful inputs, such as handwritten contents, has not been sufficiently discussed. Therefore, a remarkable question is whether handwritten information can be speculated by a malicious background application? There are two issues related to practicality and robustness that need to be addressed.

One issue is that, victims are usually *unknown* to attackers before attacks, but the speculation model is built on the training data from the *known* users. We observe that each person has his/her own handwriting habits, such as writing strength, sequence, and speed, called *chirography difference*. This chirography difference will heavily impact the accuracy of the speculation model which is trained without the victim's data [8]. The other is that the sensors' signals and noise patterns vary under different holding postures (e.g., sitting, standing), since the victim's limb jitter and handwriting strength are significantly different, called *posture variation*. Such posture variation can dramatically downgrade the speculation accuracy of a universal speculation model which is built only on a single holding posture (sitting or standing), especially when the victim's holding posture changes in the handwriting process. Moreover, posture variation should also be universal for unknown victims.

In this paper, we propose a novel character-based unrestricted-area information speculation framework, called Handwritten Information Awareness (HIAWare). The main idea of HIAWare is to track the changes of sensor signals caused by handwriting actions on the touchscreen, and recognize the patterns for each character. The above two issues were successfully solved in HIAWare. It is effective in attacking unknown victims and can be adapted to different postures.

The main contributions of our paper are as follows.

- To the best of our knowledge, HIAWare is the first work to leverage motion sensors to speculate handwritten information on mobile devices, which reveals a security threat that all shared hardware on mobile devices can be exploited for privacy leakage.
- We propose to utilize generality patterns of characters (e.g., stroke number) to alleviate the dependency on collecting training data from victims in advance and reduce the impact of chirography difference on speculation accuracy.
- We build diverse speculation models in HIAWare according to different holding postures detected by a user-independent posture-aware algorithm. Therefore, HIAWare is able to complete handwritten information speculation with a competitive accuracy in the more practical scenarios.
- We implement a HIAWare prototype using off-the-shelf mobile devices with Android platform. The comprehensive experiments demonstrate that HIAWare can speculate characters with an accuracy close to 90.0%.

We organize the rest of our paper as follows. We introduce the preliminaries in Sect. 2. Section 3 and 4 detail the design of HIAWare, and the proposed algorithms, respectively. Then, the performance of HIAWare in different conditions is shown in Sect. 5. At last, Sect. 6 summarizes related work, and Sect. 7 concludes our paper.

## 2  Preliminaries

In order to anticipate the feasibility of the solution, the targeted app, available sensors and possible threat models need to be noticed.

### 2.1  Targeted Vulnerable Apps

Although the prototype of HIAWare in this paper is implemented on Android platform, its framework can also be used on other mobile platforms, such as iOS. The targeted vulnerable apps of HIAWare, a system to speculate handwritten information, are mainly handwriting-related apps. Four application markets for Android platform are statistically analyzed. Apple App Store does not have download statistics, so the number of handwriting-related apps in American (US) and Chinese (CN) App Store is based on a third-party data set [14]. The results are shown in Table 1.

**Table 1.** Survey Of handwriting-related applications

| Markets | Downloads/Apps | Example app |
| --- | --- | --- |
| Google Play | 347,014,420 | Google handwriting input |
| AppChina | 455,663 | Chinese handwriting Recog |
| WanDouJia | 224,254 | ABC handWriting |
| AnZhi | 4,545,245 | Sogou input method |
| US Apple store | 302 | Easy writing Board |
| CN Apple store | 1,079 | NoteBook+ |

From Table 1, we can find that the total downloads of targeted apps in the Google Play is about 300 million and the number of apps in Apple Store is up to 1381 in America and China. It draws a conclusion that there are enormous targeted vulnerable apps which could potentially be compromised by HIAWare.

### 2.2  Motion Sensor Selection

The device will be shaken and rotated slightly while handwriting, and these movements will be captured by the motion sensors. Specifically, two commonly used sensors are selected: *accelerometer*, which can measure device vibration and acceleration, and *gyroscope*, which can estimate the rotation and deflection [1]. These sensors provide signals at a given frequency which contain the data in three dimensions (denoted as X, Y, and Z respectively in Fig. 1) and its corresponding timestamps. HIAWare's *Activity* collecting sensors' signals is shown in Fig. 2.

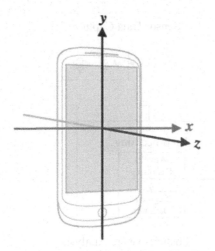

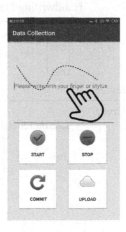

**Fig. 1.** Accelerometer and Gyroscope Axis of measurement

**Fig. 2.** *Activity* of data collection application

## 2.3   Threat Model

A malicious app is assumed that launched in application markets and pretend as a game app. To get unnoticed by the user due to unusual power consumption, the malicious app tries to stealthily run in the background, and only records sensors' signals when the user runs a handwriting-related app in the foreground. Existed work has proposed a way to determine the running app via power analysis [5, 6, 15].

Note that the signals' data is written into its own *App-specific storage*, where no storage permission is required. And the *Vanilla Android*, a system not deeply customized, only restricts the usage of mobile data in the background. However, this permission is allowed by default and so normal in game apps that the malicious app can gain the access to connect network easily for sending files to the remote server.

## 3   HIAWare Design

HIAWare consists of the following five phases: *Handwriting Detection, Sensor Data Capture, Preprocessing, Posture-Aware Analysis,* and *Character Restoration*, as shown in Fig. 3.

### 3.1   Handwriting Detection

Different from detecting unlocking screen actions, handwriting actions have no observable system broadcast [24]. Meanwhile, the other applications cannot use conventional approaches to identify whether the foreground application is a targeted app.

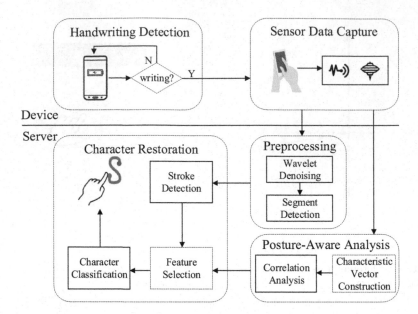

**Fig. 3.** The system architecture of HIAWare

Fortunately, different applications use different components of a device and have different usage patterns, which result in distinguishable power consumption profiles [6]. By analyzing the power consumption on mobile devices, the sensors record service can be started as soon as the handwriting actions is detected.

### 3.2 Sensor Data Capture

Once a handwriting action is detected, HIAWare uses an API provided by Android system [1] to conduct real-time recording of motion sensors' signals during the handwriting process.

HIAWare collects the signals from accelerometer and gyroscope in the meanwhile by polling and the collected data is uploaded to a remote server stealthily for analysis.

### 3.3 Preprocessing

The main tasks completed at this stage are denoising, and identifying and extracting the signal stream corresponding to each character.

**Wavelet Denoising.** The raw signals usually contain a mass of *background noise* (distinguish from the noise caused by user unexpected actions in the next paragraph). To filter the noise and restore the real signal fluctuations caused by handwriting actions, we utilize the wavelet denoising for filtering. Figure 4 shows the gyroscope's signals for handwriting actions of some characters. After wavelet denoising, the difference in the number of peaks for different characters in Fig. 4 can be observed, as the characters have different strokes.

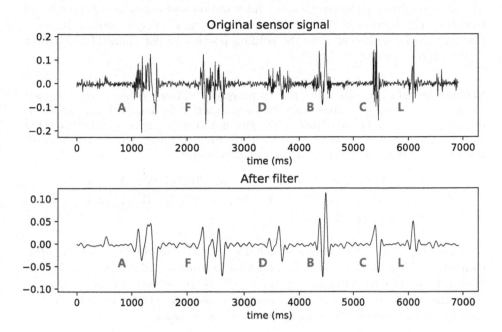

**Fig. 4.** Original vs. Wavelet Denoised sensor signals of handwriting action

**Segment Detection.** In the gap between handwriting, there is still *action noise* because of victims' arms and palms' unexpected shaking. To extract individual signal segments generated by the handwriting action of a single character, we design the modified Constant False Alarm Rate (mCFAR), a handwriting signal segment detection algorithm, to identify handwriting actions from the motion sensors' signals. The details are described in Sect. 4.1.

### 3.4 Posture-Aware Analysis

There have been many studies on human activity recognition (HAR) with motion sensors [7,16,20]. However, the features used in those former studies are not user-independent, resulting in insufficient generalization of the trained model. Here comes the challenge how to use the model trained by the known users' posture data to identify the unknown users' posture data.

To tackle this challenge, HIAWare analyzes noise signals to identify victim's holding posture based on correlation analysis of characteristics. A user-independent posture-aware algorithm is described in details in Sect. 4.2.

## 3.5  Character Restoration

In this phase, with detected handwriting signal segments, the denoised signals are grouped into candidate character sets according to the number of strokes. Then, we combine the holding posture information and candidate character sets to extract the features in the *feature selection* stage. Finally, we utilize the speculation model corresponding to the holding posture to determine the character in the *character classification* stage.

**Stroke Detection.** To solve the chirography difference issue as explained in Sect. 1, a more generality pattern of characters in handwriting signals is employed: the number of strokes (i.e., the number of peaks in a handwriting signal segment). The uppercase alphabet are divided into three clusters according to the number of strokes. The three clusters are:

$$S1 = \{`C', `G', `J', `L', `M', `N', `O', `Q', `S', `U', `V', `W', `Z'\};$$
$$S2 = \{`A', `B', `D', `G', `J', `K', `M', `N', `P', `Q', `R', `T', `X'\};$$
$$S3 = \{`A', `E', `F', `H', `I', `K', `N', `Y'\}.$$

Note that some characters are included in multiple clusters. For example, character 'A' is included in both $S2$ and $S3$ according to different writing habits. However, this situation has limited impact on the subsequent character recognition, since stroke detection aims to generate a candidate character set, which is mainly used to narrow the scope of character speculation.

**Table 2.** The type of feature extracted from the signal segment. "$\checkmark$" means that all features in the item are used, "$\phi$" means none is used, and "$-$" means partially used.

| Type | Feature | Introduction | Sit | Stand |
|------|---------|--------------|-----|-------|
| TD | Standard deviation, Maximum, Minimum, Median and Average | Calculate the four characteristics of the three coordinate axes separately | $\checkmark$ | $-$ |
| | Range | Difference between maximum and minimum | $\checkmark$ | $\phi$ |
| | Strength | Expressed by the sum of the squares of the instantaneous readings of the three axes | $\checkmark$ | $\checkmark$ |
| FD | Centroid | Indicates where the spectrum centroid is located | $\phi$ | $\checkmark$ |
| | Variance | Display the frequency density of the spectrum | $\phi$ | $\checkmark$ |
| | Skewness | Measuring the asymmetry of the spectrum | $\phi$ | $\checkmark$ |
| | Kurtosis | Describe the size of the range of changes in the spectral values | $\checkmark$ | $\checkmark$ |
| | Wiener entropy | Reflecting the flatness of the spectrum of a digital signal | $\phi$ | $\checkmark$ |

**Feature Selection.** We first extract original feature set from the time and frequency domains for a given handwriting signal segment. Then, we execute feature engineering on the segment feature vector, and generate a posture profile. The posture profile is the selected segment feature subset which makes model achieve the optimal speculation accuracy. In HIAWare, we adopt the *feature_selection* module of scikit-learn to filter redundant features from the original feature set.

Table 2 shows the type of original features extracted from the handwriting signal segment. Original features are calculated from the time domain (TD) and the frequency domain (FD). It should be clear that the feature set extracted at this time is the original feature set, and then HIAWare will filter the original feature set according to the user's holding posture to obtain the most suitable set of features for each posture.

**Character Classification.** Finally, we score the characters in the candidate set, and select the most possible character based on the score of each character. The algorithm to score the character is based on the multi-class GBDT. HIAWare can greatly reduce the impact of the victim's personal handwriting habits on the accuracy of the speculation model because of the pre-processing according to the number of strokes and postures.

## 4    Algorithm Details

In this section, the details of the algorithms in *Preprocessing* and *Posture-Aware Analysis* are explained.

### 4.1    MCFAR Algorithm

The modified Constant False Alarm Rate (mCFAR) is shown in Algorithm 1. The inputs of the algorithm are the sensor signal stream without background noise and two parameters: $t$ and $m$ (explained in the next paragraph). The output is a list of signal segments and each segment represents one character's handwriting action. The algorithm uses a polynomial fitting function to fit the action noise which can adjust the parameters automatically to generate the most appropriate fit curve. The mathematical description of our algorithm is as follows.

A stream of motion sensor signals are denoted by $D = \{d_1, d_2, ..., d_n\}$ of size $n$, where $d_i \in D$ is a set containing four-tuple values $(d_i^x, d_i^y, d_i^z, d_i^t)$: three axis reading and corresponding timestamp. Particularly, let $D^\varepsilon$ represents the projection of the $d^\varepsilon$ values in $D$, like as $D^x = \{d_1^x, d_2^x, ..., d_n^x\}$. Then, the mean values and standard deviation of $D^\varepsilon$ are denoted by $M^\varepsilon$ and $S^\varepsilon$, where

$$M^\varepsilon = \frac{1}{n} \sum_{k=1}^{n} D_k^\varepsilon \tag{1}$$

and

$$S^\varepsilon = \sqrt{\frac{1}{n} \sum_{k=1}^{n} (D_k^\varepsilon - M^\varepsilon)^2}. \tag{2}$$

**Algorithm 1.** mCFAR

---

**Input:** Original signals: $D, t, m$
**Output:** Segment signals: $D'$
1: Initialize $D', tmp$ as empty vectors;
2: $p \leftarrow F_m(D)$ // Polynomial Fitting
3: $M \leftarrow calcMean(D)$ // Calculate Mean Value
4: $S \leftarrow calcSD(D)$ // Calculate Standard Deviation
5: $flag \leftarrow False$
6: **for** $i = 1 \rightarrow length(D)$ **do**
7:      **if** $|D[i] - p[i]| > t * S$ **and** $D[i]$ is a start or end point **then**
8:          $flag \leftarrow True$
9:      **end if**
10:     **if** $flag = True$ **then**
11:         Append the element $D[i]$ to $tmp$;
12:     **end if**
13:     **if** $|D[i] - p[i]| > t * S$ **and** $D[i]$ is a start or end point **then**
14:         $flag \leftarrow False$
15:         Append the element $tmp$ to $D'$;
16:         Clear $tmp$ as empty vectors;
17:     **end if**
18: **end for**
19: **return** $D'$

---

The polynomial fitting function of $D^\varepsilon$ is $p(\varepsilon)$ (see Eq. (1)), where $F$ is a conversion function and $m$ is the degree of the fitting polynomial.

$$p(\varepsilon) = F_m(D^\varepsilon). \tag{3}$$

A valid value of handwriting signal segment is detected if

$$|D_i^\varepsilon - p(\varepsilon)| > t * S^\varepsilon, \tag{4}$$

where $t$ and $m$ are adjustable according to different granularity.

We use a sliding window of size $W$ to detect potential start and end points of a handwriting signal segment in $D_i^\varepsilon$, as shown in Fig. 5. The handwriting signal segment is detected if the distance between the start and end points are longer than a threshold $L$ in Fig. 5. $L$ is a regulable parameter chosen based on prior knowledge. Through Algorithm 1, the effect of abrupt and sharp noise can be greatly reduced.

We show the comparison between threshold-based method and our proposed mCFAR method in Fig. 6. We can observe that a low threshold will result in the redundancy of extra signals (Fig. 6(a)), while a high threshold will cause the loss of signals (Fig. 6(b)). Opposite of this, our mCFAR method can extract handwriting signal segments completely (Fig. 6(c)).

### 4.2   User-Independent Posture-Aware Algorithm

To tackle the posture variation issue, we consider two representative holding postures of mobile devices: sitting and standing, since victims usually use hand-

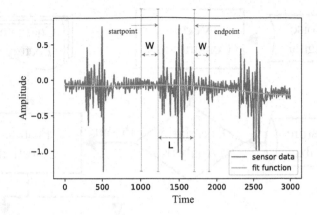

**Fig. 5.** An illustration of segment detection with accelerometer signal changes caused by handwriting actions of character '0'

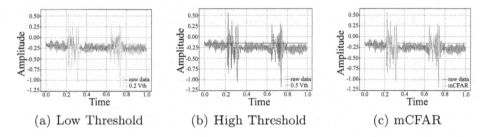

(a) Low Threshold          (b) High Threshold          (c) mCFAR

**Fig. 6.** Effect comparison of different handwritten segment detection methods

writing input under one of those two postures. Note that other holding postures can also be analyzed by our algorithm.

The feature vectors of action noise signals are constructed by commonly used digital signal processing methods, including the Power Spectral Density (PSD) and the Mel-Frequency Cepstral Coefficient (MFCC). Then, the correlation analysis is performed on the feature vectors and the prior holding posture data. The prior holding posture data contains the noise signals of the two postures we collected in advance. The type of holding posture obtained will be used to complete feature selection. The flow chart of the user-independent posture-aware algorithm is shown in Fig. 7.

**Characteristic Vector Construction.** As shown in Fig. 8, we calculate PSD and MFCC of the accelerometer's signals collected under the two holding postures. Note that the user's physical unexpected actions under the two postures can cause different action noise pattern in the signals. We can clearly observe the difference of PSD characteristics in Fig. 8(a), and MFCC characteristics in Fig. 8(b) and 8(c) under the two postures.

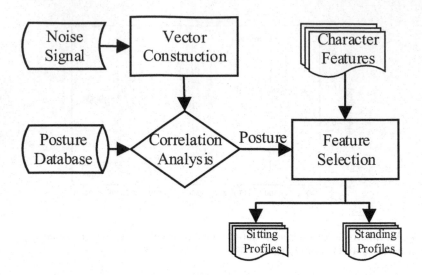

**Fig. 7.** Flow chart of user-independent posture-aware algorithm

Next, we will formalize the process of characteristic vector construction. The formal description of the noise characteristic vector for a background noise signals $N$ is $\theta_v = \{\varphi_{k0}, \xi_{k1}\}$, where $k0$ and $k1$ are the size of $\varphi$ and $\xi$, also

$$(\varphi_{k0}, \xi_{k1}) = (PSD(N), MFCC(N)),  \tag{5}$$

where $PSD$ and $MFCC$ are the functions that calculate the corresponding characteristic.

**Correlation Analysis.** In this stage, we calculate the Pearson Correlation Coefficient (PCC) between the action noise characteristic vector generated from last stage and the prior posture data. The definition of PCC is as follows:

$$\rho(X, Y) = \frac{cov(X, Y)}{\sigma_X \sigma_Y},  \tag{6}$$

where $cov(X, Y)$ is the covariance of $X$ and $Y$, $\sigma_X$ and $\sigma_Y$ are the standard deviation of $X$ and $Y$. Let us suppose the noise characteristic vectors of two postures in prior data are $\theta_{sit}$ and $\theta_{stand}$. Naturally, the correlation coefficient between $\theta_v$ and the prior holding postures signals are $\rho_s = \rho(\theta_v, \theta_{sit})$ and $\rho_{st} = \rho(\theta_v, \theta_{stand})$, respectively. We choose the holding posture as our perceived result with its correlation coefficient $\rho_{chosen} = max\{\rho_s, \rho_{st}\}$.

## 5  Performance Evaluation

In this section, the experimental setup is described firstly, and then the evaluation is conducted in the speculation accuracy and the Area Under the receiver operating characteristic Curve (AUC) under different experimental conditions.

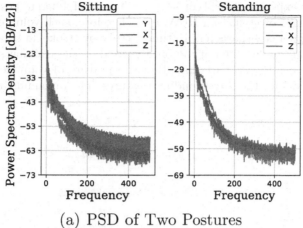

(a) PSD of Two Postures

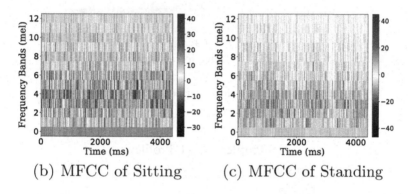

(b) MFCC of Sitting    (c) MFCC of Standing

**Fig. 8.** Example of extracted PSD and MFCC features

## 5.1 Experiment Setup

We use three Android phones: Huawei NEM-AL10, Samsung SM-G9208, and Redmi Note 4X with our app installed to collect motion sensors' signals. 12 volunteers participate in the experiments, and they are divided into three equal groups to collect signals using three devices.

Volunteers first start up our app (Fig. 2) and click the *START* button to start signal collection. There are two modes for writing characters here. One is writing characters on HIAWare's *SurfaceView*, while this stage allows the *MotionEvent* to be obtained for accurately evaluating of the performance of segment detection. The other is switching our app in the background and writing on another handwriting-related app. When the handwriting is over, volunteers then click the *STOP* button to stop signal collection. The *COMMIT* button can dump the collected data into *App-specific storage*, and the *UPLOAD* button can send the dumped files to the remote server.

Subsequent experiments will consider the impact of different postures, devices and inputs (finger or stylus) on HIAWare. Therefore, each volunteer will hand-write all the uppercase characters 15 times stroke by stroke, in two postures and two handwriting modes. 70% of the data is used as the training set, and these data are divided into two parts based on the posture of volunteers, to train a speculative model for the corresponding postures. The remaining 30% is used as the test set.

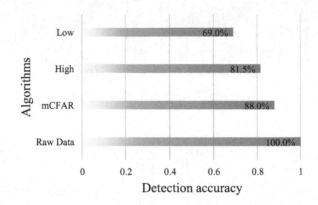

**Fig. 9.** Performance of three segment detection methods

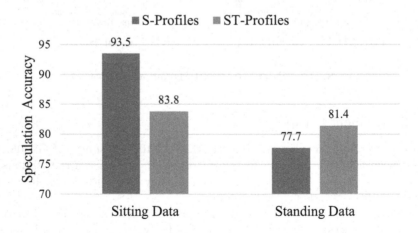

**Fig. 10.** Performance of different holding postures

## 5.2  Performance of Segment Detection

Figure 9 shows the actual detection result of three segment detection methods. It is obvious that low threshold detects the least handwriting segments, while mCFAR and high threshold's detection ability is comparable. The high threshold will cause signal loss when extracting handwriting segments from sensors' signal stream. The reason that not all segments are detected by mCFAR is that sometimes the strength of the volunteers' handwriting is so weak that it does not generate sufficient fluctuating changes. Considering the whole, the mCFAR achieved a detection rate of 88%, so that the character data for subsequent recognition accuracy experiments is based on the signal segments that are detected by the mCFAR.

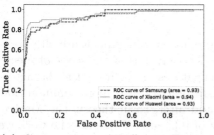
(a) Comparison of Different Devices

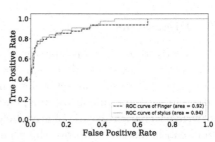
(b) Comparison on Different Inputs

**Fig. 11.** ROC curves of different experiment conditions on HIAWare

### 5.3    Performance of Different Holding Postures

Figure 10 shows the speculation accuracy under two different holding postures. From Fig. 10, we can observe that sitting data has the higher speculation accuracy, which can reach 93.5%, while the speculation accuracy of standing data is only 81.4%. This is because the sitting posture has less action noise, and relatively more signals are available for speculation model training. What's more, when using a posture profile that does not fit the current holding posture to speculate characters, the accuracy is significantly reduced.

### 5.4    Performance of Different Devices

We also investigate speculation performance of the three different devices used in our experiments with the same users. Figure 11(a) shows the ROC curves of three different devices. HIAWare has similar speculation performance on three devices with the AUC of Xiaomi as 0.94, and it is only 0.01 higher than Samsung and Huawei. In general, the discrepancy in sensor performance can lead to subtle differences in our sampling frequency and precision which may influence the speculation accuracy. However, as the training data of the speculation model continues to increase, the impact of this discrepancy could continue to decrease.

## 5.5   Performance of Different Inputs

Figure 11(b) shows the ROC curves of handwriting with the finger and the stylus. As we can see that the AUCs of the finger and stylus are 0.92 and 0.94, respectively. The AUC of stylus is 0.2 more than that of finger, which means the speculation performance with stylus is better than that with finger. This is because the physiological characteristics have a relatively weak impact on sensor signal changes when we use a stylus instead of the finger. Generally speaking, this degree of speculation performance difference is comprehensible.

## 5.6   Discussions

What needs to be stated is that the characters in this experiment are all uppercase. However, the stroke detection cannot effectively narrow the scope of speculation, since lowercase characters have more ligatures. In addition, the characters written in the experiment are independent. In practice, it is more common that the written characters are related before and after, so in theory, the Markov model can be used to improve the accuracy of inference. These are the work we need to improve in the future.

# 6   Related Work

Mobile devices' restricted-area information could be classified as two categorizations, *virtual keyboard input* and *pattern lock* [18]. Cai et al. [3] first proposed the possibility of eavesdropping virtual keyboard input via embedded sensors in smartphone. They developed TouchLogger, which can monitor the orientation signals and extract features from these signals to infer key-press information. Similar to this work, Xu et al. [21] recorded gyroscope signals to infer user input and PIN code. Ping et al. [13] proposed a method to infer even longer input. Mehrnezhad et al. [11] presented a threat of eavesdropping users' PINs by recording the sensors' signals from web page. They proposed PINlogger.js which is a JavaScript-based side channel attack embedded in a web page, recording the sensor signal changes while a user inputs the sensitive information on other web pages. All of these researches only focus on the disclosure of restricted-area information.

Currently, only few works have focused on the leakage of unrestricted-area information on touchscreens. Researchers have shown that simple touch actions including clicking, scrolling, zooming, and holding can be recognized via analyzing motion and orientation sensors' signals [12,19]. Using accelerometer and gyroscope sensors, Emanuel etal. [17] implement predicting tap locations, while Hafez [9] achieve the same functionality based on barometer sensor. However, compared with simple touch actions, handwritten information is generally more complicated to restore [8,22].

# 7   Conclusions

In this paper, we present the excogitation and evaluation of HIAWare, an efficient attack framework for handwritten information speculation on Android mobile devices, based on the motion sensor signal analysis. We utilize a generality pattern of characters (i.e., stroke number) to solve the chirography difference issue and propose a posture-aware approach to solve the posture variation issue. Moreover, we design a modified constant false alarm rate algorithm (mCFAR) to extract handwriting segments from the motion sensor signal stream, and a user-independent posture-aware algorithm which combines digital signal processing and correlation analysis. Our substantial experiments show that HIAWare can speculate the handwritten information with an accuracy close to 90.0%, which induces a significant threat against user privacy on mobile devices.

**Acknowledgments.** This research was supported in part by the National Natural Science Foundation of China under grants No. 61772383, U1836202, 62076187.

# References

1. Android Developers: Motion sensors — android developers. https://developer.android.com/guide/topics/sensors/sensors_motion
2. Aviv, A.J., Sapp, B., Blaze, M., Smith, J.M.: Practicality of accelerometer side channels on smartphones. In: Proceedings of ACSAC, pp. 41–50 (2012)
3. Cai, L., Chen, H.: Touchlogger: inferring keystrokes on touch screen from smartphone motion. In: Proceedings of HotSec (2011)
4. Chen, D., et al.: Magleak: a learning-based side-channel attack for password recognition with multiple sensors in IIoT environment. IEEE Trans. Ind. Inform. (2020)
5. Chen, J., Fang, Y., He, K., Du, R.: Charge-depleting of the batteries makes smartphones recognizable. In: Proceedings of ICPADS, pp. 33–40 (2017)
6. Chen, Y., Jin, X., Sun, J., Zhang, R., Zhang, Y.: POWERFUL: mobile app fingerprinting via power analysis. In: Proceedings of INFOCOM, pp. 1–9 (2017)
7. Chen, Z., Zhu, Q., Soh, Y.C., Zhang, L.: Robust human activity recognition using smartphone sensors via CT-PCA and Misc SVM. IEEE Trans. Ind. Inform. **13**(6), 3070–3080 (2017)
8. Du, H., Li, P., Zhou, H., Gong, W., Luo, G., Yang, P.: WordRecorder: accurate acoustic-based handwriting recognition using deep learning. In: Proceedings of INFOCOM, pp. 1448–1456 (2018)
9. Hafez, A.: Information inference based on barometer sensor in android devices. dissertation, University of Alberta (2020). https://era.library.ualberta.ca/items/15d8d051-45ab-4b1f-ba8a-005688e92f05
10. Javed, A.R., Beg, M.O., Asim, M., Baker, T., Al-Bayatti, A.H.: AlphaLogger: detecting motion-based side-channel attack using smartphone keystrokes. J. Ambient Intell. Humanized Comput. 1–14 (2020). https://doi.org/10.1007/s12652-020-01770-0
11. Mehrnezhad, M., Toreini, E., Shahandashti, S.F., Hao, F.: Stealing PINs via mobile sensors: actual risk versus user perception. Int. J. Inf. Secur. **17**(3), 291–313 (2017). https://doi.org/10.1007/s10207-017-0369-x

12. Mehrnezhad, M., Toreini, E., Shahandashti, S.F., Hao, F.: Touchsignatures: identification of user touch actions and pins based on mobile sensor data via javascript. J. Inf. Sec. Appl. **26**, 23–38 (2016)
13. Ping, D., Sun, X., Mao, B.: TextLogger: inferring longer inputs on touch screen using motion sensors. In: Proceedings of WiSec, pp. 24:1–24:12 (2015)
14. Qimai: Apple store app downloads analysis (2019). https://www.qimai.cn/
15. Qin, Y., Yue, C.: Website fingerprinting by power estimation based side-channel attacks on Android 7. In: Proceedings of TrustCom, pp. 1030–1039 (2018)
16. Quispe, K.G.M., Lima, W.S., Batista, D.M., Souto, E.: MBOSS: a symbolic representation of human activity recognition using mobile sensors. Sensors **18**(12), 4354 (2018)
17. Schmitt, E., Voigt-Antons, J.-N.: Predicting tap locations on touch screens in the field using accelerometer and gyroscope sensor readings. In: Moallem, A. (ed.) HCII 2020. LNCS, vol. 12210, pp. 637–651. Springer, Cham (2020). https://doi.org/10.1007/978-3-030-50309-3_43
18. Spreitzer, R., Moonsamy, V., Korak, T., Mangard, S.: Systematic classification of side-channel attacks: a case study for mobile devices. IEEE Commun. Surv. Tutorials **20**(1), 465–488 (2018)
19. Spreitzer, R., Kirchengast, F., Gruss, D., Mangard, S.: ProcHarvester: fully automated analysis of procfs side-channel leaks on Android. In: Proceedings of ASIACCS, pp. 749–763 (2018)
20. Wang, J., Chen, Y., Hao, S., Peng, X., Hu, L.: Deep learning for sensor-based activity recognition: a survey. Patt. Recogn. Lett. **119**, 3–11 (2019)
21. Xu, Z., Bai, K., Zhu, S.: Taplogger: inferring user inputs on smartphone touchscreens using on-board motion sensors. In: Proceedings of WiSec, pp. 113–124 (2012)
22. Yu, T., Jin, H., Nahrstedt, K.: Writinghacker: audio based eavesdropping of handwriting via mobile devices. In: Proceedings of UbiComp, pp. 463–473 (2016)
23. Zhao, R., Yue, C., Han, Q.: Sensor-based mobile web cross-site input inference attacks and defenses. IEEE Trans. Inf. Forensics Secur. **14**(1), 75–89 (2019)
24. Zhou, M., Wang, Q., Yang, J., Li, Q., Xiao, F., Wang, Z., Chen, X.: Patternlistener: cracking android pattern lock using acoustic signals. In: Proceedings of CCS, pp. 1775–1787 (2018)

# Studies of Keyboard Patterns in Passwords: Recognition, Characteristics and Strength Evolution

Kunyu Yang[1], Xuexian Hu[1(✉)], Qihui Zhang[1], Jianghong Wei[1], and Wenfen Liu[2]

[1] State Key Laboratory of Mathematical Engineering and Advanced Computing, Zhengzhou, China
xuexian_hu@hotmail.com
[2] Guilin University of Electronic Technology, Guilin, China

**Abstract.** Keyboard patterns are widely used in password construction, as they can be easily memorized with the aid of positions on the keyboard. Consequently, keyboard-pattern-based passwords has being the target in many dictionary attack models. However, most of the existing researches relies only on recognition methods defining keyboard pattern structures empirically or even manually. As a result, only those infamous keyboard patterns such as *qwerty* are recognized and many potential structures are not specified. Besides, there are limited studies focusing on the characteristics of keyboard patterns.

In this paper, we deal with the problem of recognizing and analyzing keyboard patterns in a systematic approach. Firstly, we put forward a general recognition method that can pick out keyboard patterns form passwords automatically. Next, a comprehensive study of keyboard pattern characteristics is presented, which reveals a great deal of amazing facts about the preference for passwords based on keyboard patterns, such as: (1) More than half of the pattern-based passwords are completely composed by keyboard patterns; (2) The frequency distribution of the keyboard patterns satisfies the PDF-Zipf model; (3) Users prefer to use keyboard patterns consisted by horizontal continuous keys or those characters whose physical location are on the upper left of the keyboard. We further evaluate the security of keyboard-pattern-based passwords by employing the PCFG-base cracking technique. The experimental results indicate that the keyboard patterns can reduce the security of passwords.

**Keywords:** Keyboard pattern · Password strength · Password cracking · Information security

## 1 Introduction

Passwords are used as a key to access control in every corner of the Internet service. Password security is related to users' data security and even property

© Springer Nature Switzerland AG 2021
D. Gao et al. (Eds.): ICICS 2021, LNCS 12918, pp. 153–168, 2021.
https://doi.org/10.1007/978-3-030-86890-1_9

security, but passwords are and will be an Achilles Heel in cybersecurity for the foreseeable future [2]. NIST SP800-63B [5] points out that although online services have introduced some rules to increase the complexity of passwords, the password security is still frustrating. This happens because users like to choose easy-to-remember passwords with certain structures, which greatly reduces the security of passwords [16].

Extensive researchers studied on the password characteristics and found out some password construct rules. Studies [6,11] analyzed the length of passwords and pointed out that the length of most passwords was between 6 and 10 characters. Li et al. counted the most commonly used passwords and character combinations in passwords [9]. Researches by Pearman et al. [12] and Wang et al. [14] showed that many users reused a password in different websites. Data from several studies [1,4,17] found that a large amount of personal information (e.g., name, birthday) was used in passwords. The differences between the passwords constructed by Chinese users and English users were compared by studies [10,16,20].

With the development of password security study, especially the improvement of password attack technology, network application providers have to force users to set stronger passwords than before. For example, the password length is required to be greater than 8 characters, the password must be constructed by a combination of three types of characters (i.e., digits, letters, and symbols). Therefore, many users will construct a password according to the physical position of the characters on the keyboard, which is the keyboard pattern. Passwords formed by this method can satisfy the setting requirements of passwords with ease, and the shapes formed by these characters on the keyboard are easy to remember.

Although keyboard-pattern-based passwords seem to be random, they still follow certain rules. Schweitzer et al. [13] explored the structure of keyboard patterns through visualization, and defined 11 common pattern elements (e.g., Fours, Snake, Reflected). Chou et al. [3] mapped the physical position of keys on the keyboard to a two-dimensional coordinate axis. Then three definitions of commonly used patterns (i.e., adjacent patterns, parallel patterns, and both of two) were given according to the distance between these keys. Kävrestad et al. [8] pointed out that users with high security awareness were more prone to use keyboard-pattern-based passwords. Houshmand et al. [7] introduced keyboard patterns into password attack and achieved a higher success rate than traditional methods. In addition, Wheeler et al. [19] added keyboard patterns into the password strength meter (PSM).

The above studies mainly concentrate on two aspects: (i) proposing some methods to pick out keyboard patterns from passwords; (ii) using keyboard patterns for password attack or PSM. However, most of the keyboard pattern recognition methods are rule-based. These rules are often not rich enough because they are defined empirically by researchers. Besides, few studies comprehensively analyze the characteristics of keyboard patterns.

In this paper, we propose a general method to find keyboard pattern based on the common definition of that, and extract 14.6 million keyboard-pattern-based passwords from 157 million leaked passwords. Then the systematical statistical analysis is performed, which can help us to understand users' preference in designing keyboard patterns. Finally, we employ keyboard patterns to the PCFG-based password attack model [18] and compare the attack results with the original model to evaluate the security of keyboard-pattern-based passwords. Experimental results show that the new model can guess 2.8% more passwords on average than the original.

We make the following key contributions:

- We propose a general method to recognize keyboard patterns, eliminating the need to define structure empirically.
- We comprehensively analyze the characteristics of keyboard patterns from multiple aspects, including the length distribution, the most commonly used keyboard patterns, and the frequency distribution, etc.
- We prove that using keyboard patterns can reduce passwords security through the PCFG-based password attack experiment.

## 2    General Method of Keyboard Pattern Recognition

### 2.1    Recognition Method Design

Keyboard pattern is actually physical structure (graphic) on the keyboard, but it is difficult to be found by recognizing such graphic. Although rule-based recognition methods define plenty of different rules, they have a common definition of the keyboard pattern [3, 7, 19], which we summarize as follows.

**Definition:** Keyboard pattern is a sequence of characters whose physical position in keyboard are contiguous. Contiguity refers to a duplicate key or a key next to the special key.

Based on the above definition, we propose a general keyboard pattern matching method that only focuses on the physical position of characters on the keyboard, which can avoid the problem that the structures defined by experience are not rich enough.

| Target character | Contiguous Characters |
|---|---|
| 2 | 2 @ 1 ! 3 # w W q Q |
| @ | 2 @ 1 ! 3 # w W q Q |
| s | s S a A w W e E d D x X z Z |
| ... | ... |
| Q | q Q 1 ! 2 @ w W a A |

**Fig. 1.** Contiguous characters table

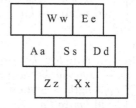

**Fig. 2.** Contiguous characters of *s*

Firstly, we remove the dictionary words from passwords in order to avoid misrecognizing them as keyboard patterns (e.g., *assw* in *password* can be regarded as a keyboard pattern). Then, we construct a table to record the contiguous characters on the keyboard (as shown in the Fig. 1). Consecutive characters of a target character include the target character itself and characters whose keys' physical position is adjacent to the target character on the keyboard. Taking the character $s$ as an example, the characters in the Fig. 2 are all consecutive characters of $s$. After that, passwords are searched to find the sequence of contiguous characters. The matched sub-strings longer than $l_1$ characters are regarded as candidate keyboard patterns. If two or more keyboard patterns are matched in a password, the positions of the keyboard patterns in the password will be considered. Two short candidate keyboard patterns can be merged into one long candidate keyboard pattern if the positions of the two in the password are consecutive. Finally, the candidate keyboard patterns whose length are longer than $l_2$ characters are the final keyboard pattern matching results. For example, a password *12#qwenjnzxcvb* is given, $l_1$ is set to 2, and $l_2$ is set to 4. First, *12#*, *qwe*, *njn*, *zxcvb* are considered as candidate keyboard patterns. Then *12#* and *qwe* can be merged because the positions in the password are adjacent. Finally, *12#qwe* and *zxcvb* are recognized as keyboard patterns in *12#qwen-jnzxcvb* because their length longer than $l_2$. In this paper, we only consider the qwerty keyboard layout without keypad, but this method can be applied to any keyboard layout by constructing a corresponding table of contiguous characters. The proposed method is described in Algorithm 1.

---

**Algorithm 1:** Keyboard pattern recognition method

**Input:** password $pw$; $l_1$; $l_2$; contiguous characters table $T$; dictionary words $D$
**Output:** keyboard pattern list $kp$

1  **if** a substring $s$ of $pw$ appears in $D$ **then**
2  $\quad$ $p \leftarrow$ remove $s$ from $pw$
3  **end**
4  candidate keyboard patterns $C_{kp} \leftarrow$ find contiguous character combinations from $p$ based on $T$
5  **for** $item_{kp}$ in $C_{kp}$ **do**
6  $\quad$ **if** $item_{kp}$.length $> l_1$ **then**
7  $\quad\quad$ $item_{kp}$.startIndex$\leftarrow$ the start index of $item_{kp}$ in $pw$
8  $\quad\quad$ $item_{kp}$.endIndex$\leftarrow$ the end index of $item_{kp}$ in $pw$
9  $\quad\quad$ add $item_{kp}$ to $kp_s$
10 $\quad$ **end**
11 **end**
12 **while** $kp_s^1$.endIndex$=kp_s^2$.startIndex for any two item in $kp_s$ **do**
13 $\quad$ $kp_l \leftarrow$ combine $kp_s^1$ and $kp_s^2$
14 $\quad$ **if** $kp_l$.length $> l_2$ **then**
15 $\quad\quad$ add $kp_l$ to $kp$
16 $\quad$ **end**
17 **end**
18 **return** $kp$

## 2.2    Recognition Results

Our empirical analysis employs 8 famous leaked password datasets which are different in terms of services, size and language. These datasets include four from Chinese website and the other four from English website. We remove the passwords that contain characters beyond 95 printable ASCII characters. Passwords with length shorter than 5 characters or longer than 20 characters are also removed. After preprocessing, these datasets have a total of 156.6 million passwords, of which there are 80.6 million unique passwords. Although these datasets are publicly available, the risk of leaking user privacy is not ruled out. Therefore, we only show aggregated statistical information, instead of analyzing for a certain password or passwords of a certain user. Table 1 summarizes the information of these 8 datasets.

**Table 1.** Information of eight datasets

| Dataset | Web service | Language | Leaked time | Original passwords | After cleaning | Removed (%) | Unique passwords |
|---|---|---|---|---|---|---|---|
| 7k7k | Gaming | Chinese | Nov. 2011 | 18, 577, 194 | 18,576,977 | 0.001 | 4, 877, 255 |
| CSDN | Programmer forum | Chinese | Dec. 2011 | 6, 374, 513 | 6,374,484 | 0.001 | 4, 006, 727 |
| Dodonew | E-commerce& gaming | Chinese | Dec.2011 | 15, 580, 010 | 15,578,470 | 0.010 | 9, 994, 170 |
| Taobao | E-commerce | Chinese | Dec. 2011 | 15, 073, 116 | 15,006,881 | 0.439 | 11, 589, 222 |
| Gmail | Email | English | Sept. 2014 | 4, 693, 896 | 4,691,609 | 0.049 | 3, 022, 756 |
| Mate1 | Online dating | English | Mar. 2016 | 27, 403, 958 | 25,570,008 | 6.692 | 11, 681, 308 |
| RockYou | Social forum | English | Dec. 2009 | 32, 382, 632 | 32,368,961 | 0.042 | 14, 195, 060 |
| Twitter | Social media | English | Nov. 2016 | 38, 564, 652 | 38,470,995 | 0.243 | 21, 194, 754 |
| Total | | | | 158, 649, 971 | 156,638,385 | 1.268 | 80, 561, 252 |

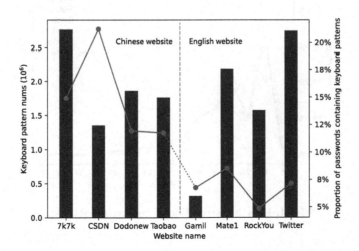

**Fig. 3.** The number of keyboard-pattern-based passwords

Keyboard-pattern-based passwords (denoted as $pw_{kp}$) are picked out from these datasets using the proposed method. $l_1$ is set to 2, and $l_2$ is set to 4. About 14.6 million $pw_{kp}$ are obtained in total, the results are shown in Fig. 3. It is apparent from this figure that Chinese passwords use keyboard patterns more than English passwords. On average, one out of every seven Chinese passwords uses keyboard patterns, while only approximately 7% of English passwords use keyboard patterns. RockYou dataset has the lowest proportion of $pw_{kp}$ at only 4.9%, while CSDN dataset has the highest proportion at about 21.2%. We then comprehensive analyze the characteristics of matched keyboard patterns in the Sect. 3.

## 3    Characteristic Analyses of Keyboard Patterns

### 3.1    Length Distribution of Keyboard Patterns

Figure 4 depicts the length distribution of keyboard patterns and passwords. We use warm colors to mark the Chinese datasets and cool colors to mark the English datasets. The length distribution of keyboard patterns is similar to that of passwords. Regardless of web service type or language, both of the most common keyboard pattern lengths and password lengths are between 6 and 10. But compared to 8-length passwords are more commonly used, the 6-length keyboard patterns are more popular. Thinking this may be because a password does not completely consist of keyboard patterns, we further analyze the length proportion of keyboard patterns in a password.

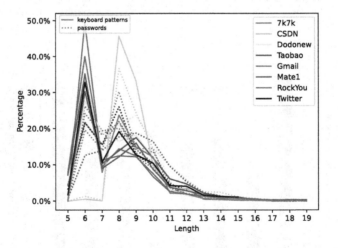

**Fig. 4.** Length distribution of keyboard patterns

Figure 5 provides the distribution of the length percentage of keyboard patterns in a password. In all datasets except the Dodonew dataset, 50.34%–67.06%

of $pw_{kp}$ are completely composed of keyboard patterns, and the remaining $pw_{kp}$ are construct by adding other elements on the basis of keyboard patterns. The most used extra elements in Chinese datasets include surname (e.g., *wang, liu*), character combinations (e.g., *abc, aa*), etc. The most used extra elements in English data include names (e.g., *john, jack*), dates (e.g., *1991, 1986*), etc. In addition, elements with the meaning of love (e.g., *woaini* and *1314* in Chinese datasets, *love* in English datasets) frequently appear in all datasets.

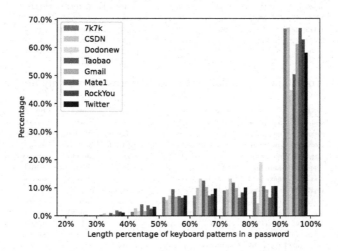

**Fig. 5.** Distribution of the length percentage of keyboard patterns in a password

## 3.2 Top Popular Keyboard Patterns

Table 2 shows the top-10 most frequent keyboard patterns in these 8 websites. *123456* is the most frequency keyboard pattern in all datasets except the CSDN dataset. The top-10 most frequent keyboard patterns are almost all continuous or repeated sequences of digits. This result may be explained by the fact that the combinations of these digits are the most commonly used passwords. What stands out in the table is that the letter sequence *qwerty* appears in the top-10 most frequent keyboard patterns in English datasets, while only *aaaaaa* appears in the Taobao dataset in Chinese datasets.

Table 3 further lists the top-10 most frequency keyboard patterns that are not entirely composed of digits. In the four Chinese datasets, the first two most popular keyboard patterns all contain the repetition of a single character (e.g., *aaaaaa*), and the first two most popular keyboard patterns of CSDN are all of that. In contrast, sequences composed of adjacent characters are used more frequently in English datasets and it is striking that *qwerty* is the most popular keyboard patterns.

From these tables, we can see that only the top-10 most popular keyboard patterns account for as high as 35.42%–55.15% of each entire dataset with Twitter being the only exception. Specifically, the proportions in Chinese datasets are

**Table 2.** Top-10 most popular keyboard patterns

| Rank | 7k7k | CSDN | Dodonew | Taobao | Gmail | Mate1 | RockYou | Twitter |
|---|---|---|---|---|---|---|---|---|
| 1 | 123456 | 123456789 | 123456 | 123456 | 123456 | 123456 | 123456 | 123456 |
| 2 | 111111 | 12345678 | 123456789 | 123456789 | 123456789 | 12345 | 12345 | 12345 |
| 3 | 123456789 | 11111111 | 111111 | 111111 | 12345 | 123456789 | 123456789 | 123456789 |
| 4 | 123123 | 123456 | 123123 | 123123 | qwerty | 1234567 | 1234567 | qwerty |
| 5 | 111222 | 00000000 | 12345 | 000000 | 12345678 | 12345678 | 12345678 | 1234567 |
| 6 | 123321 | 123123123 | 000000 | 12345 | 1234567 | 1234567890 | qwerty | 123321 |
| 7 | 12345678 | 1234567890 | 321654 | 123321 | 111111 | qwerty | 654321 | 123123 |
| 8 | 1234567 | 88888888 | 12345678 | aaaaaa | 123123 | 111111 | 000000 | 12345678 |
| 9 | 666666 | 12345 | 123321 | 1234567 | 1234567890 | 123123 | 111111 | 1234567890 |
| 10 | 12345 | 1234567 | 1234567 | 12345678 | 000000 | zxcvbnm | 123123 | 111111 |
| Sum of top-10 | 1,413,316 | 746,868 | 808,615 | 667,864 | 112,674 | 1,084,426 | 653,403 | 732,838 |
| Total passwords | 2,767,862 | 1,354,203 | 1,860,088 | 1,763,145 | 318,088 | 2,181,879 | 1,573,332 | 2,748,136 |
| % of top-10 | 51.06% | 55.15% | 43.47% | 37.88% | 35.42% | 49.70% | 41.53% | 26.67% |

**Table 3.** Top-10 most popular keyboard patterns (digits only are removed)

| Rank | 7k7k | CSDN | Dodonew | Taobao | Gmail | Mate1 | RockYou | Twitter |
|---|---|---|---|---|---|---|---|---|
| 1 | asdasd | aaaaaaaa | q123456 | aaaaaa | qwerty | qwerty | qwerty | qwerty |
| 2 | aaaaaa | qqqqqqqq | aaaaaa | zxcvbnm | zaq12wsx | zxcvbnm | zxcvbnm | Qwerty |
| 3 | zxcvbnm | qwertyuiop | qq123456 | asdasd | qwerty123 | qwertyuiop | asdfgh | qwerty123 |
| 4 | qazwsx | qq123456 | zxcvbnm | qq123456 | asdfghjkl | asdfghjkl | qwertyuiop | qwertyuiop |
| 5 | qqqqqq | asdfghjkl | asd123 | q123456 | qazwsx | asdfgh | asdfghjkl | qazwsx |
| 6 | qwerty | qazwsxedc | qwe123 | qazwsx | zxcvbnm | aaaaaa | aaaaaa | qwert |
| 7 | qq123456 | asdasdasd | q1q1q1q1 | asd123 | qwertyuiop | qwert | qazwsx | zxcvbnm |
| 8 | asd123 | qwertyui | asdasd | asdfghjkl | aaaaaa | qazwsx | qwert | qwe123 |
| 9 | qwe123 | asdfasdf | qazwsx | qwe123 | asdfgh | mnbvcxz | zxcvbn | asdfgh |
| 10 | qweqwe | qwer1234 | zxc123 | qwertyuiop | asdasd | asdfg | 123qwe | asdfghjkl |
| Sum of top-10 | 73,286 | 35,508 | 75,776 | 66,407 | 24,672 | 126,528 | 80,949 | 176,884 |
| Total passwords | 2,767,862 | 1,354,203 | 1,860,088 | 1,763,145 | 318,088 | 2,181,879 | 1,573,332 | 2,748,136 |
| % of top-10 | 2.65% | 2.62% | 4.07% | 3.77% | 7.76% | 5.80% | 5.15% | 6.44% |

slightly higher than that in English datasets. But for the top-10 most popular keyboard patterns that are not entirely composed of digits, the proportions in Chinese datasets are significantly lower than that in English datasets. This comparison shows that Chinese users prefer to use numbers to construct passwords, because the top-10 most frequent keyboard patterns are almost all digits.

### 3.3  Common Structures of Keyboard Patterns

Observing the most popular keyboard patterns, we can find that there are four main ways for users to construct keyboard patterns: (1) multiple consecutive keys starting from a certain key in the horizontal direction; (2) multiple consecutive keys starting from a certain key in the vertical direction; (3) Repetition of a certain key; (4) Combination of two or more of the above. Correspondingly, we have defined four basic structures of keyboard patterns, which are called Horizontal (Fig. 6(a)), Vertical (Fig. 6(b)), Repetition (Fig. 6(c)), and Combination (6(d)). We further analyze the frequency of each basic structure. Since keyboard patterns composed entirely by digits account for a large proportion and do not have a Vertical structure, we only count the keyboard patterns that contain letters or symbols.

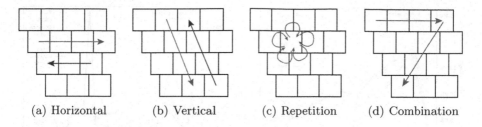

(a) Horizontal          (b) Vertical          (c) Repetition          (d) Combination

**Fig. 6.** Basic structures of keyboard patterns

Table 4 presents the proportions of each basic structure. Horizontal structure account for more than half of the whole, while the proportion of Vertical structure and Combination structure are relatively small, both accounting for less than 12%. A possible explanation for this might be that the keys on the keyboard are not completely aligned in the vertical direction, which makes it more difficult to memorize and continuous input compare to the horizontal direction. There is also a significant difference in the proportions of basic structures on different language datasets. Compared with the Chinese datasets, English datasets have more Horizontal structure but less Combined structure.

**Table 4.** Proportion of basic structures

|         | Horizontal | Vertical | Repetition | Combination |
|---------|------------|----------|------------|-------------|
| 7k7k    | 56.55%     | 7.38%    | 27.05%     | 9.02%       |
| CSDN    | 52.10%     | 7.24%    | 30.27%     | 10.38%      |
| Dodonew | 59.28%     | 5.15%    | 24.53%     | 11.04%      |
| Taobao  | 53.77%     | 6.41%    | 28.30%     | 11.51%      |
| Gmail   | 66.29%     | 10.38%   | 16.84%     | 6.48%       |
| Mate1   | 62.58%     | 5.33%    | 27.70%     | 4.39%       |
| Rockyou | 62.60%     | 7.21%    | 25.84%     | 4.35%       |
| Twitter | 67.16%     | 10.00%   | 16.33%     | 6.50%       |

## 3.4 Characters' Frequency in Keyboard Patterns

Wang et al. points out that letter distributions of passwords from diverse language datasets are obviously different [16]. We take an investigation for whether there are such differences in the character distribution of keyboard patterns, and the results are shown in the Fig. 7.

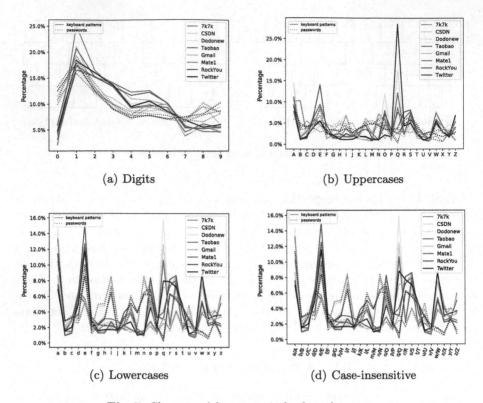

(a) Digits

(b) Uppercases

(c) Lowercases

(d) Case-insensitive

**Fig. 7.** Characters' frequency in keyboard patterns

We observe from the distribution of digital frequencies that digits 1 to 6 are used more frequently in the keyboard patterns while digit 0 is significantly lower in all passwords. For alphabetic characters, we calculate the frequency of uppercase, lowercase and case-insensitive respectively. Considering that the position of the key is the main factor that affects constructing keyboard patterns, we mainly analyze the distribution of character frequency using the result obtained by case-insensitive. The first 8 characters with the highest frequency in Chinese datasets are $a,q,w,s,d,z,e,x$. However, the most frequently occurring characters are not exactly the same in different English data sets. Generally speaking, $e$ is the most popular character, and $a,s,w,q,r,d$ are used frequently in English datasets. What stands out is that the positions of the most frequently occurring characters are mainly distributed on the left part of the keyboard, while the least frequently occurring characters (e.g., $g,m,v,b,n$) are located on the middle and bottom part of the keyboard.

In Fig. 8, we give a heatmap about the frequency of each key to show how the keys' position influences the construction of keyboard pattern more vividly. Although the most frequently used keys in different language datasets are slightly different, they are obviously concentrated in the upper left part of the keyboard, and few user show interest in symbol part of the keyboard.

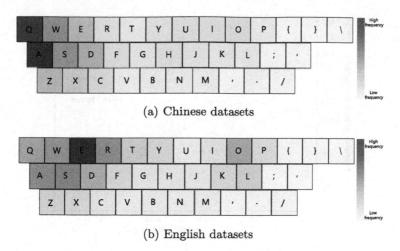

(a) Chinese datasets

(b) English datasets

**Fig. 8.** Frequency of each key

### 3.5 Frequency Distribution of Keyboard Patterns

The study researches by Wang et al. [15] has found that the distribution of real-life passwords in a dataset obeys PDF-Zipf model. The PDF-Zipf model describes the relationship between frequency and rank, which can be expressed as:

$$f_r = \frac{C}{r^s}, \tag{1}$$

where $f_r$ and $r$ are the frequency and rank of a password, $C$ and $s$ are parameters. It can be easily observed in log-log graph (10-based is used in this work).

In order to observe frequency distribution of keyboard patterns, the acquired keyboard patterns are counted and those with a frequency less than 3 are removed. All keyboard patterns are arranged in descending order of frequency. The frequency vs. the rank of keyboard patterns from different datasets are depicted in a log-log scale (Fig. 9). All lines in the figure can be approximately expressed as:

$$log(f_r) = logC - s \times log(r). \tag{2}$$

$log(f_r)$ and $log(r)$ are linearly related, $log(C)$ is the intercept and $s$ is the slope. The fitting results of each line are shown in the Table 5. All the coefficients of determination (i.e., $R^2$) are greater than 0.99, which approximately equals 1. This shows that the frequency distribution of the keyboard patterns can meet the PDF-Zipf model well.

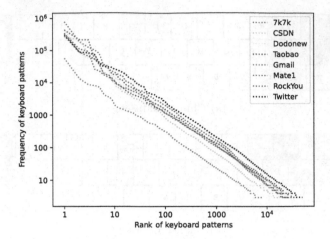

**Fig. 9.** Frequency distribution of keyboard patterns

**Table 5.** Values of parameters and coefficient of determination

|         | $s$     | $C$         | $R^2$  |
|---------|---------|-------------|--------|
| 7k7k    | 1.1204  | 326785.2554 | 0.9902 |
| CSDN    | 1.1143  | 136463.0507 | 0.9946 |
| Dodonew | 1.1160  | 263810.4820 | 0.9957 |
| Taobao  | 1.0614  | 176958.8258 | 0.9944 |
| Gmail   | 1.0741  | 39450.2511  | 0.9959 |
| Mate1   | 1.1595  | 344243.5774 | 0.9962 |
| Rockyou | 1.1551  | 304734.1164 | 0.9956 |
| Twitter | 1.1233  | 490453.4663 | 0.9964 |

## 4 Security Impacts of Keyboard-Pattern-Based Passwords

### 4.1 Method Design

We select four datasets (i.e., CSDN, Dodonew, Mate1, Twitter), which have a total of 8.1 million passwords containing keyboard patterns ($pw_{kp}$). The $pw_{kp}$ in each dataset are divided into training set and test set according to the ratio of 7 : 3. The numbers of passwords contained in each set are shown in the Table 6.

We employ the start-of-the-art password attack method (i.e., PCFG-based [18]) to evaluate the impact of keyboard patterns on passwords security. The original PCFG-based password guessing method can be divided into two steps: training and generation.

**Training:** Divide passwords into sub-strings according to character types. Then passwords in training set are parsed into basic structure by donating letters as

**Table 6.** Number of passwords in each dataset

|          | Training set | Training set (unique) | Test set | Test set (unique) |
|----------|--------------|-----------------------|----------|-------------------|
| CSDN     | 947,960      | 227,800               | 406,270  | 108,758           |
| Dodonew  | 1,302,061    | 390,050               | 558,027  | 191,908           |
| Mate1    | 1,527,315    | 321,663               | 654,564  | 157,609           |
| Twitter  | 1,923,695    | 587,494               | 824,441  | 288,577           |

$L$, digits as $D$ and symbols as $S$. For example, the password *1234zxcvbn@@* can be parsed into $D_4 L_6 S_2$. The probability of basic structure and sub-strings can be calculated by frequencies in the training set.

**Generation:** Each generating password has a probability which can be calculated by the product of the probability of basic structure and sub-strings. For example, the probability of the generating password *1234zxcvbn@@* is:

$$P(1234zxcvbn@@) = P(D_4 L_6 S_2) \times P(1234 \ in \ D_4) \\ \times P(zxcvbn \ in \ L_6) \times P(@@ \ in \ S_2). \tag{3}$$

The candidate guessing passwords can be generated in decreasing order of probability using the NEXT function.

To evaluate the change in the password security after adding keyboard patterns, we propose K-PCFG by drawing on the idea of NPC model [7]. In K-PCFG, the keyboard pattern structure (donated as K) is given the highest priority when identifying the password structure, then the rest of the password is marked in the same way as PCFG. For example, given a password *1qaz2wsxabc*, the base structure in PCFG is $D_1 L_3 D_1 L_6$, but in K-PCFG, *1qaz2wsx* is recognized as a keyboard pattern first, and then the base structure is denoted as $K_8 L_3$. The guessing flow of K-PCFG is shown as Fig. 10.

### 4.2  Evaluation Results

For each dataset, we generate 10 million candidate passwords and test them on the test set. The test includes unique matching and repeated matching, and the results are shown in the Fig. 11.

The success rate of password attack is significantly improved after adding the keyboard pattern to the PCFG method. With 1 million guesses, the candidate passwords generated by K-PCFG method can match about 2.8% more repeated passwords and 1.1% more unique passwords than that generated by the PCFG method. On the Dodonew dataset, using K-PCFG can guess at most 4.5% more repeated passwords and about 1.32% more unique passwords than the PCFG method. On the Twitter dataset, using K-PCFG can match nearly 5% more repeated passwords and 2.3% more unique passwords than the PCFG method. When the guess number exceeds 3 million on the CSDN dataset, it is noticed that

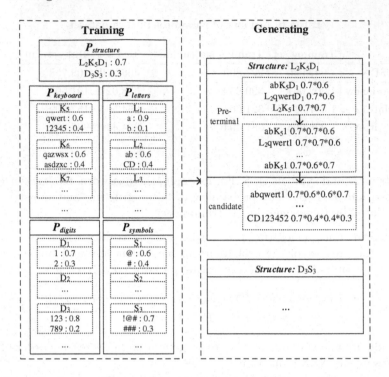

**Fig. 10.** Guessing flow of K-PCFG

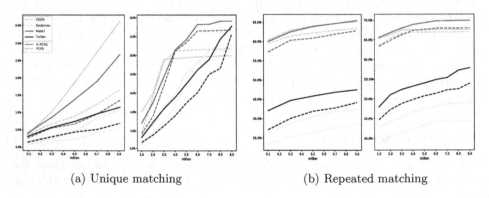

(a) Unique matching  (b) Repeated matching

**Fig. 11.** Evaluation results

the number of unique passwords that can be successfully guessed by using the K-PCFG method is starting to be lower than that by using the PCFG method, but the number of repeated passwords that can be successfully guessed by former is still higher than latter. This shows that the K-PCFG method can first generate passwords that appear more frequently in the real password dataset after adding the keyboard pattern into the PCFG method.

The experimental results reveal that adding the keyboard patterns to the password attack method can improve the attack efficiency, and it also reflects that the use of keyboard patterns in passwords reduce the security of passwords to a certain extent.

## 5    Conclusions and Suggestions

This paper mainly focuses on the research of keyboard patterns in passwords. First, a keyboard pattern recognition method is proposed, and then 14.6 million passwords containing keyboard patterns are found in 8 datasets with a total of 156.6 million passwords. We have conducted a comprehensive analysis of the obtained keyboard patterns in terms of length, most popular keyboard patterns, and frequency distribution, etc. Finally, we conduct a password guessing attack to compare the attack efficiency of the classic PCFG-based method and the K-PCFG method. The experimental results show that the use of the keyboard patterns in passwords will reduce the passwords' security. A limitation of this study is that hat our definition of keyboard patterns is not comprehensive enough. We will further explore more perfect keyboard pattern recognition methods in our future work.

Based on these studies in this paper, we put forward the following suggestions on password creation, password strength meter, and password attack. For password creation, users should completely avoid only using keyboard patterns to construct a password and using the most popular keyboard patterns. For password strength meter, attention should be paid to the influence of keyboard patterns on password strength. For a password that includes keyboard patterns, the value of strength evaluation result should be appropriately reduced. For password attack, taking into account the keyboard pattern structures in the attack method and constructing effective keyboard patterns according to the above-mentioned characteristics can improve the attack efficiency.

**Acknowledgments.** This work is supported by the National Natural Science Foundation of China (Grant Nos. 62172433, 61862011, 61872449, 61772548), and Guangxi Natural Science Foundation (Grant Nos. 2018GXNSFAA138116).

## References

1. Bonneau, J.: The science of guessing: analyzing an anonymized corpus of 70 million passwords. In: 2012 IEEE Symposium on Security and Privacy, pp. 538–552 (2012)
2. Bonneau, J., Herley, C., Van Oorschot, P.C., Stajano, F.: Passwords and the evolution of imperfect authentication. Commun. ACM **58**(7), 78–87 (2015)
3. Chou, H.C., Lee, H.C., Hsueh, C.W., Lai, F.P.: Password cracking based on special keyboard patterns. Int. J. Innov. Comput. Inf. Control **8**(1(A)), 387–402 (2012)
4. Deng, G., Yu, X., Guo, H.: Efficient password guessing based on a password segmentation approach. In: 2019 IEEE Global Communications Conference (GLOBECOM), pp. 1–6 (2019)

5. Grassi, P.A., et al.: Digital identity guidelines-authentication and lifecycle management. National Institute of Standards and Technology (2020)
6. Han, W., Xu, M., Zhang, J., Wang, C., Zhang, K., Wang, X.S.: TransPCFG : transferring the grammars from short passwords to guess long passwords effectively. IEEE Trans. Inf. Forensics Secur. **16**(pp), 451–465 (2021)
7. Houshmand, S., Aggarwal, S., Flood, R.: Next gen PCFG password cracking. IEEE Trans. Inf. Forensics Secur. **10**(8), 1776–1791 (2015)
8. Kävrestad, J., Zaxmy, J., Nohlberg, M.: Analyzing the usage of character groups and keyboard patterns in password creation. Inf. Comput. Secur. **28**(3), 347–358 (2020)
9. Li, J., Zeigler, E., Holland, T., Papamichail, D., Greco, D., Grabentein, J., Liang, D.: Common passwords and common words in passwords. In: World Conference on Information Systems and Technologies, pp. 818–827 (2020)
10. Li, Z., Han, W., Xu, W.: A large-scale empirical analysis of Chinese web passwords. In: SEC 2014 Proceedings of the 23rd USENIX Conference on Security Symposium, pp. 559–574 (2014)
11. Ma, J., Yang, W., Luo, M., Li, N.: A study of probabilistic password models. In: 2014 IEEE Symposium on Security and Privacy, pp. 689–704 (2014)
12. Pearman, S., et al.: Let's go in for a closer look: observing passwords in their natural habitat. In: Proceedings of the 2017 ACM SIGSAC Conference on Computer and Communications Security, pp. 295–310 (2017)
13. Schweitzer, D., Boleng, J., Hughes, C., Murphy, L.: Visualizing keyboard pattern passwords. Inf. Vis. **10**(2), 127–133 (2011)
14. Wang, C., Jan, S.T., Hu, H., Bossart, D., Wang, G.: The next domino to fall: empirical analysis of user passwords across online services. In: Proceedings of the Eighth ACM Conference on Data and Application Security and Privacy, pp. 196–203 (2018)
15. Wang, D., Cheng, H., Wang, P., Huang, X., Jian, G.: Zipf's law in passwords. IEEE Trans. Inf. Forensics Secur. **12**(11), 2776–2791 (2017)
16. Wang, D., Wang, P., He, D., Tian, Y.: Birthday, name and bifacial-security: understanding passwords of Chinese web users. In: SEC 2019 Proceedings of the 28th USENIX Conference on Security Symposium, pp. 1537–1554 (2019)
17. Wang, D., Zhang, Z., Wang, P., Yan, J., Huang, X.: Targeted online password guessing: an underestimated threat. In: Proceedings of the 2016 ACM SIGSAC Conference on Computer and Communications Security, pp. 1242–1254 (2016)
18. Weir, M., Aggarwal, S., de Medeiros, B., Glodek, B.: Password cracking using probabilistic context-free grammars. In: 2009 IEEE Symposium on Security and Privacy, pp. 391–405 (2009)
19. Wheeler, D.L.: zxcvbn: Low-budget password strength estimation. In: SEC 2016 Proceedings of the 25th USENIX Conference on Security Symposium, pp. 157–173 (2016)
20. Zhang, Y., Xian, H., Yu, A.: CSNN: password guessing method based on Chinese syllables and neural network. Peer-to-Peer Netw. Appl. **13**(6), 2237–2250 (2020). https://doi.org/10.1007/s12083-020-00893-7

# CNN-Based Continuous Authentication on Smartphones with Auto Augmentation Search

Shaojiang Deng, Jiaxing Luo, and Yantao Li[✉]

College of Computer Science, Chongqing University, Chongqing 400044, China
yantaoli@cqu.edu.cn

**Abstract.** In this paper, we present CAuSe, a **C**NN-based **C**ontinuous **Au**thentication on smartphones using **Au**to **A**ugmentation **Se**arch, where the CNN is specially designed for deep feature extraction and the auto augmentation search is exploited for CNN training data augmentation. Specifically, CAuSe consists of three stages of the offline stage, registration stage and authentication stage. In the offline stage, we utilize auto augmentation search on the collected data to find an optimal strategy for CNN training data augmentation. Then, we specially design a CNN to learn and extract deep features from the augmented data and train the LOF classifier after 95 features are selected by PCA in the registration stage. With the trained CNN and LOF classifier, CAuSe identifies the current user as a legitimate user or an impostor in the authentication stage. Based on our dataset, we evaluate the effectiveness of optimal strategy and the performance of CAuSe. The experimental results demonstrate that the strategy of Time-Warping(0.6)+Time-Warping(0.6) reaches the highest accuracy of 93.19% with data size 400 and CAuSe achieves the best authentication accuracy of 96.93%, respectively, comparing with other strategies and classifiers.

**Keywords:** Continuous authentication · Auto augmentation search · CNN · LOF classifier

## 1 Introduction

The mobile devices have played an essential role in our daily lives, which makes privacy protection in mobile devices extremely important, since they store a lot of private and sensitive information. Even since 2011, sales of smartphones have exceeded sales of personal computers [2]. However, due to the high-frequency usage and information interaction of these devices (e.g. smartphones), it is difficult to prevent personal information leakage and illegal access by the one-time authentication that identifies users only at the time of initial logging-in, such as personal identification numbers (PINs), passwords, voice-prints, fingerprints and face recognition. PINs face a much serious threat of online guessing and even longer PINs only attain marginally improved security [3,26]. Wang et al.

© Springer Nature Switzerland AG 2021
D. Gao et al. (Eds.): ICICS 2021, LNCS 12918, pp. 169–186, 2021.
https://doi.org/10.1007/978-3-030-86890-1_10

systematically characterized typical targeted online guessing attacks with seven sound mathematical models, each of which was based on varied kinds of data available to an attacker [27]. Biometric information cannot be acquired by direct covert observation, but once biological information is stolen, it is not naturally available to reissue [22]. For example, fingerprint recognition can be cracked by people with ulterior motives obtaining legitimate users' fingerprints left on the screen. In addition, there is a severe security and privacy threat in one-time authentication mechanisms that when a legitimate user leaves the supervision of the device after the initial authentication (the screen is unlocked), impostors can easily gain access to the device illegally.

Compared with the traditional one-time authentication mechanisms, continuous or implicit authentication approaches would provide an additional line of defense by designing a non-intrusive and passive security countermeasure [9]. The current continuous authentication mechanisms essentially use built-in sensors and accessories to frequently collect physiological or behavioral biometrics to identify the legitimacy of the user, such as voice [8], face patterns [1], touch gestures [28], typing motion [10] and gait dynamics [21]. There are two main stages for continuous authentication systems: user registration phase and continuous authentication phase. During the user registration phase, owners of mobile devices are usually asked to perform some operations to collect information to recognize the owners. During the continuous authentication phase, the system collects the user's sensor readings at regular intervals to determine whether the current user is the device owner. If the system finds that the current user is an illegal user, the system will lock the device to prevent the owner's privacy from leaking. The accelerometer, gyroscope, and magnetometer are the three most commonly used sensors for collecting behavioral biometrics without users' notice. Both accelerometer and gyroscope are motion sensors that can monitor the users' motion on the device. Magnetometer is a position sensor used to determine the physical position of the device in the true frame of reference. However, in order to obtain a high-performance continuous authentication model, it is often necessary to collect a large amount of high-quality data for training models, which costs lots of time and resources. Data augmentation methods, such as flipping, cropping, color dithering and generative adversarial networks (GANs), are very common techniques in the field of image recognition, which help cover unexplored input space, prevent overfitting and improve the generalization ability of classification model. However, there are currently few data augmentation methods specifically for time-series sensor data because time-series sensor data are quite different from image data and most of the current data augmentation methods cannot be used to create time-series data directly. Since the sufficient amount of sensor data collection needs lots of volunteers to participate, it is challenging to augment time-series sensor data. Moreover, for specific applications, artificially constructing features for time-series sensor data often requires a lot of prior expert knowledge. It is also challenging to extract features with high representation capacity on time-series sensor data.

To address the challenges of data shortage and feature contribution, we are among the first to utilize the auto augmentation search to find an optimal data augmentation strategy for CNN training and design a CNN-based deep feature extraction method consisting of feature learning and feature selection. In this paper, we present CAuSe, a CNN-based Continuous Authentication on smartphones using Auto Augmentation Search. Specifically, CAuSe consists of five modules: data collection, auto augmentation search, feature extraction, classifier training and authentication. The process of CAuSe includes three stages of the offline stage, registration stage, and authentication stage. In the offline stage, CAuSe collects time-series sensor data of the accelerometer, gyroscope, and magnetometer, and then utilizes auto augmentation search on the collected sensor data to find an optimal data augmentation strategy. In the registration stage, CAuSe applies the optimal augmentation strategy on the collected sensor data, uses the designed CNN to learn and extract deep features from the augmented data, and trains the local outlier factor (LOF) classifier after 95 deep features are selected by principal component analysis (PCA). In the authentication stage, based on the sampled sensor data, CAuSe uses the trained CNN to learn and extract features and utilizes the trained LOF classifier to conduct the authentication based on the 95 PCA-selected features. Based on our dataset, we evaluate the effectiveness of auto augmentation search and the corresponding optimal strategy and the performance of CAuSe. The experimental results demonstrate that the augmentation strategy of Time-Warping(0.6)+Time-Warping(0.6) reaches the highest authentication performance with the 93.19% accuracy, 93.77% $F_1$-score, and 3.9% EER with data size 400, and CAuSe achieves the best accuracy of 96.93% with the LOF classifier on 95 PCA-selected features, respectively, comparing with other augmentation strategies and classifiers.

The main contributions of this work are summarized as follows:

- We present CAuSe, a CNN-based continuous authentication on smartphones using auto augmentation search, leveraging the smartphone built-in accelerometer, gyroscope and magnetometer.
- We specially design a CNN for deep feature extraction and utilize the auto augmentation search to find an optimal data augmentation strategy for CNN training.
- We evaluate the effectiveness of auto augmentation search and the performance of CAuSe, and the experimental results illustrate that the searched augmentation strategy reaches the highest accuracy (93.19%) with data size 400, and CAuSe achieves the best authentication accuracy (96.93%), respectively.

The remainder of this work is organized as follows: Sect. 2 reviews the state-of-the-art on continuous authentication. We elaborate the architecture of CAuSe in Sect. 3 and evaluate the performance of the optimal strategy and CAuSe in Sect. 4. Section 5 concludes this work.

## 2    Related Work

In this section, we review the state-of-the-art of the continuous authentication systems, time-series data augmentation methods and auto augmentation methods, respectively.

### 2.1    Continuous Authentication System

In the field of continuous authentication, high-precision discrimination results are often inseparable from an efficient system framework. In recent years, researchers have creatively designed well-performed continuous authentication systems based on different data sources [20]. The mainstream continuous authentication solutions are broadly composed of two phases: registration phase and authentication phase. During the registration phase, these systems extract features from the collected datasets and train classifiers with labeled features. During the authentication phase, these systems utilize the trained classifiers to classify features that are extracted from unidentified users' data. Considering that different types of touch operations may contain quite different characteristics, the authors in [28] designed specific features for different touch operations, and then adopted the trained classifiers for authentication. Z. Sitová et al. [23] designed hand movement, orientation and grasp behavioral features based on sensor readings from smartphones, then trained and tested one-class classifiers after feature selection. Mahbub et al. [19] trained a linear SVM with statistical features obtained from face proposals that were derived from the estimated faces in their designed system. In [5], the authors proposed a continuous motion recognition system that was based on motion data from the accelerometer, gyroscope and magnetometer. They used a Siamese convolutional neural network to learn deep features, and then trained the one-class SVM with learned features of the legitimate user, to predict new observations. In [13], Li et al. proposed a two-stream convolutional neural network for feature learning in the continuous authentication system which was based on bottleneck structure of Mobilenet v2, with both time domain data and frequency domain data of the accelerometer and gyroscope as the network inputs.

Inspired by the above contributions, we design an efficient CNN-based continuous authentication system which can achieve very close performance with few sampled sensor data for training using time-series data auto Augmentation technology.

### 2.2    Time-Series Data Augmentation Method

In the image recognition field, data augmentation can be implemented by labeling the same labels for images obtained by performing operations, such as scaling, cropping, jittering and flipping on raw images. However, in the time-series data field, such as sensor data, there are few data augmentation approaches proposed. In [25], the authors were among the first to exploit geometric transformation, such as permutation, sampling, scaling, cropping and jittering, as sensor data

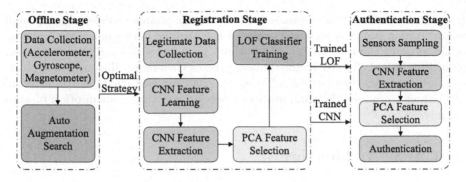

**Fig. 1.** CAuSe architecture.

augmentation approaches, which were different to those in image augmentation. DeVries et al. [7] used a sequence autoencoder to project data into feature space and investigated augmentation techniques in the feature space.

Data augmentation with generative adversarial networks (GANs) has attracted some researchers' attentions recently. Zhu et al. [31] proposed an emotion classification system using data augmentation with a cycle-consistent adversarial network (CycleGAN) and Luo et al. [17] trained a conditional Wasserstein generative adversarial network (WGAN) with electroencephalography (EEG) data to generate additional data for data augmentation. In [24], the authors investigated the possibility of using GANs to augment time-series Internet of Things (IoT) data. In [12], the author investigated five sequential data augmentation techniques (additional Gaussian noise, masking noise, signal translation, amplitude shifting, and time stretching) including sample-based and dataset-based methods to improve the intelligent fault diagnosis accuracy.

## 2.3   Auto Augmentation Method

Since the current data augmentation implementations are almost manually designed [7,25], researchers prefer to apply one or several fixed data augmentation methods based on their experience for most datasets, although there theoretically exists an optimal data augmentation method for a specific dataset. Cubuk et al. [6] first proposed the concept of auto augmentation, which automatically searched optimal augmentation policies from data to improve validation accuracy. Their search algorithm (implemented as a RNN controller based on Reinforcement) sampled thousands of policies to train a child model to measure the performance of the generalization improvement, and then updated the augmentation policy distribution with a reward signal. Despite its promising empirical performance, this scheme was difficult to apply because it was very expensive with time-consuming calculation in the whole process. Lin et al. [16] formulated the augmentation policy as a parameterized probability distribution, thus allowing the augmentation policy probability distribution parameters to be optimized along with the network parameters simultaneously. Based on a bilevel framework, this solution eliminated the need of re-training model after optimal augmentation policy search and achieved comparable performance with dozens of times faster than [6]. In [15], the authors proposed a fast auto augmentation algorithm to find effective augmentation policies via a more efficient search strategy based on density matching. Moreover, [29] proposed effective optimization algorithms to reduce the computational burden and time consumption of auto augmentation.

## 3    CAuSe Architecture

In this section, we present the architecture of CAuSe, the CNN-based continuous authentication on smartphones with auto augmentation search, as illustrated in Fig. 1. As shown in Fig. 1, CAuSe consists of three stages: the offline stage, registration stage, and authentication stage.

In the offline stage, CAuSe collects time-series sensor data and then utilizes auto augmentation search on the collected sensor data to find an optimal data augmentation strategy for CNN training data augmentation in the registration stage. First, we recruit volunteers to use smartphones equipped with sensor data collection tools to collect sensor data of the accelerometer, gyroscope and magnetometer. Then, we perform preprocessing operations on the collected time-series sensor data, and based on the preprocessed data, we conduct the auto augmentation search to obtain an optimal augmentation strategy.

In the registration stage, CAuSe applies the optimal augmentation strategy on the collected sensor data, uses the designed CNN to learn and extract deep features from the augmented data, and trains the local outlier factor (LOF) classifier after 95 deep features are selected by PCA. Specifically, the owner (the legitimate user) is required to operate on the smartphone to collect data of the accelerometer, gyroscope and magnetometer. Then, we use the optimal augmentation strategy to augment the collected sensor data including the legitimate user's for feature extraction. We specially design a CNN based on Shufflenet V2 [16] to learn and extract deep features from the augmented sensor data. 95 deep features are selected by PCA and then used to train the LOF classifier.

In the authentication stage, based on the sampled sensor data, CAuSe uses the trained CNN to learn and extract features and utilizes the trained LOF classifier to conduct the authentication based on 95 features selected by PCA. If the user is a legitimate user, CAuSe will allow the continuous usage of the smartphone and meanwhile continuously authenticate the user; otherwise, it will require the initial login inputs.

### 3.1    Data Collection and Preprocessing

**Data Collection.** The accelerometer and gyroscope are motion sensors, and they can capture the motion patterns of the devices. The magnetometer is a position sensor that records changes in the physical position of the devices. The three sensors are widely equipped on the modern smart devices. Considering the above advantages, we select the accelerometer, gyroscope, and magnetometer to collect the data for user continuous authentication.

In order to collect the sensor data for CAuSe, we recruited 88 volunteers (44 male and 44 female) to operate on 10 Samsung Galaxy S4 smartphones, each of which was installed a designed virtual keyboard. They were required to participate in 8 sessions, and they used the virtual keyboard to answer 3 questions in each session. For each answer, they entered 250 characters at least. During their operations, we collected data on the three axes of the accelerometer, gyroscope and magnetometer with a sampling rate of 100 Hz.

**Data Preprocessing.** Since the collected raw sensor data are long time-series streams, we use a sliding window to perform non-repetitive sampling, each containing 2 s-sensor data. In a sliding window, each row represents the sampled sensor data, and each column indicates the $x$, $y$, and $z$ axes of a sensor. In order to enable the time-series sensor data to be used as the inputs of a CNN with $shape = (H, W, C)$, we adaptively change the shape of the collected data. Specifically, the three sensor data are regarded as three channels ($C$), and the rows and columns of the sliding window correspond to $H$ and $W$, respectively. Ignoring the error in the sampling process and according to the sampling frequency, it can be inferred that $H = 200$.

We divide the 88 volunteers' data into three groups (88 users with 3000 windows): 68 users with 2000 windows $D_{learning}$ for CNN training, 68 users with 1000 windows $D_{positive}$ as legitimate users' testing dataset for feature extraction and classifier training, and 20 users with 3000 windows $D_{negative}$ as impostors' testing dataset for feature extraction and classifier training. $D_{learning}$ are fitted and transformed by *RobustScaler* in Python library *sklearn.preprocessing*, which ignores outliers in the dataset. $D_{positive}$ and $D_{negative}$ are transformed by the same *RobustScaler*, so that the three groups of data can be consistently normalized for data augmentation.

## 3.2 Auto Augmentation Search

**Search Space.** For images, there is spatial correlation among the pixels and other pixels around them, while for sensor data, there is temporal correlation among samples. Therefore, we design specific data augmentation strategies that consider the possible invariant geometric transformation of sensor data in time series. For each input of CNN training sensor data, we sample an augmentation strategy from the search space and apply. Each augmentation strategy is composed of two augmentation methods.

We design the candidate augmentation methods for sensor data:

1) **Rotation:** When users operate on mobile devices, the devices are likely to be flipped or rotated at a certain angle. Accordingly, the $x$, $y$, and $z$ axes of the sensors on the devices rotate at the same angle corresponding to the

Cartesian coordinate system. In order to simulate this, we design a rotation method, which rotates the $x$, $y$, and $z$ axes of the sampled sensor data by multiplying a rotation matrix to obtain angles of $(-\pi/3, -\pi/6, -\pi/12, \pi/3, \pi/6, \pi/12)$.

2) **Jittering:** Noise can be introduced in the process of sensor data collection which might be caused by environmental disturbance. Jittering function adds a noise matrix generated by a normal distribution with standard deviations of 0.05, 0.25, and 0.5 to the sampled sensor data. Note that we ignore the injection attacks in jittering augmentation [11].

3) **Scaling:** Scaling function multiplies the $x$, $y$, and $z$ axes of the sampled sensor data separately by scale factors generated by a normal distribution with standard deviations of 0.05, 0.1, and 0.2.

4) **Permutation:** Since the segmentation position of the fixed window is arbitrary for sensor data collected in a period of time, the position of the event implied in the sub-window in the whole window is meaningless. Permutation function segments the whole sample window to 4, 5, or 8 sub-windows by rows to perturb the temporal location of within-window events.

5) **Magnitude-Warping:** We sample values from a normal distribution with standard deviations of 0.2, 0.4, 0.6, feed them to *scipy.interpolate.cubicSpline* to generate three random smooth curves corresponding to $x$, $y$, and $z$ axes, and finally convolute them with the sampled sensor data.

6) **Time-Warping:** Time-Warping function utilizes the aforementioned smooth curves and one dimensional linear interpolation to perturb the temporal location smoothly.

7) **Cropping:** Cropping can diminish the dependency on event locations. In the cropping function, we randomly select different numbers of window rows (e.g. 10, 20, or 30) and set values of these selected window rows to 0.

Seven augmentation functions with specific magnitude parameters make up a total of 24 augmentation methods. In our designed augmentation strategy search space, each augmentation strategy consists of 2 augmentation methods orderly and repeatable. In other words, there are totally $24^2$ strategies in the augmentation strategy search space.

**Search Pipeline.** Inspired by Lin et al.'s work [16], we adapt distribution optimization to the continuous authentication area to search an optimal data augmentation strategy for time-series sensor data. As mentioned, since each augmentation strategy consists of two augmentation methods and there are 24 augmentation methods in total, there are $24^2$ strategies in the designed augmentation strategy search space. Thus, we first initialize a $24^2$ matrix sampled from a uniform distribution as the augmentation probability distribution $\theta$. The probability of the $k$th augmentation strategy $p_\theta$ can be formulated as:

$$p_\theta(S_k) = \frac{\frac{1}{1+e^{-\theta_k}}}{\sum_{i=1}^{K} \frac{1}{1+e^{-\theta_k}}} \tag{1}$$

where $\theta \in R^K$, and $S_k$ indicates the $k$th data augmentation strategy candidate.

Next, we perform the auto augmentation strategy search. We take an epoch $t$ of total $T$ epochs in model training process. Each input will be applied with a randomly chosen augmentation strategy for each batch $b$ of total $B$ batches. Since the validation accuracy $acc(w^*)$ of the network model is only decided by the optimal network model parameters $w^*$ and the model training process is only influenced by the augmentation strategies applied to each input, the augmentation probability distribution matrix $\theta$ is defined as a variable matrix with gradient about the network model parameters $w^*$. However, it is a tricky problem to calculate the gradient of validation accuracy $acc(w^*)$ with respect to $\theta$. To approximate the gradient, we execute the following steps four times for epoch $t$:

1) Sample and apply an augmentation strategy for each input, train the network model with augmented inputs, obtain the validation accuracy $w'$, and record the network parameters;
2) Make gradient back propagation for $\theta$, update values of $\theta$, and then clear the gradient of $\theta$;
3) Save the network parameters with the highest $w'$ as the initial network parameters for next epoch.

Based on the reinforcement learning and Monte-Carlo sampling, at the end of epoch $t$, the cumulative gradient can be approximately formulated as:

$$\nabla_\theta \Gamma(\theta) \approx \frac{1}{N} \sum_{n=1}^{N} \sum_{j=1}^{I \times B} \nabla_\theta log(p_\theta(S_{k(j),n}))acc(w,n) \tag{2}$$

where $N$ denotes the total times of network training and $acc(w, n)$ indicates the validation accuracy of the $n$th network. Network parameters with the highest validation accuracy will be broadcast to the network before the next epoch. After sufficient epochs of parameters updates, the augmentation probability distribution converges. The augmentation strategy with the highest probability is the optimal augmentation strategy we search. Note that the network model architecture is the same to the designed CNN architecture.

## 3.3 Feature Extraction

In this section, we design a CNN-based deep feature extraction method, which consists of feature learning and feature selection. In the following, we first elaborate the design of the CNN and then detail the CNN-based feature extraction.

**CNN Design.** We design the architecture of the CNN inspired by Shufflenet V2 [18], as illustrated in Table 1, for auto augmentation search, feature learning and extraction. As demonstrated in Table 1, the designed CNN is composed of a 2D convolutional layer (Conv2d), a 2D max pooling (MaxPooling2d), a stack of Shufflenet V2 units grouped into three stages (Stage 1, Stage 2, and Stage 3),

Table 1. CNN architecture.

| Layer | Output | # Kernel | KSize | Stride | Parameter | Repeat |
|---|---|---|---|---|---|---|
| Sensor | 200 × 3 × 3 | – | – | – | – | – |
| Conv2d (BN+ReLU) | 100 × 3 × 24 | 24 | 3 × 3 | (2,1) | 672 | 1 |
| MaxPooling2d | 50 × 3 × 24 | – | 3 × 3 | (2,1) | – | 1 |
| Stage 1 | 25 × 3 × 48 | 48 | – | (2,1) | 2760 | 1 |
|  | 25×3×48 | 48 | – | (1,1) | 1728 × 3 | 3 |
| Stage 2 | 13 × 3 × 96 | 96 | – | (2,1) | 8976 | 1 |
|  | 13 × 3 × 96 | 96 | – | (1,1) | 5760×7 | 7 |
| Stage 3 | 7 × 3 × 96 | 192 | – | (2,1) | 31776 | 1 |
|  | 7 × 3 × 96 | 192 | – | (1,1) | 20736 × 3 | 3 |
| Conv2d (BN+ReLU) | 7 × 3 × 1024 | 1024 | 1 × 1 | (1,1) | 197632 | 1 |
| GlobalAveragePooling2d | 1 × 1× 1024 | – | 7×3 | – | – | 1 |
| Dense | CN×1 | – | – | – | 69700 | 1 |

another Conv2d, a 2D global average pooling (GlobalAveragePooling2d), and a dense layer. We adopt BN and ReLu right after each Conv2d. In addition, Stages 1, 2, and 3 are composed of the building blocks of a basic unit followed by several basic units for spatial down sampling. 'CN' represents class number for CNN training (class_num).

**Feature Learning.** Based on the optimal strategy obtained from the offline stage, $D_{learning}$ are augmented in the registration stage. As illustrated in Table 1, with the augmented data, there are 1800 (3 sensors × 2s × $100Hz$ × 3 axes) samples in a 2s-sliding window. The first Conv2d layer with 24 filters of 3 × 3 and stride of (2,1) followed by a MaxPooling2d with kernel size of 3 × 3 and stride of (2,1), aims to make down sampling and increase channels. Then, three stages of a basic unit with stride (2,1), and several units for spatial down sampling with stride (1,1) are applied, where Stage 1 repeats 3 times of the unit for spatial down sampling, Stage 2 repeats 7 times, and Stage 3 repeats 3 times. Next, there is another Conv2d layer with 1024 filters of 1 × 1 and stride of (1,1) followed by a GlobalAveragePooling2d layer and a dense layer. The total parameters of the designed CNN are 419,228 and the second Conv2d layer contributes the most parameters (19,7632 parameters). The outputs of the GlobalAveragePooling2d are deep features learned from the sensors of the accelerometer, gyroscope and magnetometer.

**Feature Selection.** We use the principal component analysis (PCA) to select appropriate number of deep features for the classifier based on the CNN-extracted features. Based on the experiments in Sect. 4.2, PCA selects 95 deep features for the LOF classifier to conduct the authentication.

### 3.4 Authentication with LOF Classifier

With the 95 PCA-selected deep features, CAuSe utilizes the local outlier factor (LOF) classifier to identify users. LOF measures the local deviation of the data point to its neighbors, which decides whether a data point is an outlier using the anomaly score estimated by k-nearest neighbors based on a given distance metric. A data point with a substantially lower density than its neighbors will be regarded as an outlier [4].

In the registration stage, CAuSe generates the legitimate user's profile from the training data and the LOF classifier is trained by PCA-selected deep features. In the authentication stage, the trained LOF classifier classifies the PCA-selected deep features from the sampled sensor data. Based on the trained classifier and the sampled data while using the device, CAuSe authenticates the current user as a legitimate user or an impostor. If the user is a legitimate user, CAuSe will allow the continuous usage of the smartphone and meanwhile continuously authenticate the user; otherwise, it will require the initial login inputs.

## 4 Performance Evaluation

In this section, we start with experimental settings, then investigate the performance of CAuSe in terms of optimal feature number, and evaluate the effectiveness of auto augmentation search and optimal strategy, respectively.

### 4.1 Experimental Settings

**Network Model Training.** With the inputs of $D_{learning}$, 80% of the data are used for training and the rest 20% for testing, with a batch size of 128. We use the cross entropy as the loss function and the stochastic gradient descent (SGD) optimizer to update the learning rate. The initial learning rate is 0.2, and it complies with an exponential decay of decay_step = 1000 and decay_rate = 0.96. If the lowest validation loss remains for 10 continuous epochs or the network training process exceeds 150 epochs, the training process stops. The network with the lowest loss is used as the trained model.

**Auto Augmentation Strategy Search.** The parameters of the augmentation distribution initialize as a $24 \times 24$ matrix with initial values from a uniform distribution. We use Adam optimizer with learning rate 0.05, $\beta_1 = 0.9$, $\beta_2 = 0.999$, weight_decay = 0. The distribution parameters are updated 150 times in total.

**Table 2.** Accuracy (SD) % for different classifiers with varying feature numbers

| Classifier\Number | 5 | 35 | 55 | 75 | 95 | 115 | 135 | 155 | 175 | 195 |
|---|---|---|---|---|---|---|---|---|---|---|
| OC-SVM | 91.27 | **94.58** | 93.45 | 90.57 | 86.55 | 81.97 | 77.24 | 72.90 | 69.14 | 66.00 |
| | (4.10) | (2.53) | (1.98) | (1.95) | (2.13) | (2.58) | (2.78) | (3.26) | (3.68) | (4.11) |
| IF | 87.25 | 93.26 | 94.71 | 95.28 | 95.68 | 95.95 | **96.03** | 95.96 | 95.80 | 95.58 |
| | (7.17) | (4.07) | (3.32) | (2.90) | (2.49) | (2.15) | (1.92) | (1.80) | (1.78) | (1.77) |
| LOF | 80.69 | 92.51 | 94.45 | 95.40 | **96.93** | 96.79 | 96.66 | 96.38 | 95.97 | 95.77 |
| | (11.44) | (5.91) | (3.93) | (2.84) | **(1.80)** | (1.92) | (2.05) | (2.10) | (2.38) | (2.48) |

**Classifier Training.** To train the LOF classifier, we randomly select 1 legitimate user from $D_{positive}$ for 20 times. With the 1000-window data, we use 10-fold cross validation to obtain 900-window training dataset and 100-window positive testing dataset. We also randomly select 100-window from $D_{negative}$ as the negative testing dataset.

**Evaluation Metric.** We utilize three evaluation metrics: accuracy, $F_1$-score, EER to evaluate the effectiveness of CAuSe. Accuracy is the percentage ratio of the total number of correct authentication against the total number of authentication, defined as: $Accuracy = \frac{TP+TN}{TP+TN+FP+FN}$. $F_1$-score is defined as: $F_1 = \frac{2TP}{TP+FP+FN}$. EER is the point where FAR equals to FRR.

### 4.2  Feature Number and Classifier Parameter

We conduct experiments to investigate classifier selection and optimal feature number selected by PCA. We consider three classifiers of OC-SVM, IF, and LOF for classifier selection and vary feature numbers for optimal feature number. We compute the accuracy (standard deviation) of CAuSe with the three classifiers as the feature number increases from 5 to 195, as tabulated in Table 2. As shown in Table 2, the accuracy gradually increases with the feature number growing until an optimal number and then slightly decreases for all the classifiers. For OC-SVM, 35 features selected by PCA reach the best accuracy of 94.58% and for IF, 135 features achieve 96.03% accuracy. However, LOF with 95 features selected by PCA reaches the highest accuracy of 96.93% and the lowest SD of 1.80%. Therefore, we use PCA to select 95 deep features for the LOF classifier.

In addition, based on the optimal numbers of features, we utilize the grid search to seek the best parameter combinations for classifiers of the OC-SVM, IF, and LOF. We list the classifiers, number of features, and optimal parameter combination in Table 3. As shown in Table 3, the LOF classifier with 95 deep features obtains the optimal parameters of $n\_neighbors = 800$ and $p = 1$.

**Table 3.** Optimal parameter combinations

| Classifier | # Feature | Optimal parameter combination |
|---|---|---|
| OC-SVM | 35 | $\mu = 0.0001,\ \gamma = 0.015625$ |
| IF | 135 | n_estimators $= 900$ |
| LOF | 95 | n_neighbors $= 800, p = 1$ |

**Table 4.** Row and column corresponding to the optimal augmentation strategy

| Epoch | 0–2 | 3–4 | 5 | 6 | 7 | 8–10 | 11–16 |
|---|---|---|---|---|---|---|---|
| (Row, column) | (7,5) | (13,22) | (6,22) | (5,16) | (7,14) | (17,2) | (7,14) |
| Epoch | 17–19 | 20–26 | 27–45 | 46–92 | 93–107 | 108–130 | 131–149 |
| (Row, column) | (5,1) | (7,14) | (16,14) | (13,22) | (20,20) | (13,22) | (20,20) |

### 4.3   Auto Augmentation Search

We select dataset $D_{learning}^{100}$ with 100-window per user from $D_{learning}$ to conduct the evaluation of the auto augmentation search, due to the limitations of computer memory and GPU. In the auto augmentation search, we instantiate the augmentation distribution parameters as a $24\times24$ matrix and save the corresponding matrix for each epoch. Based on the saved matrices, we sum the rows of each matrix, normalize all rows for each epoch, and visualize rows varying with the epoch grows. We calculate the marginal distribution of parameters of the first augmentation method of each strategy, as illustrate in Fig. 2. As illustrated in Fig. 2, the deeper the red, the closer the probability of the method is to 1, and the deeper the blue, the closer the probability is to 0. As the search progresses, the edge probability of each method either converges to 0 or 1. When the search is complete, the edge probability of the method in rows of 4, 6, 10, 17, 18, and 21 is higher. From Fig. 2, it can be seen that during random training, the parameter values of some augmentation methods gradually increase while others gradually decrease, which indicates that some augmentation methods are abandoned while the probability of other augmentation methods is increasing.

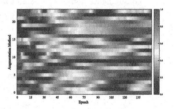

**Fig. 2.** Marginal distribution of augmentation operations. (Color figure online)

In addition, after updating the parameters of the augmentation probability distribution at the end of each epoch, we calculate the probability for each augmentation strategy by Eq. (1) and record the row and column of the corresponding optimal augmentation strategy, as shown in Table 4.

It can be seen that during the training process, with the update of the probability distribution parameters, the optimal strategy (the strategy with the highest probability) is also constantly changing, and at the end of the training, a row and a column $(20, 20)$ of the optimal strategy for local convergence can be

**Table 5.** Optimal parameter combinations

| Network | Accuracy | $F_1$-score | EER |
|---|---|---|---|
| Network without augmentation | 85.37 (7.61) | 87.54 (5.71) | 7.87 (3.59) |
| Network searched by auto augmentation | 88.88 (6.64) | 90.24 (5.29) | 6.50 (3.38) |

**Table 6.** Accuracy (SD) % on Different Strategies with Varying Data Sizes

| Strategy\Data size | 60 | 80 | 100 | 200 | 400 |
|---|---|---|---|---|---|
| No augmentation | 56.77 (6.33) | 54.67 (3.90) | 85.37 (7.61) | 90.06 (5.95) | 92.14 (5.31) |
| Rota-3+MagnWarp0.2 | 79.32 (8.18) | 81.45 (8.10) | 82.61 (7.53) | 85.10 (8.20) | 90.11 (6.02) |
| Perm8+Rotate12 | 84.34 (7.65) | 86.75(6.93) | 85.79 (7.39) | 87.49 (7.36) | 89.50 (7.02) |
| TimeWarp0.6+Perm2 | 87.99 (6.98) | 88.05 (6.43) | 89.59 (6.15) | 90.76 (6.58) | 92.47 (5.49) |
| Our strategy | 88.65 (7.40) | 88.70 (7.51) | 91.12 (5.69) | 91.89 (5.33) | 93.19 (4.85) |

**Table 7.** $F_1$ Score (SD) % on different strategies with varying data sizes

| Strategy\data size | 60 | 80 | 100 | 200 | 400 |
|---|---|---|---|---|---|
| No augmentation | 68.89 (3.34) | 68.43 (2.09) | 87.54 (5.71) | 91.16 (4.80) | 92.87 (4.37) |
| Rota-3+MagnWarp0.2 | 83.18 (5.74) | 84.68 (5.82) | 85.46 (5.52) | 87.40 (5.98) | 91.23 (4.84) |
| Perm8+Rota12 | 86.73 (5.50) | 88.55 (5.27) | 87.82 (5.50) | 89.18 (5.70) | 90.79 (5.48) |
| TimeWarp0.6+Permu2 | 89.55 (5.44) | 89.52 (5.02) | 90.76 (4.94) | 91.78 (5.27) | 93.19 (4.53) |
| Our strategy | 90.12 (5.78) | 90.16 (5.70) | 92.00 (4.70) | 92.64 (4.39) | 93.77 (4.09) |

**Table 8.** EER (SD) % on different strategies with varying data sizes

| Strategy\data size | 60 | 80 | 100 | 200 | 400 |
|---|---|---|---|---|---|
| No Augmentation | 39.12 (10.61) | 37.79 (10.32) | 7.87 (3.59) | 5.62 (2.82) | 4.65 (2.66) |
| Rota-3+MagnWarp0.2 | 10.06 (3.77) | 9.72 (4.26) | 8.62 (3.82) | 6.65 (3.24) | 5.34 (3.00) |
| Perm8+Rota12 | 9.07 (4.29) | 7.43 (3.34) | 7.90 (3.58) | 6.39 (3.31) | 5.51 (2.87) |
| TimeWarp0.6+Perm2 | 7.21 (3.80) | 6.73 (3.12) | 6.53 (3.28) | 5.24 (2.98) | 4.35 (2.73) |
| Our Strategy | 6.67 (3.60) | 6.64 (3.71) | 5.68 (3.34) | 4.99 (2.74) | 3.90 (2.48) |

obtained. It can be considered that Time-Warping (0.6) + Time-Warping (0.6) is a relatively good augmentation strategy found in our dataset in the entire search space with a CNN structure in Table 2 trained to converge. We also illustrate the continuous authentication performance of the network model trained by auto augmentation search and the network model obtained by training the same network structure without augmentation in Table 5.

### 4.4 Optimal Strategy

In the above experiments, we searched for an optimal strategy that located in the 20th row and 20th column of the probability distribution parameter matrix. The optimal strategy is a strategy composed of two identical augmentation operations Time-Warping(0.6)+Time-Warping(0.6). In order to demonstrate the superiority of the strategy, we randomly select 3 strategies from the search space to augment different size of data and compute the accuracy, $F_1$-score and EER,

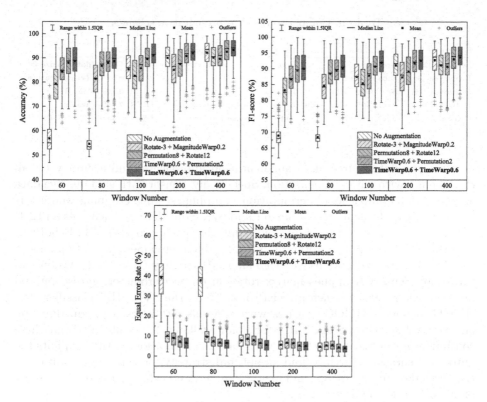

**Fig. 3.** Accuracy, $F_1$ score, and EER for different strategies with varying data sizes.

respectively. The corresponding results are tabulated in Tables 6, 7, and 8, and are plotted in Fig. 3.

We can obtain observations from Tables 6, 7, and 8, and Fig. 3:

1) When there is no data augmentation, as the data size increases, the authentication performance gradually improves, which indicates that the amount of real data is positively correlated with the authentication performance.

2) When the data size comes to 100, the EERs for the strategies of the Rotate(-3)+MagnitudeWarp(0.2) and Permutation(8)+Rotate(12) are even higher than that without data augmentation strategy, which indicates that the two strategies are relatively worse augmentation strategies.

3) On all data sizes, the strategy of Time-Warping(0.6)+Time-Warping(0.6) achieves the best authentication performance on the accuracy (93.19%), $F_1$ score (93.77%), and EER (3.9%), which proves that the optimal strategy searched by the proposed auto augmentation is optimal on different data sizes.

## 4.5   Comparison with Representative Schemes

We compare CAuSe to four representative continuous authentication schemes with data augmentation approaches, as listed in Table 9. As illustrated in Table 9,

**Table 9.** Comparison with representative schemes

| Scheme | Data source | Data augmentation approach | Accuracy |
|---|---|---|---|
| SensorAuth [13] | Acc., Gyr. | Perm., sample, scale, crop, jitter | EER: 6.29% (dataset size 200) |
| EchoPrint [30] | Face image | Rotation | BAC: 81.78% (vision features) |
| SensorCA [14] | Acc., Gyr., Mag. | Rotation | EER: 3.7% (SVM-RBF) |
| HMOG [23] | Acc., Gyr., Mag., Tou. | HMOG with tap characteristics | EER: 7.16% (walking) |
| CAuSe | Acc., Gyr., Mag. | Auto Augmentation Search | Accuracy: 96.93% (LOF) |

we show the data source, data augmentation approaches, and accuracy for all the schemes with data augmentation. Specifically, SensorAuth explores five data augmentation approaches of permutation, sampling, scaling, cropping, and jittering to create additional acccelerometer and gyroscope data and achieves an EER of 6.29% with dataset size 200 by combining the five approaches [13]. EchoPrint uses the projection matrix rotation imitating different camera poses to augment new face images and obtains 81.78% balanced accuracy (BAC) with vision features [30]. SensorCA applies matrix rotation on accelerometer, gyroscope and magnetometer data to reach an EER of 3.7% on the SVM-RBF classifier [14]. HMOG augments HMOG features with tap characteristics (e.g. tap duration and contact size) to obtain 7.16% EER for walking [23]. Different from these continuous authentication schemes with data augmentation, CAuSe exploits the auto augmentation search to find an optimal strategy for data augmentation of the accelerometer, gyroscope and magnetometer, and achieves the best accuracy of 96.93% on the LOF classifier.

## 5    Conclusion

To address the shortage of training data and improve the feature discriminability, we propose CAuSe, a CNN-based continuous authentication on smartphones using auto augmentation search, where the CNN is specially designed for deep feature extraction and the auto augmentation search is exploited for finding the optimal augmentation strategy. Although we take significant efforts to validate the effectiveness of CAuSe, there are some limitations in this work: 1) power consumption of CAuSe on smartphones, 2) impact of various attacks on CAuSe, and 3) privacy concerns on dataset collection and transportation. In future, we will consider issues of the energy, privacy and security for continuous authentication approaches.

**Acknowledgements.** This work was partially supported by the National Natural Science Foundation of China under Grant 62072061 and by the Fundamental Research Funds for the Central Universities under Grant 2021CDJQY-026.

# References

1. Abeni, P., Baltatu, M., D'Alessandro, R.: Nis03-4: Implementing biometrics-based authentication for mobile devices. In: IEEE Globecom 2006. pp. 1–5. IEEE (2006)
2. Al-Hadithy, N., Gikas, P.D., Al-Nammari, S.S.: Smartphones in orthopaedics. Int. Orthop. **36**(8), 1543–1547 (2012)
3. Bonneau, J., Herley, C., van Oorschot, P.C., Stajano, F.: Passwords and the evolution of imperfect authentication. Commun. ACM **58**(7), 78–87 (2015)
4. Breunig, M.M., Kriegel, H.P., Ng, R.T., Sander, J.: Lof: identifying density-based local outliers. SIGMOD Rec. **29**(2), 93–104 (2000)
5. Centeno, M.P., Guan, Y., van Moorsel, A.: Mobile based continuous authentication using deep features. In: Proceedings of the 2nd International Workshop on Embedded and Mobile Deep Learning, pp. 19–24 (2018)
6. Cubuk, E.D., Zoph, B., Mane, D., Vasudevan, V., Le, Q.V.: Autoaugment: learning augmentation strategies from data. In: Proceedings of the IEEE/CVF Conference on Computer Vision and Pattern Recognition, pp. 113–123 (2019)
7. DeVries, T., Taylor, G.W.: Dataset augmentation in feature space. arXiv preprint arXiv:1702.05538 (2017)
8. Feng, H., Fawaz, K., Shin, K.G.: Continuous authentication for voice assistants. In: Proceedings of the 23rd Annual International Conference on Mobile Computing and Networking, pp. 343–355 (2017)
9. Frank, M., Biedert, R., Ma, E., Martinovic, I., Song, D.: Touchalytics: on the applicability of touchscreen input as a behavioral biometric for continuous authentication. IEEE Trans. Inf. Forensics Secur. **8**(1), 136–148 (2012)
10. Gascon, H., Uellenbeck, S., Wolf, C., Rieck, K.: Continuous authentication on mobile devices by analysis of typing motion behavior. Sicherheit 2014-Sicherheit, Schutz und Zuverlässigkeit (2014)
11. Gonzalez-Manzano, L., Mahbub, U., de Fuentes, J.M., Chellappa, R.: Impact of injection attacks on sensor-based continuous authentication for smartphones. Comput. Commun. **163**, 150–161 (2020)
12. Li, X., Zhang, W., Ding, Q., Sun, J.-Q.: Intelligent rotating machinery fault diagnosis based on deep learning using data augmentation. J. Intell. Manuf. **31**(2), 433–452 (2018). https://doi.org/10.1007/s10845-018-1456-1
13. Li, Y., Hu, H., Zhou, G.: Using data augmentation in continuous authentication on smartphones. IEEE Internet Things J. **6**(1), 628–640 (2018)
14. Li, Y., Hu, H., Zhou, G., Deng, S.: Sensor-based continuous authentication using cost-effective kernel ridge regression. IEEE Access **6**, 32554–32565 (2018)
15. Lim, S., Kim, I., Kim, T., Kim, C., Kim, S.: Fast autoaugment. arXiv preprint arXiv:1905.00397 (2019)
16. Lin, C., et al.: Online hyper-parameter learning for auto-augmentation strategy. In: Proceedings of the IEEE/CVF International Conference on Computer Vision, pp. 6579–6588 (2019)
17. Luo, Y., Lu, B.L.: Eeg data augmentation for emotion recognition using a conditional wasserstein gan. In: 2018 40th Annual International Conference of the IEEE Engineering in Medicine and Biology Society (EMBC), pp. 2535–2538. IEEE (2018)
18. Ma, N., Zhang, X., Zheng, H.-T., Sun, J.: ShuffleNet V2: practical guidelines for efficient CNN architecture design. In: Ferrari, V., Hebert, M., Sminchisescu, C., Weiss, Y. (eds.) Computer Vision – ECCV 2018. LNCS, vol. 11218, pp. 122–138. Springer, Cham (2018). https://doi.org/10.1007/978-3-030-01264-9_8

19. Mahbub, U., Patel, V.M., Chandra, D., Barbello, B., Chellappa, R.: Partial face detection for continuous authentication. In: 2016 IEEE International Conference on Image Processing (ICIP), pp. 2991–2995. IEEE (2016)

20. Mosenia, A., Sur-Kolay, S., Raghunathan, A., Jha, N.K.: Caba: continuous authentication based on bioaura. IEEE Trans. Comput. **66**(5), 759–772 (2016)

21. Muaaz, M., Mayrhofer, R.: An analysis of different approaches to gait recognition using cell phone based accelerometers. In: Proceedings of International Conference on Advances in Mobile Computing & Multimedia, pp. 293–300 (2013)

22. Quan, F., Fei, S., Anni, C., Feifei, Z.: Cracking cancelable fingerprint template of ratha. In: 2008 International Symposium on Computer Science and Computational Technology, vol. 2, pp. 572–575. IEEE (2008)

23. Sitová, Z., Šeděnka, J., Yang, Q., Peng, G., Zhou, G., Gasti, P., Balagani, K.S.: HMOG: new behavioral biometric features for continuous authentication of smartphone users. IEEE Trans. Inf. Forensics Secur. **11**(5), 877–892 (2015)

24. Tschuchnig, M.E., Ferner, C., Wegenkittl, S.: Sequential IoT data augmentation using generative adversarial networks. In: ICASSP 2020–2020 IEEE International Conference on Acoustics, Speech and Signal Processing (ICASSP), pp. 4212–4216. IEEE (2020)

25. Um, T.T., et al.: Data augmentation of wearable sensor data for parkinson's disease monitoring using convolutional neural networks. In: Proceedings of the 19th ACM International Conference on Multimodal Interaction, pp. 216–220 (2017)

26. Wang, D., Gu, Q., Huang, X., Wang, P.: Understanding human-chosen pins: characteristics, distribution and security. In: 2017 ACM on Asia Conference on Computer and Communications Security (ASIA CCS 2017), pp. 372–385. ACM (2017)

27. Wang, D., Zhang, Z., Wang, P., Yan, J., Huang, X.: Targeted online password guessing: an underestimated threat. In: 2016 ACM SIGSAC Conference on Computer and Communications Security (CCS 2016), pp. 1242–1254. ACM (2016)

28. Xu, H., Zhou, Y., Lyu, M.R.: Towards continuous and passive authentication via touch biometrics: an experimental study on smartphones. In: 10th Symposium On Usable Privacy and Security (SOUPS 2014), pp. 187–198 (2014)

29. Zhang, X., Wang, Q., Zhang, J., Zhong, Z.: Adversarial autoaugment. arXiv preprint arXiv:1912.11188 (2019)

30. Zhou, B., Lohokare, J., Gao, R., Ye, F.: Echoprint: two-factor authentication using acoustics and vision on smartphones. In: MobiCom, pp. 321–336. ACM (2018)

31. Zhu, X., Liu, Y., Qin, Z., Li, J.: Data augmentation in emotion classification using generative adversarial networks. arXiv preprint arXiv:1711.00648 (2017)

# Generating Adversarial Point Clouds on Multi-modal Fusion Based 3D Object Detection Model

Huiying Wang[1,2], Huixin Shen[1,2], Boyang Zhang[1], Yu Wen[1(✉)],
and Dan Meng[1]

[1] Institute of Information Engineering, Chinese Academy of Sciences, Beijing, China
{wanghuiying,shenhuixin,zhangboyang,wenyu,mengdan}@iie.ac.cn
[2] School of Cyber Security, University of Chinese Academy of Sciences,
Beijing, China

**Abstract.** In autonomous vehicles (AVs), a critical stage of perception system is to leverage multi-modal fusion (MMF) detectors which fuse data from LiDAR (Light Detection and Ranging) and camera sensors to perform 3D object detection. While single-modal (LiDAR-based and camera-based) models are found to be vulnerable to adversarial attacks, there are limited studies on the adversarial robustness of MMF models. Recent work has proposed a general spoofing attack on LiDAR-based perception, based on the defect of ignored occlusion patterns in point clouds. In this paper, we are inspired to attack LiDAR channel alone to fool the MMF model into detecting a fake near-front object with high confidence score. We perform the first study to analyze the roubustness of a popular MMF model against the above attack and discover it is invalid due to the correction of camera. We propose a black-box attack method to generate adversarial point clouds with few points and prove the defect still exists in MMF architecture. We evaluate the attack effectiveness of different combinations of points and distances and generate universal adversarial examples at the best distance of 4m, which achieve attack success rates of more than 95% and average confidence scores over 0.9 on the KITTI validation set when the points exceed 30. Furthermore, we verify the generality of our attack and the transferability of generated universal adversarial point clouds across models.

**Keywords:** Adversarial point clouds · Adversarial attack ·
Multi-modal fusion · Autonomous vehicles · 3D object detection

## 1 Introduction

Object detection plays an important role in the visual perception system of AVs, which are equipped with multiple sensors such as LiDARs and cameras to perceive the surroundings. It is believed that fusion of multi-modal data from different sensors can obtain complementary and shared information to achieve

© Springer Nature Switzerland AG 2021
D. Gao et al. (Eds.): ICICS 2021, LNCS 12918, pp. 187–203, 2021.
https://doi.org/10.1007/978-3-030-86890-1_11

better performance and stronger robustness than a single modality. However, object detectors that rely on deep neural networks (DNNs) have been found to be vulnerable to adversarial examples [11,37], adding well-crafted malicious perturbation to inputs can deceive DNNs into making wrong predictions, such vulnerabilities can lead to catastrophic consequences in AVs.

LiDARs emit laser beams to the surface of objects to capture 360-degree high-resolution 3D information called point clouds. With the advantage of accurate depth information and reliability in poor weather or lighting conditions, LiDARs are considered as more vital sensors in AVs. Many efforts [7,8,29,32] have been paid to adversarial attacks on LiDAR-based perception, Sun et al. [29] discover that there exists ignored occlusion patterns in point clouds, making it possible to attack LiDAR-based perception by adversarial point clouds with few points. Utilizing such potential defect, they propose a general spoofing attack method by moving the occluded or distant vehicle point clouds with few points in the scene to the front, decieving the victim into believing there is an obstacle ahead. Under this circumstance, emergency braking can cause a rear-end collision and passengers injury.

While recent work [4,6,31] has shown that MMF detectors can be attacked when both LiDAR and camera channels are attacked simultaneously, it is unclear (1) whether the exploited defect in LiDAR-based perception architecture still exists in MMF-based perception architecture, (2) whether the general spoofing attack against LiDAR-based detectors mentioned above is still effective on MMF detectors and (3) how to attack the LiDAR channel of fusion models alone to achieve the goal of attacking the entire model if the above attack is invalid. We believe that the answers to these issues can make sense to improve the robustness of fusion models and better defense adversarial attacks on AVs.

In this paper, we choose AVOD [18] as our target model, a typical feature-level fusion network which is suitable for autonomous driving with fast speed and low memory usage. To answer the first two questions, we reproduce the general spoofing attack done by Sun et al. [29] and implement on AVOD. While we verify such attack does not work because the images can partially correct the false detections of point clouds, we can speculate from the experiment results that the defect still exists and it is feasible to attack the entire model by crafting adversarial point clouds with few points to attack the LiDAR channel alone.

For the third question, we propose a black-box attack method based on genetic algorithm to generate adversarial point clouds. Compared to common gradient-based white-box attacks [9,15,22], our method has stronger attack capability with no need to access the specific structures and parameters of models, and we also avoid dealing with the non-differentiable point clouds processing stage in LiDAR channel of most models, making the method simpler and easy to transfer. Then we evaluate our attack on the KITTI [14] dataset, we randomly pick up 200 scenes, on which we generate adversarial point clouds with different numbers of points from 10 to 200 in step of 10 and varying distances in front of victim ranging from 4 m to 8 m in step of 1 m, to evaluate attack effectiveness. At the optimal distance of 4 m, we generate universal adversarial point clouds, the attack success rates are stable above 95% and the average confidence scores are greater than 0.9 after the number of points exceeds 30. We further directly apply our attack method on another state-of-art MMF model EPNet [16] to make our study more general, and verify the transferability of our generated universal adversarial examples, which achieves around 50% average attack success rate.

The contributions of this paper can be summarized as follows:

- We perform the first security study to analyze the robustness of a popular MMF model against general spoofing attack on LiDAR-based models. We demonstrate that attack cannot directly generalize to fusion model for the correction of other input channel.
- To the best of our knowledge, we are the first to explore the adversarial attack against LiDAR channel on MMF model. We successfully generate adversarial point clouds with few points, proving that the defect caused by ignored occlusion patterns in point clouds still exists in MMF architecture.
- We propose a simple but effective black-box attack approach based on genetic algorithm, which can be optimized by only accessing the inputs and outputs of fusion models. Our method is easy to implement and avoids the issue of handling non-differentiable processes in gradient-based attacks.
- We provide baselines for future research. We conduct an empirical evaluation of our attack on KITTI [14] dataset, we evaluate the attack effectiveness of various combinations of point clouds and generate universal adversarial examples at the best distance of 4m, achieving attack success rates more than 95% with points exceeding 30. We also verify the generality of our attack and transferability of our generated universal adversarial point clouds.

The remainder of this paper is organized as follows, we overview the related work in Sect. 2. Section 3 details the robustness analysis of our target multi-modal fusion model against existing general spoofing attack on LiDAR-based models. Then we introduce the attack approach of generating adversarial point clouds in Sect. 4. Experimental setups and results are presented in Sect. 5. Finally we conclude the paper in Sect. 6.

## 2 Related Work

### 2.1 Multi-modal Fusion

Multi-modal fusion, or multi-sensor fusion, has been widely studied in object detection field, especially for autonomous driving. Due to the complementary and shared information, the fusion of multiple sensors is considered to achieve higher accuracy and stronger robustness in detection tasks than single sensor. AVs are safety-critical applications and the reliability of object detectors is important, therefore, they frequently utilize the fusion detection models of LiDAR and camera, with the advantages of depth information provided by point clouds and texture information captured by images. According to the different levels of fusion, multi-modal fusion can be divided into two major streams: deep fusion (feature level) and late fusion (decision level). The former is fusing the features extracted from point cloud data and image data through a certain strategy (e.g. addition, mean, concatenation), and then feed it into detection network to obtain the final detection results, e.g. MV3D [10], AVOD [18], EPNet [16]. The latter is fusing the results detected by LiDAR perception network and camera perception network respectively through a certain rule (e.g. geometric association, semantic consistency), e.g. CLOCs [23], models designed in Apollo [2] and Autoware [1]. In this work, we focus on the feature-level fusion, which is under more extensive research for its finer-grained fusion mode, from another perspective, it is also easier to grasp the defects in features to attack models.

## 2.2  Adversarial Point Clouds

Recently, with more and more mature 3D classification and detection models [19,25–27,38] proposed, the attention has been paid on the generation of adversarial examples in 3D space. Point clouds are common representation of 3D objects, existing works [4,20,21,34–36,39,40] mainly focus on generating adversarial point clouds by shifting, adding or dropping points. Xiang et al. [36] first propose an optimization algorithm with C&W framework [9] and Hausdorff/Chamfer measurements [12] to craft adversarial examples by perturbing existing points or generating new points. Liu et al. [20] extend variations of the fast gradient based method (FGSM) [15] to shift points and Zhang et al. [39] develop a variant of one-pixel attack [28] using pointwise gradient method to attach new points to the original point clouds. Besides, [35,40] opt to drop points based on point saliency maps. They all focus on point clouds data level but neglect the feasibility of reliably manufacturing adversarial examples in real world. Hence, Tsai et al. [30] extend adversarial point clouds to 3D printed physical adversarial objects.

## 2.3  Attacks on 3D Object Detection

In terms of attack methods, according to the knowledge of target model, adversarial attacks are mainly categorized into white-box and black-box. In white-box attacks, attackers have full access to the structure, parameters and other information of models while in black-box attacks, attackers can only access to the inputs and outputs, which have stronger attack ability than white-box attacks. In terms of attack effects, adversarial attacks on 3D object detection can be categorized into spoofing attack and vanishing attack, the former is to generate a fake near-front obstacle and the latter is to hide real obstacles in the scene.

**LiDAR-Based.** Cao et al. [7] are the first to implement a white-box spoofing attack on LiDAR-based perception in Apollo [2] by strategically injecting a small number of points, making the victim believe there is a vehicle ahead. Sun et al. [29] discover that current LiDAR-based detection models do not learn occlusion patterns in point clouds, allowing fake points that almost two magnitudes fewer than valid vehicle points to fool the detectors, by moving the distant or occluded vehicle point clouds with few points to the front of victim, spoofing attack can be easily implemented. Besides, [8] designs an adversarial mesh placed on the road that can evade from LiDAR-based perception system and [32] generates universal adversarial objects placed on the roof of target vehicles to make them invisible.

**MMF-Based.** There are limited studies on adversarial attacks on MMF models. Kim et al. [17] theoretically demonstrate the sensitivity of the deep fusion model to single sensor noise. Wang et al. [33] implement projected gradient descent (PGD) [22] attacks respectively on camera and LiDAR channel of simple fusion models they construct, but it is unclear whether the attack is effective on the mature and well-performed application-level fusion models. Then same as LiDAR-based attack scenario [32] where adversarial object is placed on the roof of victim, [4,31] attack image and LiDAR input modalities simultaneously by rendering a adversarial mesh with specific shape and texture to hide an existing vehicle and produce false detection. In the latest work, to make attack more practical and universal, Fang et al. [13] are committed to find a single physical-world attack vector that affects both images and point clouds, they generate adversarial 3D-printed objects based on common objects on the road, misleading AVs to fail in detecting them.

Exsiting adversarial attacks on fusion models mainly focus on attacking multiple inputs simultaneously, they are all white-box attacks that need to know the specific information inside models and have some complicated steps like rooftop approximation and sensor simulation, making their methods hard to follow. While our work is to attack the entire model by only attacking one input channel, seizing the fatal flaw of a more important sensor LiDAR in AVs to achieve a simple but effective black-box attack. In addition, the physical realization of spoofing attack explored in our work is injecting fake signals into LiDARs, which is more stealthy than the vanishing attack explored in their work of placing a real adversarial object somewhere.

## 3    Robustness Analysis

To understand the security of MMF models against general spoofing attack on LiDAR-based models, we first reproduce the attack method proposed by [29], and explore the effectiveness of directly applying it to attack the typical MMF model AVOD [18]. Firstly, We randomly extract 200 vehicle point cloud samples from KITTI [14] dataset with points within two hundred, containing 100 distant vehicle point clouds and 100 occluded vehicle point clouds. Next, we apply global transformation matrix to transform the coordinates of extracted point clouds to a near-front location, about 4–8 m in front of the victim. Then we add the transformed fake point clouds into scenes and use the merged data as the input of LiDAR channel. Finally, we run AVOD and record the detection results.

We discover that more than 95% of the fake point clouds cannot be detected or the confidence scores are lower than 0.01 (Fig. 1), and the remaining is around 0.1. Only when the confidence score exceeds 0.8, the fake point cloud is believed to be a real object by the detector. Therefore, the general adversarial attack on LiDAR-based detection models is not applicable to AVOD. A potential reason is that MMF model architecture is more robust than single-modal model architecture, the fake detections generated by cheating a single sensor can be corrected by other sensors. There are two input channels of LiDAR and camera in AVOD [18], when LiDAR perceives an object ahead but camera does not detect it, the confidence score of object will be greatly reduced.

**Fig. 1.** Results of reproduced general spoofing attack on AVOD [18] (1st row: before attack, 2nd: after attack). We map the 2d bounding boxes on images and set the confidence score threshold as 0.001 to show more objects with low confidence scores. Left value on each green box indicates the confidence score and the right is the IoU with ground truth bounding box. Specifically, the red arrows point to the fake objects and their confidence scores are lower than 0.01.

But the conclusion drew from another work is quite different from ours, Park et al. [24] exchange the image data of two scenes while keeping the point cloud data unchanged and then feed them into AVOD for detection. The results show that there is not much difference between the detections before and after the exchange, the bounding boxes of most objects are basically consistent and confidence scores are still high. To a certain extent, it illustrates that although AVOD has a symmetrical architecture, the final detections are more dependent on LiDAR.

Faced with contradictory conclusions, we observe that the confidence scores of distant or occluded objects drop a lot after exchange, even cannot be detected. We speculate that for objects with a large number of points, AVOD [18] relies more on LiDAR, while for objects with a small number of points, the weight of camera increases. We conduct two more specific experiments to prove the speculation. Under the same setting that adding point clouds extracted from one scene to the same position in another scene, the first experiment is adding the high-confidence point clouds with hundreds of points and they still output with high confidence scores (greater than 0.9), as shown in Fig. 2(a). The second one is adding the high-confidence point clouds with points within two hundred but they output with low confidence scores (less than 0.01), as shown in Fig. 2(b).

(a) With a large number of points       (b) With a small number of points

**Fig. 2.** Results of LiDAR weight comparison. Images of the upper and lower rows are original scenes and attacked scenes. (a) adds the vehicle point clouds with 569 points extracted from scene 134 to scene 454 and outputs with 0.99. (b) adds the vehicle point clouds with 65 points extracted from scene 2 to scene 4 and output with 0.01.

From the above analysis, we can conclude that blindly applying the general spoofing attack on LiDAR-based detectors to AVOD [18] is invalid, the detector with multi-modal architecture is more robust against adversarial attack. However, we discover that AVOD can output high-confidence point cloud objects in the case of (1) inputting real image data and fake point cloud data with a large number of points or (2) inputting real image data and real point cloud data with a small number of points. Between the two cases, theoretically, there are objects with few points can be detected with high confidence, overcoming the correction of images. What we do is to craft such fake point clouds to perform an adversarial attack against MMF detection models.

## 4   Generating Adversarial Point Clouds

In this section, we present a novel method for generating adversarial point clouds to attack MMF detectors, illustrated in Fig. 3. First we demonstrate the problem definition, then we introduce the perturbation rule of point clouds and the objective function to be optimized, finally we elaborate on the overall attack algorithm and optimization algorithm.

### 4.1   Problem Definition

The goal of our work is to generate adversarial point clouds with as few points as possible to fool the MMF models into detecting a fake obstacle with high confidence score in front of the victim. In the context of autonomous driving, as the distance of fake obstacle more closer, the consequence of emergency braking

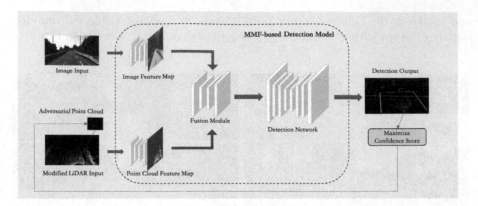

**Fig. 3.** Overview of adversarial point clouds generation pipeline. We add adversarial point cloud to LiDAR input while keeping the image input unchanged. They are processed by MMF model and output detection results. We constantly perturb the points in the direction of maxmizing the confidence score of fake object in detections to obtain the optimal adversarial point cloud eventually.

is more dangerous, so the fake obstacle is supposed to be appear near front. As the first work of attacking the LiDAR channel alone to attack the entire fusion model, we expect to explore a feasible attack method theoretically and do not consider the constraints of injecting fake point clouds into LiDAR sensors by physical equipment (e.g. photodiode, delay component, infrared laser) in real world. Besides, we target car as the category of adversarial point clouds instead of pedestrian or cyclist for the reason that cars are more common in scenes.

### 4.2 Input Perturbation

Given an image $I$ and a point cloud $P$, we keep $I$ unchanged and perturb $P$. Let $P = \{p_1,\ p_2,\ \cdots,\ p_n\}$ be an input set of N points where each point $p_i$ is represented by a vector of coordinates $p_i = [p_{i,x},\ p_{i,y},\ p_{i,z}] \in R^3$, let $\delta_i = [\delta_{i,x},\ \delta_{i,y},\ \delta_{i,z}] \in R^3$ be the perturbation vector for $p_i$. We aim to shift points to generate an adversarial point cloud $P' = \{p_1 + \delta_1,\ p_1 + \delta_1,\ \cdots,\ p_n + \delta_n\}$. In this work, we extract the distant and occluded vehicle point clouds from the dataset and transform them to the position in front of the victim as our initial fake point clouds, at the same time we get the initial fake bound boxes. For each initial fake vehicle point cloud $P = \{p_i \mid i = 1,\ \cdots,\ n\}$, its bounding box is represented by $b = [t_x,\ t_y,\ t_z,\ l,\ w,\ h]$ ($t_x,\ t_y,\ t_z$ are the bottom center coordinates and $l$, $w$, $h$ are length, width, height). We limit the perturbation of each point to the bounding box as Eq. 1:

$$p_i^{'} = [p_{i,x} + \delta_{i,x}, \ p_{i,y} + \delta_{i,y}, \ p_{i,z} + \delta_{i,z}]$$

$$s.t. \ \delta_{i,x} \in \left[ t_x - \frac{l}{2} - p_{i,x}, \ t_x + \frac{l}{2} - p_{i,x} \right]$$

$$\delta_{i,y} \in \left[ t_y - \frac{w}{2} - p_{i,y}, \ t_y + \frac{w}{2} - p_{i,y} \right]$$

$$\delta_{i,z} \in [t_z - p_{i,x}, \ t_z + h - p_{i,z}] \tag{1}$$

### 4.3 Objective Function

To make the fake vehicle appear at a specified position in front of the victim, there must be a corresponding bounding box of that vehicle output by detector and its confidence score should be above the detection threshold. Inspired from prior vanishing attack work [32] which suppresses all relevant bounding box proposals, we come up with the opposite idea that increasing the confidence score of the most probable bounding box which has the highest IoU with the initial fake bounding box. Hence, our objective is to maximize the confidence of the most relevant candidate:

$$\mathcal{L}_{adv} = (1 - IoU(b^*, \ b^{'})) \log(C_{b^{'}})$$

$$s.t. \ b^{'} = \underset{b \in B}{argmax} \, IoU(b, \ b^*) \tag{2}$$

Where $B$ is the set of all bounding boxes output by the detection model and each bounding box $b$ has a confidence score $C$. IoU denotes the Intersection over union operator, $b^*$ denotes the bounding box of initial adversarial point cloud and $b^{'}$ denotes the bounding box of the most relevant candidate.

### 4.4 Attack Method

**Attack Algorithm.** Based on the above, we propose an attack algorithm to generate adversarial point clouds on MMF detector. As detailed in Algorithm 1, given a scene, we transform an extracted distant or occluded point cloud to a specified position in front of the victim as the initial fake object, the attack iteratively searches for the perturbation of points to achieve a higher confidence score. In each iteration, the raw image and the raw scene point cloud merged with fake point cloud are as inputs and the model output the bounding boxes and confidence scores of detected objects. We then identify the most relevant candidate that has the largest IoU with initial fake bounding box. By disturbing the points in picked bounding box constantly, we generate the adversarial point cloud with a larger IoU with the initial bounding box and a higher confidence score. The attack succeeds if the confidence score exceeds threshold, otherwise enters the next iteration until the maximum number of iterations.

---

**Algorithm 1:** Generating adversarial point clouds.

---

**Input**: Raw scene point cloud $S$, raw image $I$, object detector $M$, initial fake
   point cloud $P$, initial fake bounding box $b^*$, confidence threshold $r$.
**Output**: adversarial point cloud $P'$, confidence $C_{b'}$.

1 **begin**
2     $P' \leftarrow P$
3     $iter = 0$
4     **while** $iter < maxIter$ && $C_{b'} < r$ **do**
5        $B \leftarrow M(S + P', I)$
6        $b' \leftarrow \underset{b \in B}{argmax}\, IoU(b,\ b^*)$
7        $P' \leftarrow max\,(1 - IoU(b^*,\ b'))\log(C_{b'})$
8        $iter+ = 1$
9     **end**
10     **return** $P'$, $C_{b'}$
11 **end**

---

**Optimization Algorithm.** We employ the genetic algorithm [5] to optimize
the objective function. In genetic algorithm, each individual in population represents a solution in the search space and fitness score measures how good it
is. Through the implementation of coding, selection, crossover, mutation and
other operations on the individuals for several iterations, the individual with the
highest fitness score is considered as the optimal solution to the problem. In our
case, a population of initial fake point clouds are evolved to maximize the fitness
score $\mathcal{L}_{adv}$. In each iteration, we select a new generation of population based
on roulette wheel selection [3], then preserve the candidate with the highest fitness and cross other candidates according to the crossover rate to generate new
candidates, we then add gaussian noise to points of new candidate point clouds
sampled with a mutation rate. The algorithm ends when the optimal adversarial
point cloud is found or the maximum number of iterations is reached.

## 5    Experiments

In this section, we first describe our experiment setup, including the dataset and
target fusion model, the metrics used to evaluate our attacks, and the implementation details. Then we present results and discussions on (1) attack effectiveness,
(2) universal adversarial examples, (3) attack generality, and (4)transfer attack.

### 5.1    Experiment Setup

**Dataset.** We evaluate our attack on KITTI [14] dataset, a benchmark in
autonomous driving scenarios, which contains point clouds, images, calibration
files, labels for 3D object detection. Refer to [18], we divide the trainval set into a

training set of 3712 samples for training model and a validation set of 3769 samples for evaluation. Considering the fake objects are placed in front of victims, we further remove scenes that may have conflicts with the original real near-front objects and use the selected dataset for adversarial examples generation and attack evaluation.

**Target Model.** We choose AVOD [18] as our target model, a typical two-stage object detection architecture with a region proposal network (RPN) and a second stage detection network. It uses feature extractors to generate feature maps from point clouds and images which are then shared by two subnetworks. First the feature maps are fed into RPN and fused via an element-wise mean operation after cropping and resizing, and then generate top k proposals through fully connected layers. Second projecting the proposals on feature maps and adopting similar fusion operations in RPN stage to generate final detections, including box regression, orientation estimation and category classification.

**Metrics.** We aim to generate the fake point cloud misleading the detector to believe that there is a vehicle ahead. We calculate the average confidence score (ACS) of adversarial point clouds with different numbers of points and distances to access attack effectiveness. The larger the ACS, the better the attack.

$$ACS = \frac{\sum confidence\ score}{\#\ of\ scenes}$$

Besides, object detectors often set default threshold of confidence score to filter out detected objects with low confidence. Our model considers a fake vehicle is detected if there exists a bounding box output by the detector overlaps with the initial bounding box and the confidence score is greater than 0.8. We use attack success rate (ASR) to measure the percentage at which the fake vehicle is successfully detected in scenes. The higher the ASR, the better the attack.

$$ASR = \frac{\#\ of\ success\ attack}{\#\ of\ scenes}$$

**Implementation Details.** We follow the implementation of [18] to train the target model for car class and conduct the evaluation of adversarial attacks on V100. For performing attack in a single scenario, we randomly select 200 scenes from validation set to generate adversarial point clouds with different numbers of points and different distances, then evaluate attack effectiveness. For performing attack in multiple scenarios, we generate universal adversarial point clouds on training set and evaluate on the validation set. In genetic algorithm, the population size is set to 100, the crossover operator is set to 0.8 and the mutation operator is set to 0.1, besides, we add a mutation coefficient of 0.01 to perturb the coordinates of point clouds. We repeat experiments to reduce the randomness introduced by genetic algorithm.

## 5.2   Results and Discussion

**Attack Effectiveness.** In 200 scenes, for each scene we generate adversarial point clouds with distances from 4 m to 8 m by the step of 1 m and numbers of points from 10 to 200 by the step of 10. As shown in Fig. 4(a), the ACS increases as the number of points increases under all distances, and when the number of points exceeds 30, the ACS is stable above 0.9 (0.98 at most) and the differences in attack effects get smaller. At the closest distance of 4, the adversarial point clouds with points of 10 and 20 perform better than other distances, reaching ACS of 0.69 and 0.91 respectively. When the number of points is relatively small, the closer the distance, the better the attack effect.

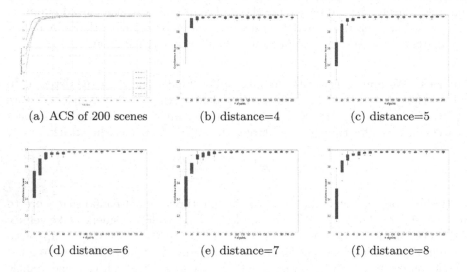

(a) ACS of 200 scenes        (b) distance=4        (c) distance=5

(d) distance=6        (e) distance=7        (f) distance=8

**Fig. 4.** The attack effectiveness of our method. (a) shows the average confidence scores of adversarial point clouds generated with different distances and number of points. (b)–(f) show the attack stability at different distances.

In addition, Fig. 4(b)–(f) shows the attack stability of generated adversarial point clouds at various distances. The data distribution is more concentrated as the length of box is shorter in box plot, which indicates the confidence scores of fake vehicles are closer and the attacks are more stable in our work. We observe that it is consistent with the trend of ACS curve, under all distances, as the number of points increases, the attack stability gets better. At the closest distance of 4, the confidence scores of adversarial point clouds with very few points (e.g. 10, 20, 30) are higher and closer than distance 5–8. Overall, the close-range attack effect and attack ability is better, we speculate the reason is that the model has better detection performance on nearby objects.

**Universal Adversarial Examples.** From the analysis of attack effectiveness, we can discover that the adversarial point clouds generated at distance of 4m perform best with the better attack effect and attack stability than other distances. Hence, at the optimal distance of 4, we generate universal adversarial point clouds with 20 groups of points and evaluate on validation set. Table 1 shows that 18 groups with different numbers of points exceeding 30 have achieved more than 95% ASR and around 0.95 ACS (caculated by successfully attacked scenes), while the adversarial point clouds with 10 points and 20 points only achieve ASR of 1.36% and 68.09%. Especially for 10 points, though it reaches 0.69 ACS (caculated by 200 scenes) in the first experiment, the confidence scores of most adversarial point clouds are less than 0.8, resulting in low ASR. Besides, with the number of points starting from 30, the ASR is basically unchanged, fluctuating between 96% and 98%, we speculate that because the adversarial point cloud with 30 points learned by our attack method is enough to express the strong features that can be detected by fusion model. A high attack success rate can be attained by fake point clouds with few points, which also proves the power of our attack method.

**Table 1.** The attack success rates and average confidence scores (caculated by successfully attacked scenes) of universal adversarial point clouds generated on AVOD [18].

| # of points | 10 | 20 | 30 | 40 | 50 | 60 | 70 | 80 | 90 | 100 |
|---|---|---|---|---|---|---|---|---|---|---|
| ASR | 1.36% | 68.09% | 95.64% | 96.78% | 98.20% | 95.72% | 97.09% | 97.00% | 96.03% | 96.29% |
| ACS | 0.93 | 0.94 | 0.95 | 0.96 | 0.98 | 0.93 | 0.97 | 0.94 | 0.94 | 0.93 |
| # of points | 110 | 120 | 130 | 140 | 150 | 160 | 170 | 180 | 190 | 200 |
| ASR | 97.47% | 96.56% | 97.79% | 96.65% | 96.65% | 96.56% | 97.47% | 96.84% | 96.75% | 97.98% |
| ACS | 0.94 | 0.96 | 0.97 | 0.95 | 0.94 | 0.95 | 0.95 | 0.94 | 0.96 | 0.95 |

Figure 5(a) shows a comparison of the detection results before and after our adversarial attack, the upper shows 2d and 3d detection bounding boxes on images and the lower shows 3d detection bounding boxes in point clouds. We can find a fake vehicle appears in front of the victim with the confidence score of 0.95, indicating the attack succeeds. Several universal adversarial point clouds with different numbers of points are presented in Fig. 5(b).

**Attack Generality.** To make our study more general, we further implement our attack on EPNet [16], a state-of-art MMF-based 3D object detection model which has superior detection performance and ranks high on KITTI [14] benchmark. First of all, we conduct robustness exploration experiments on EPNet and verify conclusions drew in Sect. 3 are not only applicable to AVOD but also to EPNet, which indicates it is feasible to craft fake point clouds with few points to attack it. Considering our attack method is black-box which only needs to combine our algorithm with the input and output of the models, it is easy to apply our method to various models. We carry out single-scenario adversarial attack at the farthest distance of 8, the attack effect is a lower bound compared

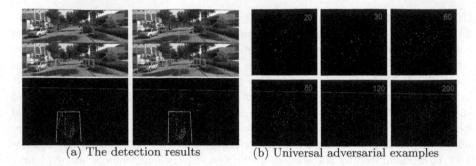

(a) The detection results        (b) Universal adversarial examples

**Fig. 5.** (a) The visualization of a successful attack in scene 1140, (b) several universal adversarial point clouds with 20, 30, 60, 80, 120 and 200 points.

to closer distances. The results show our method has a better attack effect on EPNet, with the number of points exceeding 30, the ACS of generated adversarial point clouds is stable above 0.95, but with the point of 10 and 20, the ACS is greater than 0.7 and 0.9 respectively, which can only be obtained on AVOD at the optimal distance of 4 (shown in Fig. 4(a)). From another perspective, a fusion model with better detection performance is more vulnerable to adversarial attack, probably because the model strengthens the ability of learning features from objects with few points in order to improve the detection accuracy, which makes it easier for adversaries to catch flaws in features to launch attacks. Overall, the defect of using few points to perform spoofing attack widely exsits in MMF models and our method is general to generate adversarial point clouds with good attack effect.

**Transfer Attack.** The adversarial examples generated for the same task are transferable between different models. We utilize the universal adversarial point clouds of 20 groups generated on AVOD [18] to attack EPNet [16]. Compared to AVOD, the ASR of EPNet decreases about half, mainly fluctuating between 40% and 60%, the highest is 58.83% with 20 points and the lowest is 18.26% with 70 points. While both of them are trained in the 3D object detection task, there are differences between their decision-making faces, the adversarial point clouds generated on AVOD cannot all achieve good attack performance on EPNet. We can discover that adversarial example with 10 points has better attack effect on EPNet with ASR of 18.83% probably due to a better detection performance on small objects with few points, which is consistent with our analysis in attack generality. Overall, the transferability of our adversarial examples performs well and can achieve an average attack success rate around 50%.

## 6   Conclusion

In this paper, we perform the first study to explore the adversarial attack against LiDAR channel on a popular MMF model AVOD [18] and achieve the goal of fooling the fusion model into detecting fake object ahead, which can cause rear-end collision in AVs. We first reproduce the general spoofing attack on LiDAR-based models and find that it cannot directly apply to MMF model due to the correction by other input channel. We then propose a black box attack method based on genetic algorithm to generate adversarial point clouds with few points, avoiding to deal with the issue of non-differentiable stage in models in gradient-based white-box attacks. We evaluate our attack effectiveness with different combinations of points and distances on the KITTI [14] dataset and generate universal adversarial point clouds, achieving attack success rate more than 95% when the number of points exceed 30. We further explore the attack generality and transferability of our generated adversarial examples on other representative fusion model. In the future work, we need to consider more about the physical realization in real world, and the perturbation of point clouds will be limited more strictly.

**Acknowledgement.** This work is supported by the Strategic Priority Research Program of Chinese Academy of Sciences, Grant No. XDC02010300.

# References

1. Autoware. https://www.autoware.org/
2. Baidu apollo, http://apollo.auto
3. Roulette wheel selection algorithm. https://en.wikipedia.org/wiki/Fitness_proport ionate_selection
4. Abdelfattah, M., Yuan, K., Wang, Z.J., Ward, R.: Adversarial attacks on camera-lidar models for 3d car detection. arXiv preprint arXiv:2103.09448 (2021)
5. Alzantot, M., Sharma, Y., Chakraborty, S., Zhang, H., Hsieh, C.J., Srivastava, M.B.: Genattack: practical black-box attacks with gradient-free optimization. In: Proceedings of the Genetic and Evolutionary Computation Conference, pp. 1111–1119 (2019)
6. Cao, Y., et al.: 3d adversarial object against msf-based perception in autonomous driving. In: Proceedings of the 3rd Conference on Machine Learning and Systems (2020)
7. Cao, Y., et al.: Adversarial sensor attack on lidar-based perception in autonomous driving. In: Proceedings of the 2019 ACM SIGSAC Conference on Computer and Communications Security, pp. 2267–2281 (2019)
8. Cao, Y., Xiao, C., Yang, D., Fang, J., Yang, R., Liu, M., Li, B.: Adversarial objects against lidar-based autonomous driving systems. arXiv preprint arXiv:1907.05418 (2019)
9. Carlini, N., Wagner, D.: Towards evaluating the robustness of neural networks. In: 2017 IEEE Symposium on Security and Privacy (SP), pp. 39–57. IEEE (2017)
10. Chen, X., Ma, H., Wan, J., Li, B., Xia, T.: Multi-view 3d object detection network for autonomous driving. In: Proceedings of the IEEE Conference on Computer Vision and Pattern Recognition, pp. 1907–1915 (2017)
11. Eykholt, K., et al.: Robust physical-world attacks on deep learning visual classification. In: Proceedings of the IEEE Conference on Computer Vision and Pattern Recognition, pp. 1625–1634 (2018)

12. Fan, H., Su, H., Guibas, L.J.: A point set generation network for 3d object reconstruction from a single image. In: Proceedings of the IEEE Conference on Computer Vision and Pattern Recognition, pp. 605–613 (2017)
13. Fang, J., Yang, R., Chen, Q.A., Liu, M., Li, B., et al.: Invisible for both camera and lidar: Security of multi-sensor fusion based perception in autonomous driving under physical-world attacks. arXiv preprint arXiv:2106.09249 (2021)
14. Geiger, A., Lenz, P., Urtasun, R.: Are we ready for autonomous driving? the kitti vision benchmark suite. In: 2012 IEEE Conference on Computer Vision and Pattern Recognition, pp. 3354–3361. IEEE (2012)
15. Goodfellow, I.J., Shlens, J., Szegedy, C.: Explaining and harnessing adversarial examples. arXiv preprint arXiv:1412.6572 (2014)
16. Huang, T., Liu, Z., Chen, X., Bai, X.: EPNet: enhancing point features with image semantics for 3D object detection. In: Vedaldi, A., Bischof, H., Brox, T., Frahm, J.-M. (eds.) ECCV 2020. LNCS, vol. 12360, pp. 35–52. Springer, Cham (2020). https://doi.org/10.1007/978-3-030-58555-6_3
17. Kim, T., Ghosh, J.: On single source robustness in deep fusion models. arXiv preprint arXiv:1906.04691 (2019)
18. Ku, J., Mozifian, M., Lee, J., Harakeh, A., Waslander, S.L.: Joint 3D proposal generation and object detection from view aggregation. In: 2018 IEEE/RSJ International Conference on Intelligent Robots and Systems (IROS), pp. 1–8. IEEE (2018)
19. Lang, A.H., Vora, S., Caesar, H., Zhou, L., Yang, J., Beijbom, O.: Pointpillars: fast encoders for object detection from point clouds. In: Proceedings of the IEEE/CVF Conference on Computer Vision and Pattern Recognition, pp. 12697–12705 (2019)
20. Liu, D., Yu, R., Su, H.: Adversarial point perturbations on 3d objects. arXiv preprint arXiv:1908.06062 (2019)
21. Liu, D., Yu, R., Su, H.: Extending adversarial attacks and defenses to deep 3d point cloud classifiers. In: 2019 IEEE International Conference on Image Processing (ICIP), pp. 2279–2283. IEEE (2019)
22. Madry, A., Makelov, A., Schmidt, L., Tsipras, D., Vladu, A.: Towards deep learning models resistant to adversarial attacks. arXiv preprint arXiv:1706.06083 (2017)
23. Pang, S., Morris, D., Radha, H.: Clocs: camera-lidar object candidates fusion for 3d object detection. In: 2020 IEEE/RSJ International Conference on Intelligent Robots and Systems (IROS), pp. 10386–10393. IEEE (2020)
24. Park, W.: Crafting adversarial examples on 3d object detection sensor fusion models (2020)
25. Qi, C.R., Su, H., Mo, K., Guibas, L.J.: Pointnet: Deep learning on point sets for 3d classification and segmentation. In: Proceedings of the IEEE Conference on Computer Vision and Pattern Recognition, pp. 652–660 (2017)
26. Qi, C.R., Yi, L., Su, H., Guibas, L.J.: Pointnet++: deep hierarchical feature learning on point sets in a metric space. arXiv preprint arXiv:1706.02413 (2017)
27. Shi, S., Wang, X., Li, H.: Pointrcnn: 3d object proposal generation and detection from point cloud. In: Proceedings of the IEEE/CVF Conference on Computer Vision and Pattern Recognition, pp. 770–779 (2019)
28. Su, J., Vargas, D.V., Sakurai, K.: One pixel attack for fooling deep neural networks. IEEE Trans. Evol. Comput. **23**(5), 828–841 (2019)
29. Sun, J., Cao, Y., Chen, Q.A., Mao, Z.M.: Towards robust lidar-based perception in autonomous driving: general black-box adversarial sensor attack and countermeasures. In: 29th {USENIX} Security Symposium ({USENIX} Security 20), pp. 877–894 (2020)

30. Tsai, T., Yang, K., Ho, T.Y., Jin, Y.: Robust adversarial objects against deep learning models. In: Proceedings of the AAAI Conference on Artificial Intelligence, vol. 34, pp. 954–962 (2020)
31. Tu, J., et al.: Exploring adversarial robustness of multi-sensor perception systems in self driving. arXiv preprint arXiv:2101.06784 (2021)
32. Tu, J., et al.: Physically realizable adversarial examples for lidar object detection. In: Proceedings of the IEEE/CVF Conference on Computer Vision and Pattern Recognition, pp. 13716–13725 (2020)
33. Wang, S., Wu, T., Vorobeychik, Y.: Towards robust sensor fusion in visual perception. arXiv preprint arXiv:2006.13192 (2020)
34. Wen, Y., Lin, J., Chen, K., Jia, K.: Geometry-aware generation of adversarial and cooperative point clouds (2019)
35. Wicker, M., Kwiatkowska, M.: Robustness of 3d deep learning in an adversarial setting. In: Proceedings of the IEEE/CVF Conference on Computer Vision and Pattern Recognition, pp. 11767–11775 (2019)
36. Xiang, C., Qi, C.R., Li, B.: Generating 3d adversarial point clouds. In: Proceedings of the IEEE/CVF Conference on Computer Vision and Pattern Recognition, pp. 9136–9144 (2019)
37. Xie, C., Wang, J., Zhang, Z., Zhou, Y., Xie, L., Yuille, A.: Adversarial examples for semantic segmentation and object detection. In: Proceedings of the IEEE International Conference on Computer Vision, pp. 1369–1378 (2017)
38. Yang, B., Luo, W., Urtasun, R.: Pixor: Real-time 3d object detection from point clouds. In: Proceedings of the IEEE conference on Computer Vision and Pattern Recognition. pp. 7652–7660 (2018)
39. Zhang, Q., Yang, J., Fang, R., Ni, B., Liu, J., Tian, Q.: Adversarial attack and defense on point sets. arXiv preprint arXiv:1902.10899 (2019)
40. Zheng, T., Chen, C., Yuan, J., Li, B., Ren, K.: Pointcloud saliency maps. In: Proceedings of the IEEE/CVF International Conference on Computer Vision, pp. 1598–1606 (2019)

# Source Identification from In-Vehicle CAN-FD Signaling: What Can We Expect?

Yucheng Liu[1] and Xiangxue Li[1,2,3](✉)

[1] School of Software Engineering, East China Normal University, Shanghai, China
xxli@cs.ecnu.edu.cn
[2] Shanghai Key Laboratory of Trustworthy Computing, Shanghai, China
[3] Westone Cryptologic Research Center, Beijing, China

**Abstract.** Controller Area Network (CAN) is significantly deployed in various industrial applications (including current in-vehicle network) due to its high performance and reliability. Controller area network with flexible data rate (CAN-FD) is supposed to be the next generation of in-vehicle network to dispose of CAN limitations of data payload size and bandwidth. The paper explores for the first time Electronic Control Unit (ECU) identification on in-vehicle CAN-FD network from bus signaling and the contributions are four-fold.

- Technically, we discuss the factors that might affect ECU recognition (e.g., CAN-FD controller, CAN-FD transceiver, and voltage regulator) and look into the signal ringing and its intensity where dominant states along with rising edges (from recessive to dominant states) suffice to fingerprint the ECUs. We can thereby design ECU identification scheme on in-vehicle CAN-FD network.
- For a given network topology (in terms of the stub length and the number of ECUs), we execute CAN-FD and CAN separately and one can expect considerable performance for the two kinds of protocols by using any signal characteristics (rising edges, dominant states, falling edges, and recessive states). In particular, the recognition rates by dominant states and rising edges of signals outperform significantly those by any other combinations of signal characteristics.
- As a respond to the possible transition mechanism from CAN to CAN-FD, we also allow a hybrid topology of CAN and CAN-FD, namely, there exist on the network ECUs sending purely CAN frames, ECUs sending purely CAN-FD frames, and ECUs sending both CAN and CAN-FD frames, and our suggestion on dominant states and rising edges shows robustness to source identification as expected. This shows convincing evidence on the universal applicability of our approach to forthcoming real vehicles set up by CAN-FD network.
- The proposed approach can be easily extended to intrusion detection against attacks not only initiated by external devices but also internal devices.

D. Gao et al. (Eds.): ICICS 2021, LNCS 12918, pp. 204–223, 2021.
https://doi.org/10.1007/978-3-030-86890-1_12

We hope our results could be used as a step forward and a guidance on securing the commercialization and batch production of in-vehicle CAN-FD network in the near future.

**Keywords:** Controller Area Network · CAN-FD · ECU identification

# 1  Introduction

Controller area network (CAN) is one of the most commonly used bus communication protocols between in-vehicle Electronic Control Units (ECU, similar to ordinary computer, consists of a microcontroller (MCU), some memory (ROM/RAM), input/output interface (I/O), analog-to-digital converter (A/D), and large-scale integrated circuits such as shaping and driving). It was introduced by Robert Bosch GmBH in 1983. All ECUs inside the vehicles are connected to each other through CAN bus. However, CAN protocol lacks security mechanisms, such as authentication and encryption [5]. Indeed, an adversary might easily eavesdrop on the bus, obtain all communication messages between ECUs at will, and then initiate a replay attack [23]. He can even modify the obtained messages which will be further injected into the CAN network in an attempt to control some safety-critical functions. We do see various attacks against CAN network recently [2,10,14,19,20]. In response to these attacks, researchers propose a series of countermeasures, represented by Intrusion Detection System (IDS) and Message Authentication Code (MAC). The latter is not practical however, as the length of the CAN frame data field is up to 8 bytes. And an alternative method is to use truncated MACs [22,25]. This method needs to constantly update the key, which will take up more computing resources. What's more, frequent key updates may cause malfunctions when the vehicle is moving. Fortunately, some seminal works [5,15,21] can not only detect the presence of malicious frames but also identify their sender ECUs. This is really essential for fast forensic, isolation, security patch, etc.

Robert Bosch GmBH recommends CAN FD (CAN with flexible data) [7] in 2012 to meet the requirements of modern vehicles and dispose of CAN limitations of data payload size and bandwidth. Besides compatibility with CAN, CAN-FD has the following advantages: the maximum length of the CAN-FD data field is 64 bytes; it supports variable rates (namely, a frame can use different transmission rates in different stages) and the maximum rate can reach 5Mbit/s (the maximum rate of CAN is 1Mbit/s); it can refine the load of the existing bus and increase the number of the nodes[1] on the bus.

Unfortunately, CAN-FD itself does not convey security protection either (similar to CAN) and existing attacks on CAN might also be feasible on CAN-FD. Take masquerade attack on CAN network [3] as an example. Initiating a masquerade attack and not being detected by the system, an adversary needs to stop the transmission of targeted ECU and imitate it to inject attack messages.

---

[1] As a slight abuse of terms, we use hereafter *node* and *ECU* indiscriminately.

The attack also works on in-vehicle CAN-FD network. Although an intrusion detection system based on topology verification is proposed [26] to detect attacks by using external intruding devices, it can neither detect masquerade attack nor identify the sender of the attack messages. Our proposed mechanism explores for the first time Electronic Control Unit (ECU) identification on in-vehicle CAN-FD network from bus signaling.

## 2    Background and Related Work

### 2.1    Controller Area Network

CAN uses differential signals to transmit messages. Namely, the two signals on CAN-H and CAN-L have equal amplitudes relative to 2.5 V (common mode voltage) and opposite polarities. Compared with single-ended signals, differential signals are subtracted less electromagnetic interference [15]. When the ECU sends the recessive bit (1), the voltage on CAN-H and CAN-L is about 2.5 V, so the differential voltage generated is 0 V. For the dominant bit (0), the voltages on CAN-H and CAN-L are 3.5 V and 1.5 V, respectively, and the resulting differential voltage is 2 V.

The nodes inside the CAN network communicate with each other via CAN frames. CAN frames are divided into standard frames and extended frames according to whether they contain extended identifiers. The length of the identifier of the CAN standard frame is 11 bits, and 29 bits for extended frame (including 11 bits identifier and 18 bits extended identifier). At present, most vehicles use CAN standard frames. The composition of standard frames is shown in Fig. 1(a).

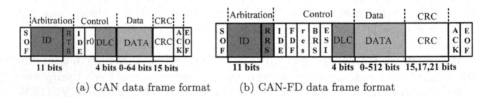

(a) CAN data frame format    (b) CAN-FD data frame format

**Fig. 1.** CAN/CAN-FD data standard frame format with 11 bit identifier.

CAN is a multi-master control bus and the bus conflicts will occur if two or more ECUs request to send data at the same time. CAN bus can detect and arbitrate these conflicts in real time by CSMA/CD [1] (Carrier Sense Multiple Access/Collision Detection) arbitration method, which supports a lossless bitwise arbitration decision process. For example, if one ECU transmits a dominant bit (0) and another ECU transmits a recessive bit (1), then there is a collision and the ECU transmitting the dominant bit gets priority.

## 2.2    Comparing CAN-FD with CAN

CAN-FD and CAN differ in the format and the length of the data frame. Compared with CAN frame, CAN-FD adds FDF (Flexible Data Rate Format), BRS (Bit Rate Switch) and ESI (Error State Indicator) fields (see Fig. 1(b)) [7]. Therein, FDF indicates whether the sent frame is a CAN frame or a CAN-FD frame and BRS stands for bit rate conversion. When the bit is a recessive bit (1), the rate is variable, and when the bit is a dominant bit (0), it is transmitted at a constant rate. ESI is an error status indicator: when ESI is a recessive bit (1), it means that the sending node is in a passive error, otherwise it is in an active error state. In addition, according to the role of different bits, CAN specification divides a frame into different fields, as shown in Fig. 1(b). And in the experiment, we set the rate of 2Mbit/s for the data field, and set the rate of 1Mbit/s in the arbitration field, control field and CRC field. The length of the CAN-FD data field is up to 64 bytes, which increases the available load.

Next we say the data rate. The maximum rate of CAN's arbitration field and data field is no more than 1Mbit/s [6]. However, CAN-FD supports variable rates, and the bit rate of its arbitration field and data field might be different. Among them, the arbitration and the ACK stages continue to use CAN2.0 specification (i.e., the highest rate does not exceed 1Mbit/s), and the data field can reach 5Mbit/s through hardware setting, or even higher.

CAN-FD is defined to be compatible with CAN at the physical layer. All CAN-FD controllers can handle a mix of CAN frames and CAN-FD frames. One might use CAN-FD controllers in conjunction with CAN controllers on in-vehicle network. Thus one might see pure CAN frames or both CAN and CAN-FD frames on the bus.

## 2.3    Related Work

Generally, we have intrusion detection systems (IDS)[2] and cryptographic solutions to strengthen in-vehicle CAN network security. Murvay and Groza [21] pioneered the methodology of studying the differences in CAN signals (sent by ECUs), which are significant for ECU identification. However, they only used the signals corresponding to the CAN frame's identifier field and did not account for the blended signals caused by the collisions between ECUs' simultaneous messages. The limitation was tackled in [5] where 18-bit identifier extension was used to fingerprint ECUs. As vehicles commonly conform to the standard specifications (e.g., ISO, SAE etc.), this scheme was howbeit impractical in real-world applications. Kneib and Huth [15] proposed Scission for in-depth analysis of CAN signals. In particular, Scission can not only detect intrusion messages, but also recognize which ECU sends the intrusion messages. For cryptographic solutions, Lin et al. [14] constructed message authentication code by sending additional messages, and the authors of [22] proposed to use truncated MACs.

For CAN-FD, security experts can pursue stronger security tricks via its higher transmission rates and larger loads. In [26], authors proposed an IDS for

---

[2] The paper focuses on signaling based IDS.

in-vehicle CAN-FD network based on topology verification. Their method uses variations of the network topology to identify intrusions initiated by external intruding devices (XIDs), but the method cannot detect attacks initiated by attackers using the vulnerability of existing ECUs in the vehicle. Woo et al. [24] proposed a security architecture for in-vehicle CAN-FD according to ISO 26262 standard. However, this method may cause GECU (gate ECU) to generate excessive load as it has to encrypt the data packets by using the targeted ECU's unique key. To relieve pressure on GECU, Agrawal et al. [1] proposed a group-based approach for the communication among different ECUs. However, their method still requires the management of a large number of keys which takes up a large amount of computing resources of the ECUs, making it beyond instant communication.

## 3    Signaling and Ringing

### 3.1    ECUs' Voltage Output Behavior

The output voltage of an ECU's regulator varies independently and differently from other ECUs' regulators, as their supply characteristics are different (e.g., different regulators' common-mode rejection ratios) [4]. Given the same power supply (i.e., a 12 V/24 V battery powering all the ECUs), one can get different output voltage of ECU regulators. On the other hand, due to the differences in the internal resistance of the CAN transceiver, the dominant voltages of CAN-H and CAN-L will be different when the dominant bit is sent. When transmitting the recessive bit, both the high and low side transistors are switched off (inside the CAN transceiver) and thus the voltages on CAN-H and CAN-L are basically the same. So the dominant voltage can be used for fingerprint ECU. Similarly, for CAN-FD, the internal components of an ECU mainly include CAN-FD controller, CAN-FD transceiver, and voltage regulator and we have the same rationale of the dominant voltages of (CAN-FD)-H and (CAN-FD)-L on the bus.

### 3.2    Ringing and Its Intensity

The impedance mismatch occurs at two points on CAN-FD bus [8,11,13], e.g., one at the junction and another at the front of the non-terminal ECUs. The non-terminal ECU causes positive reflection since its impedance can be up to several tens of $k\Omega$, significantly larger than the stub line's characteristic impedance. Conversely, the junction's impedance is lower than the stub line's characteristic impedance, resulting in negative reflection.

**From Dominant to Recessive States.** Let $n$ denote the number of ECUs connected to the junction through stub lines and ECU1 a transmitter whose signal voltage need reduce by $\Delta V$ to transfer from dominant state to recessive state. Since the dominant state's differential voltage value is approximately 2V,

$\Delta V$ has a negative polarity. As seen from Fig. 2, a total of $(n+2)$ lines are connected to the junction (i.e.,. $n$ connected stub lines and the two main bus lines). The signal transmitted from ECU1 to the junction follows $(n + 1)$ lines in parallel. Thus, the stub lines have the same impedance $\frac{Z_R}{n+1}$, where the $Z_R$'s nominal value is $120\Omega$. The reflectance ($\Gamma_d$) and transmittance ($T_d$) at the junction are calculated as:

$$\Gamma_d = \frac{\frac{Z_R}{n+1} - Z_R}{\frac{Z_R}{n+1} + Z_R} = -\frac{n}{n+2}, \ T_d = 1 + \Gamma_d = \frac{2}{n+2}. \tag{1}$$

Since $\Gamma_d$ has a negative polarity, a larger portion of the incident signal is reflected as $n$ increases, and its small part is delivered into other ECUs.

Denote $Z_{diff}$ as ECU1's differential input impedance. Now, we have ECU1's front reflectance and transmittance (i.e.,. $\Gamma_s$ and $T_s$):

$$\Gamma_s = \frac{Z_{diff} - Z_R}{Z_{diff} + Z_R}, \qquad T_s = 1 + \Gamma_s = \frac{2Z_{diff}}{Z_{diff} + Z_R}. \tag{2}$$

When the signal is at the recessive state, $Z_{diff}$ is much larger than $Z_R$. Consequently, $\Gamma_s$ has a positive polarity, and equals approximately one. Thus, ECU1's front end reflection direction is the same as the incident signal's direction, and the incident signal and reflected signals' superposition is about twice the original incident signal.

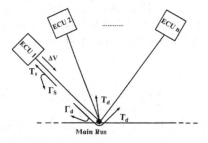

**Fig. 2.** Reflection and transmission coefficients at junction and non-terminal ECUs.

For a dominant-to-recessive transition, the negative transition signal $\Delta V$ is transmitted from ECU1 to the junction, undergoing partial transmission and reflection. The signals are transmitted to other ECUs through the junction and are partially reflected on the other ECUs' front end without changing the direction. At the ECU1's front, the signal returned from the connection is partially transmitted to ECU1. These reflections and transmissions are repeated, resulting in ringing.

**From Recessive to Dominant States.** In the transitions from recessive state to dominant state, ECU1's output impedance is very low. In the recessive state, the electrical energy is released on the network. However, when the signal transfers from recessive to dominant states, ECU1's differential output impedance becomes lowers and starts charging the network. ECU1 generates the signal of 2V, whose polarity is inverted at the junction and reflected onto ECU1. Unlike the dominant-to-recessive transition, the reflection signal is partly received at ECU1 due to the low impedance of ECU1. Since there are no reflections' repetitions, we have small ringing at the recessive-to-dominant state transition.

## 4    System Model

CAN-FD is designed to transmit large amounts of data at a faster rate and to replace CAN in future design. It has the potential to advance the current state of self-driving automobiles and add additional safety and comfort features in non-automobiles vehicles. As a respond to the possible transition mechanism from CAN to CAN-FD, we allow a hybrid topology of CAN and CAN-FD, namely, there exist on the network ECUs sending purely CAN frames, ECUs sending purely CAN-FD frames, and ECUs sending both CAN and CAN-FD frames.

As shown in Fig. 3, the network consists of two or more CAN nodes, two termination resistors, and bus lines connecting them. A twisted-wire-pair is commonly used for the bus line and its characteristic impedance is defined as **R**. The longest bus line (main bus) is terminated with the termination resistors **R** at both ends for impedance match. CAN nodes are connected to main bus through stub lines. In Fig. 3(a), the ECUs connected to the CAN-FD bus can send both CAN-FD and CAN frames. In Fig. 3(b), blue nodes represent the ECUs that can send both CAN-FD frames and CAN frames, and yellow nodes only send CAN frames. In Fig. 3(c), the ECUs connected to the CAN bus only send CAN frames.

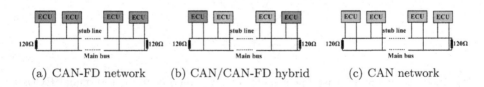

(a) CAN-FD network        (b) CAN/CAN-FD hybrid        (c) CAN network

**Fig. 3.** Network topology.

### 4.1    Threat Models

Without security protection mechanism, the in-vehicle network is vulnerable to various attacks. For example, the bus-off attack [2] can disconnect ECUs from the bus, and the masquerade attack [3] can imitate normal ECUs to inject attack messages. Since one can not determine the sender of any messages, the attacker might use related identifier to impersonate some ECU. This will seriously threaten passengers' safety. We consider two attack modes on in-vehicle CAN and CAN-FD network.

In the attack mode–known ECU, an attacker exploits the vulnerability of existing ECUs inside the vehicle. We mention that modern vehicles generally support wireless connections, such as WiFi, Bluetooth or cellular. Via these interfaces, the attackers might compromise ECUs to achieve various attacks [19, 20]. This type of attacks seems easy to implement and widespread in life (and detailed guidance could even be found freely from some online sites), and our system can detect such attacks accurately and efficiently.

In the second attack mode–unknown ECU, an attacker plugs some extra external device into the bus to send malicious messages. E.g., the device may directly access the bus through the On-Board Diagnostics (OBD)-II port[3].

### 4.2   Signal Acquisition and Preprocessing

To obtain the differential signal from CAN-FD/CAN bus prototypes, we first link the two probes of an oscilloscope to the (CAN-FD)-H/CAN-H and (CAN-FD)-L/CAN-L lines respectively. Then we use the *difference* function in the software of the oscilloscope to calculate the differential signal (CAN-FD)-H) - (CAN-FD)-L) (or (CAN-H)-(CAN-L)).

Several preprocessing steps are applied to each CAN-FD/CAN signal captured by the oscilloscope. First, all dominant states are extracted from the signals. We set a voltage threshold (=0.9V) and voltage greater than the threshold marks the start of the dominant state. The dominant states are then classified into five sets (denoted as $L_1$, $L_2$, $L_3$, $L_4$, and $L_5$) based on the number of contained bits. Let $L_i$ represent all dominant states containing exactly $i$ bits (see Fig. 4). Note that CAN-FD/CAN standards specify that a recessive bit is automatically inserted whenever five consecutive dominant bits appear in a CAN-FD/CAN signal. Thus, no dominant state can contain more than five consecutive dominant bits.

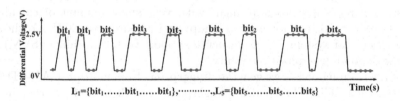

**Fig. 4.** A CAN-FD/CAN frame is divided into 5 sets.

### 4.3   Feature Extraction

We extract the statistical features from the preprocessed electrical CAN-FD/CAN signal. Due to limited computing resources of ECU, we are more

---

[3] The OBD-II port is near the dashboard interface, and the staff can understand the status of the vehicle in real time through the port.

interested in time domain features of the signal and avoid complex frequency domain conversion. Prior work also discerns the versatility of these features in ECU identification [5]. We extract 8 features for each set (see Table 1) and a total of 40 features for each electrical CAN-FD/CAN signal. As too many features might cause over fitting and computational cost in practice, we use the Relief-F [12] algorithm to weight these features. We thus get a general feature set (see Table 2). In the table, **order** column represents the order of the input features, and **feature** column represents the features selected by the algorithm, e.g., $\text{rms}(L_5^{40})$ means that rms from the set of dominant states of length 5 is selected as the first feature of the input (where 40 represents the order of this feature among all features).

**Table 1.** Vector $x$ is time domain representation of the data and $N$ its dimension.

| Feature | Description |
|---|---|
| Maximum | $Max = Max(x(i))_{i=1....N}$ |
| Minimum | $Min = Min(x(i))_{i=1....N}$ |
| Mean | $\mu = \frac{1}{N}\sum_{i=1}^{N} x(i)$ |
| Range | $R = Max - Min$ |
| Average Deviation | $adv = \frac{1}{N}\sum_{i=1}^{N} |x(i) - \mu|$ |
| Variance | $\sigma^2 = \frac{1}{N}\sum_{i=1}^{N}(x(i) - \mu)^2$ |
| Standard Deviation | $\sigma = \sqrt{\frac{1}{N}\sum_{i=1}^{N}(x(i) - \mu)^2}$ |
| Root Mean Square | $rms = \sqrt{\frac{1}{N}\sum_{i=1}^{N} x(i)^2}$ |

**Table 2.** Selected features for classification ordered by their rank

| Order | Feature | Order | Feature |
|---|---|---|---|
| 1 | $\text{rms}(L_5^{40})$ | 11 | $\max(L_1^1)$ |
| 2 | $\text{adv}(L_2^{13})$ | 12 | $\min (L_4^{26})$ |
| 3 | $\sigma^2 (L_4^{30})$ | 13 | $\text{R}(L_3^{20})$ |
| 4 | $\text{rms}(L_3^{21})$ | 14 | $\text{rms}(L_4^{32})$ |
| 5 | mean $(L_1^3)$ | 15 | $\max(L_4^{25})$ |
| 6 | $\sigma (L_4^{31})$ | 16 | $\text{adv}(L_1^5)$ |
| 7 | $\sigma^2 (L_3^{22})$ | 17 | mean $(L_2^{11})$ |
| 8 | $\sigma (L_2^{15})$ | 18 | $\text{rms}(L_3^{16})$ |
| 9 | $\text{R}(L_4^{28})$ | 19 | $\max(L_3^{17})$ |
| 10 | $\min(L_3^{18})$ | 20 | $\sigma (L_5^{39})$ |

### 4.4 Identifying ECUs

ECU identification is a multiclass classification problem and we use supervised learning to identify the source of CAN-FD/CAN signal. In particular, logistic regression (LR) is easy to implement with very small amount of calculation, which is very important for limited computing resources of ECU. To show the robustness of our system, we also execute support vector machines (SVM) algorithm with good generalization ability.

For the training phase, we generate fingerprints from multiple CAN-FD/CAN frames of each ECU. The resulting fingerprints are then used together to train the classifiers. For the testing phase, we have two types of tests. The first is to evaluate the model obtained by the training stage (i.e., whether or not it can determine the source of newly received frames), and the second is on intrusion detection.

## 5 Source Identification and Intrusion Detection

### 5.1 Experiment Setup

Our system adapts to different bus prototypes (we have three different network prototypes, see Fig. 5). Type A (see Fig. 5(a)) contains five CAN-FD nodes that can send both CAN-FD and CAN frames. Type B (see Fig. 5(b)) contains five CAN-FD nodes (the same as in Type A) and four extra CAN nodes that send purely CAN frames. Type C (see Fig. 5(c)) contains five CAN nodes that send purely CAN frames. Although the total number of ECUs in real cars might be up to 70 or even larger, in-vehicle networks are physically divided into several subnets, e.g., power-related or comfort-related. As analyzed in Sect. 3, ringing mainly exists between ECUs and junctions. Thus the rationale of fingerprinting ECUs in real cars is the same as that in our experiments. CAN protocol defines low-speed CAN and high-speed CAN. Generally speaking, high-speed CAN connects the ECUs related to the important functions of the vehicles. For example, the ECU that controls the brakes and the ECU that controls acceleration are both on high-speed CAN, and the data transmission speed of high-speed CAN is 500 kbit/s. Our CAN bus prototype takes high-speed CAN network topology.

Each CAN node that sends CAN frames consists of an Arduino UNO board and a CAN shield from Seed Studio. Each CAN shield consists of an MCP2515 controller [16] and an MCP2551 transceiver [17], and the bit rate at which they send data is 500kbit/s. For CAN-FD nodes, each one consists of a STM32F105 shield and a MCP2517FD controller [18]. MCP2517FD is known as compact, cost-effective and efficient CAN-FD controller and uses SPI interface and MCU (Microcontroller Unit) communication. In the experiments, we set the bit rate of MCP2517FD as 1Mbit/s in the arbitration phase, control phase and CRC phase, and 2Mbit/s in the data transmission phase. We mention that using signal characteristics sampled at high bit rate to identify devices is more difficult than at low bit rate. If our method shows effectiveness on the high-speed CAN-FD (and CAN), it would also function well on the low-speed CAN-FD (and CAN, respectively). To maintain the consistency of experimental environments, we require that all the stub lines, oscilloscope, and other components used in the experiments are the same in all three prototypes (except the nodes of different functions).

To simulate the in-vehicle network as realistically as possible, we use twisted pair as the communication cable in all three prototypes. Each ECU is connected to main bus through two twisted pairs (CAN-H and CAN-L) (or (CAN-FD)-H and (CAN-FD)-L). All ECUs are powered by a battery which supplies electric power to each ECU via USB ports. It is required that main bus (twisted pair as well) should be longer than any other stub line on the network (our configuration sets the length of main bus as the sum of those of stub lines). There is a 120 $\Omega$ resistor at each of the two ends of main bus. CAN-FD/CAN signals are measured by the oscilloscope PicoScope 5244D MSO with a sampling rate of 25 MS/s and a resolution of 8 bits. Two probes of the oscilloscope are connected to (CAN-FD)-H/CAN-H and (CAN-FD)-L/CAN-L respectively. For each ECU (CAN-FD or CAN node), we use 200 frames as the training set (the size of the training set could be adjusted according to the performance of the model). The machine learning library Scikit and programming software Python3 are used.

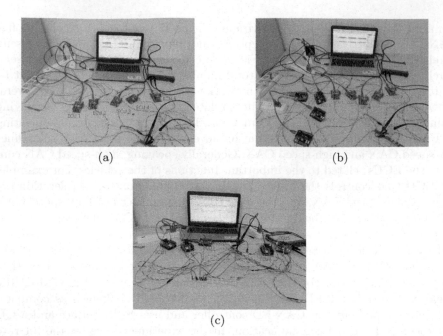

(a)                                          (b)

(c)

**Fig. 5.** Three prototypes of network topology: (a) Type A: CAN-FD nodes, (b) Type B: CAN-FD nodes and CAN nodes, (c): CAN nodes

## 5.2   Sender Identification

**Sender Identification on Pure CAN.** For Type C (see Fig. 5(c)), we consider the ringing effect. In particular, we execute SVM and LR by using dominant states and rising edges, recessive states and falling edges, and ((dominant states and rising edges) and (recessive states and falling edges)) respectively. The experimental results are shown in Table 3, Table 4, and Table 5. Each diagonal cell in the same matrix represents the accuracy of the two classification algorithms. As expected, dominant states and rising edges suffice to fingerprint ECUs.

**Using Dominant States and Rising Edges.** We then evaluate whether our system can correctly classify ECUs for Type A and Type B. Table 6 lists the confusion matrix which allows visualization of the performance of classification algorithms for 5 ECUs that send CAN-FD frames (Type A). We can see that the recognition rate of our system is sufficient to correctly recognize the ECU, and the error rate is very low. Table 7 lists the confusion matrix of 9 ECUs (Type B), of which 5 ECUs send CAN-FD frames, and the remaining 4 ECUs send CAN frames. From the result, we can see that our system can still correctly classify and recognize ECUs even in the case of a hybrid network.

**Table 3.** Confusion matrix using SVM/LR respectively for Type C and dominant states-rising edges.

|       | ECU 1 | ECU 2 | ECU 3 | ECU 4 | ECU 5 |
|-------|-------|-------|-------|-------|-------|
| ECU 1 | 99.89/99.77 | 0/0 | 0/0 | 0.11/0.23 | 0/0 |
| ECU 2 | 0/0 | 99.59/99.79 | 0/0 | 0.41/0.21 | 0/0 |
| ECU 3 | 0.14/0.46 | 0/0 | 99.76/99.54 | 0/0 | 0/0 |
| ECU 4 | 0/0 | 0/0 | 0.2/0.02 | 99.8/99.98 | 0/0 |
| ECU 5 | 0.2/0.08 | 0/0 | 0/0 | 0/0 | 99.8/99.92 |

**Table 4.** Confusion matrix using SVM/LR respectively for Type C and recessive states-falling edges.

|       | ECU 1 | ECU 2 | ECU 3 | ECU 4 | ECU 5 |
|-------|-------|-------|-------|-------|-------|
| ECU 1 | 86.52/84.66 | 0/0 | 5.23/6.01 | 8.25/9.33 | 0/0 |
| ECU 2 | 0/0 | 88.21/87.11 | 6.47/7.56 | 0/0 | 5.32/5.33 |
| ECU 3 | 14.34/11.46 | 0/0 | 85.66/88.54 | 0/0 | 0/0 |
| ECU 4 | 0/0 | 0/0 | 15.12/14.62 | 84.88/85.38 | 0/0 |
| ECU 5 | 4.32/5.01 | 0/0 | 4.66/3.84 | 5.17/6.23 | 85.85/84.92 |

**Table 5.** Confusion matrix using SVM/LR respectively for Type C and (dominant states and rising edges)-(recessive states and falling edges).

|       | ECU 1 | ECU 2 | ECU 3 | ECU 4 | ECU 5 |
|-------|-------|-------|-------|-------|-------|
| ECU 1 | 96.12/95.34 | 1.81/2.56 | 0/0 | 2.07/2.1 | 0/0 |
| ECU 2 | 4.79/5.03 | 95.21/94.97 | 0/0 | 0/0 | 0/0 |
| ECU 3 | 5.44/4.16 | 0/0 | 94.56/95.84 | 0/0 | 0/0 |
| ECU 4 | 0/0 | 0/0 | 4.12/5.02 | 95.88/94.98 | 0/0 |
| ECU 5 | 2.81/2.9 | 0/0 | 2.34/2.18 | 0/0 | 94.85/94.92 |

**Table 6.** Confusion matrix using SVM/LR respectively for Type A and dominant states-rising edges.

|       | ECU 1 | ECU 2 | ECU 3 | ECU 4 | ECU 5 |
|-------|-------|-------|-------|-------|-------|
| ECU 1 | 99.12/99.34 | 0/0 | 0/0 | 0.88/0.66 | 0/0 |
| ECU 2 | 0/0 | 99.21/99 | 0/0 | 0/0 | 0.79/1 |
| ECU 3 | 0.24/0.46 | 0/0 | 99.76/99.54 | 0/0 | 0/0 |
| ECU 4 | 0/0 | 0/0 | 0.12/0.02 | 99.88/99.98 | 0/0 |
| ECU 5 | 0.15/0.08 | 0/0 | 0/0 | 0/0 | 99.85/99.92 |

**Using Recessive States and Falling Edges.** To argue the effectiveness of our recommendation, we also consider the recognition rate if recessive edges and falling edges are used. As depicted in Sect. 3.2, ringing intensity of falling edges of signals is higher than that of rising edges. Thus recognition rate would be affected when the falling edges are used. Table 8 shows the recognition rates 81.54~86.21% for Type A. Due to space limitation, we write in the Appendix A (Table 12) the confusion matrix using SVM/LR respectively for Type B where we can see really low recognition rates (78.01~83.89%).

**Table 7.** Confusion matrix using SVM/LR respectively for Type B and dominant states-rising edges.

|       | ECU 1 | ECU 2 | ECU 3 | ECU 4 | ECU 5 | ECU 6 | ECU 7 | ECU 8 | ECU 9 |
|-------|-------|-------|-------|-------|-------|-------|-------|-------|-------|
| ECU 1 | 98.89/99.15 | 0/0 | 0/0 | 0/0 | 0.91/0.7 | 0.01/0.03 | 0/0 | 0/0 | 0.19/0.12 |
| ECU 2 | 0/0 | 98.01/99.21 | 0/0 | 1.2/0.78 | 0/0 | 0/0 | 0.79/0.01 | 0/0 | 0/0 |
| ECU 3 | 0/0 | 0/0 | 98.99/99.01 | 0.92/0.89 | 0/0 | 0/0 | 0/0 | 0/0 | 0.09/0.1 |
| ECU 4 | 0/0 | 0/0 | 0/0 | 99.29/99.11 | 0/0 | 0/0 | 0.7/0.89 | 0.01/0 | 0/0 |
| ECU 5 | 0/0 | 0/0 | 0/0 | 0/0 | 98.99/99.31 | 0/0 | 0/0 | 0/0 | 1.01/0.69 |
| ECU 6 | 1.01/0.9 | 0/0 | 0/0 | 0.01/0.1 | 0/0 | 98.98/99 | 0/0 | 0/0 | 0/0 |
| ECU 7 | 1.32/0.98 | 0/0 | 0/0 | 0/0 | 0.01/0.01 | 0/0 | 98.67/99.01 | 0/0 | 0/0 |
| ECU 8 | 0/0 | 0/0 | 0.9/0.96 | 0.01/0.03 | 0/0 | 0/0 | 0/0 | 99.09/99.01 | 0/0 |
| ECU 9 | 1.11/1.8 | 0/0 | 0/0 | 0/0.03 | 0/0 | 0/0 | 0/0 | 0/0 | 98.89/98.17 |

**Table 8.** Confusion matrix using SVM/LR respectively for Type A and recessive states-falling edges.

|        | ECU 1 | ECU 2 | ECU 3 | ECU 4 | ECU 5 |
|--------|-------|-------|-------|-------|-------|
| ECU 1 | 84.12/85.34 | 12/13.14 | 0/0 | 3.88/1.52 | 0/0 |
| ECU 2 | 0/0 | 86.21/85 | 11.79/12.78 | 2/2.22 | 0/0 |
| ECU 3 | 5.14/6.46 | 4.12/4.36 | 82.76/81.54 | 3.51/3.96 | 4.47/3.68 |
| ECU 4 | 0/0 | 15.82/16.62 | 0/0 | 84.18/83.38 | 0/0 |
| ECU 5 | 0/0 | 12.32/12.01 | 2.93/3.17 | 0/0 | 84.75/84.82 |

**Table 9.** Confusion matrix using SVM/LR respectively for Type A and (dominant states and rising edge)-(falling edges and recessive states).

|        | ECU 1 | ECU 2 | ECU 3 | ECU 4 | ECU 5 |
|--------|-------|-------|-------|-------|-------|
| ECU 1 | 94.32/95.24 | 3.36/3.14 | 0/0 | 0/ 0 | 2.32/1.62 |
| ECU 2 | 0/0 | 93.21/94.21 | 5.78/5.01 | 0/0 | 1.01/0.78 |
| ECU 3 | 5.14/1.46 | 0/0 | 93.76/94.54 | 1.1/0.45 | 0/0 |
| ECU 4 | 0/0 | 5.2/6.33 | 0/0.09 | 94.8/93.58 | 0/0 |
| ECU 5 | 5.05/5.15 | 0.2/0.23 | 0/0 | 0/0 | 94.75/94.62 |

**Using (Dominant States and Rising Edges) and (Recessive States and Falling Edges).** We also compare the execution rates when the system uses (dominant states and rising edges) and (Recessive States and falling Edges). Table 9 and Table 10 show the results of Type A and Type B respectively, both lower than that using dominant states and rising edges.

**Table 10.** Confusion matrix using SVM/LR respectively for Type B and (dominant states and rising edges)-(recessive states and falling edges).

|        | ECU 1 | ECU 2 | ECU 3 | ECU 4 | ECU 5 | ECU 6 | ECU 7 | ECU 8 | ECU 9 |
|--------|-------|-------|-------|-------|-------|-------|-------|-------|-------|
| ECU 1 | 93.89/ 94.15 | 5.98/ 5.51 | 0.13/ 0.34 | 0/0 | 0/0 | 0/0 | 0/0 | 0/0 | 0/0 |
| ECU 2 | 0/0 | 92.01/ 93.21 | 0/0 | 0/0 | 6.01/ 5.89 | 0/0 | 0/0 | 1.98/ 0.9 | 0/0 |
| ECU 3 | 0/0 | 0/0 | 94.01/ 93.1 | 0/0 | 5.53/ 6.01 | 0/0 | 0/0 | 0.46/ 0.89 | 0/0 |
| ECU 4 | 3.9/4.01 | 0/0 | 0/0 | 95.29/ 95.11 | 0.81/ 0.88 | 0/0 | 0/0 | 0/0 | 0/0 |
| ECU 5 | 0/0 | 0/0 | 5.8/6.91 | 0/0 | 93.99/ 92.31 | 0/0 | 0/0 | 0/0 | 0.21/ 0.78 |
| ECU 6 | 6.01/6.4 | 0/0 | 0/0 | 2.1/1.5 | 0/0.01 | 91.98/ 92.09 | 0/0 | 0/0 | 0/0 |
| ECU 7 | 5.32/4.91 | 0/0 | 0/0 | 0/0 | 1.08/ 1.01 | 0/0 | 93.67/ 94.01 | 0/0 | 0/0 |
| ECU 8 | 0/0 | 0/0 | 0.9/ 0.2 | 0.01/0.03 | 5.9/6.86 | 0/0 | 0/0 | 93.09/ 92.01 | 1.01/ 1.82 |
| ECU 9 | 1.11/1.8 | 0/0 | 0/0 | 1.01/ 0.03 | 0/0 | 0/0 | 0/0 | 5.1/6.01 | 93.89/ 92.17 |

## 5.3   Detecting Known/Unknown ECUs

The proposed ECU identification scheme is readily extended to intrusion detection system on in-vehicle CAN-FD network and the resulting IDS can not only detect attacks initiated by external devices but also internal devices. The recognition rate can be up to 99%. Due to space limitation, we write the evaluation in the Appendix B and C.

In practice, one can deploy the offline trained models on dedicated ECU which is inserted to the bus. Main function of the exact ECU is to monitor the traffic silently and detect anomaly. In the ECU, a digital signal processor (DSP chip, a microprocessor especially suitable for digital signal processing operations) can be integrated to establish the function of an oscilloscope: collect signals in real time and pass them to the model for detection.

## 6   Discussions

**Sample Rate.** We duplicate the experiments at various sample rates to inspect our system's effectiveness, especially for Type B. Note that at different sample rate one will be at different position of sample sizes (which might convey tight relationship with system performance). Fortunately, our approach manifests robustness as expected (due to the contribution of rising edges and dominant states). Table 11 shows the average identification and false positive rates at the sample rates 10~25MS/s (1000 frames for each ECU).

**Comparable Performance Between Type A and Type C.** For a given network topology (in terms of the stub length and the number of ECUs), one may note considerable performance for Type A (CAN-FD) and Type C (CAN) by

**Table 11.** LR Performance at various sample rates.

| Sample rate(MS/s) | 10 | 15 | 20 | 25 |
|---|---|---|---|---|
| Identification rate | 97.11 | 98.95 | 99.01 | 99.15 |
| False positive rate | 2.89 | 1.05 | 0.99 | 0.85 |

using any signal characteristics (rising edges, dominant states, falling edges, and recessive states). In fact, Type C could obtain generally a tiny little better recognition rate than Type A. On the one hand, CAN-FD supports data size up to 512 bits, a drastically larger number than that (64 bits) in CAN specification, thus the cumulative effect of ringing for Type A might be more powerful than for Type C. On the other hand, CAN-FD provides variable transmission rate and our experiments specify the bit rate 2 Mbit/s for the data field of CAN-FD frames and 1Mbit/s for other fields (e.g., arbitration field, control field and CRC field), whereas CAN frame (Type C) regulates the rate of 500kbit/s. Namely, our experiments have the bit width 2000ns in a CAN frame, and 1000ns in the non-data field of and 500ns in the data field of a CAN-FD frame. Now, it is more likely for Type A (than Type C) that ringing of recessive states functions unceasing (even though the bit itself was already completed on the network)[4] and thus involves the coming dominant states before it attenuates to be unnoticeable.

**Applicability to CAN-FD Network in Real Vehicles.** The controllers used in our evaluation conform to ISO11898-1:2015 and support CAN-FD [18]. We also take into account the possible transition mechanism from CAN to CAN-FD (i.e., Type A and Type B). Our results show expressive evidence on the universal applicability of our approach to forthcoming real vehicles set up by CAN-FD network. We do hope our results could be used as a step forward and a guidance on securing the commercialization and batch production of in-vehicle CAN-FD network in the near future.

**Environmental Factors.** The electrical characteristics of CAN signals may remain unchanged for several months [21]. However, in actual vehicles, changes in the internal temperature of the vehicle will affect the characteristics of electrical signals. A typical example is that the voltage output may deviate from 0.012 V to 0.026 V [15] when we start the vehicle from a cooled turn-off engine to a warmed-up engine. This situation may also exist for the CAN-FD network. Howbeit, the length of CAN FD frame is greater than 512 bits, and the number of dominant states contained would be much likely greater than that in CAN frame. We might thus expect an acceptable impact of temperature changes on signal characteristics (and further on the system). Precise assessment is left as one of the future work.

---

[4] It is already reported [8,9] that for CAN-FD protocol, high-speed data phase and low-speed arbitration phase challenge the same ringing surrounds (as ringing does not depend on the transmission rate), and ring of some recessive bit might not converge until criterion and interfere with the next dominant bit.

**Battery/ECU Aging.** Generally speaking, the service life of car battery is of 3 to 5 years and its real usage duration is also related to the driver's driving habits. Battery aging might affect the characteristics of the electrical signals. The same problem exists on CAN network. For now, however, we can not track the impact of battery aging on the system by simulating CAN-FD nodes and car battery as there is no CAN-FD vehicle for real driving. We hope we can explore the interesting topic in the coming future. On the other hand, ECU has a relatively long service life and the aging process is really slow. It is thus rational not to consider the impact of ECU aging on electrical signals.

**Limitation of the Model.** Our method can detect compromised ECUs by monitoring CAN bus. It can determine whether particular frame on the bus originates from some ECU that is allowed to commit the corresponding identifier. If not, the system will issue a warning. Otherwise said, an attack will be detected once a known ECU professes some message identifier affiliated with another normal ECUs. However, if a known ECU abuses its own identifier (that is permitted under normal circumstances) to launch some attack, our system cannot recognize the attack.

**Acknowledgement.** The work was supported by Shanghai Municipal Education Commission (2021-01-07-00-08-E00101), the National Natural Science Foundation of China (Grant No. 61971192), and the National Cryptography Development Fund (Grant No. MMJJ20180106).

# A    Source Identification on Type B and Recessive States-Falling Edges

As depicted in Sect. 3.2, ringing intensity of falling edges of signals is higher than that of rising edges. Thus recognition rate would be affected when the falling edges are used. Table 12 show the results for Type B (and Table 8 for Type A) and we can see really low recognition rates.

**Table 12.** Confusion matrix using SVM/LR respectively for Type B and recessive states-falling edges.

|  | ECU 1 | ECU 2 | ECU 3 | ECU 4 | ECU 5 | ECU 6 | ECU 7 | ECU 8 | ECU 9 |
|---|---|---|---|---|---|---|---|---|---|
| ECU 1 | 79.89/ 78.15 | 15.98/ 16.51 | 3.12/ 4.32 | 0/0 | 0/0 | 0/0 | 0/0 | 0/0 | 1.01/ 1.02 |
| ECU 2 | 0/0 | 80.01/ 79.21 | 0/0 | 0/0 | 16.01/ 17.99 | 0/0 | 0/0 | 3.78/ 2.8 | 0/0 |
| ECU 3 | 0/0 | 0/0 | 78.01/ 79.1 | 0/0 | 18.53/ 17.01 | 0/0 | 0/0 | 3.73/ 3.89 | 0/0 |
| ECU 4 | 16.01/ 15.99 | 0/0 | 0.01/ 0.19 | 80.29/ 80.11 | 3.6/ 3.71 | 0/0 | 0/0 | 0/0 | 0/0 |
| ECU 5 | 0/0 | 0/0 | 16.48/ 15.91 | 0/0 | 78.99/ 79.31 | 0/0 | 1.32/ 1.01 | 0/0 | 3.21/ 3.77 |
| ECU 6 | 15.01/ 14.98 | 0/0 | 0/0 | 3.1/ 3.25 | 0.91/ 0.76 | 80.98/ 81.01 | 0/0 | 0/0 | 0/0 |
| ECU 7 | 15.32/ 15.91 | 0/0 | 0/0 | 0/0 | 1.01/ 1.1 | 0/0 | 83.67/ 82.99 | 0/0 | 0/0 |
| ECU 8 | 0/0 | 0/0 | 15.91/ 14.99 | 2.01/ 2.18 | 5.9/ 6.86 | 0/0 | 0/0 | 80.09/ 81.01 | 1.99/ 1.82 |
| ECU 9 | 14.11/ 15.01 | 0/0 | 0/0 | 1.01/ 1.99 | 0/0 | 0/0 | 0/0 | 0.99/ 0.83 | 83.89/ 2.17 |

**Table 13.** Confusion matrix of the IDS using SVM

| Support vector machines | | | |
|---|---|---|---|
| Prototype | True | Predicted | |
|  |  | No Attack | Yes |
| CAN-FD | No Attack | 99.38 | 0.62 |
|  | Yes | 1.5 | 98.5 |
| CAN-FD And CAN | No Attack | 99.01 | 0.99 |
|  | Yes | 1.18 | 98.82 |
| CAN | No Attack | 99.58 | 0.52 |
|  | Yes | 0.99 | 99.01 |

**Table 14.** Confusion matrix of the IDS using LR

| Logistic regression | | | |
|---|---|---|---|
| Prototype | True | Predicted | |
|  |  | No Attack | Yes |
| CAN-FD | No Attack | 99.85 | 0.42 |
|  | Yes | 1.88 | 98.12 |
| CAN-FD And CAN | No Attack | 99.11 | 0.89 |
|  | Yes | 1.89 | 98.11 |
| CAN | No Attack | 99.44 | 0.56 |
|  | Yes | 0.89 | 99.11 |

# B    Detecting Known ECUs

For Type C (Fig. 5(c)), we assume that ECU 1 is normal and the attackers can use other ECUs to send messages with the same identifier as ECU 1. We collect a total of 500 frames, of which 300 are used as attack frames and the rest as normal frames. As shown in Table 13, we achieve a detection rate of 99.01%. For Type A (Fig. 5(a)), we use the same assumptions and operations as for Type C and achieve a detection rate of 98.5% (see Table 13). For Type B (see Fig. 5(b)), we regard ECU 7, ECU 8 and ECU 9 as attackers (equipped with the ability of sending both CAN and CAN-FD frames). We collect 1000 frames, of which 600 are used as attack frames and the rest are normal. Table 14 shows the results with comparable performance to Type A and Type C.

## C    Detecting Unknown ECUs

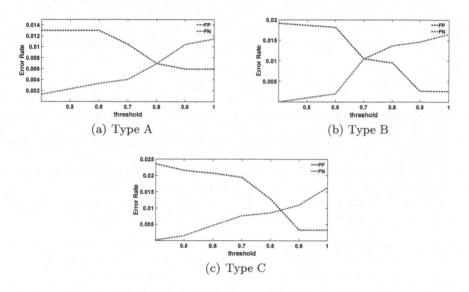

(a) Type A                    (b) Type B

(c) Type C

**Fig. 6.** Error rates at varying thresholds.

For unknown ECUs, we adopt a threshold-based method to extend our model. For Type A, we first remove ECU 5 and obtain about 500 frames from the remaining ECUs. These data are used to train a new model. Then we plug ECU 5 back to the network and sample a total of 600 frames now. The obtained model is used to classify the newly collected data and Fig. 6(a) shows the False Positive (FP) and False Negative (FN) rates for different threshold values. The recognition rate can be up to 99.36% at threshold = 0.8. For Type B, we remove ECU 8, use the remaining ECUs to train a new model, and then plug ECU 8 back to the network. We collect now a total of 1,000 data which will be classified by the obtained model. Figure 6(b) shows FP and FN vs threshold, and the recognition rate is 99% at threshold = 0.7. For Type C, we use similar method and Fig. 6(c) shows FP and FN vs threshold. We see the 99.1% recognition rate at threshold = 0.83.

# References

1. Agrawal, M., Huang, T., Zhou, J., Chang, D.: CAN-FD-Sec: improving security of CAN-FD protocol. In: Hamid, B., Gallina, B., Shabtai, A., Elovici, Y., Garcia-Alfaro, J. (eds.) CSITS/ISSA -2018. LNCS, vol. 11552, pp. 77–93. Springer, Cham (2019). https://doi.org/10.1007/978-3-030-16874-2_6
2. Cho, K., Shin, K.G.: Error handling of in-vehicle networks makes them vulnerable. In: Proceedings of ACM CCS, pp. 1044–1055 (2016)
3. Cho, K., Shin, K.G.: Fingerprinting electronic control units for vehicle intrusion detection. In: 25th USENIX Security Symposium, pp. 911–927 (2016)
4. Cho, K., Shin, K.G.: Viden: attacker identification on in-vehicle networks. In: Proceedings of 2017 ACM CCS, pp. 1109–1123. ACM (2017)
5. Choi, W., Jo, H.J., et al.: Identifying ECUs using inimitable characteristics of signals in controller area networks. IEEE Trans. Veh. Technol. 67(6), 4757–4770 (2018)
6. GmbH, R.B.: CAN Specifcation Version 2.0 (1991)
7. GmbH, R.B.: CAN with Flexible Data-Rate (2012)
8. H. Mori, Y.S., et al.: Novel ringing suppression circuit to increase the number of connectable ECUs in a linear passive star CAN. In: International Symposium on Electromagnetic Compatibility - EMC EUROPE, pp. 1–6 (2012)
9. Islinger, T., Mori, Y.: Ringing suppression in can fd networks. CAN Newsletter (2016)
10. Karl, K., Alexei, C., et al.: Experimental security analysis of a modern automobile. In: IEEE Symposium on Security and Privacy, pp. 447–462 (2010)
11. Kim, G., Lim, H.: Ringing suppression in a controller area network with flexible data rate using impedance switching and a limiter. IEEE Trans. Veh. Technol. 68(11), 10679–10686 (2019)
12. Kononenko, I.: Estimating attributes: analysis and extensions of RELIEF. In: Bergadano, F., De Raedt, L. (eds.) ECML 1994. LNCS, vol. 784, pp. 171–182. Springer, Heidelberg (1994). https://doi.org/10.1007/3-540-57868-4_57
13. Lim, H., Kim, G., et al.: Quantitative analysis of ringing in a controller area network with flexible data rate for reliable physical layer designs. IEEE Trans. Veh. Technol. 68(9), 8906–8915 (2019)
14. Lin, C., Sangiovanni-Vincentelli, A.L.: Cyber-security for the controller area network (CAN) communication protocol. In: 2012 ASE International Conference on Cyber Security, pp. 1–7. IEEE Computer Society (2012)
15. Marcel, K., Christopher, H.: Scission: signal characteristic-based sender identification and intrusion detection in automotive networks. In: Proceedings of the 2018 ACM Conference on Computer and Communications Security, pp. 787–800 (2018)
16. Microchip-Corporation: Stand-Alone CAN Controller With SPI Interface (2005)
17. Microchip-Corporation: MCP2551 High-Speed CAN Transceiver (2007)
18. Microchip-Corporation: Externa CAN FD Controller with SPI Infertface (2017)
19. Miller, C., Valasek, C.: Adventures in automotive networks and control units. Def Con 21(260–264), 15–31 (2013)
20. Miller, C., Valasek, C.: Remote exploitation of an unaltered passenger vehicle. Black Hat USA 2015(S 91) (2015)
21. Pal-Stefan, M., Bogdan, G.: Source identification using signal characteristics in controller area networks. IEEE Signal Process. Lett. 21(4), 395–399 (2014)
22. Schweppe, H., Roudier, Y., et al.: Car2x communication: securing the last meter-a cost-effective approach for ensuring trust in car2x applications using in-vehicle symmetric cryptography. In: 2011 IEEE VTC Fall, pp. 1–5 (2011)

23. Tobias, H., Jana, D.: Sniffing/replay attacks on can buses: A simulated attack on the electric window lift classified using an adapted cert taxonomy. In: Proceedings of the 2nd workshop on embedded systems security (WESS), pp. 1–6 (2007)
24. Woo, S., Jo, Hyo Jin, A.O.: A practical security architecture for in-vehicle CAN-FD. IEEE Trans. Intell. Transp. Syst. **17**(8), 2248–2261 (2016)
25. Woo, S., Jo, H.J., et al.: A practical wireless attack on the connected car and security protocol for in-vehicle CAN. IEEE Trans. Intell. Transp. Syst. **16**(2), 993–1006 (2015)
26. Yu, T., Wang, X.: Topology verification enabled intrusion detection for in-vehicle CAN-FD networks. IEEE Commun. Lett. **24**(1), 227–230 (2020)

# EmuIoTNet: An Emulated IoT Network for Dynamic Analysis

Qin Si[1,2], Lei Cui[2], Lun Li[2(✉)], Zhenquan Ding[2], Yongji Liu[2], and Zhiyu Hao[2]

[1] School of Cyber Security, University of Chinese Academy of Sciences,
Beijing, China
[2] Institute of Information Engineering, Chinese Academy of Sciences, Beijing, China
{siqin,cuilei,lilun,dingzhenquan,liuyongji,haozhiyu}@iie.ac.cn

**Abstract.** Dynamic analysis of IoT firmware is an effective method to discover security flaws and vulnerabilities. However, limited by emulation methods concentrating on a single IoT device, it is challenging to find security issues hidden in communication channels. This paper presents EmuIoTNet, a tool capable of automatically building an emulated IoT network for dynamic analysis. First, EmuIoTNet prepares an emulated hardware environment to emulate a number of devices for firmware. Then, it employs network virtualization tools to setup two types of networks, IntraNet and InterNet, which connect emulated devices, companion applications, and cloud endpoints to support many communication protocols. Meanwhile, it reconfigures the IP address of emulated devices at will to support simultaneous operations of multiple users. The experimental results show that EmuIoTNet can automatically build various emulated networks and facilitate security analysis in communication channels.

**Keywords:** IoT network · IoT security · Device emulation · Network emulation · Firmware analysis

## 1 Introduction

With over billions of IoT (internet of things) devices being placed in many aspects of our lives, e.g., home, hospital, and industry environments, the IoT devices face increasing attacks on them [5,25]. The attackers leverage exploitable vulnerabilities in the device to gain the privilege, build botnets, steal private data, perform DDoS attack, etc., posing severe threats to the devices [6,22,28].

Firmware, which sits between the hardware and the outside world, is reported to be potentially vulnerable and exploitable. To discover the security flaws and vulnerabilities, dynamic analysis of firmware, including security testing, vulnerability detection and discovery, is one promising approach [7,8,12,24,26]. It

Supported in part by the National Natural Science Foundation of China under Grant 61972392 and Grant 62072453 and in part by the Youth Innovation Promotion Association of the Chinese Academy of Sciences under Grant 2020164.

© Springer Nature Switzerland AG 2021
D. Gao et al. (Eds.): ICICS 2021, LNCS 12918, pp. 224–240, 2021.
https://doi.org/10.1007/978-3-030-86890-1_13

launches the firmware, carefully monitors the runtime behavior of firmware, and finally perceives malicious operations on-the-fly. As the first step, launching the firmware is fundamental to perform dynamic analysis. Many methods have been proposed in this literature recently. They intend to provide an independent environment (i.e., an emulated IoT device) including common hardware (e.g., CPU, memory) and specific hardware (e.g., NVRAM) so that the firmware can run successfully, e.g., Firmadyne [7], Avatar [26], Avatar$^2$ [21], P$^2$IM [14]. With device emulation, a lot of vulnerabilities have been detected and discovered [23,29].

Unfortunately, emulating only a single IoT device is inadequate to expose security threats in IoT environments, especially when suffering wide attack surfaces. Take the home IoT network as an example, as shown in Fig. 1, the IoT devices are controlled by a companion application (mobile or desktop) and connect to the cloud endpoints (i.e., public websites) for requesting specific services. In such a networked scenario, there exist many security issues in the communication channels due to insecure protocols, unclosed ports, and old vulnerable firmware. For example, a recent study [5] reports 20 of 27 channels are susceptible to MITM (man in the middle) attack. Therefore, compared to emulating a single device, emulating an IoT network can provide more opportunities for dynamic analysis, particularly in communications.

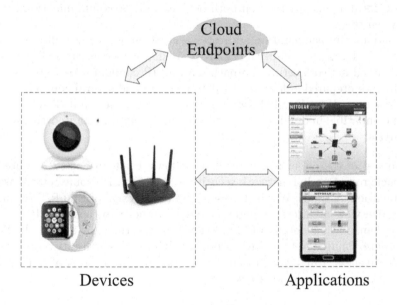

**Fig. 1.** Home based IoT network.

In this paper, we present EmuIoTNet to automatically setup an emulated IoT network for dynamic analysis. EmuIoTNet leverages full-system emulation to prepare common hardware and proposes an adaptive NVRAM simulator to allow NVRAM operations, so that a number of firmware can be launched and network reachable. Based on the emulated devices, it proposes an automated network setup approach to build intra-network (IntraNet) and inter-network (InterNet), thereby allowing the connection between emulated devices, companion applications and cloud endpoints. Meanwhile, EmuIoTNet designs an IP reconfiguration approach which sets the given address for each node. Consequently, it supports many communication protocols, e.g., HTTP, HTTPS, DNS, etc. Compared to existing works that focus on emulating a single device, EmuIoTNet provides the ability of emulating an IoT network, which facilities dynamic analysis on not only a single device but also the communication channels between IoT device and companion application, and between IoT device and cloud endpoints.

We have implemented EmuIoTNet based on Firmadyne, which is a full system emulation platform for IoT devices. EmuIoTNet allows more firmware to be launched and network reachable than Firmadyne. Moreover, it supports automated network setup for analyzing HTTP, HTTPS, DNS, SSH, and Telnet traffics. We employ several security analysis tools such as Wireshark, sslsplit, and Nmap to analyze the emulated networks, and successfully find many weaknesses in both the firmware and the communication channels. To summarize, this paper makes the following contributions.

– First, we present EmuIoTNet, an automated tool to build emulated IoT networks for dynamic analysis on both devices and the communication channels between them.
– Second, our implementation addresses several challenges in building IoT networks, including an adaptive NVRAM simulator to emulate IoT devices, an automated network setup approach, and an IP configuration approach.
– Third, we prepared a dataset of 1,026 firmware and several companion applications to evaluate EmuIoTNet. The results show that EmuIoTNet is scalable and compatible in building IoT networks, and is useful in supporting dynamic analysis.

The rest of the paper is organized as follows. First, in Sect. 2, we present a brief background on IoT network security and summarize some related work of IoT emulation methods. We then present the basic design of EmuIoTNet, including the overview architecture and challenges we need to tackle with in Sect. 3. The implementation details of EmuIoTNet is described in Sect. 4. In Sect. 5, we evaluate EmuIoTNet in three aspects to show its scalability, compatibility and effectiveness. Finally, we discuss some existing limitations of our proposed framework and conclude our work in Sect. 6.

## 2   Background and Related Work

### 2.1   Security Issues in IoT Network

IoT devices are facing increasing attacks on them. For example, Kaspersky reported 102 million attacks on IoT devices in the first 6 months of 2019, compared with just 12 million attacks in the first half of 2018 [3]. The devices are suffering a wide range of attacks, including Denial of Service, Botnet, Device hijacking, and Man-in-the-middle attack, making IoT security an important issue [4].

Firmware is the onboard software that sits between the hardware and the outside world in IoT devices. Generally, it is composed of boot-up software, operating system, drivers, applications, and web servers. Firmware is known to be vulnerable, mainly because the associated devices have low computational power and hardware limitations that don't allow for built-in security features [1]. For example, due to the heavy computational overhead of encryption, the users' passwords are stored in plain text, and many web services in firmware are still using insecure HTTP rather than HTTPS. Besides, even if a vulnerability has been reported, the firmware may not be updated and repaired in time, leaving the affected devices still in danger.

Firmware analysis is essential in discovering the security issues in IoT devices. However, recent studies report that IoT suffers a wide range of attacks. They not only attack the device (vulnerabilities, weak authentications) but also attack the companion application (gain-privilege, information leakage) and communications (lack of encryption, MITM). Therefore, analysis of only a firmware is insufficient to find threats such as MITM attack in the network environment.

## 2.2 IoT Emulation Methods

Dynamic analysis is one promising method in firmware analysis. Generally, it launches the firmware, monitors the runtime behavior, and employs analysis tools to detect security issues [9, 21, 23]. Device emulation is necessary and essential for dynamic analysis. Generally, it refers to preparing the environment (software or hardware) so that the firmware can be launched, run continuously, and be network reachable.

Current device emulation methods fall into two categories, full software emulation and hybrid emulation. Full software emulation uses software to emulate the function of hardware required for firmware execution, including CPU, memory, disk, etc. [11,12,20]. For example, Firmadyne [7] employs QEMU to emulate CPUs of different architectures to support Linux-based firmware. It is totally software-based and thus shows well scalability and compatibility. HALucinator [10] decouples the hardware from the firmware by providing high-level replacements for HAL functions (a process termed High-Level Emulation–HLE). It locates the library functions of firmware, and then after binary analysis, it provides generic implementations of the functions in a full-system emulator, which enables re-host firmware and allows the virtual device to be used normally. $P^2IM$ [14] models the I/O behaviors of the processor-peripheral interfaces with a generic processor emulator (QEMU), which is the first to enable peripheral-oblivious emulation of MCU devices. And in turn, it allows MCU firmware to be dynamically tested with high code coverage, at scale, and without hardware dependence. PRETENDER [15] creates models of peripherals automatically by recording the interactions between the original hardware and the firmware. It allows for the execution of the firmware in a fully-emulated environment and the models are interactive, stateful, and transferable. However, due to diversity of IoT devices, many firmware cannot be successfully launched when specific hardware is missing, e.g., camera lens, Bluetooth chip, etc.

The hybrid emulation method mitigates this problem by combining software emulation and physical hardware. It first launches the firmware in a software emulation environment (mainly the CPU) and then forwards specific instructions to the physical hardware, so that the firmware can run successfully [16–19,24]. For example, Avatar [26] is a framework that enables complex dynamic analysis of embedded devices by orchestrating the execution of an emulator together with the real hardware. By injecting a specific software proxy in the embedded device, Avatar can execute the firmware instructions inside the emulator while channeling the I/O operations to the physical hardware. It leverages the real hardware to handle I/O operations, but extracts the firmware code from the embedded device and emulates it on an external machine. Taking a hybrid approach, Avatar overcomes the limitation of full software based firmware emulation. Avatar$^2$ [21], a successor of Avatar, is a dynamic multi-target orchestration process framework that aims to achieve interoperability between different dynamic binary analysis frameworks, debuggers, emulators, and actual physical devices. Avatar$^2$ enables analysts to organize different tools in complex topologies and then transfers the execution of binary code from one system to another. It transfers the internal state of the device/application automatically, as well as fowards the configuration of the I/O and memory access to the physical peripherals or emulated targets. This method provides more opportunities for firmware execution with the assistance of physical hardware, especially the one that cannot be emulated. However, it requires experienced engineers to develop interfaces for specific hardware, which is non-trivial and greatly limits the scalability of firmware analysis.

Many studies have employed device emulation to perform dynamic analysis and successfully discovered dozens of vulnerabilities [13,27]. For example, FirmFuzz [23] is an automated device-independent emulation and dynamic analysis framework for Linux-based firmware images. It provides targeted and deterministic bug discovery within the firmware image by employing a grey box-based generational fuzzing approach coupled with static analysis and system introspection. Firm-AFL [29] is a high-throughput greybox fuzzer for IoT firmware. By combining system mode emulation and user-mode emulation, it proposes a novel technique called augmented process emulation, providing high compatibility as system-mode emulation and high throughput as user-mode emulation. Nevertheless, the emulation of only a single device makes them unable to perform dynamic analysis on companion application and communication. Consequently, they fail to discover many aspects of security issues hidden in IoT networks.

## 3    Basic Design of EmuIotNet

### 3.1    Design Goals

EmuIoTNet intends to build emulated IoT networks for supporting dynamic analysis, especially on communication between IoT devices, companion applications (mobile or desktop), and cloud endpoints. Considering that network scenarios are diverse due to various vendors, firmware, and protocols, EmuIoT-Net is designed to satisfy the following goals.

First, a wide range of IoT devices spanning different vendors. Due to the large number and variety of firmware, it is necessary to support a wide range of IoT device vendors (e.g., Netgear, D-Link, TP-Link), device types (e.g., router, switch, camera), and firmware versions.

Second, various types of communications. EmuIoTNet provides connections between different components for analyzing various communication traffics. 1) Connection between application and IoT device to support Telnet, HTTP, HTTPS, UPnP. 2) Connection between IoT device and cloud endpoint to support DNS, NTP, HTTP. 3) Connection between application and cloud endpoint to support DNS, HTTPS.

Third, automated network setup. EmuIoTNet enables multiple users to deploy and operate IoT networks simultaneously. A user simply chooses IoT devices (actually the firmware) and companion application, provides the network topology, and leaves the complicated network setup to EmuIoTNet.

### 3.2 Overview Architecture

The overview architecture of EmuIoTNet is illustrated in Fig. 2. As can be seen, it consists of three components, i.e., Controller, Device Emulator, and Network Configurator.

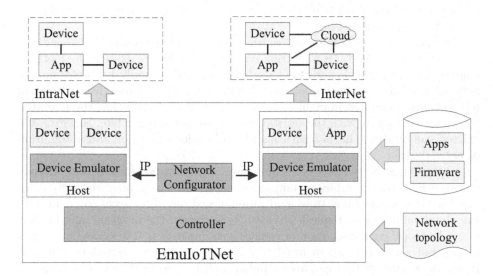

**Fig. 2.** Overview of EmuIoTNet.

**Controller.** When using EmuIoTNet, the user first feeds a network topology which describes i) the nodes within the network (mainly the devices and the companion applications), ii) how the nodes are connected, and iii) whether they access the cloud endpoints. Following the topology, Controller determines the IP address for each node, requests the firmware and applications from the database, and employs Device Emulator to emulate the node, which is either an emulated IoT device for running the firmware or a companion application for operating the emulated IoT device. Finally, it leverages Network Configurator to configure IP for nodes so that the IoT devices, applications, and cloud endpoints are connected.

**Device Emulator.** It is responsible for emulating two types of nodes, i.e., IoT device and companion application. For IoT device, Device Emulator prepares the environment so that the firmware (mainly the services inside, e.g., web server) can be launched and runs continuously. For application, it takes charge of emulating desktop node or mobile node where the application can be deployed, e.g., Netgear Genie that manages Netgear routers, TP-Link that manages TP-Link devices.

**Network Configurator.** It takes charge of connecting IoT devices, applications, and cloud endpoints so that communication protocols, including HTTP, HTTPS, DNS, NTP, UPnP, can be supported. In addition, it is responsible for configuring the IP address on-demand, especially when multiple users simultaneously deploy IoT networks.

### 3.3 Challenges

Although the idea of building IoT networks is simple, there exist several challenges to realize EmuIoTNet.

First, only a small number of firmware can be network reachable using existing emulation methods. E.g., Firmadyne, a full software emulation method, enables 8,591 of 23,035 firmware successfully launched, yet only 1,971 of them are network reachable [7]. Such a low rate definitely limits the capability of building IoT networks. The hybrid emulation method, such as Avatar, has the potential to run more firmware with the assistance of physical hardware [26]. However, the existence of hardware hurts both scalability of device emulation and automation of network setup.

Second, there exist many types of communication needs for different analysis requirements. For example, a companion application may operate many emulated devices at the same time, and an emulated device expects to connect to the cloud endpoints to request services. Accordingly, the analysis may be performed on local communication or remote communication. Therefore, it requires automated network setup to build IntraNet (a local and isolated network) or InterNet (a network that can access public websites) for different requirements.

Third, the firmware often stores the default IP address in a hard-coded manner, which causes many emulated devices to share the same address, for example, 192.168.0.1 and 192.168.1.1 in many Netgear and OpenWrt routers, causing in IP conflicts in building networks. When multiple users deploy networks concurrently, the IP range of these networks should be different. Although manual configuration is feasible, it is tedious and requires network experience.

## 4    Implementation Details

This section describes how to tackle the challenges above. We start by presenting the emulation approach to emulate IoT devices and companion applications. Then, we explain how to set up two network modes. Finally, we introduce how we reconfigure the IP of the emulated IoT devices.

### 4.1    IoT Device Emulation

EmuIoTNet leverages Firmadyne, a full software emulation method, to emulate a single IoT device for a given firmware. As mentioned before, although Firmadyne enables many firmware to run successfully, the success rate is still low. There exist many factors that limit Firmadyne's ability. Of them, the missing or imperfect NVRAM device is one key factor. In detail, many drivers and applications in firmware request execution-related parameters from NVRAM. If they fail to read the value, the application or even the device would stop execution. Firmadyne hard-codes a few commonly used parameters in a simulated NVRAM. Meanwhile, it attempts to read values from several file locations, e.g., */etc./nvram.default*. Unfortunately, these files are often missing, and the hard-coded values cannot satisfy the diverse requirements from a larger amount of firmware, making many firmware crash when requested parameters are not provided.

To mitigate this problem, we implement NvramSim, an adaptive NVRAM simulator capable of providing requested parameters on demand. As shown in Fig. 3, NvramSim is mainly composed of two modules, Nvram Library and Nvram Configuration.

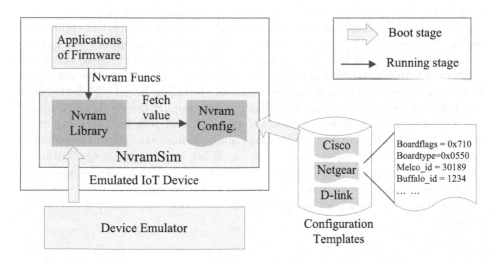

**Fig. 3.** Workflow of NvramSim.

Nvram Library provides interfaces related to NVRAM functions, such as *nvram_get, nvram_set, nvram_get_buffer*. It is statically loaded during the OS boot stage so that those functions issued by the firmware will be redirected to Nvram Library. Subsequently, the associated functions in Nvram Library will operate Nvram Configuration, a file storing parameters that are requested by the firmware, e.g., *boardflags, boardtype* of Netgear. Compared to the hardcoded manner, NvramSim decouples the relatively unified interfaces from diverse configurations. Therefore, the required parameters can be provided at will and modified adaptively without interfering with the interfaces.

Preparing the configuration requires manual effort. Fortunately, the firmware of the same vendor are likely to share similar or even the same parameters. Inspired by this, we can prepare a template configuration for each vendor at prior. Upon device emulation given a firmware, we simply place the associated template configuration and the nvram library into the firmware, as shown in the Boot stage in Fig. 3. Thereafter, when called during the Running stage, the NVRAM functions will fetch the parameters from the configuration and return them to the firmware, thereby enabling the successful execution of the firmware. We have prepared template configurations for many vendors, such as Netgear, TP-Link, D-Link, etc., which help improve the number of emulated IoT devices a lot. It is also feasible for an user to generate a template configuration for a new vendor by googling the NVRAM configurations.

### 4.2   Companion Application Emulation

Companion application is used to control the associated devices. To deploy an application in IoT networks, we first install the application in QEMU virtual machine (VM) configured with desktop OS (e.g., Windows 7) or mobile OS (e.g., Android) and then launch it to connect the emulated device.

Note that it's complicated to automatically install or launch the application unless the operating system is customized. Fortunately, an application intends to manage a series of IoT devices, e.g., TP-Link is compatible with dozens of routers. Therefore, the number of applications is much less than the number of devices. Motivated by this, we adopt the idea of template-based VM deployment used in many cloud platforms. Specifically, we manually create the VM, install the OS and the companion application at prior, and prepare the setup steps such as user registration and application configuration. Afterward, the prepared VM is stored as an application template. When a companion application is desired upon network setup, EmuIoTNet deploys the VM from the associated application template, configures the network on-demand, and starts the VM. Finally, the user launches the application to connect the device for dynamic analysis.

It's worth noting that the installation of Android in QEMU VM is more complicated than Windows, mainly because Android requires specific hardware support such as camera lens and sensors, which are not provided by QEMU. Fortunately, Android-x86, an open-source project, offers a complete solution for running Android on common QEMU platforms [2]. Thus, in EmuIoTNet, we install Android-x86 in QEMU VM to run mobile applications.

### 4.3   Network Models

EmuIoTNet provides two modes of networks, i.e., IntraNet and InterNet, that connect the companion application, emulated devices, and cloud endpoints. Therefore, it supports dynamic analysis on many protocols involved in IoT networks.

**IntraNet.** It is an isolated network that provides local connection between the companion application and the emulated devices, so that Telnet, UPnP, HTTP, and HTTPs traffic between the two can be analyzed. We employ two representative network virtualization tools, Open vSwitch and VXLAN, to set up IntraNet. More specifically, suppose a 3-node network is to be deployed, where $App_1$ and $Dev_1$ are placed on $Host_1$ and $Dev_3$ is placed on $Host_2$, as shown in Fig. 4. We first leverage Open vSwitch to establish a virtual switch on each host, e.g., $vSwitch_1$ and $vSwitch_3$, and connect the network nodes to the associated switch, e.g., $App_1$ and $Dev_1$ to $vSwitch_1$. Then, we use VXLAN to set up a tunnel ($tunnel_1$) between $vSwitch_1$ and $vSwitch_2$, so that the nodes connected to the two switches can communicate with each other. Finally, $App_1$, $Dev_1$ and $Dev_3$ form an isolated network even they are placed on disparate hosts.

**InterNet.** It allows the connection among three components, so that the traffic outside to the cloud endpoints, e.g., DNS, NTP, HTTP, can also be analyzed. To accomplish this, we extend IntraNet with NAT (network address translation). As shown in Fig. 4, $Dev_2$, $App_2$ and $Dev_4$ have formed an IntraNet via $tunnel_2$. Then, we add a gateway $GW$ to $vSwitch_4$, which serves as the gateway of this network. Meanwhile, we use *iptables* to NAT the traffic so that the traffic sent from these nodes will be transmitted to the physical network interface of $Host_2$ and finally reach the cloud endpoints. Eventually, these nodes are able to access cloud endpoints.

Note that EmuIoTNet currently only supports wired communication. Therefore the nodes using other communication types, such as Wifi, Bluetooth, Zigbee, cannot be connected. We leave the support of them as our future work.

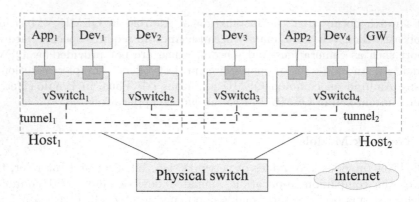

**Fig. 4.** IoT network setup.

### 4.4 IP Configuration

IP configuration sets the desired IP addresses to a node when deploying networks. IP configuration for a VM configured with desktop OS or mobile OS has been well studied. Therefore, this paper focuses on automatically configuring IP addresses for emulated IoT devices. EmuIoTNet presents two methods for IP configuration, i.e., dynamic configuration and static rewriting.

**Dynamic Reconfiguration.** It reconfigures the IP address after the emulated device and its services have been started. Specifically, it executes a configuration script, which involves setting the IP address, adding a route to the gateway, and setting DNS nameserver (only necessary for InterNet). For example, many Linux-based systems place services initialization-related scripts in directory */etc./rc.d*, and execute them in order by the assigned priorities. Therefore, we can place the configuration script in this directory with a lower priority. In this way, the script can be executed automatically, and the reconfiguration result is not affected by other scripts.

**Static Rewriting.** It rewrites the textual configurations or binary applications to set the IP address, netmask, route and gateway. Some firmware store network configuration in a plain text, e.g., */etc./network/interfaces*. In this case, we use string replacement to modify the configuration. However, some firmware hard-code this information in binary applications, which makes rewriting non-trivial since it requires careful operations to ensure binaries' integrity. To achieve this, we design a binary rewriting method, which is simple and easily adapts to binaries on different architectures, such as mipsel, mipseb and armel. Specifically, we open the binary application in binary format and search the old hard-coded IP. Once found, if the old IP is longer than the new assigned IP, then the old IP is replaced directly with the new IP in the binary form. Otherwise, we check $k$ bytes following the old IP, where $k$ is the length of new IP minus the length of old IP. If these bytes are all 0s, then the replacement can be performed safely. Or else, we request Network Configurator to re-assign a shorter IP address, e.g.,

10.0.0.0/24 instead of 192.168.0.0/16 which is used in most firmware, to ensure the new IP can be filled without damaging the binary.

Dynamic reconfiguration is easy to implement. However, it may fail when some shell commands are missing (e.g., *route*), or the */etc./rc.d* directory does not exist. Besides, some firmware may perform improperly if the IP address assigned with dynamic reconfiguration is inconsistent with the hard-coded address. Therefore, static rewriting will be employed to tackle these issues.

## 5   Evaluation

We evaluate EmuIoTNet in three aspects. First, how many firmware can be successfully launched and network reachable, i.e., scalability. Second, how many types of IoT networks can be built, i.e., compatibility. Third, whether it helps to perform analysis, i.e., effectiveness.

To answer these questions, we prepare a dataset from Firmadyne dataset, containing 1,026 firmware from 5 IoT vendors, i.e., TP-Link, D-Link, OpenWrt, Netgear, Tomato by Shibby. These vendors are selected since they release a large account of firmware. Meanwhile, we download 7 associated companion applications, i.e., TP-LINK (desktop and mobile), Mydlink Lite (desktop and mobile), Netgear Genie (desktop and mobile), Openwrt on Android (mobile). Then, we employ EmuIoTNet to emulate devices, build emulated networks, and perform analysis on networks.

### 5.1   Scalability in Device Emulation

EmuIoTNet employs Firmadyne to emulate a single device. Thus, it is expected to show well scalability and compatibility benefitted from Firmadyne. Moreover, it improves Firmadyne with NvramSim in device emulation. We first compare the two in terms of three gradual metrics, i.e., how many firmware can be successfully launched, network reachable, and port opened. Here, a firmware is launched successfully if its initial services and applications can be started. Following Firmadyne, we use whether the IP can be inferred as the sign. Moreover, it is network reachable if the IP address gets pinged, and it is port opened if any of the following ports is opened, i.e., 21 of FTP, 22 of SSH, 23 of Telnet, 53 of DNS, 80 and 8080 of HTTP, 123 of NTP, 443 of HTTPS, 1900 and 5000 of UPnP.

As illustrated in Table 1, EmuIoTNet launches more firmware than Firmadyne and enables more emulated devices to be network reachable and port opened.

**Table 1.** Comparison of device emulation.

| Vendor | Count | Extracted | Firmadyne | | | EmuIoTNet | | |
|---|---|---|---|---|---|---|---|---|
| | | | $L^*$ | NR | $PO$ | L | NR | PO |
| D-Link | 113 | 59 | 42 | 24 | 22 | 44 | 33 | 31 |
| TP-Link | 184 | 68 | 48 | 42 | 40 | 48 | 42 | 40 |
| Tomato | 102 | 102 | 6 | 6 | 0 | 36 | 36 | 36 |
| Netgear | 282 | 113 | 29 | 17 | 8 | 49 | 33 | 29 |
| OpenWrt | 345 | 328 | 90 | 52 | 0 | 116 | 63 | 0 |
| Total | 1,026 | 670 | 215 | 141 | 70 | 293 | 207 | 136 |

*L (launched), NR(network reachable), PO(port opened).

For example, for 282 Netgear firmware of which 113 are extracted[1], Firmadyne successfully launches 29 and enables 17 to be network reachable. As a comparison, EmuIoTNet launches 49 firmware, and 33 of them are network reachable. Meanwhile, it increases the number of firmware whose ports are opened from 8 to 29. The improvement is contributed to NvramSim, which provides the required parameters upon initialization of services and applications. Thus, many services and applications can start and run successfully. Note that EmuIoTNet does not improve Firmadyne for TP-Link firmware, mainly because the parameters required by TP-Link firmware have been hard-coded in the simulated NVRAM of Firmadyne. In total, EmuIoTNet improves Firmadyne by 36.3%, 47.1%, and 94.3% respectively in three metrics. Consequently, it provides opportunities for performing dynamic analysis on more firmware.

## 5.2 Compatibility in Network Setup

We employ EmuIoTNet to build networks that connect the companion application, emulated devices, and cloud endpoints. Note that most applications interact with devices via HTTP interface (e.g., port 80), thereby ignoring other channels. Thus, we also deploy SSH client and Telnet client as applications to establish more communication channels. In addition, although many vendors have released companion applications to control the devices, an application may only work with a certain series of devices and firmware versions. E.g., Netgear Genie currently supports 18 Netgear devices, which is far smaller than the total number of released devices. Considering that many devices, e.g., routers, provide control-related HTTP interface that allows both applications and web browsers to interact with, we therefore install Chrome browser in VM and use the browser as the companion application to connect emulated devices.

When building a network, we only use one device and one application. If the two can communicate with each other, then the network is built successfully.

---

[1] Some firmware cannot be extracted if they are encrypted or do not contain a valid file system.

Table 2 lists the number of networks via various communication channels. As can be seen, of 33 Netgear emulated devices that are network reachable, 21 are connected with HTTP applications. Therefore, these 21 networks support the analysis on HTTP traffic. Meanwhile, 2 and 14 emulated devices can be connected with SSH and Telnet clients respectively, and 28 emulated devices request DNS from public sites. OpenWrt devices are unable to build networks since all their ports are closed. Despite this, the results in Table 2 demonstrate that EmuIoTNet can build a variety of networks to support analysis on many communication protocols.

**Table 2.** Number of networks via different communications.

|          | Count | SSH | Telnet | HTTP | DNS |
|----------|-------|-----|--------|------|-----|
| D-Link   | 33    | 0   | 18     | 27   | 2   |
| TP-Link  | 42    | 0   | 1      | 26   | 0   |
| Tomato   | 36    | 32  | 24     | 23   | 33  |
| Netgear  | 33    | 2   | 14     | 21   | 28  |
| OpenWrt  | 63    | 0   | 0      | 0    | 0   |

### 5.3 Dynamic Analysis on Networks

We perform analysis on the emulated devices and networks. Some important results are listed below.

1) **DNS services.** 63 emulated devices use open DNS services, i.e., 8.8.8.8 of Google, 208.67.222.222/220 and 208.67.220.220/222 of OpenDNS. The use of open DNS services is vulnerable to DNS cache poisoning. DNS cache poisoning is a user-end method of DNS spoofing. The user system will record the fraudulent IP address in the local memory cache, causing DNS to recall bad sites. In this way, users are at risk of personal data leakage, malware infection, and halted security updates.

2) **HTTP protocol.** 97 emulated devices use HTTP protocol and 7 of them do not use TLS/SSL sessions (HTTPS). HTTP is insecure since its data is not encrypted, and it can be intercepted by attackers to gather data. Built on top of HTTP, HTTPS uses SSL or TLS to provide encryption processing data, verify the identity of the other party, and protect data integrity.

3) **UPnP.** 15 emulated devices open 1900 or 5000 port (UPnP), which originally designed to find other smart devices and connect to them automatically. However, the use of open UPnP may allow an attacker to gain control of multiple devices once a single device has been hijacked. The vulnerabilities in the UPnP protocol can also lead the router to direct the network to another remote address (rather than the local IP address) by malicious programs or make the device at the risk of being exploited by attackers to perform DDoS attack.

4) **NTP.** 55 emulated devices use NTP protocol for synchronizing the time. It's known that NTP is vulnerable to NTP Reply Flood Attack, DDoS attack, MITM attack, etc.
5) **MITM.** We use the emulated networks to detect whether the communication is susceptible to MITM attack. MITM attack means a perpetrator positions himself in a conversation between a user and an application—either to eavesdrop or to impersonate one of the parties, making it appear as if a normal exchange of information is underway, aiming to steal personal information. We use Wireshark and sslsplit to inspect HTTP and HTTPS traffic and test their susceptibility to MITM attack. Of the 20 communications we inspected, 3 are susceptible to a MITM attack.

## 6    Discussion and Conclusion

This paper presents EmuIoTNet, a tool of automatically building emulated IoT networks for dynamic analysis. Compared to existing works on a single IoT device emulation, EmuIoTNet provides connections between IoT devices, applications, and cloud endpoints, enabling many network traffic types to be analyzed. A set of results show that EmuIoTNet is scalable and compatible in building IoT networks and effectively support dynamic analysis.

There exist two main limitations in EmuIoTNet. First, limited by Firmadyne, EmuIoTNet currently only emulates wired routers and switches yet fails for wireless devices and devices that require specific hardware (e.g., camera lens). Second, it requires manual operations for dynamic analysis, including tool setup, traffic capture and analysis. In the future, we will work towards these two directions. We plan to explore other emulation methods, e.g., Avatar, to emulate more devices. In addition, we plan to integrate several automated testing tools, e.g., fuzzing tools, to support automated analysis.

## References

1. Anatomy of an iot malware attack. https://developer.ibm.com/articles/iot-anatomy-iot-malware-attack/
2. Android-x86. https://www.android-x86.org/
3. Iot attacks up significantly in first half of 2019. https://www.darkreading.com
4. Sonicwall: Encrypted attacks, IoT malware surge as global malware volume dips. http://blog.sonicwall.com
5. Alrawi, O., Lever, C., Antonakakis, M., Monrose, F.: Sok: security evaluation of home-based IoT deployments. In: S&P, pp. 1362–1380 (2019)
6. Antonakakis, M., et al.: Understanding the mirai botnet. In: USENIX Security Symposium, pp. 1093–1110 (2017)
7. Chen, D.D., Woo, M., Brumley, D., Egele, M.: Towards automated dynamic analysis for Linux-based embedded firmware. NDSS **16**, 1–16 (2016)
8. Chen, J., et al.: Iotfuzzer: discovering memory corruptions in IoT through app-based fuzzing. In: NDSS (2018)

9. Chipounov, V., Kuznetsov, V., Candea, G.: S2e: a platform for in-vivo multi-path analysis of software systems. Acm Sigplan Notices **46**(3), 265–278 (2011)
10. Clements, A.A., et al.: Halucinator: firmware re-hosting through abstraction layer emulation. In: 29th USENIX Security Symposium (USENIX Security 20), pp. 1201–1218 (2020)
11. Costin, A., Zaddach, J., Francillon, A., Balzarotti, D.: A large-scale analysis of the security of embedded firmwares. In: USENIX Security Symposium, pp. 95–110 (2014)
12. Costin, A., Zarras, A., Francillon, A.: Automated dynamic firmware analysis at scale: a case study on embedded web interfaces. In: AsiaCCS, pp. 437–448 (2016)
13. Davidson, D., Moench, B., Ristenpart, T., Jha, S.: Fie on firmware: finding vulnerabilities in embedded systems using symbolic execution. In: 22nd USENIX Security Symposium (USENIX Security 2013), pp. 463–478 (2013)
14. Feng, B., Mera, A., Lu, L.: P 2 IM: scalable and hardware-independent firmware testing via automatic peripheral interface modeling. In: USENIX Security Symposium (2020)
15. Gustafson, E., et al.: Toward the analysis of embedded firmware through automated re-hosting. In: 22nd International Symposium on Research in Attacks, Intrusions and Defenses (RAID 2019), pp. 135–150 (2019)
16. Kammerstetter, M., Burian, D., Kastner, W.: Embedded security testing with peripheral device caching and runtime program state approximation. In: 10th International Conference on Emerging Security Information, Systems and Technologies (SECUWARE) (2016)
17. Kammerstetter, M., Platzer, C., Kastner, W.: Prospect: peripheral proxying supported embedded code testing. In: Proceedings of the 9th ACM Symposium on Information, Computer and Communications Security, pp. 329–340 (2014)
18. Koscher, K., Kohno, T., Molnar, D.: SURROGATES: enabling near-real-time dynamic analyses of embedded systems. In: 9th USENIX Workshop on Offensive Technologies (WOOT 15) (2015)
19. Li, H., Tong, D., Huang, K., Cheng, X.: Femu: a firmware-based emulation framework for soc verification. In: CODES+ISSS, pp. 257–266 (2010)
20. Magnusson, P.S., et al.: Simics: a full system simulation platform. Computer **35**(2), 50–58 (2002)
21. Muench, M., Nisi, D., Francillon, A., Balzarotti, D.: Avatar2: a multi-target orchestration platform. Workshop Binary Anal. Res. **18**, 1–11 (2018)
22. Sha, L., Xiao, F., Chen, W., Sun, J.: Iiot-sidefender: detecting and defense against the sensitive information leakage in industry IoT. World Wide Web **21**(1), 59–88 (2018)
23. Srivastava, P., Peng, H., Li, J., Okhravi, H., Shrobe, H., Payer, M.: Firmfuzz: automated IoT firmware introspection and analysis. In: IoT S&P, pp. 15–21 (2019)
24. Talebi, S.M.S., Tavakoli, H., Zhang, H., Zhang, Z., Sani, A.A., Qian, Z.: Charm: facilitating dynamic analysis of device drivers of mobile systems. In: USENIX Security Symposium, pp. 291–307 (2018)
25. Yu, T., Sekar, V., Seshan, S., Agarwal, Y., Xu, C.: Handling a trillion (unfixable) flaws on a billion devices: rethinking network security for the internet-of-things. In: Hot Topics in Networks, pp. 1–7 (2015)
26. Zaddach, J., Bruno, L., Francillon, A., Balzarotti, D., et al.: Avatar: a framework to support dynamic security analysis of embedded systems' firmwares. NDSS **14**, 1–16 (2014)

27. Zhang, L., Chen, J., Diao, W., Guo, S., Weng, J., Zhang, K.: Cryptorex: large-scale analysis of cryptographic misuse in IoT devices. In: 22nd International Symposium on Research in Attacks, Intrusions and Defenses (RAID 2019), pp. 151–164 (2019)
28. Zhang, Z.K., Cho, M.C.Y., Shieh, S.: Emerging security threats and countermeasures in IoT. In: Proceedings of the AsiaCCS, pp. 1–6 (2015)
29. Zheng, Y., Davanian, A., Yin, H., Song, C., Zhu, H., Sun, L.: Firm-afl: high-throughput greybox fuzzing of IoT firmware via augmented process emulation. In: USENIX Security Symposium, pp. 1099–1114 (2019)

# Software Security

# ACGVD: Vulnerability Detection Based on Comprehensive Graph via Graph Neural Network with Attention

Min Li[1,2], Chunfang Li[1,2], Shuailou Li[1,2], Yanna Wu[1], Boyang Zhang[1], and Yu Wen[1(✉)]

[1] Institute of Information Engineering, Chinese Academy of Sciences, Beijing, China
{limin1,lichunfang,lishuailou,wuyanna,zhangboyang,wenyu}@iie.ac.cn
[2] School of Cyber Security, University of Chinese Academy of Sciences, Beijing, China

**Abstract.** Vulnerability is one of the main causes of network intrusion. An effective way to mitigate security threats is to find and repair vulnerabilities as soon as possible. Traditional vulnerability detection methods are limited by expert knowledge. Existing deep learning-based methods neglect the connection between semantic graphs and cannot effectively deal with the structure information. Graph neural network brings new insight into vulnerability detection. However, benign nodes on the graph account for a large proportion, resulting in vulnerability information could be disturbed by them. To address the limitations of existing vulnerability detection approaches, in this paper, we propose ACGVD, a vulnerability detection method by constructing a graph network with attention. We first combine multiple semantic graphs together to form a more comprehensive graph. We then adopt the Graph neural network instead of the sequence-based model to automatically analyze the comprehensive graph. In order to solve the problem that the vulnerability information could be covered up, we add a double-level attention mechanism to the graph model. We also add a novel classification layer to extract the high-level features of the code. To make the experiment more realistic, the model is trained over the latest published real world dataset. The experiment results demonstrate that compared with state-of-the-art methods, our model ACGVD achieves 5.01%, 13.89%, and 8.27% improvement in accuracy, recall and F1-score, respectively.

**Keywords:** Vulnerability detection · Graph neural network · Attention · Comprehensive graph

## 1 Introduction

The existence of vulnerabilities allows attackers to access to or destroy the system without authorization, leading to great economic losses for users and endangering national security [23,33]. Although there have been many vulnerability detection tools, the number of new vulnerabilities is still emerging in an endless stream

© Springer Nature Switzerland AG 2021
D. Gao et al. (Eds.): ICICS 2021, LNCS 12918, pp. 243–259, 2021.
https://doi.org/10.1007/978-3-030-86890-1_14

[4, 5, 11] due to the high complexity of software writing process. As a result, it is necessary to find and repair vulnerabilities earlier to minimize the loss [16].

Traditional vulnerability detection methods can be categorised into static analysis [2, 18, 28], dynamic analysis [9, 13, 19, 27, 34] and hybrid analysis [15, 31]. However, these methods rely on expert knowledge and known vulnerability related patterns, resulting in low coverage of vulnerability types. The emergence of deep learning has alleviated the reliance on human experts. Deep learning can automatically extract vulnerability feature information from training samples. But this method has the following problems. The program is either treated as a simple natural language sequence [14] or a single semantic graph [8, 17, 20–22, 24–26]. The former treatment of program completely loses the structural information of the code. The latter neglects the relationship between semantic graphs despite it considers the structural information. For example, the semantic Control Flow Graph (CFG) cannot detect data related vulnerabilities, syntax vulnerabilities cannot be identified by CFG and Data Flow Graph (DFG). Therefore, only when we consider theses semantic graphs together can we detect more types of vulnerabilities. Besides, the semantic graph is directly put into the sequence-based model for classification in deep learning. The sequence-based models are not applicable to non-Euclidean structures on graphs. Because each node in a graph has a different number of neighbors.

To cope with above challenges, we propose a more comprehensive graph which combines multiple semantic graphs. Thus, we can incorporate more sufficient syntax and semantic information of vulnerabilities. In addition, the emergence of graph neural network offers a feasible solution to process the non-Euclidean structures [36]. But the graph model is easy to be disturbed by considerable benign nodes, leading to vulnerability information cannot be fully explored.

```
1 static int tcp_set_msgfds(CharDriverState *chr, int *fds, int num)
2 {
3     TCPCharDriver *s = chr->opaque;
4     /* clear old pending fd array */
5     g_free(s->write_msgfds);
6     if (num) {
7         s->write_msgfds = g_malloc(num * sizeof(int));
8         memcpy(s->write_msgfds, fds, num * sizeof(int));
9     }
10    s->write_msgfds_num = num;
11    return 0;
12 }
```

(a)

```
1 static int tcp_set_msgfds(CharDriverState *chr, int *fds, int num)
2 {
3     TCPCharDriver *s = chr->opaque;
4     /* clear old pending fd array */
5     if (s->write_msgfds) {
6         g_free(s->write_msgtds);
7     }
8     if (num) {
9         s->write_msgfds = g_malloc(num * sizeof(int));
10        memcpy(s->write_msgfds, fds, num * sizeof(int));
11    }
12    s->write_msgfds_num = num;
13    return 0;
14 }
```

(b)

**Fig. 1.** An example of the *Double Free* vulnerability.

As shown in Fig. 1(a), this is a potential *Double Free* vulnerable code from Qemu dataset, due to it releases the heap without first determining whether $'s-> write\_msgfds'$ has been freed. The patched code is shown in Fig. 1(b). The vulnerability exists in line 5 of the code. Lines 1 and 3 of the code have data dependence on the vulnerability, while lines 6, 7, 8, 9, 10 and 11 are non-vulnerable lines. With the increase of the number of function lines, more non-vulnerable lines corresponding to the benign nodes on graph will occur, which bring much interference to judgment. Consequently, the node representation of the whole graph vector tends to be benign, and the vulnerability information could be covered. Hence, we design a node-level attention mechanism to pay more attention to those vulnerability related nodes, when the number of nodes is relatively large.

In addition, in some cases, a few semantic graphs in the comprehensive graph are more important than other semantic graphs for vulnerability detection. As shown in Fig. 1, DFG contains more information about the flow direction of vulnerability related data $'s- > write\_msgfds'$. Therefore, DFG is more important for vulnerability identification, while the rest of the semantic graph only provides auxiliary information for vulnerability detection. Hence, our work also proposes a path-level attention mechanism, believing it can automatically give more attention to semantic graphs that are more relevant to the vulnerability.

Finally, in order to separate the vulnerable features from the non-vulnerable features, we adopt a novel classifier model that is more suitable for our needs and conduct experiments on the latest public dataset.

The main contributions of this paper are as follows:

- We combine multiple semantic graphs to form a more comprehensive code graph representation. By taking into account all the semantic graphs (i.e., CFG and DFG), we can detect more types of vulnerabilities than single semantic graph.
- In order to make better use of the structural information of the code, we use the graph neural network to learn the vector representation of the code automatically. So the long-range dependencies induced by using the same variable or function in distant locations can be treat as a relationship between neighbors. And we propose a novel classifier layer, which can extract more high-level code features after graph networks.
- To alleviate the problem that the vulnerability information in the graph is easy to be concealed, the node-level and path-level attention mechanism are designed to extract the vulnerability information. The double-level attention mechanism can learn attention weights on different parts of graph, effectively paying more attention to those vulnerable related information.
- Finally, we evaluate ACGVD on the latest published real-world dataset, and the results prove the effectiveness of our scheme.

The rest of this paper is organized as follows: Sect. 2 reviews the related work of this paper. Section 3 details the design of the proposed system. In Sect. 4, we introduce the implementation environment of the experiment. In Sect. 5, we provide the experiment results and provide case studies to demonstrate how our model performed compared with other methods. In Sect. 6, we analyze the threats to performance of our work. Section 7 concludes the paper.

## 2   Related Work

The traditional static analysis methods include inference method [2] and rule-based method [18,28]. Inference method [2] is to conclude all possible behaviors in the process of program execution to eliminate the existence of vulnerabilities, but this method is limited by the generalization of behavior types. Rule-based method [28] uses pre-defined rules to observe matching degrees between programs and rules. However, the recognition accuracy depends on the quality of rules [31].

Different from static analysis, dynamic analysis method can monitor the behavior of program during running. Dynamic analysis methods include symbolic execution [19], fuzzing [34], and taint analysis [27]. Ghaffarian et al. [15] by putting specific test cases to determine whether the program has vulnerabilities during the program running. Newsome et al. [27] generated path constraint conditions to judge path reachability by collecting branch conditions and related variables during program execution. In order to make full use of the advantages of dynamic analysis and static analysis, a hybrid analysis method is proposed in [15], but it is easily limited by the two methods.

Work [29,30] regard the program as a text sequence. Scandariato et al. [30] directly uses the Continuous Bag-Of-Words (CBOW) technology in natural language processing technology to express the program as a collection of related tokens as the feature of the vulnerability, ignoring the sequence feature of the code. Dam et al. [14] use Long Short-Term Memory (LSTM) technology to capture the context of code, although this method can capture the sequence relationship between codes better, it ignores the structural features between codes. Li et al. [20] take the structural information between codes into consideration, the program is first transformed into Abstract Syntax Tree (AST), and then four rules are used to extract nodes to obtain intermediate code- and Semantics-based Vulnerability Candidate (iSeVC). After that, it is converted into sequence tokens and input into the sequence model Recurrent Neural Network (RNN) for classification. Harer et al. [17] obtain the Control Flow Graph (CFG) according to the control flow relationship in the code, and make the vector representation of fixed size by hand, which is input into the Random Forest (RF) for learning later. Li et al. [21] use Program Dependence Graph (PDG) as a code feature representation, PDG increases data dependence on the basis of CFG. This method improves the detection rate of vulnerabilities related to both data and control. Although [17,20,21] consider the structure information of the code by using the semantic graph, they do not make full use of the connection between these semantic graphs. At the same time, in order to reduce the complexity of calculation, they first convert these graphs into sequences, and then input them into the sequence-based classifier for classification. This method cannot deal with the structure information perfectly.

Zhou [36] propose a new *Conv* module for the first time, using graph neural network to learn code features, no longer dependent on the sequence-based model. On the basis of the work of [36], Chakraborty et al. [10] focus on solving the problem of dataset imbalance. But different from our work, all their work do not consider the vulnerability information could be interfered by benign information.

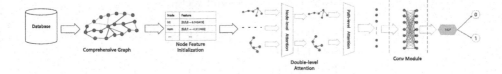

**Fig. 2.** Overview of the ACGVD vulnerability detection framework.

# 3 ACGVD Pipeline

## 3.1 Overview of ACGVD

This work focuses on vulnerability detection at the individual function level. Given a function, we need to accurately identify whether it contains vulnerabilities. If it contains vulnerabilities, then the label is '1', otherwise, the label is '0'. We will give a brief introduction to the ACGVD pipeline as shown in the Fig. 2.

First, we use the Joern[1] tool to transform the samples in the database into comprehensive graphs, and then the initial node feature representations of the structure graph are obtained by Word2vec[2] model. Next, we use a double-level attention mechanism to update the graph node information. Finally, the node information is aggregated and input into convolution module to get the high-level features of the code. And a MLP module is used to compress the dimension of high-level features to judge whether the attributes of the sample contain vulnerabilities.

## 3.2 Comprehensive Graph Representation

In the field of vulnerability detection, the semantic graph representation of code is various. In fact, whether building AST graph, CFG graph or PDG graph, programs need to be transformed into different types of nodes and edges first, and then connected to construct graphs. To cover more vulnerability information, we use 12 connection relationships between nodes to construct a more comprehensive code graph as shown in Table 1.

$IS\_AST\_PARENT$ edge represents that AST graph is being constructed. AST is an intermediate representation in the process of program compilation, which stores the syntax information of functions. While $IS\_AST\_PARENT$ edge can construct the structures of the program, it is unable to reason about program behavior without understanding the semantics of the structures on AST. Semantic information can be obtained by the data flow and control flow information. $FLOWS\_TO$ edge describes the order in which code statements are executed, and the conditions that need to be met for a particular path of execution to be taken. $REACHES$ edge represents the data dependency between variables. $Use\text{-}Def$ edge captures the variable to the sub-tree that uses the defined value.

---

[1] https://joern.readthedocs.io/en/latest/index.html.
[2] http://radimrehurek.com/gensim/models/word2vec.html.

**Table 1.** Edge types and vector.

| Edge type | Vector |
|---|---|
| IS_AST_PARENT | 1 |
| FLOWS_TO | 3 |
| USE | 5 |
| DEF | 4 |
| REACHES | 6 |
| CONTROLS | 7 |
| DOM | 9 |
| POST_DOM | 10 |
| IS_FUNCTION_OF_AST | 11 |
| IS_FUNCTION_OF_CFG | 12 |
| IS_CLASS_OF | 2 |
| DECLARES | 8 |

We use the Joern tool to convert each sample from the dataset into its corresponding comprehensive graph. Under the *REACHES* connection relationship, the constructed semantic graph of the code in Fig. 1 can be represented as shown in Fig. 3. The final Comprehensive Graph of the Fig. 1 can be seen in Fig. 4. For better visualization, Fig. 4 only shows a comprehensive graph representation composed of three edges.

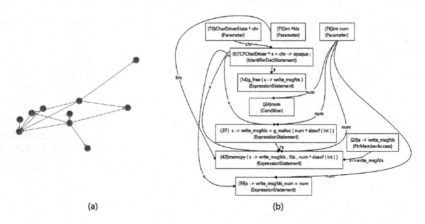

(a)                                        (b)

**Fig. 3.** Semantic graph of the *REACHES* connection relationship. (a) is a simplified version of the semantic graph which nodes with only Key values; (b) corresponds the Key values to the code, where the number in brackets are the Key values in (a).

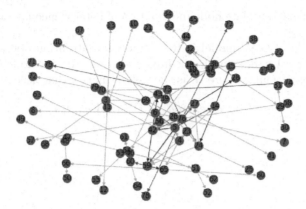

**Fig. 4.** A comprehensive graph representation composed of three edges: the *REACHES* represented by blue edge, *FLOWS_TO* represented by red edge and *IS_AST_PARENT* represented by yellow edge. (Color figure online)

### 3.3   Node Feature Initialization

For each node in the graph, as shown in Fig. 3(b), it contains two parts of information: code and node type. Since edge types are exhaustible, we replace them with a preset number vector as shown in Table 1. These vectors are transformed into one-hot form later. The node types are generated from AST by the Joern.

In order to make better use of the sequence information of the code, the word2vec model is used to learn the vector representation of the node code. The comprehensive graph can be expressed as: $G = (\nu, \varepsilon)$, where $\nu$ represents the node set of the graph, and $\varepsilon$ represents the different edge set of the graph. Concretely, the edge type is represented as $\varepsilon_j$, $j \in K$, $K = 12$. The node feature of each node is represented as $h_i = [t_i, c_i]$, $i \in N$, where $t_i$ refers to the node type vector, $c_i$ refers to the node code vector, and $N$ is the total number of nodes in the graph.

### 3.4   Double-Level Attention Mechanism

**Node Level Attention.** Considering that in a semantic graph, the benign nodes account for a large proportion. When we update the state vectors by aggregating all incoming messages from node neighbors, the vectors tend to benign due to that the most information it learned is non-vulnerable. We use the attention mechanism to imitate manually analyzing vulnerabilities. By giving more weight to vulnerability related nodes, the node vector can contain more vulnerable related features.

This paper uses the Graph Attention Networks (GAT) model [32] to implement the node-level attention mechanism. However, GAT cannot process heterogeneous graphs. Therefore, we first divide heterogeneous graphs into multiple homogeneous graphs according to edge types. And then the GAT model is adopted to learn and update node information of heterogeneous graphs as shown in Fig. 5.

The key step of GAT model is to perform self-attention operation on each node in the graph. Under $\varepsilon_k$ path, the importance of node $j$ to node $i$ is calculated by the following formula, where $W$ is the weight matrix that needs to be learned.

$$e_{ij}^{\varepsilon_k} = attention_{node}(Wh_i, Wh_j; \varepsilon_k) \tag{1}$$

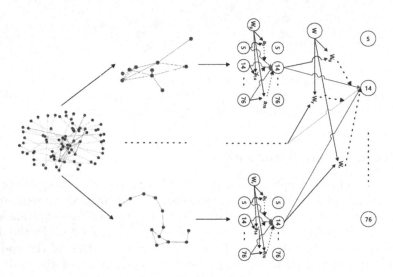

**Fig. 5.** Double-level attention mechanism.

The attention coefficient is normalized and then combined with its corresponding features to get the updated node features as follows:

$$\tilde{h}_i^{\varepsilon_k} = \sigma\left( \sum_{j \in N_i^{\varepsilon_k}} (soft\max_j(e_{ij}^{\varepsilon_k}) \cdot Wh_j) \right) \tag{2}$$

**Path Level Attention.** Majority of vulnerabilities are subtle, they only can be spotted with a join consideration of the composite semantic graphs. However, there are still a small number of vulnerabilities much related to some certain semantic graphs. In response to this situation, we propose a path-level attention mechanism. The graph model can automatically pay more attention to venerability related semantic graphs and weaken the weight of non-vulnerable semantic graphs.

This paper adopts the semantic-level attention [35] to realize the path-level attention mechanism. In order to learn the importance of each path, a layer of Multilayer Perceptron (MLP) is applied to transform the embedding of a specific path nonlinearly, and the similarity between semantic embedding is measured by the path level vector q. Finally, the result will be averaged.

$$\omega_{\varepsilon_k} = \frac{1}{|\nu|} \sum_{i \in \nu} q^T \cdot \tanh(W \cdot \tilde{h}_i^{\varepsilon_k} + b) \tag{3}$$

Normalization is given:

$$\beta_{\varepsilon_k} = \frac{\exp(\omega_{\varepsilon_k})}{\sum_{k=1}^{K} \exp(\omega_{\varepsilon_k})} \tag{4}$$

Integrating different path information, the final embedding is as follows:

$$\tilde{h} = \sum_{i=1}^{K} \beta_{\varepsilon_k} \cdot \tilde{h}_{\varepsilon_k} \tag{5}$$

### 3.5  Classifier Module

The goal of [35] is the node level classification task. Different from their work, the input sample of graph model in this paper is function-level code. So the task is a graph-level classification. The general method of graph classification is to add a MLP layer at the end of graph network. First, the embedded representation of nodes learned by the neural network $H_i$ and the initialized node features $h_i$ are connected together, and next they are input into the MLP for predicting the labels of samples. As shown below:

$$\tilde{y} = sigmoid(\sum_{i \in N} MLP(E_i)) \tag{6}$$

where

$$E_i = [H_i, h_i] \tag{7}$$

Different from the general graph classification work, the connected nodes are convolved at two layers in this work to obtain the high-level features which are more related to vulnerabilities. Finally, the learned features are fed into the multi-layer fully connected neural network for binary classification.

$$y_1 = MAXPOOL\_1(RELU(CONV\_1(E_i))) \tag{8}$$

$$y_2 = MAXPOOL\_1(RELU(CONV\_2(y_1))) \tag{9}$$

$$\tilde{y} = sigmoid(\sum MLP(y_2^T)) \tag{10}$$

If the final value of $\tilde{y}$ is 1, it means that the function contains vulnerabilities. Otherwise, the function is benign.

# 4    Experiments

## 4.1    Datasets

The quality of datasets has a great impact on experimental results. The existing public datasets for vulnerability detection are mostly derived from Juliet Test Suite [12], NVD [3], and SARD [7]. The Juliet Test Suite and the SARD are synthetic data, which are artificially synthesized using known vulnerability patterns and do not have authenticity. NVD are semi-synthetic data, it cannot fully capture the complexity of the real world. Considering this situation, we use the latest datasets FFmpeg and Qemu in [36] as our dataset to ensure the authenticity and certainty of our experimental data. The dataset information is shown in Table 2.

**Table 2.** Datasets.

| Dataset | Graphs | Vul graphs | Non_Vul graphs |
|---------|--------|------------|----------------|
| FFmpeg | 6716 | 3420 | 3296 |
| Qemu | 15645 | 6648 | 8997 |
| FFmpeg+Qemu | 22361 | 10068 | 12293 |

## 4.2    Implementation Details

We use Pytorch 1.5.0 with Cuda version 10.2 to implement our method. We ran our experiments on Nvidia Tesla V100 PCIe 32 GB GPU, Intel(R) Xeon(R) Gold 5115 CPU @ 2.40 GHz with 128 GB ram. In order to eliminate the randomness of the experiment, we perform 15 independent experiment runs. And the final result adopts the median value.

In the stage of word2vec, we encode the node code into vector with 100 dimensions, with node features of 69 dimensions and edge types of 12 dimensions. Batch size is 128; learning rate is 0. 0001. Other parameter configuration of the model is shown in Table 3.

**Table 3.** Parameter settings.

| Model | Parameter | Value |
|---|---|---|
| Hierarchical attention module | Input size | 169 |
| | Hidden size | 200 |
| | Out size | 100 |
| | Num of heads | 4 |
| | Dropout | 0.75 |
| | Learning rate | 0.0001 |
| Classification layer | Conv filter | (1,3) (1,1) |
| | Pool filter | (1,3) (2,2) |
| | Dropout | 0.2 |
| | Activation function | RELU |
| | Number of hidden layers | 4 |

### 4.3 Evaluation Metrics

This paper uses four popular evaluation metrics: Accuracy, Precision, Recall, and F1-score. Although there is no inevitable correlation between Precision and Recall according to the calculation formula, these two metrics are often restricted each other. In general, if the Precision is high, Recall could be low. The F1 score is used to weigh the two metrics comprehensively.

## 5 Experiments Study

### 5.1 How Effective Is ACGVD When Compared with the Traditional Static Analysis Tools?

Traditional vulnerability analysis methods include static analysis, dynamic analysis and hybrid analysis. Since we focus on source code level vulnerability identification, we will compare with static analysis methods.

**Table 4.** Comparison between ACGVD and static analysis tools.

| Method | Accuracy (%) | Precision (%) | Recall (%) | F1 (%) |
|---|---|---|---|---|
| Flawfinder [1] | 54.32 | 49.76 | 14.65 | 22.64 |
| RATS [6] | 53.86 | 46.05 | 6.65 | 11.61 |
| ACGVD | **63.58** | **57.12** | **76.62** | **65.45** |

We compare ACGVD with open source static analysis tools, such as Flatfinder [1] and RATS [6]. These tools have been widely used in the field of

C/C++ source code vulnerability detection. The experimental results are shown in Table 4.

From Table 4 we can see that although Flatfinder and RATS perform well on accuracy (about 50%), the recall rate is very low (less than 20%). The reason is that static analysis tools mistakenly judge many benign samples as vulnerability samples. Compared with these tools, our model ACGVD has a better accuracy rate and a better recall rate.

### 5.2    How Effective Is ACGVD When Compared with Deep Learning Method Based on Single Semantic Graph?

On the one hand, in order to make full use of the vulnerability information contained in each semantic graph, we combine multiple semantic graphs together to make a comprehensive graph. On the other hand, instead of the traditional sequence-based model, we use graph neural network to aggregate the structure information of learning code. We compare ACGVD with the state-of-the-art deep learning methods based on single semantic graph and sequence model. The experimental results are shown in Table 5.

Table 5. Comparison between ACGVD and deep learning method.

| Method | Accuracy (%) | Precision (%) | Recall (%) | F1 (%) |
|---|---|---|---|---|
| Russel [29] | 58.13 | 54.04 | 39.5 | 45.62 |
| VulDeePecker [22] | 53.58 | 47.36 | 28.7 | 35.2 |
| SySeVR [21] | 52.52 | 48.34 | 65.96 | 56.03 |
| ACGVD | **63.58** | **57.12** | **76.62** | **65.45** |

We can see that, whether the code is treated as a natural language sequence [29], a gadeget fragment [22], or the PDG graph [21] and then input into the sequence-based model (e.g., CNN, RNN, BLSTM) for classification, the experiment result is not good as our model. Concretely, BGNN4VD achieves 11.06% higher on accuracy, 8.78% higher on recall, 10.66% higher on precision, and 9.42% higher on F1-measure than SySeVR which perform best in the three deep learning methods we compared. This shows the advantage of using comprehensive graph and graph neural network, which can consider more vulnerability information and make better use of code structure information.

Here we focus on the analysis of why the method of [22] performs the worst on our dataset. [22]'s motivation is a little similar to ours. They hold that the proportion of vulnerable code in a program is too low, resulting vulnerable and non-vulnerable code is hardly distinguishable. In order to locate the vulnerability lines more accurately, they represent the program as a more sophisticated representation than a function. Therefore, they propose the concept of code gadgets according to CFG and DFG information, and then classify gadgets by using sequence-based model. They rely on some expert knowledge in the process of extracting gadgets, which makes them only identify vulnerabilities related to library/API calls. Considering more types of vulnerabilities and building more accurate gadgets have become their limitations, which are also the main reason for their poor performance in our experiment.

### 5.3 How Effective Is ACGVD When Compared with Graph Neural Network Method Without Attention Mechanism?

Since Devign [36] did not publish their implementation. And there is little information about the configuration of parameters. We encountered difficulties in reproduce the process. Chakraborty et al. [10] reproduces the work of Devign [36], and opens all experimental configurations. We quote the experimental results, which are shown in Table 6. ACGVD outperforms Devign [36] in all performance metrics. ACGVD's accuracy, precision, recall, and F1-scores are 5.01%, 3.52%, 13.89%, and 8.27% higher respectively than Devign [36]. This proves the superiority of ACGVD.

Chakraborty et al. [10] is different from what we pay attention to. They mainly solve the imbalance problem on the basis of [36]. They divide the vulnerability detection problem into two supervised learning stages. In the first stage, graph neural network is used to study the high-level features of the code. In the second stage, smote oversampling and representation learning is used to sample for binary classification. Since solving the problem of imbalanced datasets is not our concern, it seems unreasonable to compare our work with theirs. Nevertheless, we still compared with the work of [10], and the experimental results show that our scheme still has advantages as shown in Table 6.

**Table 6.** Comparison between ACGVD and graph neural network method without attention mechanism.

| Method | Accuracy (%) | Precision (%) | Recall (%) | F1 (%) |
|---|---|---|---|---|
| Devign [36] | 58.57 | 53.6 | 62.73 | 57.18 |
| Reveal [10] | 62.51 | 56.85 | 74.61 | 64.42 |
| CGVD | 62.14 | 55.82 | 76.25 | 64.46 |
| ACGVD | **63.58** | **57.12** | **76.62** | **65.45** |

In order to verify the validity of the double-level attention mechanism, we also compare ACGVD with the CGVD which only adds the node-level attention mechanism. The results showed that comparing with CGVD, ACGVD shows significant improvements on vulnerability detection capability. Specifically, ACGVD's accuracy, precision and recall are 1.44%, 1.3%, and 0.37% higher respectively than CGVD. This proves the necessity of using both node-level and path-level attention mechanism simultaneously.

### 5.4    What Is the Impact of Modifying the Classifier on the Experiment?

This work replaces the traditional classifier in graph classification method which directly inputting the node features into the MLP for classification. Instead, a novel classifier with convolution layer is designed which is more helpful for the vulnerability detection. To understand the contribution of our classifier module, we create a variant of ACGVD with traditional classifier which is called ACGVD_Tra. Table 7 shows the performance metrics for the above setup. Overall, ACGVD's accuracy, precision, recall, and F1 are 3.26%, 2.84%, 1.54%, and 2.44% improvement than ACGVD_Tra. We contend that, since the convolution layer is added after the graph neural network, more abundant features about the vulnerability are learned. The result indicates that our classifier improves the performance of vulnerability prediction.

**Table 7.** Comparison between ACGVD and ACGVD_Tra.

| Method | Accuracy (%) | Precision (%) | Recall (%) | F1 (%) |
|---|---|---|---|---|
| ACGVD_Tra | 60.32 | 54.28 | 75.08 | 63.01 |
| ACGVD | **63.58** | **57.12** | **76.62** | **65.45** |

## 6    Threats Factors

One the one hand, the vulnerability samples which are more relevant to some certain semantic graphs accounts for a small portion in the dataset. As a result, the path-level attention mechanism cannot significantly improve the experimental result. We believe that the performance of our model would be better if Zhou [36] published their Linux Kernel dataset. The source of this insight is the special case in [36]: on the Linux kernel dataset, the accuracy of model recognition only based on AST graph is better. On the other hand, our dataset is full of function-level code. This leads to the poor performance of cross-function vulnerability identification, which directly affects the detection accuracy of our model.

Another threat to our study is the design of model parameters. A slight change of each parameter could have a great impact on the experimental results. However, there are many adjustable parameters in the model. In this paper, when the experimental results prove the validity of our model, we fix the parameters. Although we cannot confirm whether the current parameter choose is the optimal solution of the model or not, ACGVD can still achieve best performance.

## 7    Conclusion

In this paper, we propose a more comprehensive graph, which considers multiple semantic graphs together. Compared with the previous deep learning method which only considers single semantic graph, we can cover more vulnerability types. In order to learn the structural information of the code better, we abandon the traditional deep learning method based on sequence model. We propose to learn the node vector representation of code directly by graph neural network, and then add a novel classification layer for end-to-end training. Considering that the proportion of vulnerable codes is small and the vulnerability information could be disturbed by benign code information, we add a node-level and a path-level attention mechanism at the same time. We conduct experiments on the latest public vulnerability dataset. The experimental results show that ACGVD achieves a better performance than the state-of-the-art methods. In the future, we plan to adopt the algorithm to automatically select the optimal parameters of the model. In the future, we will also focus on file-level vulnerability detection to identify cross-function vulnerabilities better.

**Acknowledgement.** This work is supported by the Strategic Priority Research Program of Chinese Academy of Sciences, Grant No. XDC02010300.

# References

1. Flawfinder. http://www.dwheeler.com/flawfinder
2. Infer static analyzer. https://fbinfer.com/
3. National vulnerability database. https://nvd.nist.gov
4. National vulnerability database (2019). https://nvd.nist.gov
5. Record-breaking number of vulnerabilities disclosed in 2017: Report (2017). https://www.securityweek.com/record-breaking-number-vulnerabilities-disclosed-2017-report
6. Rough audit tool for security. https://code.google.com/archive/p/rough-auditing-tool-for-security/
7. Software assurance reference dataset. https://samate.nist.gov/SRD/index.php
8. Ban, X., Liu, S., Chen, C., Chua, C.: A performance evaluation of deep-learnt features for software vulnerability detection. Concurrency Comput. Pract. Exp. **31**(19), e5103 (2019)
9. Cao, K., Jing, H.E., Fan, W.Q., Huang, W.: PHP vulnerability detection based on stain analysis. J. Commun. Univ. China (Sci. Technol.) (2019)
10. Chakraborty, S., Krishna, R., Ding, Y., Ray, B.: Deep learning based vulnerability detection: are we there yet? arXiv preprint arXiv:2009.07235 (2020)
11. Chen, Z., Zou, D., Li, Z., Jin, H.: Intelligent vulnerability detection system based on abstract syntax tree. J. Cyber Secur. **4**, 1–13 (2020)
12. Choi, M.j., Jeong, S., Oh, H., Choo, J.: End-to-end prediction of buffer overruns from raw source code via neural memory networks. arXiv preprint arXiv:1703.02458 (2017)
13. Dai, H., Murphy, C., Kaiser, G.: Configuration fuzzing for software vulnerability detection. In: 2010 International Conference on Availability, Reliability and Security, pp. 525–530. IEEE (2010)
14. Dam, H.K., Tran, T., Pham, T., Ng, S.W., Grundy, J., Ghose, A.: Automatic feature learning for vulnerability prediction. arXiv preprint arXiv:1708.02368 (2017)
15. Ghaffarian, S.M., Shahriari, H.R.: Software vulnerability analysis and discovery using machine-learning and data-mining techniques: a survey. ACM Comput. Surv. (CSUR) **50**(4), 1–36 (2017)

16. Guo, J., Wang, Z., Li, H., Xue, Y.: Detecting vulnerability in source code using CNN and LSTM network (2021)
17. Harer, J.A., et al.: Automated software vulnerability detection with machine learning. arXiv preprint arXiv:1803.04497 (2018)
18. Lee, M., Cho, S., Jang, C., Park, H., Choi, E.: A rule-based security auditing tool for software vulnerability detection. In: 2006 International Conference on Hybrid Information Technology, vol. 2, pp. 505–512. IEEE (2006)
19. Li, H., Kim, T., Baterdene, M., Lee, H.: Software vulnerability detection using backward trace analysis and symbolic execution. Int. J. Comput. Biol. Drug Des. 6(6), 255–62 (2013)
20. Li, Z., Zou, D., Xu, S., Chen, Z., Zhu, Y., Jin, H.: Vuldeelocator: a deep learning-based fine-grained vulnerability detector. IEEE Trans. Dependable Secure Comput. (2021)
21. Li, Z., Zou, D., Xu, S., Jin, H., Zhu, Y., Chen, Z.: SySeVR: a framework for using deep learning to detect software vulnerabilities. IEEE Trans. Dependable Secure Comput. (2021)
22. Li, Z., et al.: Vuldeepecker: a deep learning-based system for vulnerability detection. arXiv preprint arXiv:1801.01681 (2018)
23. Lin, G., Wen, S., Han, Q.L., Zhang, J., Xiang, Y.: Software vulnerability detection using deep neural networks: a survey. Proc. IEEE 108(10), 1825–1848 (2020)
24. Lin, G., Zhang, J., Luo, W., Pan, L., Xiang, Y.: Poster: vulnerability discovery with function representation learning from unlabeled projects. In: Proceedings of the 2017 ACM SIGSAC Conference on Computer and Communications Security, pp. 2539–2541 (2017)
25. Lin, G., et al.: Cross-project transfer representation learning for vulnerable function discovery. IEEE Trans. Industr. Inf. 14(7), 3289–3297 (2018)
26. Ndichu, S., Kim, S., Ozawa, S., Misu, T., Makishima, K.: A machine learning approach to detection of Javascript-based attacks using AST features and paragraph vectors. Appl. Soft Comput. 84, 105721 (2019)
27. Newsome, J.: Dynamic taint analysis for automatic detection, analysis, and signature generation of exploits on commodity software. Chin. J. Eng. Math. 29(5), 720–724 (2005)
28. Pewny, J., Schuster, F., Bernhard, L., Holz, T., Rossow, C.: Leveraging semantic signatures for bug search in binary programs. In: Proceedings of the 30th Annual Computer Security Applications Conference, pp. 406–415 (2014)
29. Russell, R., et al.: Automated vulnerability detection in source code using deep representation learning. In: 2018 17th IEEE International Conference on Machine Learning and Applications (ICMLA), pp. 757–762. IEEE (2018)
30. Scandariato, R., Walden, J., Hovsepyan, A., Joosen, W.: Predicting vulnerable software components via text mining. IEEE Trans. Softw. Eng. 40(10), 993–1006 (2014)
31. Semasaba, A.O.A., Zheng, W., Wu, X., Agyemang, S.A.: Literature survey of deep learning-based vulnerability analysis on source code. IET Softw. 14, 654–664 (2020)
32. Veličković, P., Cucurull, G., Casanova, A., Romero, A., Lio, P., Bengio, Y.: Graph attention networks. arXiv preprint arXiv:1710.10903 (2017)
33. Votipka, D., Stevens, R., Redmiles, E., Hu, J., Mazurek, M.L.: Hackers vs. testers: a comparison of software vulnerability discovery processes. In: IEEE Symposium on Security and Privacy (2018)
34. Wang, T., Wei, T., Gu, G., Zou, W.: TaintScope: a checksum-aware directed fuzzing tool for automatic software vulnerability detection. In: 2010 IEEE Symposium on Security and Privacy, pp. 497–512. IEEE (2010)

35. Wang, X., et al.: Heterogeneous graph attention network. In: The World Wide Web Conference, pp. 2022–2032 (2019)
36. Zhou, Y., Liu, S., Siow, J., Du, X., Liu, Y.: Devign: effective vulnerability identification by learning comprehensive program semantics via graph neural networks. arXiv preprint arXiv:1909.03496 (2019)

# TranFuzz: An Ensemble Black-Box Attack Framework Based on Domain Adaptation and Fuzzing

Hao Li[1], Shanqing Guo[1(✉)], Peng Tang[1], Chengyu Hu[1,2],
and Zhenxiang Chen[3]

[1] School of Cyber Science and Technology, Shandong University, Qingdao, China
202020883@mail.sdu.edu.cn, guoshanqing@sdu.edu.cn
[2] Key Laboratory of Network Assessment Technology, CAS (Institute of Information Engineering, Chinese Academy of Sciences), Beijing 100093, China
[3] University of Jinan, Jinan, China

**Abstract.** A lot of research effort has been done to investigate how to attack black-box neural networks. However, less attention has been paid to the challenge of data and neural networks all black-box. This paper fully considers the relationship between the challenges related to data black-box and model black-box and proposes an effective and efficient non-target attack framework, namely TranFuzz. On the one hand, TranFuzz introduces a domain adaptation-based method, which can reduce data difference between the local (or source) and target domains by leveraging sub-domain feature mapping. On the other hand, TranFuzz proposes a fuzzing-based method to generate imperceptible adversarial examples of high transferability. Experimental results indicate that the proposed method can achieve an attack success rate of more than 68% in a real-world CVS attack. Moreover, TranFuzz can also reinforce both the robustness (up to 3.3%) and precision (up to 5%) of the original neural network performance by taking advantage of the adversarial re-training.

**Keywords:** Domain adaptation · AI security · Fuzzing · Black-box attack

## 1 Introduction

Recently, Deep Neural Networks (DNNs) have been applied to many realistic AI systems, such as image classification [1]. However, due to the catastrophic overfitting or underfitting problem, the DNN-based systems always show a very vulnerable behavior in many corner cases [2]. In other words, if the DNN models are not tested in particular corner cases (which is referred to as the adversarial example, i.e., clean data that adds a well-designed noise), there would be devastating consequences. Thus, like traditional software testing, it is particularly important to systematically test and check the quality of the DNN-based models.

© Springer Nature Switzerland AG 2021
D. Gao et al. (Eds.): ICICS 2021, LNCS 12918, pp. 260–275, 2021.
https://doi.org/10.1007/978-3-030-86890-1_15

There exists plenty of works to test the quality of DNN-based systems by taking advantage of the model attack method [2, 3, 20]. These works always belong to white-box attacks, where the adversary usually first draws upon knowledge of the structure and parameters of the target model and then injects some imperceptible perturbation into a test example to create an adversarial example, and attacks the victim model. Nevertheless, in real-world scenarios, the target victim model is always a black box, where an attacker cannot access a complete knowledge of the target model. This will increase the attack difficulty. How to successfully attack the target model under the black-box condition is a very challenging issue.

To solve the above problem, the researchers have proposed two kinds of adversarial black-box attack methods: i) *Query-based attack* method [5, 15], where the target black-box model is treated as an optimization problem, and the attackers use query prediction information (for example, a probability value) as an instruction to generate adversarial examples. Although the query-based attack method usually has a high attack success rate, it does take a very high number of requests, which will incur a higher cost [6]. ii) *Transfer-based attack* method [4, 20], where the adversary ought to construct a comparable model as the local substitution to the target, and construct highly transferable adversarial examples that can successfully attack the local model. Then, these adversarial examples are transferred to attack the target black-box model. In this paper, we mainly focus on the transfer-based attack.

Unfortunately, the existing transfer-based attack methods [4, 20] usually only consider that the network architecture of the target model is a black box, but do not consider the target training data is a black box, and it is assumed that the training data of the source and target models are following the same data distribution [20]. However, it is not the case for practical application. As a result, these methods often suffer from low attack success rates or poor transfer efficiency in the black-box attack task. To remedy the above deficiency, in our black-box attack problem, we summarize the following two key challenges:

**C1: Target Training Data Is Black-Box.** Due to commercial confidentiality, the training data used in the target model will not be publicly available, and only some of the test/validation data examples may be provided to developers with some special scenes (e.g., adversarial competition[1]). This will lead to a different data distribution between training data used in the local (or source) model and that used in the target model. How to exploit the limited test/validation data examples of the target model to achieve a successful attacking purpose is a critical challenge.

**C2: Target Model's Network Is Black-Box.** Developing an effective adversarial example always requires that the attacker has complete information about

---

[1] https://github.com/tensorflow/cleverhans/tree/master/examples/nips17_adversarial_competition/dataset.

the target model, such an attack method is also known as a white-box attack. However, the attacker is generally unaware of what kind of neural network architecture is implemented in the target model. How to successfully tackle the target model under the $C2$ is another major challenge.

In this paper, we fully consider and explore the relationship between challenges $C1$ and $C2$. On the one hand, to address the challenge of different data distributions, we propose a methodology based on domain adaptation to reduce the difference between the data of the source and target domains by using subdomain mapping. Based on this method, we can implement a transfer-based attack. On the other hand, to counter the challenge of the model's black-box ($C2$), we illustrate an adversarial example (AE) generation framework based on neuron coverage to measure the logical runtime of the DNN model. Within our fuzzing framework, we also propose a novel ensemble-based seed mutation strategy to improve AEs attack transferability. The strategy introduces a small change in input mutations and to maximizes the expected difference between the original and the adversarial example. Certainly, there exist some transfer-based adversarial attack methods [3, 20] which are used in iterative gradient attack methods [20]. But they do not have a guide for exposing incorrect corner case behaviors. This can result in incorrect DNN behaviors remaining unexplored after thousands of iterations (low transferability caused by overfitted issues [3]).

In the end, we design a black-box model attack framework, namely, TranFuzz, which combines the domain adaptation method with the fuzzing strategy. Evaluation experiments show that the proposed TranFuzz method is best able to achieve an attack success rate of over 68% in the real-world Cloud Vision Service (CVS) scenario.

**Summary of Contributions** – The major contributions to this paper are shown as follows:

- We propose a black-box attack framework, namely, TranFuzz, which can generate highly transferable adversarial examples by interconnecting the domain adaptation-based local alternative model construction method and fuzzing-based method, respectively. To the best of our knowledge, it is the first work to combine domain adaptation and fuzz methods against the black-box model.
- TranFuzz[2] takes full account of the challenges of the data black-box and the neural network black-box. To create highly transferable AEs, we propose an ensemble-based seed mutation strategy, which can rapidly and efficiently trigger objective functions in our fuzzing framework. The experimental results show that the average attack success rate of TranFuzz can exceed 10%, compared to the state-of-the-art baselines.
- In five real-world Cloud Vision Services (i.e., Aliyun, Baidu, Tencent, Azure, and Clarifai) attacking scenes, the TranFuzz can better perform over 68% attack success rate. Furthermore, our proposed method can also enhance the robustness of the victim's model with adversarial training.

---

[2] https://github.com/lihaoSDU/ICICS2021.

**Organization** – The remainder of this paper is organized as follows. We present an overview methodology of the TranFuzz framework in Sect. 2. In Sect. 3, we illustrate details of the evaluation experiments and results. Section 4 highlights the related work of the paper and we conclude our work in Sect. 5.

## 2 Methodology

In this section, we first introduce an overview of the TranFuzz (Sect. 2.1), and then we illustrate the local model construction method based on domain adaptation to break the barrier of the data black-box challenge (Sect. 2.2). Finally, we generate optimal adversarial examples with high transferability by presenting a fuzzing-based method to address the neural network black-box challenge (Sect. 2.3).

### 2.1 Overview of TranFuzz

TranFuzz takes full account of the unique nature of the data black-box challenge co-existing with the model black-box challenge in the realistic application scene. In this paper, we design an effective adversarial example generation framework, named TranFuzz. The TranFuzz framework is depicted in Fig. 1. In TranFuzz, we first develop an algorithm based on a deep sub-domain adaptation network (DSAN) to construct a local substitute model. Afterward, we manufacture an adversarial example of high transferability on the strength of the mutation-based fuzzing strategy.

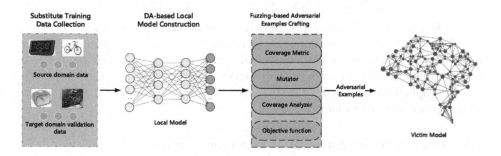

**Fig. 1.** Framework of TranFuzz.

### 2.2 Domain Adaptation-Based Local Model Construction

In the data black-box challenge, we assume that the attacker cannot get the target model's training data. Only unsupervised validation/test data can be accessed. To resolve the problem mentioned above and construct a local replacement model, in this section, TranFuzz uses a deep subdomain adaptation network

(DSAN) algorithm [9] with a certain improvement of the DSAN pseudo-labeling part. DSAN uses a classification loss function and an adaptive loss function to close the huge gap between the data of the source and target domain. We formulate the objective function of the DSAN as:

$$min_{f} \frac{1}{n_s} \sum_{i=1}^{n_s} J(f(x_i^s), y_i^s) + \sum_{l \in L} LMMD_l(p, q) \tag{1}$$

where $J(\cdot, \cdot)$ is a cross-entropy function, $f(\cdot)$ is the predict function, $n_s$ is the number of a source domain's samples, and $x^s$ and $y^s$ corresponding to the source domain's samples with the label. $LMMD(\cdot, \cdot)$ is the *local maximum mean difference* function to calculate the loss in the process of subdomain domain adaptation. The $l$ is an active layer in the subdomain distribution $L$. $p$ and $q$ are the data distributions of the source and target domains. To address the challenge of the data black box, our optimization goal is to minimize Eq. 1 under the conditions of different data distributions between $p$ and $q$.

From Eq. 1, DSAN leverages an algorithm based on domain adaptation networks (DANs) [10] and designs a local maximum mean difference as the difference metric between source and target domains. To compute the LMMD and reduce the data distribution difference between source and target domains. In our proposed method, we leverage a query-based strategy that adopts the target victim model as a benchmark to predict the target samples. For each sample, a single request is necessary for our proposed method. The improved method can significantly enhance the generalization capabilities of the local surrogate model's construction. Formally, the proposed method is formalized in Eq. 2.

$$LMMD_l(p, q) \stackrel{\triangle}{=} \mathbf{E}_c || \mathbf{E}_{p^{(c)}} [\phi(x^s)] - \mathbf{E}_{q^{(c)}} [f^t(x^t)] ||^2 \tag{2}$$

where $\mathbf{E}[\cdot]$ is the mathematical expectation function, $c$ is the different classes (e.g. labels), $x^t$ is the target domain's example. $\phi(\cdot)$ means the feature mapping function. In this paper, we use a universal function of the *Gaussian Kernel* as a mapping function between source and target domains. Our proposed approach can be applied to any neural network architecture to construct a local model with a promising performance.

### 2.3  Fuzzing-Based Adversarial Examples Crafting

In this section, we leverage the coverage-based fuzzing method to fuzz our local substitution model and generate high transferability adversarial examples. In the following chapter, we first describe the coverage gauge to guide our fuzzing neural network framework. Then we also elucidate the fuzzing objective function. Second, we describe our coverage analysis method (also called coverage analyzer) of the TranFuzz. Finally, we propose a new comprehensive mutation strategy to generate highly transferable adversarial examples.

**Definitions of the Neuron Coverage and Objective Functions.** The following are several definitions that are used in the fuzzing framework.

**Neuron Coverage.** TranFuzz exploits neuron coverage (NC) as our fuzzing coverage criteria, proposed by DeepXplore [2]. Neuron coverage is a metric for testing the comprehensiveness of the DNN model. NC can also calculate how many neurons are at least activated once during the current process. The formula of the neuron coverage is shown below:

$$NCov(TS, seed) = \frac{|\{n_i | \forall seed \in TS, f(n_i, seed) > th\}|}{K} \tag{3}$$

where $TS$ is a set of test seeds $\{ts_1, ..., ts_n\}$. We suppose all neurons of the model as $N = \{n_1, ..., n_K\}$, $K$ is the number of neurons in the model, $th$ is the fuzzing threshold to be considered as an activated neuron (in this paper we define the value as zero). $f(\cdot)$ is the function that allows you to send back the output value of the neuron. It is worth mentioning that in our proposed TranFuzz framework, we did not intentionally pursue a higher neuron coverage as our optimization objective. We took neuron coverage as a guide instruction metric to discover more exceptional adversarial examples that can crash the local substitute model.

**Objective Functions.** TranFuzz fully considers the high transferability and human imperceptibility as the objective fuzzing functions to craft adversarial examples. If an adversarial example complies with the objective function constraint, the fuzzing process will jump out of the execution loop.

On the one hand, TranFuzz loosens the differential testing objective function used in [2] and proposed a novel function, specifically,

$$obj_{DX} : O_{f_1(x)} \cup O_{f_2(x)} \cup ... O_{f_k(x)} = 1, \tag{4}$$

where $f(\cdot)$ is the predicted function of local models, $\cup$ is the union function, $k$ is the number of local models. If $f_i(x)$ is not equal to the true label of $x$, then $O_{f_i(x)} = 1$. In addition, the differential testing method requires several local DNN models with the same prediction task, which will increase training costs. Unlike the differential testing method above-mentioned, TranFuzz only needs a local substitution model as our fuzzing framework input ($k = 1$ in Eq. 4). Accordingly, one of the fuzzing objective functions $obj_{TF}^1$ in TranFuzz describes as following:

$$obj_{TF}^1 : f(x_{adv}) \neq true\ label\ of\ x_{adv} \tag{5}$$

On the other hand, to generate imperceptible adversarial examples, the structural similarity between adversarial examples and the original examples is another objective function used in the fuzzing framework. We introduce an Average Structural Similarity (ASS) [11] as the similarity metric. To deduce structural changes, ASS captures pixel intensity patterns, especially among adjacent pixels. ASS can also measure the brightness and contrast of the image which affects the perceived quality of the image. The formalization of the objective function based on structural similarity is

$$obj_{TF}^2 : ASS(x_{avd}, x) > \tau \tag{6}$$

where $\tau$ is the pre-setting threshold value of ASS, we set it as 0.96 in our evaluation experiments. In the end, we combine $obj_{TF}^1$ and $obj_{TF}^2$ as our objective functions. If these conditions are triggered, the current fuzzing loop will shut down.

**Coverage Analyzer.** In the coverage parser section, if the adversarial example $x_{adv}^i$ has not satisfied the objective functions, the coverage analyzer will randomly select unfuzzed network layer neurons. After that, the coverage analyzer will split a new fuzzing path. Next, the coverage analyzer will calculate the current neuronal loss values and gradient values $grads^i$. The coverage analyzer will combine the $x_{adv}^i$ and $grads^i$ as the *Mutator* inputs. If the $x_{adv}^{i+1}$ can reach the objective functions, the coverage analyzer will update the neuron coverage and break the current fuzzing loop.

**Mutator.** Mutator is the schedule against too many fuzzing execution iterations. The mutator can also craft adversarial examples of high transferability and imperceptibility. In this section, TranFuzz proposes a novel ensemble-based mutation method that leverages multiple perturbation strategies to generate adversarial examples. The formal representation is $x_{adv} = x + \delta$, where $\delta$ is the optimal adversarial perturbation.

The mutator is based on the gradient value which is computed by the local surrogate model output layer and the hidden layer. To generate adversarial perturbation $\delta_{dx}$, we adopt the occlusion strategy described by DeepXplore to simulate the camera lens that may be accidentally or deliberately occluded. In contrast to DeepXplore's strategy, we are implementing a smaller occlusion adversarial perturbation $\delta_{DX} = occlusion_{i:i+m,j:j+n}$ (smaller rectangle that has $m*n$ pixels, and $(i, j)$ is the coordinate of a pixel) and operating randomly in multiple seed positions.

Moreover, to improve the success of the attack under the premise of imperceptibility, TranFuzz does not implement general mutation methods with various fuzzers, e.g., image blurring, image contrast adjusting, image brightness adjusting [13]. The TranFuzz proposed a novel method based on the scale of images [12] to transform adversarial perturbations. The mutator first leverages the cumulative distribution function (CDF) to calculate the equalization of the image perturbation histogram. The formalization of the histogram equalization function $(T(\cdot))$ is as follows:

$$T(r_k) = \frac{G-1}{MN} \sum_{j=0}^{k} n_j, k = 0, 1, ..., G-1 \tag{7}$$

where $MN$ is the sum of pixels, $r_k$ is the gray level of the image, $\sum_{j=0}^{k} n_j$ is the number of $r_k$, and $G$ is the number of possible gray levels of the image.

After equalizing the histogram, we adopt the linear interpolation method to insert the initial Gaussian derived noise from the gradient. Then we calculated the adversarial perturbation $\delta_{TF}$. To avoid invalid values in $\delta_{TF}$, the mutator also quantifies the adversarial perturbations [14]. Finally, we describe the optimal adversarial perturbation as $\delta = \delta_{DX} + \delta_{TF}$.

## 3 Evaluation

In this section, we first introduce specific details of the data sets adopted in our evaluation experiments, then the attack configurations are also depicted in this section (Sect. 3.1). After that, we attack both the non-robustness (Sect. 3.2) and robustness (Sect. 3.3) of the black-box models that leverage the adversarial examples generated by TranFuzz. We also compare our method with eight state-of-the-art baseline methods. In addition, five different business Cloud Vision services are conducted as black-box victim targets in real-world scenarios (Sect. 3.4). We also analyze the positive impacts of the proposed method on the target model defensibility by leveraging adversarial retraining (Sect. 3.5).

### 3.1 Experimental Setting

**Datasets.** To build black-box data with experimental domain adaptation environments, we use two different image data sets (namely, Office-31 [16] and Office-Home [19]), which are often the benchmark dataset in the domain adaptation field. In our evaluation experimental setting, to simulate the $C1$ challenge, different categories are regarded as the source and target domains (specifically, *Amazon* and *Webcam* of the Office-31, *Product* and *RealWorld* of the Office-Home), respectively. All domain adaptation data are downloaded from the open-source site[3].

**Attack Configurations.** The settings for black-box model attacks and evaluation baselines are described in this section.

**Black-Box Model Setting.** ResNet50, AlexNet, VGG19, and DenseNet121 are different models of neuronal network structures. We conducted the above-mentioned models in our black-box attack evaluation experiments. Moreover, to build the experimental configuration about the model black-box under the $C2$ challenge, we use the ResNet50 neural network as the source domain model. The other three models are as the target domain attacked model. For example, under the $C2$ challenge, on the one hand, the source domain model with its training data is the ResNet50 and *Webcam*. On the other hand, the target domain model with its training data is defined as DenseNet121 and *Amazon*. While the two different models all have 31 different classes (`backpack`, `bike`, etc.), the training data of the models are different and obeys the $C1$ challenge.

---

**Baselines.** 1) In the non-robustness black-box model attack, we leverage eight different state-of-the-art attack methods as the baselines in our comparison experiments. Specifically, five white-box attack methods (namely, DDN [14], PGD [4], FGSM [20], L-BFGS [17], C&W [18]), and three black-box attack methods (ZOO [15], Pixel Attack (PA) [21], and Spatial Transformation (ST) [22]). 2) In the robustness black-box model attack task, we use adversarial training algorithms to build robust models as the victim models. The compared baselines are the same as the above-mentioned in step 1). The details of the adversarial training algorithms are Fast is Better than Free (FBF) [25] and Madry's Protocol [4]. 3) To demonstrate the effectiveness of our proposed approach in real-world black-box attack scenarios, we are also attacking five state-of-the-art commercial Cloud Vision Services (CVS), namely, Aliyun[4], Baidu[5], Tencent[6], Azure[7], and Clarifai[8].

**Implementation Details.** 1) In our evaluation experiments, all the training iterations numbers are set as 200. 2) Also, implementation information of the eight baseline methods are depicted in the AdverTorch [23] and ART [24]. It is worth noting that the Spatial Transformation parameters of `max_translation` and `max_rotation` are all equal to 30 degrees (which is consistent with the [22]). Other employment parameters are all set to default. 3) In the implementation of adversarial training, we conduct the ART tool to re-train the robustness models (`AdversarialTrainerFBF` and `AdversarialTrainerMadryPGD`). The maximum number of training iterations also is set to 200. The maximum perturbation parameter with its step is set to 0.3 and 0.1, respectively. 4) Considering the cost of the commercial Cloud Vision Services (e.g., one thousand API access will need to be 3$ for Aliyun), we randomly select 50 adversarial examples to attack the five CVS which are generated by TranFuzz. The detailed description is shown in Sect. 3.4. Among the five CVS, we access the provided API of Aliyun, Baidu, and Clarifai. In addition, the Tencent and Azure attacks are making use of web browsers upload manually. 5) We randomly divided the target domain data into the train (80%) and test (20%) parts. The train part is for training the target victim model and the test part is for constructing the source local substitute model. 6) In the experimental evaluation, we perform Attack Success Rate[9] to evaluate our proposed framework.

We provide a summary of our trained DNN-based models of the target domain in Table 1.

---

[4] https://vision.aliyun.com/imagerecog.

[5] https://ai.baidu.com/tech/imagerecognition/general.

[6] https://ai.qq.com/product/visionimgidy.shtml.

[7] https://azure.microsoft.com/en-us/services/cognitive-services/computer-vision.

[8] https://www.clarifai.com/label.

[9] The Attack Success Rate is the proportion of adversarial examples misclassified by the target DDN [14].

**Table 1.** Summary of the target DNN-based models

| Target model type | Dataset | Index | Architecture | Train & test data | Testing accuracy | Training iterations |
|---|---|---|---|---|---|---|
| Non-robust | Office-31 | 1 | AlexNet | Amazon | 77.53% | 200 |
| | | 2 | AlexNet | Webcam | 95.91% | 200 |
| | | 3 | VGG19 | Amazon | 83.62% | 200 |
| | | 4 | VGG19 | Webcam | 91.81% | 200 |
| | | 5 | DenseNet121 | Amazon | 79.97% | 200 |
| | | 6 | DenseNet121 | Webcam | 93.57% | 200 |
| Non-robust | Office-Home | 7 | AlexNet | Product | 80.22% | 200 |
| | | 8 | AlexNet | RealWorld | 69.6% | 200 |
| | | 9 | VGG19 | Product | 85.68% | 200 |
| | | 10 | VGG19 | RealWorld | 80.4% | 200 |
| | | 11 | DenseNet121 | Product | 85.68% | 200 |
| | | 12 | DenseNet121 | RealWorld | 82.85% | 200 |
| Robust (FBF) | Office-31 | 13 | DenseNet121 | Amazon | 62.89% | 200 |
| | | 14 | DenseNet121 | Webcam | 83.04% | 200 |
| Robust (Madry's Protocol) | Office-31 | 15 | DenseNet121 | Amazon | 52.88% | 200 |
| | | 16 | DenseNet121 | Webcam | 76.02% | 200 |

## 3.2 Black-Box Attack Against Non-robustness Model

This section primarily describes our proposed method to attack the non-robustness black-box model's performance. Two different image datasets (Office-31 and Office-Home) were implemented in the experiments, mentioned in Sect. 3.1. Details of the comparison experimental results are given in Table 2 and Table 3.

On the one hand, we use ResNet50 as the source neural network to train a local substitute model for attacking other three different networks. Training data of the local substitute model differs from the target model on the promise of C1 challenge. From Table 2, TranFuzz can better achieve more than 33.9% and 37.5% attack success rates on *Webcam* and *Amazon* data, respectively. On average, TranFuzz can perform a Top-1 attack success rate (specifically, 25.6%) compared to the other baseline methods. On the other hand, from Table 3, comparison experiments also use ResNet50 as the source neural network. The TranFuzz can mislead the target victim model over 31.3% and 46.9% on *RealWorld* and *Product* data, respectively. For the average attack success rate, TranFuzz is also capable of making the Top-1 success rate (28.5%) compared to other baseline methods.

From Table 2 and 3, we observe that AlexNet is not robust against the nine different black-box attack methods, compared with the DenseNet121 and VGG19. This demonstrates that the defender should design a more robust and complex network structure to enhance the DNN-based model's performance. Furthermore, the C&W attack is one of the most effective and widely used among the primary attacks. From the evaluation experiments of non-robustness black-box attack, our proposed method can surpass the C&W method over 2.38% and 6.4% in Office-31 and Office-Home datasets. In addition, the L-BFGS attack uses L-BFGS to minimize the distance of the original and perturbed images.

**Table 2.** The success rate of attacks against the non-robustness black-box model in the Office-31 dataset

| Source model | Source data | Target model | Target data | Attack | | | | | | | | |
|---|---|---|---|---|---|---|---|---|---|---|---|---|
| | | | | DDN | PGD | FGSM | L-BFGS | C&W | ZOO | PA | ST | TranFuzz |
| ResNet50 | Amazon | AlexNet | Webcam | 12.9% | 19.2% | 22.4% | 21.64% | 25.25% | 1.92% | **34.1%** | 33.9% | 33.9% |
| | | VGG19 | | 9.9% | 14.6% | 17.4% | 15.35% | 20.05% | 2.44% | 17.9% | **26.9%** | 19.3% |
| | | DenseNet121 | | 16.9% | 25.6% | 18.8% | 22.22% | **27.49%** | 1.22% | 11.9% | 11.7% | 18.1% |
| | Webcam | AlexNet | Amazon | 10.6% | 19.9% | 23.4% | 10.28% | 20.04% | 2.93% | 22.3% | 35% | **37.5%** |
| | | VGG19 | | 10.6% | 11.7% | 12.9% | 10.62% | 20.9% | 1.18% | 11.9% | **23.7%** | 20.7% |
| | | DenseNet121 | | 11.1% | 19.3% | 18.1% | 14.12% | 25.61% | 2.34% | 8.4% | 16.2% | **24.2%** |
| **Average attack success rate** | | | | 12.0% | 18.4% | 18.8% | 15.71% | 23.22% | 2.01% | 17.8% | 24.5% | **25.6%** |

**Table 3.** The success rate of attacks against the non-robustness black-box model in the Office-Home dataset (DN121: DenseNet121)

| Source model | Source data | Target model | Target data | Attack | | | | | | | | |
|---|---|---|---|---|---|---|---|---|---|---|---|---|
| | | | | DDN | PGD | FGSM | L-BFGS | C&W | ZOO | PA | ST | TranFuzz |
| ResNet50 | Product | Alexnet | RealWorld | 13.0% | 18.0% | 11.8% | 17.93% | 21.49% | 2.1% | 21.5% | 30.7% | **31.3%** |
| | | VGG19 | | 14.5% | 22.3% | 19.6% | 13.92% | 20.71% | 2.8% | 16.4% | **29.9%** | 24.8% |
| | | DN121 | | 13.4% | **26.9%** | 19.9% | 13.25% | 21.27% | 1.9% | 11.0% | 21.7% | 20.7% |
| | RealWorld | AlexNet | Product | 13.5% | 23.1% | 24.8% | 16.29% | 23.72% | 2.1% | 15.0% | 36.7% | **46.9%** |
| | | VGG19 | | 14.9% | 18.5% | 19.2% | 14.1% | **24.15%** | 1.9% | 10.0% | 23.5% | 22.5% |
| | | DN121 | | 12.1% | **30.4%** | 20.9% | 10.27% | 21.23% | 0.1% | 8.1% | 20.9% | 25.1% |
| **Average attack success rate** | | | | 13.6% | 23.2% | 19.4% | 14.29% | 22.1% | 1.8% | 13.7% | 27.2% | **28.5%** |

The TranFuzz method also can be better than L-BFGS under the premise of C1 and C2 (specifically, 9.9% in Office-31, 14.2% in Office-Home).

Besides, from the results of the experiment, we also observe that the Spatial Transformation (ST) method has a promising effect on the local black-box model attack under $C1$ and $C2$ challenges, but is still weaker than TranFuzz (4.2% lower than us). From the adversarial examples generated by ST, we conclude that the ST method is not similar to the original natural structure of the images. Hence, from the ST [22] evaluation result, the adversarial example will have a partial loss compared to the original due to image rotations. Consequently, the target black-box model cannot predict the successful adversarial examples which trade by spatial transformation. While the adversarial examples generated by TranFuzz can retain the original ASS leverages our proposed image mutating strategy (examples of the AE can be found on our website mentioned before).

### 3.3   Black-Box Attack Against Robustness Model

In our evaluation experiments, to achieve the proposed robust networks as a black-box victim model, we implement commonly adversarial training methods, specifically, Fast Is Better Than Free (FBF) [25] and Madry's Protocol [4]. Furthermore, the neural network structure of the target model is also set as DenseNet121, and the source model is ResNet50. The data set that we have implemented in this section is the same as in Table 2. Detailed information on the deployment and implementation of robust models can be found in Sect. 3.1, Table 1.

**Table 4.** The success rate of attacks against the robust black-box model

| Adversarial trainer | Source data | Target data | Attack | | | | | | | | |
|---|---|---|---|---|---|---|---|---|---|---|---|
| | | | DDN | PGD | FGSM | L-BFGS | C&W | ZOO | PA | ST | TranFuzz |
| FBF [25] | Amazon | Webcam | 38.0% | 9.9% | 11.7% | 9.64% | 15.32% | 11.7% | 48.3% | 56.7% | **74.8%** |
| | Webcam | Amazon | 6.4% | 3.0% | 1.9% | 7.33% | 18.58% | 0.5% | 24.8% | 51.0% | **53.8%** |
| Madry's Protocol [4] | Amazon | Webcam | 5.3% | 1.2% | 1.8% | 19.89% | 22.81% | 1.2% | 31.7% | **64.3%** | 53.8% |
| | Webcam | Amazon | 4.9% | 4.0% | 3.0% | 4.18% | 8.88% | 2.1% | 6.8% | 9.9% | **13.6%** |
| **Average attack success rate** | | | 18.1% | 8.9% | 9.1% | 10.26% | 16.4% | 8.6% | 29.4% | 44.4% | **48.6%** |

Table 4 shows the black-box attacks result against two distinct robustness models between TranFuzz and the other eight baseline methods. According to the table, TranFuzz can achieve a maximum attack success rate of 74.85%, and the proposed method also is able to accomplish an average attack success rate of 48.6%. From the evaluated experiment results, we can conclude that our proposed method is optimal compared with others. Additionally, we also observe that Madry's and FBF's defense methods are effective in resisting gradient attacks (i.e., FGSM-based attacks like PGD, L-BFGS, and FGSM). Specifically, in PGD, FGSM, and L-BFGS, the worst one only reaches 1.2% attack success rate, and the average attack success rate only achieves 8.9%, 9.1%, and 10.26%, respectively. The attack success rate has dropped by more than half compared to Table 2, Sect. 3.2. Besides, the C&W method can achieve a success rate of 16.4%, which is also lower than the proposed TranFuzz method. Nevertheless, Madry's protocol and FBF defense methods are unable to effectively defend Spatial Transformation and TranFuzz methods. The reason is that TranFuzz performs an ensemble-based AEs generation method to enhance transferability. But the Spatial Transformation algorithm adopts the technique of image transformation in space that will decrease the imperceptible performance of the image. In addition, the TranFuzz method can also exceed the ST method on a mean attack success rate of 4.2%.

Accordingly, based on the above-mentioned investigation, we have put forward some hypotheses and conjectures for the defense strategy here. The defender should consider various algorithms to generate adversarial examples (e.g., gradient-based, space-based, and color-based changes strategy) under the process of building an adversarial trainer. In our evaluation experiments, we also use adversarial examples generated by TranFuzz as the robust retraining data to defend against other attack methods. The implementation details are illustrated in Sect. 3.5.

## 3.4    Black-Box Attack Against Cloud Vision Services

In this section, we focus on the black-box attack in real-world scenarios. The attacking targets are five different businesses Cloud Vision Services (Aliyun, Baidu, Tencent, Azure, and Clarifai). Considering the cost issue mentioned in Sect. 3.1, we randomly select 50 images from the Office-Home data set and develop adversarial examples using our proposed approach. It should be noted that we define a new attack success metric: if the Top-1 prediction tag (which access from Cloud Vision Services response) is different between the adversarial example and the original natural one, we consider the attack as successful.

On the one hand, we call the API provided by Aliyun, Baidu, and Clarifai and return the detection result. After that, the experimental results of the attack success rate are being calculated. On the other hand, we also take advantage of a method that manually uploads the picture via the web browser (Tencent and Azure), and we also record the response detection results on each of the adversarial examples. Ultimately, the success rate for the attack on Aliyun, Baidu, Tencent, Azure, and Clarifai is 19/50, 18/50, 34/50, 13/50, and 8/50 respectively. From the CVS detection results, our proposed method can perform a 68% higher attack success rate on Tencent, and can also do 16% to attack Clarifai even though it is the worst.

### 3.5   Adversarial Defending

In this section, we demonstrate that TranFuzz can also enhance the robustness of the target model. To meet this objective, we are focusing on an adversarial training strategy based on additional data. We retrain the victim model from scratch on the union of the TranFuzz crafted adversarial examples with the original natural images. In this section, to retrain the target neural network, we implement the Office-Home dataset and DenseNet121 shown in Table 3. The maximum number of training iterations is also set to 200.

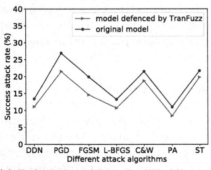
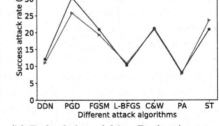

(a) Defended model in *RealWorld*'s data detection.

(b) Defended model in *Product*'s data detection.

**Fig. 2.** Comparison with success attack rate before and after TranFuzz defend in seven different attack methods.

In the evaluation experiments, we defend against other Top-7 different attacking methods (DDN, PGD, FGSM, L-BFGS, C&W, PA, and ST). The source model also is the ResNet50 and the defense results are illustrated in Fig. 2. From Fig. 2(a), our proposed model defending method is implemented on the DenseNet121 that can hamper more than 3.3% average success rate on the non-target attacks, compared with the non-defense original model. Furthermore, from Fig. 2(b), the retrained model can also improve robustness performance by around 1% on average in the *Product* data.

Moreover, our proposed retraining method can also improve the detection accuracy of the model over the clean data. We perform the defended network to predict the clean data in *RealWorld* and *Product*, respectively. Further investigation shows that our retrained model can achieve classification accuracy over 87.2% and 92.27%, which are improving more than 5% on average compares to the original model (the classification accuracy of the original model is 82.85% and 85.68%, as shown in Index 11 and Index 12 of Table 1).

## 4 Related Work

In this section, we list several related works with TranFuzz, specifically, attack methods and adversarial defenses, and DNN-based model fuzzing techniques.

### 4.1 Adversarial Attacks and Defenses

**Black-Box Attacks.** The transfer-based strategy is an extremely important black-box attack method, and several types of research [3,4] were proposed based on the transfer attack method. Su et al. [21] analyzed an attack situation under extreme conditions and proposed an adversarial perturbation based on differential evolution to perform a single-pixel attack. The results of the experiment show that the reported method can modify the output of the model and only change one pixel of the image. In addition, ZOO [15] is a query-based black box attack method, and they exploit the non-derivative optimization strategy and symmetrical injury difference to estimate the Hessian gradient matrix. The method does not need to obtain the gradient information of the target model. Engstrom et al. [22] proposed an attack method based on spatial transformation, which makes it possible to study the vulnerability of neural network classifiers by carrying out image rotation and translation operations.

**Adversarial Training.** Adversarial training is a data enhancement strategy to improve the robustness of the model. Madry et al. [4] proposed a min-max optimization framework using the projected gradient descent method to generate conflicting samples as augmentation data. This method first finds several examples by adopting the PGD and then uses these examples as the adversarial training data to decrease the training loss. Wong et al. [25] adopt a weaker, lower-cost adversarial strategy to form a robust model. The method combines the fast gradient sign and random initialization methods in adversarial training. The results of the experiment have shown that it has effective performance with lower cost compared to the PGD-based adversarial training method.

## 4.2 DNN Model Fuzzers

DeepXplore [2] is a fuzzing-based method to verify the DNN system. The method first proposed Neuron Coverage as a coverage metric for guiding the DNN model's testing. DeepXplore uses differential testing and generates some test inputs to identify the incorrect behavior of the deep learning system without the necessary manual operations. In addition, to effectively mutate test inputs, [13] proposed eight mutation strategies that include neuronal network weight-based mutation, neuron-based mutation, and layer-based mutation.

## 5    Conclusion

In this paper, we fully consider the relationship between the challenges of data black-box. Based on that, we proposed a non-targeted black-box attack framework. The evaluation experiment results show that our proposed framework can address both the non-robustness and robustness black-box attack tasks. In addition, TranFuzz can perform over 68% attack success rate against real-world Cloud Vision Services. Moreover, by taking advantage of the adversarial training strategy with data augmentation, TranFuzz can also strengthen the robustness of the original model.

**Acknowledgment.** This research was supported by National Natural Science Foundation of China (No. 62002203), Major Scientific and Technological Innovation Projects of Shandong Province, China (No. 2018CXGC0708, No. 2019JZZY010132), Shandong Provincial Natural Science Foundation (No. ZR2020MF055, No. ZR2020LZH002, No. ZR2020QF045), The Fundamental Research Funds of Shandong University (No. 2019GN095), The Open Project of Key Laboratory of Network Assessment Technology, Institute of information engineering, Chinese Academy of Sciences (No. KFKT2019-002).

## References

1. He, K., Zhang, X., Ren, S., Sun, J.: Deep residual learning for image recognition. In: CVPR 2016, pp. 770–778 (2016)
2. Pei, K., Cao, Y., Yang, J., Jana, S.: DeepXplore: automated Whitebox testing of deep learning systems. In: SOSP 2017, pp. 1–18 (2017)
3. Xie, C., et al.: Improving transferability of adversarial examples with input diversity. In: CVPR 2019, pp. 2730–2739 (2019)
4. Madry, A., Makelov, A., Schmidt, L., Tsipras, D., Vladu, A.: Towards deep learning models resistant to adversarial attacks. CoRR abs/1706.06083 (2017)
5. Bhagoji, A.N., He, W., Li, B., Song, D.: Exploring the space of black-box attacks on deep neural networks. In: European Conference on Computer Vision (2019)
6. Suya, F., Chi, J., Evans, D., Tian, Y.: Hybrid batch attacks: finding black-box adversarial examples with limited queries. In: USENIX Security Symposium 2020, pp. 1327–1344 (2020)
7. Pan, S.J., Yang, Q.: A survey on transfer learning. IEEE Trans. Knowl. Data Eng. **22**(10), 1345–1359 (2010)

8. Wang, M., Deng, W.: Deep visual domain adaptation: a survey. Neurocomputing **312**, 135–153 (2018)
9. Zhu, Y., Zhuang, F., Wang, J., et al.: Deep subdomain adaptation network for image classification. IEEE Trans. Neural Netw. Learn. Syst. **32**, 1713–1722 (2020)
10. Long, M., Cao, Y., Wang, J., Jordan, M.I.: Learning transferable features with deep adaptation networks. In: ICML 2015, pp. 97–105 (2015)
11. Wang, Z., Bovik, A.C., Sheikh, H.R., Simoncelli, E.P.: Image quality assessment: from error visibility to structural similarity. IEEE Trans. Image Process. **13**(4), 600–612 (2004)
12. Xiao, Q., Chen, Y., Shen, C., Chen, Y., Li, K.: Seeing is not believing: camouflage attacks on image scaling algorithms. In: USENIX Security Symposium 2019, pp. 443–460 (2019)
13. Hu, Q., Ma, L., Xie, X., Yu, B., Liu, Y., Zhao, J.: DeepMutation++: a mutation testing framework for deep learning systems. In: ASE 2019, pp. 1158–1161 (2019)
14. Rony, J., Hafemann, L.G., Oliveira, L.S., Ayed, I.B., Sabourin, R., Granger, E.: Decoupling direction and norm for efficient gradient-based L2 adversarial attacks and defenses. In: CVPR 2019, pp. 4322–4330 (2019)
15. Chen, P.-Y., Zhang, H., Sharma, Y., Yi, J., Hsieh, C.-J.: ZOO: zeroth order optimization based black-box attacks to deep neural networks without training substitute models. In: AISec@CCS 2017, pp. 15–26 (2017)
16. Saenko, K., Kulis, B., Fritz, M., Darrell, T.: Adapting visual category models to new domains. In: Daniilidis, K., Maragos, P., Paragios, N. (eds.) ECCV 2010. LNCS, vol. 6314, pp. 213–226. Springer, Heidelberg (2010). https://doi.org/10.1007/978-3-642-15561-1_16
17. Szegedy, C., et al.: Intriguing properties of neural networks. In: ICLR (Poster) (2014)
18. Carlini, N., Wagner, D.A.: Towards evaluating the robustness of neural networks. In: IEEE Symposium on Security and Privacy, pp. 39–57 (2017)
19. Venkateswara, H., Eusebio, J., Chakraborty, S., Panchanathan, S.: Deep hashing network for unsupervised domain adaptation. In: CVPR 2017, pp. 5385–5394 (2017)
20. Goodfellow, I.J., Shlens, J., Szegedy, C.: Explaining and harnessing adversarial examples. In: International Conference on Learning Representations (2015)
21. Jiawei, S., Vargas, D.V., Sakurai, K.: One pixel attack for fooling deep neural networks. IEEE Trans. Evol. Comput. **23**(5), 828–841 (2019)
22. Engstrom, L., Tran, B., Tsipras, D., Schmidt, L., Madry, A.: Exploring the landscape of spatial robustness. In: ICML 2019, pp. 1802–1811 (2019)
23. https://github.com/BorealisAI/advertorch
24. https://github.com/Trusted-AI/adversarial-robustness-toolbox
25. Wong, E., Rice, L., Zico Kolter, J.: Fast is better than free: revisiting adversarial training. In: ICLR 2020 (2020)

# Software Obfuscation with Non-Linear Mixed Boolean-Arithmetic Expressions

Binbin Liu[1(✉)], Weijie Feng[1], Qilong Zheng[1], Jing Li[1], and Dongpeng Xu[2(✉)]

[1] University of Science and Technology of China, Hefei, China
{robbertl,fengwj}@mail.ustc.edu.cn, {QLZheng,lj}@ustc.edu.cn
[2] University of New Hampshire, Durham 03824, USA
dongpeng.xu@unh.edu

**Abstract.** Mixed Boolean-Arithmetic (MBA) expression mixes bitwise operations (e.g., AND, OR, and NOT) and arithmetic operations (e.g., ADD and IMUL). It enables a semantic-preserving program transformation to convert a simple expression to a difficult-to-understand but equivalent form. MBA expression has been widely adopted as a highly effective and low-cost obfuscation scheme. However, state-of-the-art deobfuscation research proposes substantial challenges to the MBA obfuscation technique. Attacking methods such as bit-blasting, pattern matching, program synthesis, deep learning, and mathematical transformation can successfully simplify specific categories of MBA expressions. Existing MBA obfuscation must be enhanced to overcome these emerging challenges.

In this paper, we first review existing MBA obfuscation methods and reveal that existing MBA obfuscation is based on "linear MBA", a simple subset of MBA transformation. This leaves the more complex "non-linear MBA" in its infancy. Therefore, we propose a new obfuscation method to unleash the power of non-linear MBA. Non-linear MBA expressions are generated from the combination or transformation of linear MBA rules based on a solid theoretical underpinning. Comparing to existing MBA obfuscation, our method can generate significantly more complex MBA expressions. To present the practicability of the non-linear MBA obfuscation scheme, we apply non-linear MBA obfuscation to the Tiny Encryption Algorithm (TEA). We have implemented the method as a prototype tool, named *MBA-Obfuscator*, to produce a large-scale dataset. We run all existing MBA simplification tools on the dataset, and at most 147 out of 1,000 non-linear MBA expressions can be successfully simplified. Our evaluation shows *MBA-Obfuscator* is a practical obfuscation scheme with a solid theoretical cornerstone.

**Keywords:** Software obfuscation · Mixed Boolean-Arithmetic expression · Expression transformation

## 1 Introduction

Software obfuscation [26] performs a semantics-preserving transformation to hide the implementation of a program, which leads the program is hard to understand

© Springer Nature Switzerland AG 2021
D. Gao et al. (Eds.): ICICS 2021, LNCS 12918, pp. 276–292, 2021.
https://doi.org/10.1007/978-3-030-86890-1_16

and analyze. Many obfuscation techniques have been developed in the literature to hide the behavior of code in different ways, *Mixed Boolean-Arithmetic*(MBA) obfuscation is such a prominent example. Zhou et al. [32] define MBA expression that mixes bitwise operations(e.g., $\neg, \wedge, \oplus, \ldots,$) and arithmetic operations(e.g., $+, -, \times$). MBA obfuscation transforms a simple expression like $x+y$ to a complex but equivalent expression with mixed bitwise and arithmetic operators. Multiple academic tools and industry projects [6,13,17,21,23] have embedded MBA obfuscation into their products.

The wide application of MBA obfuscation has attracted researchers to explore how to recover the initial form of the MBA expression. Guinet et al. [11] presents a tool, *Arybo*, which normalizes MBA expressions to bit-level symbolic representation with only $\oplus$ and $\wedge$ operations. Eyrolles et al. [9] simplify MBA expressions by a pattern matching method. Blazytko et al. [4] apply program synthesis techniques [12] to learn the underlying semantics of obfuscated code and generate another simpler but equivalent expression. Feng et al. [10] introduce a novel solution based on deep learning, named NeuReduce, to simplify MBA expression. Liu et al. [18] prove a hidden two-way transformation feature and present a novel technique to simplify MBA expression. These methods propose a great challenge to the MBA obfuscation technique. However, we note that existing MBA obfuscation rules are generated from linear MBA expressions, because non-linear MBA research is still at an early stage and related non-linear MBA translation rules are rare.

To improve the resilience of the MBA obfuscation technique, we explore a new research direction: non-linear MBA obfuscation, which is the relative complement of linear MBA expression in the MBA obfuscation area. Firstly, multiple methods are demonstrated to create unlimited non-linear MBA expressions, whose correctness is guaranteed based on the basic math rules. Next, we present a practical application of the non-linear MBA obfuscation technique on the Tiny Encryption Algorithm(TEA), which hides the key and transforms the original operation into another different format. We have implemented the method as an open-source tool named *MBA-Obfuscator*. Given a simple expression as input, *MBA-Obfuscator* generates a related complex non-linear MBA expression. To the best of our knowledge, *MBA-Obfuscator* is the first tool to generate diversified non-linear MBA expressions.

To demonstrate the strength of *MBA-Obfuscator*, we evaluate it on a comprehensive dataset containing 1,000 diversified linear and related non-linear MBA expressions. The evaluation demonstrates that existing deobfuscation methods cannot effectively simplify non-linear MBA expressions. On the other hand, the overhead to use non-linear MBA expression is low. Our evaluation results show that the non-linear MBA technique is an practical obfuscation scheme.

In summary, we make the following key contributions:

- We propose how to use linear MBA rules to generate non-linear MBA expression and guarantee its correctness.
- We discuss one concrete application of non-linear MBA expression, which obfuscates the Tiny Encryption Algorithm(TEA).

– Our large-scale evaluation shows that *MBA-Obfuscator* outperforms existing linear MBA obfuscation in terms of better resilience, potency, and cost. *MBA-Obfuscator*'s source code and the benchmark are available at https://github.com/nhpcc502/MBA-Obfuscator.

The rest of the paper is structured as follows. Section 2 illustrates the background of existing MBA obfuscation and simplification methods. Section 3 introduces our methods to generate infinite non-linear MBA expressions. Next, we present one practical application (Sect. 4) of non-linear MBA obfuscation on the Tiny Encryption Algorithm(TEA). Finally, we give the details on our evaluation results (Sect. 5) and conclude (Sect. 6)

## 2   Preliminaries

In this section, we provide background on Mixed-Boolean-Arithmetic (MBA) transformation and deobfuscation techniques. Zhou et al. [31,32] formally present how to generate unlimited linear MBA expressions. Since MBA rules are applied as an information hiding technique, it has already generated follow-up deobfuscation research to simplify MBA expressions, including bit-blasting, pattern matching, program synthesis, deep learning, and mathematical transformation approaches.

### 2.1   MBA-Based Obfuscation

Zhou et al. [31,32] firstly define an MBA expression as the mixture usage of bitwise operations ($\vee$, $\wedge$, $\oplus$, $\neg$) and integer arithmetic operations($+$, $-$,$\times$). The formal definition of polynomial MBA expression is denoted as follows:

**Definition 1.** *A polynomial MBA expression is:*

$$\sum_{i \in I} a_i * (\prod_{j \in J_i} e_{i,j}(x_1, \ldots, x_t)),$$

*where $a_i$ are constants, $e_{i,j}$ are bitwise expressions of variables $x_1, \ldots, x_t$ over $B^n$, and $I, J_i \subset Z$, $\forall i \in I$. $a_i * (\prod_{j \in J_i} e_{i,j}(x_1, \ldots, x_t))$ is called a term.*

**Definition 2.** *A linear MBA expression is a polynomial MBA expression of the form:*

$$\sum_{i \in I} a_i * e_i(x_1, \ldots, x_t),$$

*where $a_i$ are constants, $e_i$ are bitwise expressions of variables $x_1, \ldots, x_t$ over $B^n$, and $I \subset Z$. $a_i * e_i(x_1, \ldots, x_t)$ is called a term.*

Examples of linear and polynomial MBA expression are shown as follows:

$$x + (x \wedge y) + y - 2 * (x \oplus y) + (x \vee y),$$
$$x + y * (x \oplus y) - 2 * (\neg x \vee y) * (\neg x) - 1.$$

In addition, Zhou et al. [32] present an approach, which uses truth tables(linearly dependent column vectors) to generate a linear MBA identity, as shown in Example 1. They propose a formal theoretical foundation to guarantee the correctness of this method to build unlimited linear MBA rules.

*Example 1.* The 0,1-matrix

$$
M = \begin{matrix} x & y & x \oplus y & \neg(x \vee y) & -1 \\ \begin{pmatrix} 0 & 0 & 0 & 1 & 1 \\ 0 & 1 & 1 & 0 & 1 \\ 1 & 0 & 1 & 0 & 1 \\ 1 & 1 & 0 & 0 & 1 \end{pmatrix} \end{matrix}, \vec{v} = \begin{pmatrix} 1 \\ 1 \\ 1 \\ 2 \\ -2 \end{pmatrix},
$$

$$
\Rightarrow M\vec{v} = [0,0,0,0,0]^{T}.
$$

It produces a linear MBA identity:

$$
x + y + (x \oplus y) + 2 * \neg(x \vee y) - 2 * (-1) = 0.
$$

This linear MBA identity can be transformed into multiple MBA rules:

$$
x + y = -(x \oplus y) - 2 * \neg(x \vee y) - 2,
$$
$$
2 = -x - y - (x \oplus y) - 2 * \neg(x \vee y).
$$

Example 1 presents how to transform $x + y$ into a much more complex form: $-(x \oplus y) - 2 * \neg(x \vee y) - 2$, another example of the code before and after linear MBA obfuscation is shown in Fig. 1.

Unfortunately, Zhou et al. [32] only propose the formation of polynomial MBA expression, without any further research. Therefore, all existing MBA obfuscation work is mainly based on the linear MBA rules. For example, Cloakware [17], Irdeto [13], and Quarkslab [23] apply MBA transformation in their commercial products. Tigress[1] [5], an academic C source code obfuscation tool, applies MBA rules to encode integer variables and expressions [6,7]. Mougey and Gabriel [21] use MBA rules to do instruction substitution in a Digital Rights Management (DRM) system. Recently, Blazy and Hutin [3] strengthen the CompCert C compiler [16] to generate programs with formally verified MBA obfuscation rules. Xmark applies MBA obfuscation to hide the static signatures of software watermarking [19]. ERCIM News reported in 2016 that the MBA obfuscation technique had been used in malware compilation chains [1].

## 2.2   MBA Deobfuscation

After the MBA obfuscation technique was proposed, researchers have explored how to simplify MBA expressions. Eyrolles's PhD thesis [8] is the first work to explore this subject at full length. She focuses on simplifying MBA expressions by

---

[1] https://tigress.wtf/index.html.

```
int fun(int x,int y){
    int res;
    res = x + y;

    return res;
}
```

```
int fun(int x,int y){
    int res;
    res = 5*(x&y)+4*(x&~y)-5*(x|y)+2*x+6*~(x|~y);

    return res;
}
```

(a) Original program.                    (b) Linear MBA obfuscated program.

**Fig. 1.** An example of MBA obfuscation for x + y.

applying various analyzing tools. Her experiments show that popular computer algebra software such as Maple [20], SageMath [25], Wolfram Mathematica [29], and Z3 [22] fail to simplify MBA expression, because existing reduction rules only work either on pure bitwise expressions or pure arithmetic expressions. Furthermore, LLVM compiler optimization passes [15] have a very limited effect on MBA simplification. Therefore, Eyrolles et al. [9] propose a pattern matching method to simplify MBA expressions. Multiple MBA rules are hard-coded in the tool, named SSPAM. However, the pattern matching method only detects and simplifies MBA expressions by a range of fixed patterns. Such an approach cannot reduce generic MBA rules.

Guinet et al. [11] create a tool, *Arybo*, which reduces MBA expressions to bit-level symbolic representation with only ⊕ and ∧ operations. One of the drawbacks is the performance penalty caused by the bloated size of bit-level expressions. Biondi [2] presents an algebraic approach to simplify MBA expression, reducing the complexity of MBA obfuscation, but the proposed method strongly depends on the specific MBA rules.

Blazytko et al. [4] apply program synthesis [12] guided by Monte Carlo Tree Search (MCTS) to do code deobfuscation. It produces input-output samples from the obfuscated code and then learns the semantics based on those input-output pairs. Then it automatically generates another simpler but equivalent expression. Due to the non-determinism and sampling mechanism of program synthesis, their method cannot guarantee the correctness of the simplification result.

Feng et al. [10] introduce a novel solution, named NeuReduce, to simplify MBA expression. They train NeuReduce using MBA rules based on the sequence to sequence neural network models. The input of NeuReduce is a character string format of complex MBA expression, and the related output is the simplification expression.

Liu et al. [18] investigate the mathematical mechanism of MBA expression and prove a hidden two-way transformation feature in the MBA obfuscation. They transform all bitwise expressions to specific MBA forms and then perform arithmetic reduction to simplify MBA expression.

# 3 Non-Linear MBA Expression Generation

In this section, we focus on how to generate unlimited non-linear MBA expressions. First, we present how to create polynomial MBA expressions from the linear MBA rules. Next, multiple methods are applied to produce non-linear MBA rules.

## 3.1 Polynomial MBA Expression

Using Proposition 1, a polynomial MBA expression can be generated based on the existing linear MBA expression.

**Proposition 1.** *Let $E$ be an expression, $n$ be positive integer,*

- $E \equiv \bar{E}$, $\bar{E}$ *is a linear MBA expression,*
- $E_k$ *is a linear MBA expression, $E_k = 1$, $k = 1, 2, \ldots, n$,*
- $E' = \bar{E} * E_1 * E_2 * \ldots * E_n$,

*Then*
$E' \equiv E$, *and $E'$ is a polynomial MBA expression.*

By Proposition 1, any simple expression can be transformed into a complex polynomial MBA expression, as shown in Example 2.

*Example 2.* For a simple expression $x + y$, we have
$x + y = (x \vee y) + (x \wedge y)$, $1 = (x \wedge y) - y - (x \vee \neg y)$,
then

$$
\begin{aligned}
x + y &= ((x \vee y) + (x \wedge y)) * ((x \wedge y) - y - (x \vee \neg y)) \\
&= (x \vee y) * (x \wedge y) + (x \wedge y) * (x \wedge y) - (x \vee y) * y \\
&\quad - (x \wedge y) * y - (x \vee y) * (x \vee \neg y) - (x \wedge y) * (x \vee \neg y).
\end{aligned}
$$

However, Proposition 1 exposes a potential drawback: one polynomial MBA expression generated by Proposition 1 has a fixed pattern, which can be used to simplify the polynomial MBA expression through basic algebra laws (i.e., commutation, association, and distribution laws). Firstly, a reverse engineer can carefully recover the original linear MBA expression by factoring. Then, the linear MBA expressions can be simplified with existing linear MBA simplification tools. For instance, the polynomial MBA expression in Example 2 can be simplified as follows:

$$
((x \vee y) + (x \wedge y)) * ((x \wedge y) - y - (x \vee \neg y)) = (x + y) * 1 = x + y.
$$

This flaw shows that the strength of polynomial MBA expression is the same as linear MBA expression. In order to address this issue, we firstly introduce the concept of *0-equality*, whose formal definition is given in Definition 3. Example 3 presents how to generate a *0-equality*.

**Definition 3.** *Let $E_1$ be an expression, $E_2$ and $\bar{E}_2$ be linear MBA expressions,*

- *the coefficients in $E_1$ are randomly reversed or not, to construct a new expression $\bar{E}_1$,*
- *a sub-expression of $E_2$ forms part of $\bar{E}_2$, and $\bar{E}_2 \equiv 0$,*

*Then $E_1 * E_2$'s 0-equality is $Z_{E_1 E_2} = \bar{E}_1 * \bar{E}_2 \equiv 0$.*

*Example 3.* For expressions
$E_1 = (x \vee y) + (x \wedge y)$, $E_2 = (x \wedge y) - y - (x \vee \neg y)$, we have
$\bar{E}_1 = (x \vee y) - (x \wedge y)$, $\bar{E}_2 = (x \wedge y) - y + (\neg x \wedge y) = 0$,
$\Rightarrow Z_{\bar{E} E_1} = \bar{E}_1 * \bar{E}_2 = ((x \vee y) - (x \wedge y)) * ((x \wedge y) - y + (\neg x \wedge y)) = 0$.

Through the *0-equality*, Proposition 2 transforms a polynomial expression generated by Proposition 1 into another format, which has broken the fixed pattern introduced by Proposition 1. One detailed instance is shown in Example 4.

**Proposition 2.** *Let $E$ be an expression,*

- *$E = \bar{E}$, $\bar{E}$ is a linear MBA expression,*
- *$E_k$ is a linear MBA expression, $E_k = 1$, $k = 1, 2, \ldots, n$,*
- *$E' = (\ldots((\bar{E} * E_1 + Z_{\bar{E} E_1}) * E_2 + Z_{\bar{E} E_1 E_2}) * \ldots) * E_n + Z_{\bar{E} E_1 E_2 \ldots E_n}$,*

*Then*
*$E' \equiv E$, and $E'$ is a polynomial MBA expression.*

*Example 4.* For an expression $E = x + y$ and Example 3, we have

$$\begin{aligned}
x + y &= \bar{E} * E_1 + Z_{\bar{E} E_1} \\
&= ((x \vee y) + (x \wedge y)) * ((x \wedge y) - y - (x \vee \neg y)) \\
&\quad + ((x \vee y) - (x \wedge y)) * ((x \wedge y) - y + (\neg x \wedge y)) \\
&= 2 * (x \vee y) * ((x \wedge y) - y) - (x \vee \neg y) * ((x \vee y) + (x \wedge y)) \\
&\quad + (\neg x \wedge y) * ((x \vee y) - (x \wedge y)).
\end{aligned}$$

### 3.2  MBA-related Rules

So far, a simple expression can be transformed into a complex polynomial MBA expression. Next, we demonstrate how to generate a new non-linear MBA expression based on existing MBA rules.

**Recursively Apply MBA Rules.** This method firstly transform the simple expression into a linear MBA expression, then recursively apply MBA rules to convert the related linear MBA expression into a complex non-linear MBA format, as seen in Example 5.

*Example 5.*

$$\begin{aligned}
x + y &= y - \neg x - 1 \Leftarrow x + y = (x \vee y) + (x \wedge y) \\
&= ((y - \neg x) \vee -1) + ((y - \neg x) \wedge -1) \Leftarrow x + y = y + (x \wedge \neg y) + (x \wedge y) \\
&= (((-\neg x) + (y \wedge \neg(-\neg x)) + (y \wedge (-\neg x))) \vee -1) \\
&\quad + (((-\neg x) + (y \wedge \neg(-\neg x)) + (y \wedge (-\neg x))) \wedge -1).
\end{aligned}$$

**Replace Sub-expression With MBA Expression.** This method replaces part of the original expression with an equivalent non-linear MBA rule, as shown in Example 6.

*Example 6.*

$$
\begin{aligned}
2 =& (-3 * (x \vee y) - 1 - 3 * (\neg x \wedge \neg y)) * (-x - \neg x) \\
& + (3 * (x \vee y) + 1) * (-x + (x \vee y) + (x \wedge \neg y)) \\
=& 3 * (\neg x \wedge \neg y) * (x + \neg x) + (3 * (x \wedge y) + 1) * (\neg x + 1 + 3 * (x \vee y)) \\
\Rightarrow x + y =& -(x \oplus y) - 2 * \neg(x \vee y) - 2 \\
=& -(x \oplus y) - 2 * \neg(x \vee y) - 3 * (\neg x \wedge \neg y) * (x + \neg x) \\
& - (3 * (x \wedge y) + 1) * (\neg x + 1 + 3 * (x \vee y)).
\end{aligned}
$$

**Linear Combination of MBA Expression.** A new MBA expression can be generated from the linear combination of existing MBA obfuscation rules. More specifically, we add a non-linear MBA expression which is equal to 0 to the original expression, as seen in Example 7.

*Example 7.*

$$
\begin{aligned}
x =& (x \wedge y) + (x \vee y) - y, \\
0 =& x * y - (x \wedge y) * (x \vee y) - (x \wedge \neg y) * (\neg x \wedge y), \\
\Rightarrow x =& (x \wedge y) + (x \vee y) - y + x * y - (x \wedge y) * (x \vee y) - (x \wedge \neg y) * (\neg x \wedge y).
\end{aligned}
$$

---

**Algorithm 1.** MBA Expression Generation

---

1: *Input*: Simple expression $E$, Flag $F$.
2: **function** MBA-OBFUSCATOR($E$, $F$)
3:     Generate a new linear MBA expression $\bar{E}$ that equals to input $E$.
4:     **if** $F$.polynomial **then**
5:         Generate a new polynomial MBA expression $E'$ based on Proposition 2.
6:     **else if** $F$.recursively **then**
7:         Generate multiple MBA rules that equal to $x + y$.
8:         Apply the rules to transform $\bar{E}$ into $E'$.
9:     **else if** $F$.replace **then**
10:         Generate multiple MBA rules that equal to the sub-expression of $\bar{E}$.
11:         Apply the rules to transform $\bar{E}$ into $E'$.
12:     **else if** $F$.combination **then**
13:         Generate a polynomial MBA expression $Z$ that equals to 0.
14:         $E' = \bar{E} + Z$.
15:     **end if**
16:     return $E'$.
17: **end function**

---

We integrate all methods described above into Algorithm 1. The algorithm takes a simple expression $E$ and flag $F$ as input and returns its related complex non-linear MBA expression based on the flag $F$. One random seed is contained in the algorithm, as the one in Definition 3, to generate a randomized obfuscated expression. Note that Algorithm 1 can transform a constant into a complex non-linear MBA expression, as shown in Example 6.

# 4   Case Study

In this section, we present how to apply the MBA obfuscation technique on the Tiny Encryption Algorithm(TEA) to protect the key and algorithm, since MBA obfuscation technique can hide constant or complicate operations.

In cryptography, the Tiny Encryption Algorithm (TEA) is a block cipher designed by David Wheeler and Roger Needham [28]. The encryption routine code in C of TEA is shown in Fig. 2. Given the feature of the symmetric encryption algorithm, the respective decryption routine is similar to the encryption routine. Since its simple structure and low cost, TEA has been widely applied to many scenarios [14, 24, 27, 30].

The whole obfuscation process is shown in Fig. 3. Firstly, the key is transformed into a function that outputs a constant value, as shown in Fig. 3a. Next, the operation in the TEA is replaced with related complex MBA expression, which is shown in Fig. 3b. For every one operation, MBA-Obfuscator generates the related complex and equivalent MBA expression. After that, the original algorithm is replaced with the obfuscation routine, and the entire obfuscation code is seen in Fig. 3c. Considering the randomness in Algorithm 1, every expression generated by MBA-Obfuscator is different from each other.

```c
void TEAencrypt(uint32 v[2], uint32 k[4]) {
    uint32 v0=v[0], v1=v[1], sum=0, i;                    /*set up*/
    uint32 delta=0x95136ac5;              /*a key schedule constant*/
    uint32 k0=k[0], k1=k[1], k2=k[2], k3=k[3];              /*key*/

    /*basic cycle*/
    for (i = 0; i < 32; i++) {
        sum += delta;
        v0 += ((v1<<4) + k0) ^ (v1 + sum) ^ ((v1>>5) + k1);
        v1 += ((v0<<4) + k2) ^ (v0 + sum) ^ ((v0>>5) + k3);
        }

    v[0]=v0;
    v[1]=v1;
    }
```

**Fig. 2.** The TEA encryption procedure implemented in C programming language.

```
uint32 keyfunction(uint32 x, uint32 y) {
    //let the key be 0x12345678
    uint32 key = MBA-Obfuscator(0x12345678, Flag);

    return key;
}
```

(a) Hiding the encryption key.

```
uint32 ob_operation(uint32 x, uint32 y, uint32 sum,
                    uint32 (*kf0)(uint32, uint32),
                    uint32 (*kf1)(uint32, uint32)) {
    uint32 k0 = kf0(x, y);
    uint32 k1 = kf1(x, y);
    uint32 y1 = y << 4;
    uint32 z0 = MBA-Obfuscator(y1 + k0, Flag);
    uint32 z1 = MBA-Obfuscator(y + sum, Flag);
    uint32 y2 = y >> 5;
    uint32 z2 =  MBA-Obfuscator(y2 + k1, Flag);
    uint32 res = MBA-Obfuscator(z0 ^ z1 ^ z2, Flag);

    return res + x;
}
```

(b) Obfuscating the encryption operations.

```
void TEAencrypt_ob(uint32 v[2],uint32 (*kf)(uint32, uint32)[4]) {
    uint32 v0=v[0], v1=v[1], sum=0;                    /*set up*/
    uint32 delta=0x95136ac5;           /*a key schedule constant*/
    uint32 kf0=kf[0], kf1=kf[1];                          /*key*/
    uint32 kf2=kf[2], kf3=kf[3];

    /*basic cycle*/
    for (i = 0; i < 32; i++){
        sum = MBA-Obfuscator(sum + delta, Flag);
        v0 = ob_operation(v0,v1,sum,kf0,kf1);
        v1 = ob_operation(v1,v0,sum,kf2,kf3);
    }

    v[0]=v0;
    v[1]=v1;
}
```

(c) TEA encryption routine after Obfuscation.

**Fig. 3.** The steps of running MBA-Obfuscator on TEA encryption.

# 5    Implementation and Evaluation

We implement Algorithm 1 in a prototype tool called *MBA-Obfuscator*. *MBA-Obfuscator* accepts a simple expression input and generates a random related non-linear MBA expression. The whole prototype is written in 1,900 lines of Python code. We leverage the `NumPy` library for generating a linear MBA expression, and the `SymPy` library for arithmetic operation. *MBA-Obfuscator* also includes utilities for measuring the quantitative metrics for MBA expressions, such as counting MBA alternation and the number of terms.

We conduct a set of experiments to seek for checking the capability of *MBA-Obfuscator*. In particular, we design experiments to answer the following research questions.

1. **RQ1:** Is *MBA-Obfuscator* able to resist related methods simplifying MBA expression? *(resilience)*
2. **RQ2:** How complex is a non-linear MBA expression? *(potency)*
3. **RQ3:** How much overhead does *MBA-Obfuscator* introduce? *(cost)*

As the answer to RQ1, we apply existing deobfuscation tools to simplify non-linear MBA expressions. To address RQ2, we calculate the complexity metrics such as the number of MBA alternation. In response to RQ3, we study *MBA-Obfuscator*'s performance data such as running time and memory usage.

## 5.1    Experimental Setup

**Dataset.** Note that the use of non-linear MBA expression is by simply substituting where the linear MBA rule is used. Therefore we check the capacity of linear MBA expression and the related non-linear MBA expression generated by *MBA-Obfuscator*. We collect 1,000 linear MBA expressions and related ground truth (correctly simplified form) from existing works [4,5,8,18,31,32]. Next, we use *MBA-Obfuscator* to generate a non-linear MBA expression that is equivalent to the ground truth. Therefore, we get the Dataset including 1,000 MBA expressions. Every sample in the dataset is a 3-tuple: $(G, L, NL)$. $G$ is a simple expression, named ground truth. $L$ is the related complex linear MBA expression. $NL$ is the related complex non-linear MBA expression. One example is shown in Fig. 4.

```
Linear MBA rule:
   x + y = 3*(x&y)+8*(x&~y)-1*y-7*~(x&y)+7*~(x|y)+9*~(x|~y)
Non-Linear MBA rule:
   x + y = 2*(x|y)*((x&y)-y)-(x+y)*(x|~y)+(~x&y)*(x^y)
```

**Fig. 4.** The linear and non-linear MBA expressions for the ground truth expression x + y.

**MBA Complexity Metrics.** We use the following metrics to measure MBA complexity. For these complexity metrics, a larger value indicates a more complex MBA expression.

1. **MBA Alternation.** MBA alternation is to count the number of operations that connect arithmetic operation and bitwise operation. For example, in $(x \wedge y) + 2 * (x \vee y)$, the $+$ represents an MBA alternation operation, because its left operator is a bitwise operator, and its right operator is an arithmetic operator.
2. **Number of Terms.** How many terms are included in an MBA expression.
3. **MBA Length.** The string length of an MBA expression is measured as the MBA length.

**MBA Deobfuscation Tools.** We collect and check existing, state-of-the-art MBA deobfuscation tools, as described in Sect. 2.2. Arybo[2], MBA-Blast[3], NeuReduce[4], SSPAM[5], and Syntia[6] focus on simplifying MBA expression by bit-blasting, mathematical transformation, machine learning-based, pattern matching, and program synthesis. Arybo is a tool for transforming MBA expression to a symbolic representation at the bit-level written in Python. MBA-Blast is a novel technique to simplify MBA expressions to a normal simple form by arithmetical reduction. SSPAM (Symbolic Simplification with Pattern Matching) is a Python tool for simplifying MBA expression. Syntia is a program synthesis framework for synthesizing the semantic of obfuscated code. It produces input-output pairs from the obfuscated rules and then generates a new simple expression based on these pairs. NeuReduce is a string to string method based on neural networks to learn and reduce complex MBA expressions automatically. We download and apply those tools to simplify MBA expressions for checking the strength of the related obfuscation technique.

**Machine Configuration.** Our experiments were performed on an Intel Xeon W-2123 4-Core 3.60GHz CPU, with 64GB of RAM, running Ubuntu 18.04.

## 5.2    Deobfuscation on Non-Linear MBA Expression

In this evaluation, we check the *resilience* of *MBA-Obfuscator*, which refers to the robustness of an obfuscation tool for an automatic deobfuscator. Eyrolles's PhD thesis [8] states that reverse engineer focuses on recovering the initial form of the expression, meaning simplifying the MBA expression. Her experiments show that existing symbolic software cannot simplify MBA expression because math reduction rules only work on pure Boolean expressions(e.g., normalization and constraint solving), or on pure arithmetic expressions(e.g., the algebra laws). So far, no publicly known methods, including both static and dynamic methods, can effectively simplify MBA expressions. This fact attracts researchers to

---

[2] https://github.com/quarkslab/arybo.
[3] https://github.com/softsec-unh/MBA-Blast.
[4] https://github.com/nhpcc502/NeuReduce.
[5] https://github.com/quarkslab/sspam.
[6] https://github.com/RUB-SysSec/syntia.

develop multiple tools to simplify MBA expression, such as Arybo, MBA-Blast, NeuReduce, SSPAM, and Syntia. Therefore, we test those deobfuscation tools on linear and related non-linear MBA expression. After simplification, we use the Z3 solver [22] library[7] to check whether every simplified result is equivalent to the original expression.

**Table 1.** Deobfuscation results. ✓ means equivalent(Z3 returns a UNSAT solution), ✗ means not equivalent(Z3 returns an SAT solution), – means time out(deobfuscation tools cannot return a result in 1 h), "Ratio" indicates the ratio of outputs passing equivalence checking, "Time" reports the average processing time (seconds) that each tool takes to process an MBA sample in the ✓ column.

| Method | Linear MBA | | | | | Non-Linear MBA | | | | |
|---|---|---|---|---|---|---|---|---|---|---|
| | ✓ | ✗ | −− | Ratio(%) | Time | ✓ | ✗ | −− | Ratio(%) | Time |
| Arybo | 569 | 0 | 431 | 56.9 | 1936.5 | 84 | 0 | 916 | 8.4 | 2304.9 |
| MBA-Blast | 1,000 | 0 | 0 | 100.0 | 0.06 | 147 | 853 | 0 | 14.7 | 0.09 |
| NeuReduce | 756 | 244 | 0 | 75.6 | 0.06 | 1 | 999 | 0 | 0.0 | 0.06 |
| SSPAM | 386 | 356 | 258 | 38.6 | 1465.7 | 103 | 192 | 705 | 10.3 | 2132.1 |
| Syntia | 97 | 903 | 0 | 9.7 | 29.7 | 98 | 902 | 0 | 9.8 | 29.8 |

**Table 2.** The complexity distribution of the MBA expressions in the Dataset.

| Metrics | Linear MBA | | | Non-Linear MBA | | |
|---|---|---|---|---|---|---|
| | Min | Max | Average | Min | Max | Average |
| MBA Alternation | 2 | 99 | 30.6 | 5 | 92 | 30.3 |
| Number of Terms | 3 | 99 | 31.4 | 6 | 79 | 27.4 |
| MBA Length | 17 | 1498 | 438.0 | 47 | 1140 | 391.0 |

Table 1 shows the deobfuscation result on the Dataset. All existing deobfuscation tools can simplify part or all of the linear MBA expressions. However, up to 14.7% of non-linear MBA expressions can be simplified by these tools in 1 h. Arybo only simplifies 84 out of 1,000 samples, because it suffers from severe performance penalties when it normalizes all operators to algebraic normal form. We observe that MBA-Blast performs well on 2-variable polynomial MBA expressions. However, it has a limited simplification effect on other non-linear MBA expressions. Thus, MBA-Blast can simplify 147 non-linear MBA expressions. Since NeuReduce is trained on the dataset of linear MBA expressions, it cannot simplify non-linear MBA expressions. SSPAM can simplify 103 out of 1,000 non-linear MBA expression. Limited by the nature of program synthesis, Syntia only returns 98 correct results for non-linear MBA expressions.

[7] https://github.com/Z3Prover/z3.

## 5.3    Complexity of Non-Linear MBA Expression

In the second evaluation, we seek to check the capability of *potency* after obfuscation by MBA expressions. *Potency* represents how *complex* or *unreadable* the obfuscated rules are to a reverse engineer. Table 2 shows the complexity results on the Dataset. From Definition 2, the complexity of a linear MBA expression mainly depends on the number of terms in the expression. However, a non-linear MBA expression's complexity depends on multiple factors, such as the number of terms and MBA alternation. Overall, the non-linear MBA expression is slightly simpler than the linear MBA expression from the "Average" column.

**Table 3.** *MBA-Obfuscator*'s performance for generating a non-linear MBA expression with different complexity. "Time" reports the time (seconds) that *MBA-Obfuscator* generates a non-linear MBA expression 10,000 times.

| # Of Terms | Time(Second) | Memory(MB) |
|---|---|---|
| 10 | 103.5 | 0.01 |
| 50 | 259.4 | 0.01 |
| 100 | 381.1 | 0.01 |
| 200 | 534.2 | 0.02 |

**Table 4.** Run-time overhead on non-linear MBA expressions with different complexity. The timing result is the time to run the MBA expression 10,000 times repeatedly.

| # of Terms | Time (Second) | | |
|---|---|---|---|
| | Linear MBA | Non-Linear MBA | Ratio(%) |
| 10 | 0.5 | 1.5 | 300.0 |
| 50 | 3.6 | 8.2 | 227.8 |
| 100 | 8.1 | 22.4 | 276.5 |
| 200 | 18.3 | 57.9 | 316.4 |

## 5.4    Cost of Non-Linear MBA Expression

As the last experiment, we study the *cost* of *MBA-Obfuscator*: performance overhead representing the cost for generating an MBA expression, and run-time overhead that refers to the cost when the obfuscated code is running. Table 3 presents the time and memory cost when *MBA-Obfuscator* generates a non-linear MBA expression with different complexity measured by the number of terms. *MBA-Obfuscator* is effective because it relies on the existing linear MBA rules.

For run-time overhead, we integrate MBA expressions into a C program, and then use gcc to compile it with the -O2 option. Table 4 shows that run-time overhead introduced by non-linear MBA rules is about 3X as long as the related linear MBA rules. However, the average run-time overhead for a non-linear MBA rule is low, which is less than 0.01 s (0.0058 s).

# 6 Conclusion

Mixed Boolean-Arithmetic (MBA) expression, which mixes both bitwise and arithmetic operations, can be applied to complicate a simple expression. Our work is the first research to propose a non-linear MBA obfuscation challenge. We investigate the class of MBA expression, and demonstrate how to generate infinite non-linear MBA expressions based on existing linear MBA expressions. Furthermore, we present a practical application of the non-linear MBA obfuscation technique on obfuscating the Tiny Encryption Algorithm(TEA). We develop a prototype tool *MBA-Obfuscator*, a novel non-linear MBA obfuscation technique. Our large-scale experiment demonstrates that *MBA-Obfuscator* is effective and efficient—existing deobfuscation tools may be available to simplify polynomial MBA expressions but hardly for other non-linear MBA expressions, and the cost of applying non-linear MBA obfuscation is low. Developing *MBA-Obfuscator* not only advances the application of the MBA obfuscation technique, but also develops a dataset for future research on MBA deobfuscation direction.

**Acknowledgments.** We would like to thank our shepherd Roland Yap and anonymous paper reviewers for their helpful feedback. We also thank team members from Anhui Province Key Laboratory of High Performance Computing (USTC) and the UNH SoftSec group for their valuable suggestions.

# References

1. Biondi, F., Josse, S., Legay, A.: Bypassing Malware Obfuscation with Dynamic Synthesis. https://ercim-news.ercim.eu/en106/special/bypassing-malware-obfuscation-with-dynamic-synthesis (July 2016)
2. Biondi, F., Josse, S., Legay, A., Sirvent, T.: Effectiveness of synthesis in concolic deobfuscation. Comput. Secur. **70**, 500–515 (2017)
3. Blazy, S., Hutin, R.: Formal verification of a program obfuscation based on mixed boolean-arithmetic expressions. In: Proceedings of the 8th ACM SIGPLAN International Conference on Certified Programs and Proofs (CPP 2019) (2019)
4. Blazytko, T., Contag, M., Aschermann, C., Holz, T.: Syntia: synthesizing the semantics of obfuscated code. In: Proceedings of the 26th USENIX Security Symposium (USENIX Security 2017) (2017)
5. Collberg, C., Martin, S., Myers, J., Nagra, J.: Distributed application tamper detection via continuous software updates. In: Proceedings of the 28th Annual Computer Security Applications Conference (ACSAC 2012) (2012)
6. Collberg, C., Martin, S., Myers, J., Zimmerman, B.: Documentation for Arithmetic Encodings in Tigress. http://tigress.cs.arizona.edu/transformPage/docs/encodeArithmetic

7. Collberg, C., Martin, S., Myers, J., Zimmerman, B.: Documentation for Data Encodings in Tigress. http://tigress.cs.arizona.edu/transformPage/docs/encodeData

8. Eyrolles, N.: Obfuscation with Mixed Boolean-Arithmetic Expressions: Reconstruction, Analysis and Simplification Tools. Ph.D. thesis, Université Paris-Saclay (2017)

9. Eyrolles, N., Goubin, L., Videau, M.: Defeating MBA-based obfuscation. In: Proceedings of the 2016 ACM Workshop on Software PROtection (SPRO 2016) (2016)

10. Feng, W., Liu, B., Xu, D., Zheng, Q., Xu, Y.: Neureduce: Reducing mixed boolean-arithmetic expressions by recurrent neural network. In: Proceedings of the 2020 Conference on Empirical Methods in Natural Language Processing: Findings, pp. 635–644 (2020)

11. Guinet, A., Eyrolles, N., Videau, M.: Arybo: Manipulation, canonicalization and identification of mixed boolean-arithmetic symbolic expressions. In: Proceedings of GreHack 2016 (2016)

12. Gulwani, S., Polozov, O., Singh, R.: Program Synthesis. Found. Trends in Program. Lang. **4**(1–2), 1–119 (2017)

13. Irdeto: Irdeto Cloaked CA: a secure, flexible and cost-effective conditional access system. www.irdeto.com (2017)

14. Israsena, P.: Securing ubiquitous and low-cost rfid using tiny encryption algorithm. In: 2006 1st International Symposium on Wireless Pervasive Computing, pp. 4-pp. IEEE (2006)

15. Lattner, C., Adve, V.: LLVM: a compilation framework for lifelong program analysis & transformation. In: Proceedings of the International Symposium on Code Generation and Optimization: Feedback-directed and Runtime Optimization (CGO 2004) (2004)

16. Leroy, X.: Formal verification of a realistic compiler. Commun. ACM **52**, 107–115 (2009)

17. Liem, C., Gu, Y.X., Johnson, H.: A compiler-based infrastructure for software-protection. In: Proceedings of the 3rd ACM SIGPLAN Workshop on Programming Languages and Analysis for Security (PLAS 2008) (2008)

18. Liu, B., Shen, J., Ming, J., Zheng, Q., Li, J., Xu, D.: Mba-blast: unveiling and simplifying mixed boolean-arithmetic obfuscation. In: 30th USENIX Security Symposium (USENIX Security 2021) (2021)

19. Ma, H., Jia, C., Li, S., Zheng, W., Wu, D.: Xmark: dynamic software watermarking using collatz conjecture. IEEE Trans. Inf. Forensics Secur. **14**, 2859–2874 (2019)

20. MapleSoft: The Essential Tool for Mathematics. https://www.maplesoft.com/products/maple/ (2020)

21. Mougey, C., Gabriel, F.: DRM obfuscation versus auxiliary attacks. In: REcon Conference (2014)

22. Moura, L.D., Bjørner, N.: Z3: an efficient SMT solver. In: Proceedings of the 14th International Conference on Tools and Algorithms for the Construction and Analysis of Systems (TACAS) (2008)

23. Quarkslab: Epona Application Protection v1.5. https://epona.quarkslab.com (July 2019)

24. Rahim, R., et al.: Tiny encryption algorithm and pixel value differencing for enhancement security message. Int. J. Eng. Technol. **7**(2.9), 82–85 (2018)

25. Sagemath: SageMath. http://www.sagemath.org/ (2020)

26. Schrittwieser, S., Katzenbeisser, S., Kinder, J., Merzdovnik, G., Weippl, E.: Protecting software through obfuscation: Can it keep pace with progress in code analysis? ACM Comput. Surv. (CSUR) **49**(1), 1–37 (2016)

27. Suwartadi, E., Gunawan, C., Setijadi, A., Machbub, C.: First step toward internet based embedded control system. In: IEEE 5th Asian Control Conference, vol. 2, pp. 1226–1231 (2004)
28. Wheeler, D.J., Needham, R.M.: TEA, a tiny encryption algorithm. In: Proceedings of the 2nd International Workshop on Fast Software Encryption (1994)
29. WOLFRAM: WOLFRAM MATHEMATICA. http://www.wolfram.com/mathematica/ (2020)
30. Zafar, F., Olano, M., Curtis, A.: Gpu random numbers via the tiny encryption algorithm. In: Proceedings of the Conference on High Performance Graphics, pp. 133–141 (2010)
31. Zhou, Y., Main, A.: Diversity via Code Transformations: A Solution for NGNA Renewable Security. The National Cable and Telecommunications Association Show (2006)
32. Zhou, Yongxin, Main, Alec, Gu, Yuan X., Johnson, Harold: Information hiding in software with mixed boolean-arithmetic transforms. In: Kim, Sehun, Yung, Moti, Lee, Hyung-Woo (eds.) WISA 2007. LNCS, vol. 4867, pp. 61–75. Springer, Heidelberg (2007). https://doi.org/10.1007/978-3-540-77535-5_5

# VIRSA: Vectorized In-Register RSA Computation with Memory Disclosure Resistance

Yu Fu[1,2], Wei Wang[1,3](✉), Lingjia Meng[1,2], Qiongxiao Wang[1,2], Yuan Zhao[4], and Jingqiang Lin[5,6](✉)

[1] State Key Laboratory of Information Security, Institute of Information Engineering, CAS, Beijing 100089, China
{fuyu,wangwei}@iie.ac.cn
[2] School of Cyber Security, University of Chinese Academy of Sciences, Beijing 100089, China
[3] Data Assurance and Communication Security Research Center, CAS, Beijing, China
[4] Ant Group, Hangzhou, China
[5] School of Cyber Security, University of Science and Technology of China, Hefei 230027, Anhui, China
linjq@ustc.edu.cn
[6] Beijing Institute, University of Science and Technology of China, Beijing, China

**Abstract.** Memory disclosure attacks give adversaries access to sensitive data in memory, posing a serious threat to the security of cryptographic systems. For example, the plain private key in RAM is exposed to the attacker during RSA operation. In this paper, we propose a register-based RSA system with high efficiency, called VIRSA, so that CRT-enabled 2048-bit RSA is entirely carried out on CPU registers. The private key and the intermediate results during the calculation process are all stored in registers, and will not appear in memory, which effectively prevents memory disclosure attacks. The input RSA parameters are encrypted by an AES key. The AES key is stored in the privileged debug registers. For performance, we use AVX-512F instruction set to accelerate the RSA calculation. We adopt vector instructions to implement 1024-bit Montgomery multiplication and make use of redundant representation to solve the carry propagation problem. Experiments on Intel Xeon Silver 4208 CPU shows that VIRSA achieves a performance factor of 0.8 compared to the OpenSSL RSA implementation, which outperforms existing approaches such as PRIME. Furthermore, we make use of the windowing method to improve the RSA performance. The precomputed table is encrypted by the AES key to ensure security. The performance of VIRSA using the fixed windowing method slightly exceeds OpenSSL, achieving a performance factor of 1.02.

This work was partially supported by Shandong Province Key Research & Development Plan/Major Science & Technology Innovation Project (Grant No. 2020CXGC010115), and National Natural Science Foundation of China (Grant No. 61772518).

D. Gao et al. (Eds.): ICICS 2021, LNCS 12918, pp. 293–309, 2021.
https://doi.org/10.1007/978-3-030-86890-1_17

**Keywords:** Memory disclosure · RSA · Register · AVX-512 · Montgomery multiplication · Vector instruction · Redundant representation

# 1 Introduction

Memory disclosure attacks, like Cold-boot attacks [1] and DMA attacks, pose a threat to RSA [2] systems although RSA is considered computationally secure. Attackers obtain the sensitive data in RAM during the RSA private key operation to destroy the security of RSA systems in the practical application. Adversaries exploit the *remanence effect* of RAM to launch cold-boot attacks. Since the memory content stops fading away at low temperatures after loosing power and is readable once the memory chip is powered on again, attackers can freeze the memory chip and move the chip into another prepared machine. DMA attacks use direct memory access through high-speed peripheral ports like Firewire [3] and PCI [4] to access memory space.

An effective method to resist memory disclosure attacks is to implement cryptographic operations within CPUs [5–14] because attacks on CPU chips are costly and hard to accomplish. Register-based [5–11] schemes have been proposed to implement such cryptography system. For register-based RSA system, PRIME [8] and RegRSA [6] use 256-bit AVX registers to store RSA private key and finish private key operations.

Before AVX-512 registers appeared, the storage space in registers was limited. We need to balance the computation speed and storage consumption. Scalar operations save storage space but reduce performance, while vector method improves performance but takes up more space. Redundant representation [15] is a major method for RSA vector implementations. PRIME [8] makes use of 256-bit AVX vector registers to realize RSA system for the first time. However, due to the limitation of register space, PRIME abandons CRT and the windowing method, resulting in low performance. PRIME puts the plaintext of some intermediate results in RAM. RegRSA [6] adopts CRT and the windowing method. However, RegRSA uses scalar instructions to calculate RSA and uses AVX vector registers only for data storage instead of the SIMD feature. Besides, RegRSA symmetrically encrypts the precomputed table and CRT intermediate results, then stores them in RAM.

Both the number and bit-width of AVX-512 [16] registers are twice those of AVX registers. AVX-512 register provides more register space, and processes twice as much data as AVX register in parallel by executing vector instructions. We use redundant representation to solve the carry overflow problem during vector operations. Based on AVX-512 registers, we improve the security and performance of RSA system.

In this paper, we propose VIRSA, a secure and efficient 2048-bit RSA system based on AVX-512 registers. Also, we use AVX-512F instruction subset to optimize performance. Our contributions are:

1. On the AVX-512 capable processor, we complete CRT-enabled RSA system entirely in registers without relying on memory. In the whole process of calculation, all the sensitive data is held in registers and will not appear in memory in any form. VIRSA effectively resists memory disclosure attacks.
2. We use AVX-512F vector instructions to optimize Montgomery multiplication. Compared to the optimal scalar implementation with MULX instruction, our execution time is shortened to 87%.
3. The performance of VIRSA is 8.26 times higher than PRIME, and is 80% of OpenSSL. When using the windowing method, the performance is 10.9 times higher than PRIME, and is 102% of OpenSSL. That means VIRSA achieves acceptable performance while offering protection against memory disclosure.

The rest of this paper is organized as follows. Section 2 briefly introduces the background information about VIRSA. Section 3 shows how to design VIRSA. Section 4 describes the implementation highlights of our system. We evaluate the performance and security of VIRSA in Sect. 5. Section 6 shows the related work. Section 7 concludes this paper.

## 2    Preliminaries

In this section, we will present the basic knowledge about VIRSA. We first give RSA-related algorithms and then give a description of AVX-512. We also introduce how to use redundant representation to solve carry propagation problems in vector operations.

### 2.1    RSA-Related Algorithms

Since the modular multiplication is the major operation of RSA and directly determines the efficiency of RSA, we decide to use Montgomery multiplication to finish modular multiplication to avoid time-consuming division operations. We choose the CIOS (Coarsely Integrated Operand Scanning) in [17], shown in Algorithm 1, to perform Montgomery multiplication. By using Algorithm 1, we get the result $S = A \times B \times R^{-1}$, where $R = 2^{1024}$.

---

**Algorithm 1.** CIOS

---

**Input:** $0 \leq A, B < M < R$, M is modules, $B = \sum_{i=0}^{n-1} b_i 2^{iw}$, $R = 2^{nw}$, $2^w$ is radix, n
　　　　 is digits number, $gcd(M, R) = 1$, $\mu = -M^{-1} mod\ 2^w$

**Output:** $S = A \times B \times R^{-1} (mod\ M)$

**Operation:**

1:　S = 0
2:　for i = 0 to n-1
3:　　　$S = S + A \times b_i$
4:　　　$q = S[0] \times \mu\ mod\ 2^w$
5:　　　$S = S + M \times q$
6:　　　$S = S/2^w$
7:　end for
8:　if $S \geq M$ then
9:　　$S = S - M$
10:　end if
11:　return S

---

We use the Montgomery exponentiation [18], shown in Algorithm 2, to accomplish modular exponentiation operations. Montgomery exponentiation will call the lower-level Montgomery multiplication.

---

**Algorithm 2.** Montgomery Exponentiation

---

**Input:** $a = (a_{l-1}, \ldots, a_0)_2$, $d = (d_{l-1}, \ldots, d_0)_2$, $n = (n_{l-1}, \ldots, n_0)_2$, $r = 2^l$

**Output:** $a^d\ mod\ n$

**Operation:**

1:　$result = 1$
2:　$base = a \times r\ (mod\ n)$
3:　for $i = 0$ to $l - 1$ do
4:　　　if $d_i = 1$ then $result = Montgomery\ Multiplication(result, base)$;
5:　　　$base = Montgomery\ Multiplication(base, base)$
6:　end
7:　return $Montgomery\ Multiplication(result, 1)$

---

CRT method [18] is able to transform 2048-bit Montgomery exponentiation into two 1024-bit Montgomery exponentiations. We use CRT, shown in Algorithm 3, to improve the performance of RSA.

---

**Algorithm 3.** CRT Operations

---

**Input:** $p$, $C_p = C\ mod\ p$, $d_p = d\ mod\ (p-1)$, $q$, $C_q = C\ mod\ q$, $d_q = d\ mod\ (q-1)$,
　　　　 $q^{-1}\ mod\ p$

**Output:** Plaintext M

**Operation:**

1:　$M_p = C_p{}^{d_p}\ mod\ p$
2:　$M_q = C_q{}^{d_q}\ mod\ q$
3:　$h = q^{-1}(M_p - M_q)\ mod\ p$
4:　$M = M_q + hq$

---

## 2.2   AVX-512

Intel Advanced Vector Extensions (AVX) [19] provides a set of registers that support SIMD instructions. AVX registers are 256-bit width, and process eight 32-bit or four 64-bit data with a single instruction. AVX capable processors have sixteen 256-bit AVX registers, represented as YMM0 – YMM15. The lower 128-bit of AVX registers are SSE registers, represented as XMM0 – XMM15.

Compared with AVX registers, AVX-512 register expands the bit-width from 256-bit to 512-bit, also doubles the number of AVX registers. AVX-512 capable processors have thirty-two 512-bit registers, represented as ZMM0 – ZMM31. Each ZMM register processes eight 64-bit or sixteen 32-bit data simultaneously with a single instruction. The lower 256-bit of ZMM registers are YMM registers. The AVX-512 instruction set consists of several subsets. Different processors may support distinct subsets. For example, Skylake-SP processors only support AVX-512F, AVX-512CD, AVX-512VL, AVX-512DQ, AVX-512BW subsets. AVX-512F subset is supported by all AVX-512 capable CPUs. Thus, we use AVX-512F instructions to accelerate cryptographic operations, which is available for all AVX-512 capable CPUs and has broad applicability.

## 2.3   Redundant Representation

Redundant representation is used to handle complex carry propagation problems in vector instructions. It is easy to deal with carry propagation in scalar operations. In 64-bit operating system, scalar instruction ADD adds two 64-bit data stored in GPRs. If the sum of the two integers exceeds 64-bit, then CF is set to 1 to save the carry. For scalar multiplication, MUL instruction is used to multiply two 64-bit integers. The lower 64-bit of the result is stored in RAX register, and the higher 64-bit is stored in RDX in default. However, dealing with carry propagation in vector instructions is difficult. We take an AVX-512 instruction [20,21] VMULPD as an example. The detail of VMULPD is shown in Table 1. We find that only the lower 64-bit of the result is retained and the higher 64-bit is discarded in every multiplication.

**Table 1.** Details of VMULPD instruction.

| Instruction Details of VMULPD |
| --- |
| **Instruction:** VMULPD zmm1, zmm2, zmm3 |
| **Description:** Multiply packed double-precision (64-bit) floating-point elements in zmm1 and zmm2, and store the results in zmm3 |
| **Operation:** For j = 0 to 7<br>    i = j * 64<br>    zmm3[i+63 : i] = zmm1[i+63 : i] * zmm2[i+63 : i]<br>ENDFOR<br>zmm3[MAX : 512] = 0 |

Redundant representation [15] solves carry overflow in vector operations. The core idea is to put smaller data into large container, so that the computation result will not produce carry. For example, if the two integers are less than

$2^{32}$, the product is less than $2^{64}$. The overflow is avoided. Suppose an n-bit integer A, written in $2^{64}$ radix as an $l$-digits integer, where $l = \lceil n/64 \rceil$ and every 64-bit digit $0 \le a_i \le 2^{64}$. We choose a positive integer m satisfies m < 64 and we use radix $2^m$ to represent integer A with k-digits, $k = \lceil n/m \rceil > l$ and $A = \sum_{i=0}^{k-1} x_i \times 2^{m \times i}$, $x_i$ satisfies $0 \le x_i \le 2^m$. Then, A is able to be written in $2^m$ redundant representation. If we select $2^m$ redundant representation, we only fill the low m-bit in 64-bit container, and the high (64-m) bits are filled with 0. By choosing an appropriate m, we ensure that carry overflow will not occur in vector instructions. Finally, we use Algorithm 4 in [15] to transform the result into $2^m$ redundant representation, then transform the $2^m$ redundant representation to regular $2^{64}$ representation.

---

**Algorithm 4.** Redundant-to-$2^m$

**Input:** U in redundant representation with k digits
**Output:** U in radix $2^m$ representation
**Operation:**
1:    temp = 0
2:    for i = 0 to k-1
3:        temp = temp + $u_i$
4:        $v_i$ = temp mod $2^m$
5:        temp = temp / $2^m$
6:    End for
7:    Return V

---

## 3   System Design

In this section, we present the core idea of designing VIRSA.

### 3.1   Securing Critical Data and Operations

To protect the cryptographic system against memory disclosure attacks, we need to ensure both the keys and private key operations are secure.

**Protecting Keys.** We put an AES master key in debug registers. Attackers are not allowed to access debug registers during RSA operations to prevent AES key disclosure. RSA private key is encrypted by the AES key. The encrypted private key only appears in RAM in the input phase. The plain private key only appears in registers.

**Secure Private Key Operations.** In our 2048-bit RSA system, the whole operations, including modular multiplication, modular exponentiation and CRT, are register-based. When we use CRT acceleration only, any sensitive data during RSA operations is stored in registers. There is no memory interaction. When we use both CRT and the windowing method [18] to improve performance, we load the AES-encrypted precomputed table in memory in the form of ciphertext. In order to avoid context switch during RSA computations, we implement RSA module atomically in Linux kernel space, so that the data in registers will not be exchanged to RAM.

## 3.2 Improving Performance

In this paper, VIRSA is designed based on ZMM registers. The ZMM register is 512-bit width, and processes eight 64-bit elements in parallel by using CPU vector instructions, so as to improve the performance.

The carry overflow problems can be solved by using redundant representation when we use vector instructions. For CRT-enabled 2048-bit RSA, the large integer operations involve 1024-bit Montgomery multiplication and 1024-bit large integer multiplication. Montgomery multiplication [22] is the core of RSA computation and directly determines the performance of private key operations. For 1024-bit integer, we take $m = 28$ to ensure no overflow happens during Montgomery multiplication (Sect. 4.1 explains why we choose 28). When $m = 28$, $k = \lceil 1024/28 \rceil = 37$. The number of ZMM registers required is q, satisfying $q = \lceil 37/8 \rceil = 5$. We use ZMM registers to load 1024-bit integer in the form of $2^{28}$ redundant representation, so that we can apply AVX-512F subset to operate 8 digits by SIMD instructions to improve performance. Figure 1 shows the $2^{28}$ redundant representation form of a 1024-bit integer A.

The windowing method [18] during Montgomery exponentiation further improves the performance of RSA computation, because it reduces the times of Montgomery multiplication. However, the windowing method needs space in RAM to store the precomputed table. To ensure security, we use the AES key to encrypt the precomputed table. In this paper, we both analyze the performance of 2048-bit RSA with and without the windowing method.

A[1023 : 0]

| ZMM4: | $0^{64}$ | $0^{64}$ | $0^{64}$ | $0^{48}$\|A[1023:1008] | $0^{36}$\|A[1007:980] | $0^{36}$\|A[979:952] | $0^{36}$\|A[951:924] | $0^{36}$\|A[923:896] |
|---|---|---|---|---|---|---|---|---|
| ZMM3: | $0^{36}$\|A[895:868] | $0^{36}$\|A[867:840] | $0^{36}$\|A[839:812] | $0^{36}$\|A[811:784] | $0^{36}$\|A[783:756] | $0^{36}$\|A[755:728] | $0^{36}$\|A[727:700] | $0^{36}$\|A[699:672] |
| ZMM2: | $0^{36}$\|A[671:644] | $0^{36}$\|A[643:616] | $0^{36}$\|A[615:588] | $0^{36}$\|A[587:560] | $0^{36}$\|A[559:532] | $0^{36}$\|A[531:504] | $0^{36}$\|A[503:476] | $0^{36}$\|A[475:448] |
| ZMM1: | $0^{36}$\|A[447:420] | $0^{36}$\|A[419:392] | $0^{36}$\|A[391:364] | $0^{36}$\|A[363:336] | $0^{36}$\|A[335:308] | $0^{36}$\|A[307:280] | $0^{36}$\|A[279:252] | $0^{36}$\|A[251:224] |
| ZMM0: | $0^{36}$\|A[223:196] | $0^{36}$\|A[195:168] | $0^{36}$\|A[167:140] | $0^{36}$\|A[139:112] | $0^{36}$\|A[111:84] | $0^{36}$\|A[83:56] | $0^{36}$\|A[55:28] | $0^{36}$\|A[27:0] |

**Fig. 1.** $2^{28}$ redundant representation of a 1024-bit integer A.

### 3.3  System Architecture

To protect VIRSA from memory disclosure, we integrate RSA module into Linux kernel space and complete RSA computation in register space. The user processes pass RSA parameters encrypted by the AES key to kernel RSA module through system call. We read the AES-encrypted RSA parameters into ZMM registers, then use the AES key stored in debug registers to decrypt RSA parameters. We get the plaintext of RSA parameters and perform RSA computation in register space. We divide RSA module into three levels, the highest level is 2048-bit CRT, the middle level is 1024-bit Montgomery exponentiation and the lowest level is 1024-bit Montgomery multiplication. High-level operations call low-level operations. The system architecture is shown in Fig. 2.

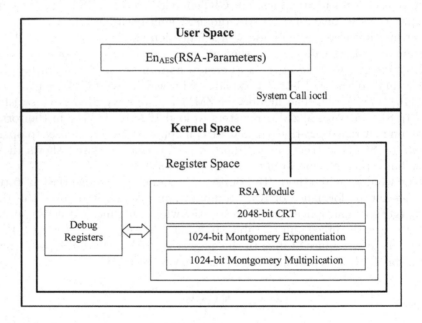

**Fig. 2.** System architecture of VIRSA.

## 4  Implementation Highlights

In this section, we describe implementation highlights of VIRSA based on AVX-512 registers. We use assembly language to control registers. We also integrate RSA module into Linux kernel space to access debug registers and ensure atomicity of RSA module.

## 4.1   Montgomery Multiplication

We use CIOS, shown in Algorithm 1, to perform 1024-bit Montgomery multiplication. We transform the 1024-bit inputs A, B, M into $2^{28}$ redundant representation to avoid carry propagation in CIOS operations and apply vector instructions in AVX-512F to speed up computation. In step 3 and step 5 of Algorithm 1, we use VPBROADCASTQ to fill 64-bit data into all elements of ZMM registers, then use vector multiplication instruction VPMULUDQ and vector addition instruction VPADDQ. In step 6, the right-shift vector instruction VALIGNQ is selected. The mod operation in step 4 is realized by scalar instruction AND. Table 2 describes the main AVX-512F instructions we used.

We use AVX-512F instructions and get the vectorized implementation of CIOS algorithm refer to [15]. The vectorized CIOS is shown in Algorithm 5. We finish Montgomery multiplication (Algorithm 5) completely in registers and use assembly language to control registers.

**Table 2.** Explanation of AVX-512F instructions used in CIOS algorithm.

| Instruction | Description |
|---|---|
| **vpbroadcastq** zmm, r64 | Broadcast 64-bit integer to all elements of ZMM. |
| **vpmuludq** zmm1, zmm2, zmm3 | Multiply the low unsigned 32-bit integers from each packed 64-bit element in ZMM1 and ZMM2, and store the unsigned 64-bit results in ZMM3 |
| **vpaddq** zmm1, zmm2, zmm3 | Add packed 64-bit integers in ZMM1 and ZMM2, and store the results in ZMM3 |
| **valignq** zmm1, zmm2, zmm3, imm8 | Concatenate ZMM1 and ZMM2 into a 128-byte immediate result, shift the result right by imm8 64-bit elements, and store the low 64 bytes (8 elements) in ZMM3 |

---

**Algorithm 5.** Vectorized implementation of CIOS with AVX-512F instructions

**Input:** A, B, M, in radix $2^{64}$. $\mu$.

**Output:** $S = A \times B \times R^{-1} (mod\ M)$

**Conventions:**

We choose m=28, k=37, q=5.

$A_i$, $M_i$, $X_i$, $T$, represent 512-bit width ZMM registers.

B is consist of $b_0$, $b_1$, ..., $b_{k-1}$.

$X_i[j]$ is the $j^{th}$ 64-bit element of ZMM register $X_i$.

**Operation:**

1:   Transform A, B, M from radix $2^{64}$ to radix $2^m$

2:   $x_0 = 0$, $X_q, \ldots, X_0 = 0$

3:   $a_0 = A\ mod\ 2^m$ (digit 0 of A)

4:   $m_0 = M\ mod\ 2^m$ (digit 0 of M)

5:   Put digits 1, 2, ..., (k-1) of A into ZMM registers $A_1, \ldots, A_q$

6:   Put digits 1, 2, ..., (k-1) of M into ZMM registers $M_1, \ldots, M_q$

7:   for i = 0 to k-1

8:       $x_0 = x_0 + a_0 \times b_i$

9:       $T = Broadcast\ b_i$ // using VPBROADCASTQ instruction

10:      for j = 1 to q

11:          $X_j = X_j + A_j \times T$ // using VPMULUDQ and VPADDQ

12:      $y_0 = x_0 \times \mu\ mod\ 2^m$

13:      $x_0 = x_0 + m_0 \times y_0$

14:      $T = Broadcast\ y_0$ // using VPBROADCASTQ instruction

15:      for j = 1 to q

16:          $X_j = X_j + M_j \times T$ // using VPMULUDQ and VPADDQ

17:      $x_0 = x_0 \gg m$

18:      $x_0 = x_0 + X_1[0]$

19:      $X_q, \ldots, X_1 = X_q, \ldots, X_1 \gg 64$ // using VALIGNQ instruction

20:  Using Algorithm 4 to convert $X_q, \ldots, X_1, x_0$ from redundant representation to $2^m$ radix representation. We get result S after Algorithm 4.

21:  Transform S from $2^m$ representation to regular $2^{64}$ representation.

22:  If $S \geq M$, $S = S - M$

23:  End if

24:  Return S

---

We choose m = 28 instead of other integers. Assuming m is larger than 28, for example, m is 29. Then, carry overflow will occur once during CIOS operation. If m is less than 28, m is 27. Then, 38 digits are needed to restore a 1024-bit integer, leading to extra computational overhead. In conclusion, the best value of m in this paper is 28.

## 4.2   Montgomery Exponentiation

We use Algorithm 2 to perform Montgomery exponentiation entirely in CPU registers to resist memory disclosure.

Since the windowing method [18] reduces the times of Montgomery multipli-cation, we can also use the windowing method to speed up Montgomery expo-nentiation operations. For 2048-bit RSA, 6-bit windows size is the best option. However, 6-bit fixed windowing method needs 8KB to store the precomputed table, which exceeds the storage space of registers. So we allocate 8KB memory to store precomputed table. The table is not allowed to be stored in plaintext. We use the AES key stored in debug registers to encrypt the precomputed table and load it into RAM in ciphertext. When data in table is needed, the encrypted data is loaded into ZMM registers and then is decrypted. We use *kmalloc* function to allocate 8KB memory on kernel heap and use function *kfree* to free memory after RSA computation. If we use kernel stack space to store the 8KB table, a stack-overflow error will happen because the Linux kernel stack space is not enough.

## 4.3    CRT

We complete CRT operations (shown in Algorithm 3) in registers and keep all the sensitive data generated during CRT computation, $M_p$, $M_q$, $h$ and $h \times q$, in ZMM registers. The last step of CRT involves large integer multiplication. Multiplication of two 1024-bit integers yields 2048-bit result. Although 1024-bit integer multiplication will not cause a significant impact on performance (since it is only executed once), we also use redundant representation to perform vector operations in Algorithm 6.

## 4.4    RSA Module

RSA module starts from receiving RSA parameters. Parameters, including CRT parameters and Montgomery parameters, are encrypted by the AES key before RSA computation. The encrypted parameters only appear in memory in the input phase and then are decrypted in registers. We finish RSA module in register space without memory interaction. RSA module is consist of CRT, Montgomery exponentiation and Montgomery multiplication. The high level calls the low level operation by sending parameters. The low level returns computation results to high level. Algorithm 7 shows the operations in RSA module.

**Algorithm 6.** Vectorized implementation of 1024-bit integer multiplication

**Input:** A, B, in radix $2^{64}$.

**Output:** $S = A \times B$

**Conventions:**

$A_i$, $R_i$, $T$, represent 512-bit width ZMM registers.

B is consist of $b_0, b_1, \ldots, b_{k-1}$.

$R_i[j]$ is the $j^{th}$ 64-bit element of ZMM register $R_i$.

$r_i : r$ is a pointer to an array of 64-bit values, $i$ is the index.

**Operation:**

1:     Transform A, B from radix $2^{64}$ to radix $2^m$
2:     $R_q, \ldots, R_0 = 0$
3:     Put digits 0, 1, . . . , (k-1) of A into ZMM registers $A_1, \ldots, A_q$
4:     for i = 0 to k-1
5:         $T = Broadcast\ b_i$ // using VPBROADCASTQ instruction
6:         for j = 1 to q
7:             $R_j = R_j + A_j \times T$ // using VPMULUDQ and VPADDQ
8:         store $R_1[0]$ at $r_i$
9:         $R_q, \ldots, R_1 = R_q, \ldots, R_1 \gg 64$ // using VALIGNQ instruction
10:    Store $R_q, \ldots, R_1$ starting at $r_k$
11:    Using Algorithm 4 to convert $r_i$ from redundant representation to $2^m$-radix.
       We get result S after Algorithm 4.
12:    Transform S from $2^m$ representation to regular $2^{64}$ representation.
13:    Return S

**Algorithm 7.** 2048-bit RSA module operation

**Input:** AES-encrypted RSA parameters (CRT and Montgomery parameters).
$En(p, C_p, d_p, q, C_q, d_q, q^{-1} \ mod\ p, R^2 \ mod\ p, R^2 \ mod\ q, -p^{-1} \ mod\ 2^r, -q^{-1} \ mod\ 2^r)$

**Output:** Plaintext M

**Operation:**

1:     Passing encrypted parameters $p, C_p, d_p, R^2 \ mod\ p, -p^{-1} \ mod\ 2^r$ from RAM to
       registers, then decrypt these parameters in register space.
2:     In register space, calculate $M_p = C_p^{d_p} \ mod\ p$, store the result in ZMM registers.
3:     Passing encrypted parameters $q, C_q, d_q, R^2 \ mod\ q, -q^{-1} \ mod\ 2^r$ from RAM to
       registers, then decrypt these parameters in register space.
4:     In register space, calculate $M_q = C_q^{d_q} \ mod\ q$, store the result in ZMM registers.
5:     Calculate $M = M_q + [(M_p - M_q) \times (q^{-1} \ mod\ p) \ mod\ p] \times q$ in registers.
6:     Return M

## 4.5   Building Execution Environment

We integrate RSA module into Linux kernel and compile it. We regard RSA module as a char module and provide interfaces for user processes. Processes in userspace access RSA module by *ioctl* system call. The passed RSA parameters are encrypted by the AES key. The AES key is stored in debug registers dr0 – dr3, that can only be accessed by ring 0 privilege. In order to protect VIRSA from memory disclosure attacks, we need to satisfy the following two security prerequisites.

**Atomicity.** Before executing RSA module, we call *preempt_disable* to suspend kernel preemption and call *local_irq_save* to forbid interrupts. So data in registers will not appear in RAM by context switch. We use *preempt_enable* to restore kernel preemption and use *local_irq_restore* to enable interrupts after finishing RSA module.

**AES Key Protection.** We use the existing method TRESOR [7] to produce and protect the AES key. Before any user process startup, we input the password to derive the AES key. The generated AES key is placed in debug registers. The way for userspace processes to access debug registers is through *ptrace* system call. We modify these system functions to prevent attackers from accessing debug registers.

## 5    Evaluation

In this section, we evaluate the performance and security of VIRSA. The experimental platform is Intel Xeon Silver 4208 CPU (2.2 GHz frequency), 16GB memory. The operating system is Ubuntu 18.04 64-bit.

### 5.1    Performance

We evaluate the performance improvement from two aspects, including Montgomery multiplication and 2048-bit RSA private key operation.

**Montgomery Multiplication.** We apply vector instructions to accomplish Montgomery multiplication based on AVX-512 registers, and compare its performance with optimal scalar implementation. The scalar implementation uses MULX instruction to improve performance. We run vector implementation and scalar implementation respectively and compare the performance difference from two aspects of execution time and CPU cycles. The execution time and cycles of each round are the average results of 10 million trials. We find that the performance of vector implementation is better than scalar implementation in every round, with 13% performance improvements. The scalar implementation achieves 87% performance of our vector implementation. Table 3 shows the time ($\mu$s) and CPU cycles when executing Montgomery multiplication.

**Table 3.** Performance of scalar and vector Montgomery multiplication implementation.

| | First round | | Second round | | Third Round | | Average | |
|---|---|---|---|---|---|---|---|---|
| | Times | Cycles | Times | Cycles | Times | Cycles | Times | Cycles |
| Scalar [6] implementation | 0.396 | 869 | 0.396 | 874 | 0.393 | 861 | 0.396 | 868 |
| Our vector implementation | 0.345 | 757 | 0.346 | 759 | 0.345 | 761 | 0.345 | 759 |
| Vector/Scalar | 87% | 87% | 87% | 87% | 88% | 88% | 87% | 87% |

**RSA Private Key Operations.** We complete VIRSA to resist memory disclosure attacks and accelerate RSA computation by vector instructions. We compare our vector implementation with OpenSSL and PRIME. We run our ASM code in kernel space and run OpenSSL (version 1.1.1) in user space. By comparing the execution time and CPU cycles, we find that when we abandon the windowing method, our performance is 80% of OpenSSL. When we adopt the windowing method, our performance is 102% of OpenSSL. That means VIRSA achieves acceptable performance while offering protection against memory disclosure. Table 4 shows our experimental data based on RSA computation. In Table 4 and Table 5, we use Register-CRT to represent we finish CRT-enabled RSA system completely within registers. We use Win-CRT to represent we finish CRT-enabled RSA system with the windowing method.

We also compare our running results with PRIME in Table 5. PRIME [8] states that the performance of PRIME is 8.6% of OpenSSL. In Table 5, we find that PRIME achieves 10.8% of Register-CRT and 8.4% of Win-CRT. That means the performance of Register-CRT is improved by 8.26 times and the performance of Win-CRT is improved by 10.9 times compared to PRIME.

**Table 4.** Performance comparison between Register-CRT, Win-CRT and OpenSSL.

|  | Register-CRT | Win-CRT | OpenSSL | OpenSSL/Register-CRT | OpenSSL/Win-CRT |
|---|---|---|---|---|---|
| Execution time | 4.6 ms | 3.6 ms | 3.7 ms | 80% | 103% |
| Cycles | 10117276 | 7965110 | 8110468 | 80% | 102% |

**Table 5.** Performance comparison between Register-CRT, Win-CRT and PRIME.

|  | PRIME/OpenSSL | PRIME/Register-CRT | PRIME/Win-CRT |
|---|---|---|---|
| Performance | 8.6% | 10.8% | 8.4% |

### 5.2 Security

**Resistance to Memory Disclosure Attacks.** Memory disclosure attacks, like Cold-boot attacks and DMA attacks, allow adversaries to partially or entirely acquire memory contents. To effectively resist these attacks, we implement VIRSA, a register-based CRT-enabled RSA system. When we abandon the windowing method, the RSA module is executed entirely in register space. All private data, including RSA parameters and intermediate results are stored in registers during running RSA module. No private information appears in RAM. When we use the windowing method, the 8KB precomputed table is encrypted by the AES key and placed in memory as the form of ciphertext. The AES key is not allowed to be accessed. To ensure the atomicity of RSA module, we disable kernel preemption and interrupts to avoid context switch. Thus, the attacker will not get any sensitive information from memory.

**Resistance to Cache Timing Side-Channel Attacks.** When using the fixed windowing method, our system is resistant to cache timing side-channel attacks [23–25]. We complete CIOS (in Algorithm 1) with always performing the final subtraction. Besides, the Montgomery multiplication in step 4 (in Algorithm 2) is always executed. Thus, no side-channel timing based on execution flow happens because there is no branch in instruction paths. Every time when we load data from precomputed table, we read the table as a whole to stop attackers deducing the exponents. Thus, no side-channel timing based on data access happens. When we abandon the windowing method, all the data during RSA computation is stored in registers and will not appear in cache, which obviously resists cache-based timing attacks.

### 5.3 Discussions

Compared with symmetric cryptography, the implementation of register-based asymmetric cryptography requires more register space. Fortunately, the amount of available register space is gradually increasing, from 128-bit XMM registers to 256-bit YMM registers and then 512-bit ZMM registers. In RegRSA based on 256-bit YMM registers, the total available register space has reached 704-byte. In VIRSA based on 512-bit ZMM registers, it is 2240-byte including sixteen 64-bit general purpose registers (GPRs), eight 64-bit MM registers and thirty-two 512-bit ZMM registers. Those registers are enough to complete RSA, ECC and corresponding acceleration algorithms. Thanks to the improvement of register space and vector instructions, the performance of RSA, from [11] based on XMM registers, to PRIME and RegRSA based on YMM registers, to VIRSA based on ZMM registers in this paper, is also improving. At the same time, using registers to implement cryptographic system can prevent physical attacks in board-level and memory disclosure caused by system vulnerabilities. However, current register-based RSA schemes are almost performed in Linux kernel, which is difficult to implement and deploy. How to run cryptographic systems within CPUs in user space is worth considering.

In this paper, we use $2^{28}$ redundant representation to finish 2048-bit RSA because $2^{28}$ is suitable for AVX-512F subset and AVX-512F is suitable for all AVX-512 capable CPUs. In fact, on some processors like Cannon Lake processors and Ice Lake processors, the best option is AVX-512IFMA subset since AVX-512IFMA is able to be used to realize RSA with $2^{52}$ redundant representation.

## 6  Related Work

Both register-based [5–11] and cache-based [12–14] cryptography schemes are implemented within CPUs and have memory disclosure resistance. For register-based symmetric cryptosystem, AESSE [9], Amnesia [10] and TRESOR [7] keep the AES keys in registers and complete AES symmetric cryptographic operations entirely in registers. As for register-based asymmetric cryptosystem, the work in [11] utilizes SSE XMM registers to store RSA key. PRIME [8] uses AVX

YMM registers to implement RSA computation without CRT and the windowing method. RegRSA [6] utilizes AVX YMM registers to store sensitive data during RSA operations and uses scalar instructions to implement 2048-bit private key operations instead of using vector instructions. The study in [5] designs and implements an ECC system in register space.

Cache-based schemes use cache to store private data and implement cryptography computations outside RAM, which is Cache-as-RAM. FrozenCache [12] uses CPU caches to store keys instead of RAM. Copker [13] stores RSA private key and intermediate states only in CPU caches and registers during RSA private key operations to avoid memory disclosure. Mimosa [14] uses hardware transactional memory (HTM) to protect sensitive information and uses caches to store data. However, Mimosa relies on special CPU hardware features.

## 7 Conclusion

In this paper, we present VIRSA, which implements 2048-bit CRT-enabled RSA entirely in CPU registers. Thus, all private data only exists in registers and will not appear in RAM to effectively resist memory disclosure attacks. We apply AVX-512 vector instructions to improve performance. Montgomery multiplication directly determines the performance of RSA computation. So we use AVX-512F instructions to optimize Montgomery multiplication. We also use the windowing method to reduce the number of Montgomery multiplication and put the AES-encrypted precomputed table in memory. We carry out experiments on Intel Xeon Silver 4208 CPU, and find that the performance is 80% of OpenSSL without the windowing method. When using the windowing method, the performance is 102% of OpenSSL.

In the future, we will make use of AVX-512IFMA instructions to realize RSA with $2^{52}$ redundant representation. We will also consider the impact of Hyper-Threading for our RSA computation. Similarly, vector operations can speed up the performance of ECC algorithm. We plan to use vector instructions and redundant representation to finish a high-performance ECC system in prime field with memory disclosure resistance.

## References

1. Halderman, J.A., et al.: Lest we remember: cold-boot attacks on encryption keys. Commun. ACM **52**(5), 91–98 (2009)
2. Rivest, R.L., Shamir, A., Adleman, L.: A method for obtaining digital signatures and public-key cryptosystems. Commun. ACM **21**(2), 120–126 (1978)
3. Böck, B., Austria, S.B.: Firewire-based physical security attacks on windows 7, EFS and BitLocker. Secure Business Austria Research Lab (2009)
4. Carrier, B.D., Grand, J.: A hardware-based memory acquisition procedure for digital investigations. Digit. Investig. **1**(1), 50–60 (2004)
5. Yang, Y., Guan, Z., Liu, Z., Chen, Z.: Protecting elliptic curve cryptography against memory disclosure attacks. In: Hui, L.C.K., Qing, S.H., Shi, E., Yiu, S.M. (eds.) ICICS 2014. LNCS, vol. 8958, pp. 49–60. Springer, Cham (2015). https://doi.org/10.1007/978-3-319-21966-0_4

6. Zhao, Y., Lin, J., Pan, W., Xue, C., Zheng, F., Ma, Z.: RegRSA: using registers as buffers to resist memory disclosure attacks. In: Hoepman, J.-H., Katzenbeisser, S. (eds.) SEC 2016. IAICT, vol. 471, pp. 293–307. Springer, Cham (2016). https://doi.org/10.1007/978-3-319-33630-5_20

7. Müller, T., Freiling, F.C., Dewald, A.: TRESOR runs encryption securely outside RAM. In: USENIX Security Symposium, vol. 17, p. 103 (2011)

8. Garmany, B., Müller, T.: PRIME: private RSA infrastructure for memory-less encryption. In: Proceedings of the 29th Annual Computer Security Applications Conference, pp. 149–158 (2013)

9. Müller, T., Dewald, A., Freiling, F.C.: AESSE: a cold-boot resistant implementation of AES. In: Proceedings of the Third European Workshop on System Security, pp. 42–47 (2010)

10. Simmons, P.: Security through amnesia: a software-based solution to the cold boot attack on disk encryption. In: Proceedings of the 27th Annual Computer Security Applications Conference, pp. 73–82 (2011)

11. Parker, T.P., Xu, S.: A method for safekeeping cryptographic keys from memory disclosure attacks. In: Chen, L., Yung, M. (eds.) INTRUST 2009. LNCS, vol. 6163, pp. 39–59. Springer, Heidelberg (2010). https://doi.org/10.1007/978-3-642-14597-1_3

12. Pabel, J.: Frozen cache (2009). Blog: http://frozenchache.blogspot.com

13. Guan, L., Lin, J., Luo, B., Jing, J.: Copker: computing with private keys without RAM. In: NDSS (2014)

14. Guan, L., Lin, J., Luo, B., Jing, J., Wang, J.: Protecting private keys against memory disclosure attacks using hardware transactional memory. In: 2015 IEEE Symposium on Security and Privacy, pp. 3–19. IEEE (2015)

15. Gueron, S., Krasnov, V.: Software implementation of modular exponentiation, using advanced vector instructions architectures. In: Özbudak, F., Rodríguez-Henríquez, F. (eds.) WAIFI 2012. LNCS, vol. 7369, pp. 119–135. Springer, Heidelberg (2012). https://doi.org/10.1007/978-3-642-31662-3_9

16. Intel advanced vector extensions 512 (avx-512). https://www.intel.com/content/www/us/en/architecture-and-technology/avx-512-overview.html

17. Koc, C.K., Acar, T., Kaliski, B.S.: Analyzing and comparing montgomery multiplication algorithms. IEEE Micro **16**(3), 26–33 (1996)

18. Koc, C.K.: High-speed RSA implementation. Technical report, TR-201, RSA Laboratories (1994)

19. Lomont, C.: Introduction to intel advanced vector extensions. Intel white paper 23 (2011)

20. Intel: Intel 64 and IA-32 architectures software developer's manual volume 2 (2A, 2B, 2C & 2D): Instruction set reference, A-Z (2020)

21. Intel intrinsics guide. https://software.intel.com/sites/landingpage/Intrinsics Guide/

22. Montgomery, P.L.: Modular multiplication without trial division. Math. Comput. **44**(170), 519–521 (1985)

23. Bernstein, D.J.: Cache-timing attacks on AES (2005)

24. Bonneau, J., Mironov, I.: Cache-collision timing attacks against AES. In: Goubin, L., Matsui, M. (eds.) CHES 2006. LNCS, vol. 4249, pp. 201–215. Springer, Heidelberg (2006). https://doi.org/10.1007/11894063_16

25. Brumley, D., Boneh, D.: Remote timing attacks are practical. Comput. Netw. **48**(5), 701–716 (2005)

# Informer: Protecting Intel SGX from Cross-Core Side Channel Threats

Fan Lang[1,2], Wei Wang[1,3(✉)], Lingjia Meng[1,2], Qiongxiao Wang[1,2], Jingqiang Lin[4,5], and Li Song[1,2]

[1] State Key Laboratory of Information Security, Institute of Information Engineering, Chinese Academy of Sciences, Beijing 100089, China
wangwei@iie.ac.cn
[2] School of Cyber Security, University of Chinese Academy of Sciences, Beijing 100089, China
[3] Data Assurance and Communication Security Research Center, CAS, Beijing 10089, China
[4] School of Cyber Security, University of Science and Technology of China, Hefei 230027, Anhui, China
[5] Bejing Institute, University of Science and Technology of China, Bejing, China

**Abstract.** As one of the major threats facing Intel SGX, side-channel attacks have been widely researched and disclosed as actual vulnerabilities in recent years, which can severely harm the integrity and confidentiality of programs protected by SGX. Most existing defense schemes are built based on the assumption that the adversary launches attacks from the same core as the victim, which however have been proved insufficient by newly-emerged cross-core side-channel attacks (e.g. CrossTalk). We present Informer, a defensive approach for SGX against side-channel attacks launched from any location, whether the adversary resides in the same physical CPU core as the victim or not. Informer achieves this goal by creating dummy threads that temporarily monopolize all CPU cores when security-critical codes are being executed, which breaks the essential concurrent execution condition of side-channel attacks. A key challenge is to ensure all those threads are scheduled exclusively to occupy all CPU cores even within an untrusted OS. Informer can defend against side-channel attacks from any core, and only incurs 22% performance overhead in OpenSSL. An additional mechanism is designed to reduce the impact on the operating system, as well as an optional extension to reduce the performance overhead brought to other programs.

**Keywords:** Software Guard Extension · Side-channel attack · Cross-core

This work was supported by the National Key R&D Program of China (Award No. 2020YFB1005800) and National Natural Science Foundation of China (Grant No. 61772518).

# 1    Introduction

With the increasing code size of operating systems and the emergence of more and more applications, system vulnerabilities and malicious threats from unknown applications have become more serious. Therefore, a solid and reliable trusted execution environment becomes particularly important. The Software Guard eXtension (SGX) [1–3] is currently a hardware-level security extension that has a wide range of applications and is highly researched. SGX is a set of CPU instructions, which is applied on the x86 processor to provide a trusted execution environment called enclave for user mode processes. SGX can provide integrity and confidentiality guarantee for programs running inside enclaves. Even the privileged applications, including a potentially malicious operating system or hypervisor cannot break it. However, SGX technology is still found to have security vulnerabilities. The most serious security threat to SGX is information leakage through the side-channel. In recent years, attacks on SGX scenarios using or indirectly using side-channel mainly include traditional cache time side-channel attacks, page fault side-channel attacks, and the most popular micro-architectural data sampling (MDS) attack, such as Meltdown [4] and Spectre [5]. More worrying is that in the security assumptions of SGX, the applications in the privileged mode may have malicious behaviors, which can reduce the noise in the side-channel attacks through many privileged behaviors to make the above attacks easier to launch.

Whether it is a traditional side-channel attack or an MDS attack, two necessary conditions are required: (1) The adversary and the victim must share some necessary resources. For example, in a cache side-channel attack, the adversary must share cache resources with the victim. (2) Both must be executed concurrently. The adversary needs to execute simultaneously with the victim in order to obtain the runtime state of the victim. Many existing researches [6–12] have focused on how to isolate the resources used by the victim, making it impossible for the adversary to analyze the shared resources. In addition, there are some researches [13–15] focus on the conditions of concurrent running, trying to create shadow threads and using effective detection mechanisms to prohibit concurrent execution of the program when the victim program is running critical operations. Currently, the attacks are mainly divided into three types: (1) The adversary achieves concurrent attacks from the same logical core by continuously interrupting the victim thread. (2) Attacks at the sibling core. (3) The adversary launches an attack from a completely different physical core, we call it cross-core attack. The existing schemes like Racing [14] and Varys [13] leverage extra thread to occupy the sibling core and leverage cache access time to defend against side-channel attack from sibling core. And they have 36.4% and 15% performance overhead respectively. But they can not defend against the cross-core attack. Recently, attacks from different physical cores such as Crosstalk [16] pose a great threat. Therefore, a defense scheme that can effectively deal with side-channel attacks from all cores is essential. Although E-SGX [15] can resist cross-core threats, it only supports single thread scenario and imposes a performance overhead of over 47.2%.

In this work, we propose Informer, an efficient scheme against various side-channel attacks. Informer monopolizes the whole processor during security-critical execution to break the concurrent execution condition of side-channel attack. We design an attack detection method to ensure whether the OS is truthfully scheduling SGX threads. And our solution occupies all idle cores, which can defend against side-channel attack from all cores. Different from E-SGX, we specifically set a control thread to be responsible for safety detection instead of the computing thread, which greatly improves the performance of the computing thread itself, and also allow multiple computing threads to execute concurrently on different cores.

Since thread scheduling depends on the OS, it is a significant challenge to ensure all the logical cores has been occupied with an untrusted or even malicious OS. In fact, Under the protection of Informer, if the adversary tries to launch a side-channel attack, he has to make an SGX thread Asynchronous Enclave Exit (AEX) and resume it afterwards, which will take a fixed number of CPU cycles. Therefore, we leverage it as a time threshold to create a control thread to challenge threads on other cores and detect whether the AEX occurs on itself periodically, which will prevent malicious scheduling of the operating system.

In addition, we also optimize the operating system blocking problem caused by our scheme and design an extension to reduce the exclusivity to other programs. We implement Informer based on the Intel Linux SGX SDK and apply it to ECDSA signature and RSA signature to protect the private key calculation part with about 39% and 22% performance overhead respectively.

- We propose Informer, a defensive approach against side-channel attacks from any core without trusting system software.
- We design a mechanism to reduce the impact on the OS and propose an additional optimized extension to reduce the performance overhead brought to other programs.
- We implement Informer on OpenSSL with 22% performance overhead.
- We propose an extension that can reduce Informer exclusivity.

## 2   Backgroud

### 2.1   Intel SGX

As an extension of the instruction set architecture, SGX provides a trusted execution environment for applications from the hardware level. SGX uses the Preserved Random Memory (PRM) as the memory page of the enclave and encrypts the PRM through Memory Encryption Engine (MEE) technology. Only when the data enters processors will it be decrypted, and it is always stored in memory in the form of ciphertext and measured externally to ensure Confidentiality and integrity of data. In addition, processor provides users with attestation services for enclave codes and data through the keys preset in processors.

When a program is running inside an enclave, the processor core on which it is located will enter enclave mode. In order to allow the core to switch between enclave mode and normal mode normally, SGX provides EENTER and EEXIT instructions to allow the core to enter and exit enclave mode. However, when a program running inside an enclave is interrupted, the core cannot exit normally. In this case, the processor will replace it with an asynchronous enclave exit (AEX). In order to protect the running state and data information in the enclave, The important information such as the contents of the register is securely saved inside the enclave (in the thread's GPRSGX region of State Save Area (SSA) frame) [17] and then replaced by a synthetic state before the core exits the enclave mode. An AEX process takes a fixed number of CPU cycles, and we call this fixed number of CPU cycles **an AEX duration.**

After the AEX, the ERESUME instruction can resume the program to re-enter the enclave mode and load the content of the SSA area to restore the running state. Regarding the SSA, it is a thread-local storage area of enclave memory. It saves the contents of various registers when the program is running inside an enclave, and important information when an exception or interrupt occurs. The organization of the GPRSGX domain includes RIP field and EXIT-INFO field. The RIP field stores the value of the instruction register when an AEX occurs. This field is always updated when the AEX occurs.

## 2.2  Hyper-Threading

Hyper-threading technology [18] is a synchronous multi-threading technology implemented by Intel, which allows a physical core to execute multiple (usually two) control flows concurrently. In this case, a physical core is actually divided into multiple logical cores. The architectural-state storage sections, APIC and some CPU internal buffers are duplicated among logical cores. In addition, other resources such as L1/L2 cache, TLB, bus interface, and branch prediction unit are all competitively shared. In fact, it is precisely because of these completely shared resources that CPU will have many attack threats from the sibling core. At the same time, the performance of programs will be greatly affected by occupying resources from the sibling core.

## 2.3  Side-Channel Attacks and Related Attacks

Launching a cache-time side-channel attack [19–24] needs to concurrently learn the access pattern of the victim's program control flow to the cache in a fine-grained manner, and then leverages the condition of shared cache resources to obtain cache side-channel information. At present, typical attacks include flush+reload, prime+probe, and so on. In addition to cache side-channel attacks, there are also researches on side-channel attacks using page table information [25,26]. These researches obtain side-channel information by modifying the page table entry to observe page conditions.

In addition to the traditional side-channel attack researches, the MDS attacks such as Meltdown have emerged in recent years. These attacks take advantage of the out-of-order execution or speculative execution to access some originally inaccessible data, and leak it to the cache before the instruction rolls back.

Then, they exploit the cache time side-channel attack to steal information. In subsequent researches, more in-depth MDS attacks such as Zomebieload [27], RIDL [28], and Cacheout [29] are proposed. They leak temporary information in various caches inside the CPU. Although Intel subsequently released patches for these vulnerabilities, but with little effect. In fact, the initial MDS attack did not work against SGX. The reason is that they did not consider the restrictions of SGX on page access. However, by bypassing the protection mechanism of SGX, MDS attacks such as Foreshadow [30] and Sgaxe [31] were derived as meltdown–type attacks on SGX scenarios. Although the internal buffers of the CPU used by them to disclose secrets are different, the principles they use are similar essentially. At present, most MDS attacks are carried out from the same core or sibling core where the victim is located, so most of the current defense schemes are also aimed at attacks from the same physical core.

### 2.4 Cross-Core Threats

At present, most defenses against side-channel attacks only focus on the protection of the L1 cache which is shared by sibling cores, because cross-core attacks are very difficult and have a huge workload. However, attacks from the non-sibling logical cores have also appeared. It is aimed at some internal buffers shared by multiple physical cores, such as LLC, staging buffers, etc. Liu *et al.* [32] and Disselkoen *et al.* [33] each designed an LLC based cache side-channel attacks to extract key information. The latest MDS attack, Crosstalk, leaks information directly from any cores, exploiting staging buffers. It has a great impact on the security mechanism of SGX. For these cross-core attacks, SGX is unable to resist using existing mechanisms. Therefore, a universal defense scheme that allows SGX programs to resist attacks from any core side channel is urgent.

## 3   Threat Model

We assume the standard SGX threat model. The adversary takes complete control of privileged software, in particular, the operating system. He can suspend or wake up any thread at any time, and at the same time modify the CPU affinity of the thread. At the same time, the adversary can read, write, delay, lock, and replay any memory area except the enclave. However, he cannot directly compromise the CPU, steal its internal fuse key, and cannot directly access the enclave by bypassing Page Miss Handler (PMH) and MEE. In addition, we assume that the operating system will provide sufficient execution time for the SGX program, regardless of denial of service provision. The adversary can exploit all resources except the enclave to launch attacks from any location, including

- The adversary attacks from the same logical core by frequently interrupting the victim program and alternately executing with the victim program.
- By running concurrently on the same physical core, The adversary continuously monitors and leaks the victim's operating status from the sibling core.
- By running malicious programs on any physical core, The adversary continuously monitors and reveals the victim's operating status from any core.

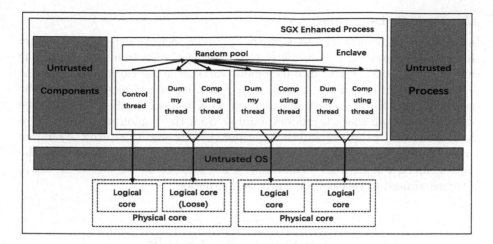

**Fig. 1.** The architecture of Informer. The control thread occupies a logical core alone, and the remaining cores are alternately occupied by computing and dummy threads.

## 4    Design

### 4.1    Architecture Overview

Informer mitigates side-channel attacks from any location by introducing a challenge-response style detection mechanism shortly monopolizing all cores during security-critical operations. Figure 1 illustrates an overview of Informer, which includes four main components: a control thread, n-1 dummy threads, n-1 computing threads (n is the number of CPU logic cores) and a random number pool. Among them, the dummy threads and the computing threads are paired one to one, and each pair occupies a logical core. The computing threads are responsible for performing security-critical operations. The dummy threads occupy the idle cores when the computing threads are performing security-critical operations. The control thread occupies a core for a long time, and is responsible for detection to ensure that all cores are monopolized by the dummy threads and the computing threads during the security-critical operations. Any malicious actions against informer threads, such as interruption or suspension, will be discovered by the control thread. And the random number pool provides random numbers for the control thread as the challenge content.

These components together serve a detection mechanism, in which the computing threads and the dummy threads respond to periodic challenges initiated by the control thread, acting as "informers". In this paper, the term "informer thread" is used to refer to either a computing thread or a dummy thread. Besides, even shortly occupying all the cores will also affect the running of the OS. For this reason, we also design a "response delay tolerance" mechanism, allowing the OS to work on a core that has lower detection sensitivity but still has security guarantees.

## 4.2    Periodic Detection

In Informer, the control thread performs periodic security detection to ensure that the adversary cannot attack the computing thread from any core. The control thread sets a cyclic auto-increment variable as the clock, and uses it to divide the detection period. As shown in Fig. 2, in each detection period, the control thread will first check whether an AEX has occurred. After that, the control thread will select a random number from the random number pool to challenge informer threads, and the informer threads need to return the random number within the period as a challenge response. When the detection fails, the control thread will generate an alarm. The measures to be taken after an alarm is triggered need to be based on the actual situation.

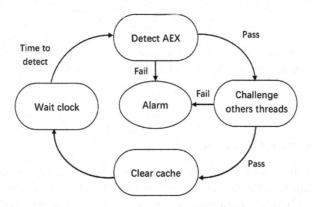

**Fig. 2.** Attack detection state transition diagram of the control thread.

The length of the detection period directly affects the security strength and reliability of Informer. When the detection period is too long, the adversary can suspend any thread on any core to launch an attack. After the attack succeeds, the suspended thread can be resumed in time to respond to the challenge, which fails the detection mechanism indeed. And when the length of the detection period is too short, it may cause false alarms. From this point of view, the setting of the period length is very important. Because Informer occupies all cores, the adversary has to suspend the enclave thread on one of them to initiate an attack, which will cause an AEX to occur and take an AEX duration. Therefore, setting the detection cycle to an AEX duration enables the control thread to effectively perform safety detection on other threads. However, through experiments we found that there is still room for optimization in the setting of the detection cycle. In fact, if the adversary wants to launch an attack, in addition to suspending an enclave thread before the attack, he also needs to resume the suspended enclave thread after the attack ends in Informer scheme. The process of resume is basically the reverse operation of AEX, so it will take a similar time to an AEX. Therefore, setting the detection cycle to about twice an AEX duration is enough for the control thread to detect malicious behavior in time.

### 4.3   AEX Monitoring

We leverage the method used in the Racing [14] solution to determine whether the control thread has the AEX behavior by monitoring the changes in the content of the SSA. Specifically, when the enclave thread occurs an AEX, it will affect the content of the SSA inspected, such as the RIP field. Therefore, we can detect whether the AEX has occurred by setting the RIP field to 0 and periodically observing whether the content of the RIP field is still 0. In Informer, except for the control thread itself, all other threads are inspected by the control thread. Therefore, the detection of the AEX is only for the control thread itself. We set the control thread to the highest priority in FIFO mode and do not allow it to be suspended. Therefore, once an AEX is detected, it is considered that the system has suffered a malicious attack.

### 4.4   Response Delay Tolerance

The model proposed in Informer will affect the normal operation of the OS. Therefore, we propose the "response delay tolerance" mechanism as part of Informer. The main idea is setting the sibling logical core of the core hosting the control thread as a loose core. The priority of the informer thread on the loose core is turned down, so that the OS can perform the necessary execution on the core. At the same time, Informer can tolerate cases in which the informer thread on the loose core cannot respond to the challenge timely, to a certain extent. Although this enables the OS to run relatively normally, this mechanism undoubtedly also brings security risks to Informer. The adversary can perform side-channel attacks on this loose core without being discovered. Currently, disclosed cross-core attacks such as Crosstalk perform MDS attacks through the staging buffer shared by all cores and leverage L1 cache to perform side-channel attacks to steal secrets. Therefore, we let the control thread periodically refresh the L1 cache. Since the L1 cache is shared on the same physical core, the adversary can no longer obtain side-channel information from the L1 cache of the loose core. To enable the "response delay tolerance" mechanism, Informer requires the application computer to support hyper-threading technology.

## 5   Implementation and Analysis

### 5.1   Implementation

Informer needs to make the control thread judge whether an AEX has occurred on its own by monitoring the content of the SSA. The SGX SDK does not provide such an interface, so we modify the SDK shown in Listing 1.1. In the initialization phase, the control thread will set the content of the RIP field to 0. When an AEX occurs, this field will be changed to store the contents of the RIP register, which will be detected by the control thread.

In terms of specific implementation, we apply Informer to the RSA signature and ECDSA signature in SGX-OpenSSL library to protect the private key calculation part. In the OpenSSL, the private key calculation part of the RSA signature and the ECDSA signature are performed in two functions, BN_mod_exp_mont and EC_POINTS_mul, respectively. The BN_mod_exp_mont function is mainly responsible for Montgomery modular exponentiation using the private key during the RSA signature process. The EC_POINTS_mul function is mainly responsible for the use of the private key for multiplying points during the ECDSA signature process. We modify the source code of OpenSSL to set the flags at the start and end of the two functions to represent the start and end of the protected operation, and add code fragments to respond to the challenge in all low_level_function calls of the two functions. In addition, at the start and end flags, the computing thread will also record the clock of the current control thread and compare them to ensure that the control thread is working normally. Once the two functions in OpenSSL are executed after recompilation, the protection mechanism of the solution starts to work and periodically challenges the thread that executes the calculation. And since all low_level_functions are added with challenge response code, the challenge can be responded to in time.

**Listing 1.1.** set SSA RIP

```
void SGXAPI set_ssa_ip (){
    thread_data_t* td = get_thread_data();
    ssa_gpr_t* ssa_gpr = reinterpret_cast<ssa_gpr_t
        *>(td->first_ssa_gpr);
    ssa_gpr -> REG(ip) = 0;
}
```

In addition, in order to allow the OS to run normally, we lower the priority of the informer thread of the same physical core as the control thread so that the system threads can preempt the core for normal scheduling. In order to ensure that the adversary cannot launch side-channel attacks through this core, the control thread periodically refreshes the L1 data cache and instruction cache shared with the physical core. Since both the data cache and the instruction cache have a capacity of 32 KB, and they are all virtual address indexes, we only need to prepare a 32 KB continuous data and instruction space and continuously access it to refresh the L1 cache.

### 5.2   Analysis

A key point of Informer is the definition of the detection period. The length of the detection period should be set to no more than an AEX duration. In this way, before the adversary completes an attack and resumes normal operations, he can be discovered. In E-SGX [15], the AEX duration was tested. It is between 8000 and 9000 CPU cycles. Therefore, it is safe to set the detection cycle to 8000 CPU cycles, but frequent detection will cause too many false positives. In addition to suspending the SGX thread, the attacker launching an attack also needs to resume the suspended thread. The resume operation is almost the reverse operation of the suspend operation, which will take almost the same time. We conducted 200 rounds of tests and found that the average time it takes to suspend and resume an SGX thread is 19817 CPU cycles. It is slightly larger than twice the AEX duration, which is in line with our speculation. Therefore, it is safe and reliable to set the detection cycle to around 19817 CPU cycles.

The attacks can be launched from any positions. First, if the adversary suspends the control thread to attack, it will cause the control thread to generate an AEX. This behavior will be detected by the AEX monitoring. In addition, in order to prevent the attack from suspending the control thread for a long time and causing the alarm mechanism to fail, the computing thread will compare the clock of the control thread at the beginning and end of its operation. If the clocks are consistent, it indicates that the control thread is not working properly during the security-critical operations and thus triggers an alarm. Second, if the adversary suspends the computing thread itself or another informer thread, it takes longer to suspend and resume the informer thread than a detection cycle of the control thread. The suspended informer thread will not be able to respond to the challenge of the control thread in time, and an alarm will be triggered.

**Table 1.** Attack detection. We conduct three rounds of attacks from three different locations (core 3, core 4 and core 5) and record the detection rate. The core 4 is the loose core.

|  | Core 1 | Core 2 | Core 3 | Core 4 | Core 5 | Core 6 | Core 7 |
|---|---|---|---|---|---|---|---|
| Round 1 | C | C | C/A | D(L) | D | D | D |
| Detection rate | 0 | 0 | 99.97% | 0 | 0 | 0 | 0 |
| Round 2 | C | C | C | D/A(L) | D | D | D |
| Detection rate | 0.06% | 0.03% | 0.04% | 99.95% | 0 | 0 | 0 |
| Round 3 | C | C | C | D(L) | D/A | D | D |
| Detection rate | 0.12% | 0.03% | 0.03% | 0 | 99.96% | 0 | 0 |

Computing: A computing thread.
Dummy: A dummy thread.
Attack: An attack thread.
Loose: Loose core.

# 6    Evaluation

Our experiments were conducted on an ThinkCentre with an SGX-enabled i7-6700 processor and 16 GB DRAM. The processor has 4 cores per CPU and 2 threads per core (with hyper-threading enable), whose maximum frequency is 3.4 GHz. The size of EPC was 128 MB. The operating system was Ubuntu 16.04 with Linux kernel version 4.15.0.132. We use GCC 5.4.0 to compile the source code including the Linux SGX SDK and OpenSSL.

## 6.1   Security Evaluation

Regardless of page table side-channel attacks, cache side-channel attacks, or MDS attacks that leverage cache side-channel attacks, if the adversary wants to launch an attack against Informer, it will cause the thread of the protected enclave to hang. We conduct security evaluation of scenarios where the informer threads are suspended in various locations relative to computing threads. In addition, we also conduct a special analysis of the informer thread on the loose core responding to challenges, as well as the cache access of the loose core.

**Table 2.** We conduct two rounds of testing, respectively, setting different cores (core 4 and core 5) as the loose core and recording the detection rate. The core 4 and the core of the control thread belong to the same physical core. The cycles of "delay response tolerance" is limited to 3.

|  | Core 1 | Core 2 | Core 3 | Core 4 | Core 5 | Core 6 | Core 7 |
|---|---|---|---|---|---|---|---|
| Round 1 | C | C | C | D(L) | D | D | D |
| Detection rate | 0.05% | 0.04% | 0.09% | 0 | 0 | 0 | 0 |
| Round 2 | C | C | C | D | D(L) | D | D |
| Detection rate | 0.05% | 0.03% | 0.09% | 0 | 0.01% | 0 | 0 |

Computing: A computing thread.
Dummy: A dummy thread.
Loose: Loose core.

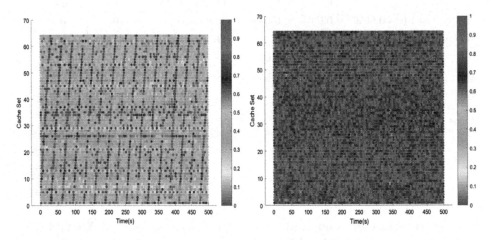

**Fig. 3.** L1 cache access time of loose core. The left is unrefreshed. The right is refreshed and no visible information is exposed.

***Validating the Effectiveness of Detection Mechanism.*** We leverage Informer to protect the private key calculation part of the ECDSA signature process in OpenSSL. We let the three cores perform the ECDSA signature operation 1000 times at the same time and launch the cache side-channel attack from three locations: (1) the same logical core as a certain computing thread. (2) the sibling logical core of a certain computing thread on the same physical core. (3) a different physical core. The detection cycle of the control thread is set to 19000 CPU cycles. As shown in the Table 1, the unaffected dummy thread can always respond to the challenge of the control thread in a timely manner. The unaffected calculation threads are constantly doing signature operations. The operations include numerous other calculations besides the private key calculation, which leads to frequent switching with the dummy thread, resulting in a few false alarms. In the worst case, the Informer will have a false alarm rate as low as 0.1%. The attacked core frequently suspends the informer thread due to the attacking program, so that the informer thread on the core cannot respond to the challenge of the control thread in time, resulting in numerous alarms. The detection rate is as high as 99.9%.

Since we set the sibling logical core of the core hosting the control thread as a loose core, the informer thread running on it is allowed to be interrupted shortly by the OS. Because it cannot respond to the challenge of the control thread timely, we have implemented a "response delay tolerance" strategy for this thread, which allows it to respond to the challenge within a certain delay. Specifically, we found through testing that when the delay is smaller than 3 detection periods, it will not trigger an alarm. In addition, since the informer thread and the control thread are in the same physical core, they share resources such as L1 cache and TLB, and can respond to challenges faster than the informer threads on other physical cores. As shown in the Table 2, within the limit of 3 detection period, When the informer thread and the control thread are not running on the same physical core, there is a phenomenon that it cannot respond to challenges in time and alarm. If the two threads are in the different physical core for a long time, the probability of generating an alarm will increase. If the adversary only temporarily schedules during the attack, frequent thread scheduling will also generate numerous alarms.

*Residual Cache Leakage of the Loose Core.* In addition, consider the case when the adversary initiates a page table side-channel attack similar to controlled-channel attacks during the suspension of the informer thread on the loose core. Such an attack will cause numerous AEXs in the computing thread, the thread will not be able to respond to the challenge in time and thus trigger an alarm. When the adversary leverages the cache to perform a side-channel attack, as shown in the Fig. 3, because the control thread refreshes the L1 cache periodically, the original L1 cache access information will no longer be observed. The adversary is unable to obtain the L1 cache side-channel information. We also evaluate Informer's resistance to last-level cache(LLC) attacks. Because the LLC is Physical Index Physical Tag(PIPT), the side-channel attacks against LLC cannot directly select consecutive addresses as the eviction set like L1 cache. Instead, the appropriate pages position should be selected first as the eviction set according to the access time. Our experimental device has 8192 LLC sets, so we must construct an LLC eviction set containing 8192 cache sets firstly. We conduct 20 experiments to build the LLC expulsion sets on each of three scenarios: the idle state, the original OpenSSL runtime, and the OpenSSL protected by Informer runtime. Compared with the idle state, when the original OpenSSL was running, there was a 10% failure rate in constructing the LLC eviction sets. When the OpenSSL protected by Informer was running, the LLC eviction sets cannot be constructed successfully.

## 6.2   Performance Evaluation

***Performance in OpenSSL.*** In order to verify the practicability of our scheme, we mainly design two sets of experiments. Respectively, Informer is applied to the ECDSA signature and RSA signature in the OpenSSL. We leverage the security mechanism of Informer to protect their private key calculation part and record its performance overhead. At the same time, because Informer supports multi-threading scenarios, we record the performance overhead under different concurrency during these two sets of experiments.

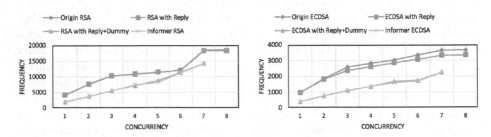

**Fig. 4.** The performance of the original OpenSSL and the OpenSSL with Informer. The horizontal axis is the number of cores used by the computing thread, and the vertical axis is the number of runs in 100 s. On the left is the performance of the RSA signature algorithm. On the right is the performance of the ECDSA signature algorithm. Both have the highest performance when the amount of concurrency is highest. At this time, the performance overhead is 22% and 39% respectively.

Since our experimental environment is a 4-core 8-thread CPU, and does not involve network transmission and reception processing, the target program performs best when 8 logical cores are in use. We measure the performance overhead of the original RSA signature and ECDSA signature programs from a single thread to 8 threads. In Informer, the control thread needs to occupy a single core for a long time, so the maximum number of cores that can actually be used for calculation is 7, so we measure the performance of the protected RSA signature and ECDSA signature when the amount of concurrency is from 1 to 7. As shown in Fig. 4, whether it is an RSA signature or an ECDSA signature, the original performance and the performance under protection are improved with the increase in the number of concurrency. Compared with the original program, under the protection mechanism of Informer has a certain performance overhead, about 22% and 39%. The performance overhead is mainly caused by the dummy threads, which cause frequent thread switching.

***Performance Overhead of the OS.*** In addition, since Informer will monopolize all CPUs when running, this will affect the operating system operation and scheduling, even if we set a loose core. We use perf to evaluate the impact of Informer on the operating system. As shown in Fig. 5, compared with the original OpenSSL, Informer increases the CPU occupancy rate of the protected program by 8.87%, and at the same time imposes a burden of 1.58% on the system scheduling process. But the overall impact on the operation of the system is not significant, only the CPU usage of the swapper process responsible for occupying idle time is greatly reduced, which is irrelevant to the normal operation of the operating system.

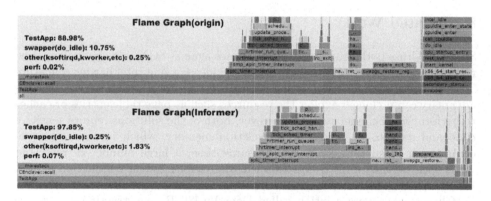

**Fig. 5.** Flame Graph. The TestApp is the target program, and most of its runtime is spent on switching in and out of the enclave. The swapper process is a process that runs when the operating system is idle. It mainly performs do_idle operations and has no practical meaning.

## 7   Extension

Informer occupies all the cores when the computing thread runs security-critical operations, which will affect the operation of other programs. So we design an additional extension to allow other programs to execute concurrently to improve core utilization. As shown in Fig. 6, other SGX programs can replace the dummy threads to respond to the challenge of the control thread and execute concurrently with the protected computing thread. As in different enclaves, the communication security in response to the challenge cannot be guaranteed. So we let the protected program share the session key (sk) and the random number pool (random[n], n is the size of the random number pool) with other SGX programs through SGX local attestation or sharing author identity during the initialization phase. During each challenge, the control thread will generate a random number R and share it with other running SGX programs. The other programs will randomly pick a number (random[x]) from the pool and leverage Eq. 1 to generate a reply to the control thread. The control thread uses the reverse Eq. 1 to obtain random numbers for comparison to determine whether the challenge is correctly responded.

$$reply = (x||random[x]) \oplus sk \oplus R \tag{1}$$

We have implemented an SGX applet as a "co-current program", which continuously run SHA-256 in its own enclave and constantly respond to the challenge of the control thread. Compared to the original program, our solution generate about 10% performance overhead, which is mainly due to fact that the applet needs to continuously accept the challenge of the control thread and respond to the challenge through calculation.

## 8   Related Work

Shih and Gruss both propose defense schemes against SGX side-channel attacks, namely T-SGX [6] and Cloak [7]. Their basic idea is using Intel hardware feature TSX to cover the side-channel information of the SGX programs so that the adversary can no longer exploit the resources such as page table entries or caches to perform side-channel attacks. However, both of the solutions require sensitive code and data to be placed in transaction memory, which means additional compilation to support TSX. At the same time, a large number of read and write operations within a transaction will also cause frequent transaction aborts, which seriously affects efficiency.

Chen et al. gave a solution called Déjà Vu [8]. By constructing an enclave with the ability to query the execution time of the application, each control node in the operation flow chart queries the running time of the application. If a side-channel attack occurs, the cryptographic computing thread in the enclave generates an AEX, which will cause the observation time at the next control node to be significantly longer. Déjà Vu leverages this to determine whether an

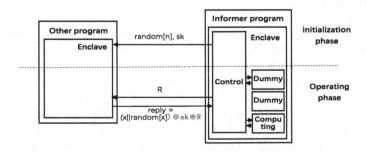

**Fig. 6.** Interaction process of challenge response between other programs and the protected program

attack has occurred. However, this solution cannot defend against attacks from non-hyperthreaded cores.

In addition, there are some solutions similar to the idea of our scheme, which also prevent side-channel attacks by preventing the adversary from executing concurrently. Among them, varys [13] and Racing [14] eliminate the side-channel attacks from sibling cores by occupying the sibling core with a thread and using the cache access time to ensure that the OS cannot schedule it to other cores. However, it is not effective for cross-core attacks. Although E-SGX [15] can defend against cross-physical core attacks, it only supports single-threaded programs and has a large performance overhead, which is not practical.

## 9    Discussion

*Cache Refreshing.* We only refresh the L1 cache on the sibling core where the control thread is located, but the side-channel information existing in the L2 cache and LLC has not been completely cleared. This is mainly because the number of operations that the control thread can do in a detection cycle is limited. Refreshing the L2 cache and LLC requires a larger memory collection and longer refresh time, which cannot be done in a detection cycle. And refreshing the L1 data cache and instruction cache also has a greater impact on the lower-level cache. To a certain extent, it also has a defensive effect on attacks from various levels of cache.

*Core Utilization.* Racing [14] first proposes the idea of occupying the core to defend side-channel attacks. But it only considers attacks from hyper-threading and requires a shadow thread to occupy the sibling core. Varys [13] mention that the threads responsible for the calculation can be paired to occupy a physical core, which makes more efficient use of the cores. It also proposes an extension scheme against LLC cache side-channel attacks, but it needs hardware support. E-SGX [15] provides more comprehensive protection against attacks from various cores. However, the overall detection needs to be completed by the thread responsible for the computing, so it only supports single-threaded programs, and the remaining cores cannot be used normally. What's more, it affects the normal operation of the OS Seriously. In our scheme, only the control thread will occupy a core for a long time, and the remaining cores can be used normally. In Sect. 6.1, we evaluated the security of the loose core when a dummy thread is running on it. Nevertheless, when computing thread is running on the loose core, since it generates a small amount of false alarms, the control thread can no longer determine whether the loose core is attacked based on whether there is an alarm. Therefore, in this case, an alarm threshold for the loose core needs to be measured as a criterion for judging whether the loose core have been attacked.

## 10   Conclusion

We present Informer, a scheme that can make the SGX program resist cross-core side-channel attacks. Informer defends against most side-channel attacks by monopolizing all CPU cores temporarily. We apply it to the OpenSSL. Our evaluation shows that Informer is secure against side-channel attacks with a detection rate of as high as 99.9%, and only incurs 22% and 39% performance overhead in RSA signature and ECDSA signature respectively. At the same time, through an additional mechanism and extension, Informer has a small impact on the operating system and other programs.

## References

1. Hoekstra, M., Lal, R., Pappachan, P., Phegade, V., Del Cuvillo, J.: Using innovative instructions to create trustworthy software solutions, p. 11 (2013)
2. McKeen, F., et al.: Innovative instructions and software model for isolated execution. In: HASP@ ISCA 10 (2013)
3. Costan, V., Devadas, S.: Intel SGX explained. IACR Crypt. ePrint Arch. **2016**, 86 (2016)
4. Lipp, M., et al.: Meltdown. ArXiv e-prints (2018)
5. Kocher, P., et al.: Spectre attacks: exploiting speculative execution. ArXiv e-prints (2018)
6. Shih, M.W., Lee, S., Kim, T., Peinado, M.: T-SGX: eradicating controlled-channel attacks against enclave programs. In: Proceedings of the 2017 Annual Network and Distributed System Security Symposium (NDSS), San Diego, CA (2017)
7. Gruss, D., Lettner, J., Schuster, F., Ohrimenko, O., Haller, I., Costa, M.: Strong and efficient cache side-channel protection using hardware transactional memory. In: USENIX Security Symposium (2017)
8. Chen, S., Zhang, X., Reiter, M.K., Zhang, Y.: Detecting privileged side-channel attacks in shielded execution with Déjà Vu. In: Proceedings of the 2017 ACM on Asia Conference on Computer and Communications Security, pp. 7–18. ACM (2017)
9. Kuvaiskii, D., et al.: SGXBOUNDS: memory safety for shielded execution. In: Proceedings of the Twelfth European Conference on Computer Systems, pp. 205–221. ACM (2017)
10. Crane, S., Homescu, A., Brunthaler, S., Larsen, P., Franz, M.: Thwarting cache side-channel attacks through dynamic software diversity. In: NDSS, pp. 8–11 (2015)
11. Seo, J., et al.: Sgx-shield: enabling address space layout randomization for sgx programs. In: Proceedings of the 2017 Annual Network and Distributed System Security Symposium (NDSS), San Diego, CA (2017)
12. Tromer, E., Osvik, D.A., Shamir, A.: Efficient cache attacks on aes, and countermeasures. J. Cryptol. **23**(1), 37–71 (2010)
13. Oleksenko, O., Trach, B., Krahn, R., Silberstein, M., Fetzer, C.: Varys: protecting SGX enclaves from practical side-channel attacks. In: 2018 USENIX Annual Technical Conference, USENIX ATC 2018, Boston, MA, USA, 11–13 July 2018, pp. 227–240 (2018). https://www.usenix.org/conference/atc18/presentation/oleksenko
14. Chen, G., et al.: Racing in hyperspace: closing hyper-threading side channels on SGX with contrived data races. In: 2018 IEEE Symposium on Security and Privacy, SP 2018, Proceedings, San Francisco, California, USA, 21–23 May 2018, pp. 178–194. IEEE Computer Society (2018). https://doi.org/10.1109/SP.2018.00024

15. Lang, F., et al.: E-SGX: effective cache side-channel protection for intel SGX on untrusted OS. In: Wu, Y., Yung, M. (eds.) Inscrypt 2020. LNCS, vol. 12612, pp. 221–243. Springer, Cham (2021). https://doi.org/10.1007/978-3-030-71852-7_15
16. CrossTalk. http://cve.mitre.org/cgi-bin/cvename.cgi?name=CVE-2020-0543
17. Intel: Intel Software Guard Extensions Programming Reference (2014). reference no. 329298–002US
18. Guide, P.: Intel® 64 and ia-32 architectures software developer's manual (2016)
19. Brasser, F., Müller, U., Dmitrienko, A., Kostiainen, K., Capkun, S., Sadeghi, A.R.: Software grand exposure: SGX cache attacks are practical, p. 33. arXiv preprint arXiv:1702.07521 (2017)
20. Schwarz, M., Weiser, S., Gruss, D., Maurice, C., Mangard, S.: Malware guard extension: Using SGX to conceal cache attacks. In: Polychronakis, M., Meier, M. (eds.) DIMVA 2017. LNCS, vol. 10327, pp. 3–24. Springer, Cham (2017). https://doi.org/10.1007/978-3-319-60876-1_1
21. Götzfried, J., Eckert, M., Schinzel, S., Müller, T.: Cache attacks on intel sgx (2017)
22. Yarom, Y., Falkner, K.: Flush+reload: a high resolution, low noise, l3 cache side-channel attack. In: Usenix Conference on Security Symposium (2014)
23. Gruss, D., Maurice, C., Wagner, K., Mangard, S.: Flush+flush: a fast and stealthy cache attack. In: Detection of Intrusions and Malware, and Vulnerability Assessment - 13th International Conference, DIMVA 2016, San Sebastián, Spain, 7–8 July 2016, Proceedings, pp. 279–299 (2016). https://doi.org/10.1007/978-3-319-40667-1_14
24. Osvik, D.A., Shamir, A., Tromer, E.: Cache attacks and countermeasures: the case of AES. In: Topics in Cryptology - CT-RSA 2006, The Cryptographers' Track at the RSA Conference 2006, San Jose, CA, USA, 13–17 February 2006, Proceedings, pp. 1–20 (2006). https://doi.org/10.1007/11605805_1
25. Xu, Y., Cui, W., Peinado, M.: Controlled-channel attacks: deterministic side channels for untrusted operating systems. In: 2015 IEEE Symposium on Security and Privacy (SP), pp. 640–656. IEEE (2015)
26. Wang, W., Chen, G., Pan, X., Zhang, Y., Wang, X., Bindschaedler, V.: Leaky cauldron on the dark land: Understanding memory side-channel hazards in sgx. In: Conference on Computer and Communications Security: Proceedings of the Conference on Computer and Communications Security. ACM Conference on Computer and Communications Security (2019)
27. Schwarz, M., et al.: Zombieload: cross-privilege-boundary data sampling. In: The 2019 ACM SIGSAC Conference (2019)
28. van Schaik, S., et al.: RIDL: rogue in-flight data load. In: 2019 IEEE Symposium on Security and Privacy, SP 2019, San Francisco, CA, USA, 19–23 May 2019, pp. 88–105. IEEE (2019). https://doi.org/10.1109/SP.2019.00087
29. van Schaik, S., Minkin, M., Kwong, A., Genkin, D., Yarom, Y.: Cacheout: leaking data on intel cpus via cache evictions. CoRR abs/2006.13353 (2020). https://arxiv.org/abs/2006.13353
30. Bulck, J.V., et al.: Foreshadow: extracting the keys to the intel SGX kingdom with transient out-of-order execution. In: 27th USENIX Security Symposium, USENIX Security 2018, Baltimore, MD, USA, 15–17 August 2018, pp. 991–1008 (2018). https://www.usenix.org/conference/usenixsecurity18/presentation/bulck
31. van Schaik, S., Kwong, A., Genkin, D., Yarom, Y.: Sgaxe: how sgx fails in practice (2020). http://cacheoutattack.com/files/SGAxe.pdf

32. Liu, F., Yarom, Y., Ge, Q., Heiser, G., Lee, R.B.: Last-level cache side-channel attacks are practical. In: 2015 IEEE Symposium on Security and Privacy (SP), pp. 605–622. IEEE Computer Society, Los Alamitos (2015). https://doi.org/10.1109/SP.2015.43
33. Disselkoen, C., Kohlbrenner, D., Porter, L., Tullsen, D.M.: Prime+abort: a timer-free high-precision L3 cache attack using intel TSX. In: 26th USENIX Security Symposium, USENIX Security 2017, Vancouver, BC, Canada, 16–18 August 2017, pp. 51–67 (2017). https://www.usenix.org/conference/usenixsecurity17/technical-sessions/presentation/disselkoen

# Internet Security

# Towards Open World Traffic Classification

Zhu Liu[1,2], Lijun Cai[1(✉)], Lixin Zhao[1], Aimin Yu[1], and Dan Meng[1]

[1] Institute of Information Engineering, Chinese Academy of Sciences, Beijing, China
{liuzhu,cailijun,zhaolixin,yuaimin,mengdan}@iie.ac.cn
[2] School of Cyber Security, University of Chinese Academy of Sciences,
Beijing, China

**Abstract.** Due to the dynamic evolution of network traffic, open world traffic classification has become a vital problem. Traditional traffic classification methods have achieved success to a certain extent but failed with unknown traffic detection due to the assumption of a closed world. Existing techniques on unknown traffic detection suffer from an unsatisfactory accuracy and robustness because they lack design according to the hierarchical structure of network flows. Meanwhile, the diverse flow patterns in the same attacks and the similar flow patterns from different attacks lead to the existence of hard examples, which degrades the classification performance. As a solution, we present a Siamese Hierarchical Encoder Network for traffic classification in an open world setting. We import a hierarchical encoder mechanism which mines the potential sequential and spatial characteristics of traffic deeply and adopt the siamese structure with a new designed complementary loss function which focuses on mining hard paired examples and quickens the convergence. Both of the key designs conjointly learn the intra-class compactness and inter-class separateness in the feature space to set aside more space for unknown traffic. Our comprehensive experiments on real-world datasets covering intrusion detection and malware detection indicate that SHE-Net achieves excellent performance and outperforms the state-of-the-art methods.

**Keywords:** Open world · Traffic classification · Hierarchical encoder

## 1 Introduction

With the advent of information technology and network intercommunication, more attention has been paid to network management and cyberspace security owing to the explosion and evolution of network traffic data volume [3,21]. In network management, to properly prioritize different applications across the limited bandwidth, most QoS mechanisms have a traffic classification module to first recognize which application the packets belong to. In cyberspace security, traffic classification, especially for unknown traffic detection towards open world, can support security analysts in their effort to identify and classify attack behaviors to ensure the safety of equipment and information. Thus, most researchers bend their minds to traffic classification by machine learning methods [2,5,15,27].

© Springer Nature Switzerland AG 2021
D. Gao et al. (Eds.): ICICS 2021, LNCS 12918, pp. 331–347, 2021.
https://doi.org/10.1007/978-3-030-86890-1_19

There are two main machine learning approaches to cope with this challenge: supervised methods and unsupervised methods. The supervised methods train data-driven classifiers on known traffic samples and achieve satisfactory results, but the results do not take into account the samples that non-existence in training [8]. Consequently, once a previously unseen sample is sent to the classifier, it may be misclassified as the predefined class and causes a high false alarm rate. With the dynamic and evolution of the traffic, this becomes a major bottleneck in a traffic classification system. In contrast, unsupervised methods, such as clustering, are typically used to deal with the unknown by gathering the unlabeled samples from the same class in the feature space [22,27]. Unfortunately, it may not work well in high-dimensional traffic data and result in poor precision, which places restrictions on its practical usefulness. Thereafter, most researches on traffic classification are shifting to deep learning.

The Deep learning method has made promising progress in many fields for its excellent capability of feature extraction. Recently, it also has made a breakthrough in traffic classification and unknown traffic detection. A recurrent neural network (RNN) is used to extract sequential features for traffic classification, but it is bounded by a closed world and cannot solve the unknown traffic detection problem [12,24]. RNN-based attention network is employed for traffic classification, whereas the network is time-consuming, especially when the input sequence is too long [10]. Besides, a convolutional neural network (CNN) applied to the studies [5,29] is to discover unknown traffic by learning the spatial feature. However, with the lack of consideration of the intra-class diversity and inter-class similarity, the feature extractor confused by hard examples has adverse effect on detection results. In addition, all of the above methods lack design according to hierarchical structure of traffic flows and merely extract a single feature (e.g., sequential features or spatial features), and thereby these facts highlight the necessity of building a robust classifier towards open world traffic classification.

Under the comprehensive consideration, we present a new model named SHE-Net for traffic classification in the open world, which learns deep-level sequential and spatial features from the raw network flows rather than manually designed features. Factoring the structure of the network flows, the major structure of the SHE-Net is a hierarchical encoder that mirrors the hierarchical structure of network flows from bytes to packets to flows. Firstly, the byte encoder is composed of semantic embedding and position embedding to learn the byte embedding. Subsequently, the packet encoder is the self-attention network, which handles the input sequences with different flow lengths and learns the sequential features among bytes within a packet. Finally, the flow encoder is the convolution network to acquire the spatial features of a whole network flow. Inspired by metric learning, these hierarchical encoders are integrated into the siamese network. Moreover, except for the contrastive loss function ($L_{con}$), we design a new complementary loss function ($L_{com}$) for the siamese network to accelerate convergence and mine the hard paired examples. Both $L_{con}$ and $L_{com}$ jointly supervise the model to learn compacted intra-class features and separated inter-class features to set aside more space for unknown instances. After that, we utilize the thresholds and the Support Vector Machine (SVM) classifier in open world traffic classification.

In summary, we make the following contributions:

- We propose a SHE-Net model towards open world traffic classification by incorporating the sequential and spatial network into the siamese network without any statistical feature design.
- We design a new complementary loss function ($L_{com}$) which can accelerate the model learning and hard paired data mining, in association with the contrastive loss function ($L_{con}$), to learn the discriminative features.
- We conduct extensive experiments on four traffic datasets, inclusive of intrusion datasets and malware datasets. The results show that our methods can achieve excellent performance on open world traffic classification.

## 2    Related Work

### 2.1    Closed World Traffic Classification

Many of the existing traffic classification methods are limited to the closed-world premise. That is, no new classes appear in testing.

In the early studies, some algorithms seek to solve the problem at packet level features [1,9,11,13,18,26]. The port-based method [18] identifies the application type by matching the port number in network packets. However, this method is failed in the situation that dynamic changes of the allocated port bring by new technology. The payload-based method [9,11] avoids the dynamic port problem by inspecting the specific application signature in the packet payload for matching, but it involves user privacy and needs to constantly update the protocol feature library. Yun et al. propose the Securitas and extract the statistical protocol message formats by clustering n-grams with the same semantics [26]. Others focus on flow level features [4,7]. Gil et al. use the C4.5 decision tree and time-related features like flow duration, flow bytes per second, and arrival intervals [7]. However, such approaches eventually fail to put into practice due to the error-prone and time-consuming statistical feature design.

Multiple research working on traffic classification is currently based on the deep learning [12,16,24,28]. Wang et al. present the HAST-IDS system by using CNN and RNN to learn the low-level spatial features and high-level sequential features [24]. Zhao et al. design a new objective function to solve the problem of intra-class diversity and inter-class similarity [28]. A Deep Packet framework, with a stacked autoencoder and CNN, is employed to integrate both feature extraction and packet classification phases into one system [16]. Although these methods are efficient and time-saving in deep feature extraction, they are all confined to the closed set and cannot be applied to unknown traffic detection.

### 2.2    Open World Unknown Traffic Detection

The real world is open and dynamic, and in many situations, the model cannot expect it sees everything in training, which makes open-world unknown traffic detection intractable.

There are some flow level approaches to cope with this issues [5,15,27,29]. Zhang et al. propose a parameter-optimized scheme named RTC by combining supervised and unsupervised machine learning techniques to get an additional confidence score and judge the probability that a test sample is seen or unseen [27]. Inspired by the RTC, studies in [5,15] use the deep learning methods to obtain embedding representations and set critical values between known class and unknown class. Zhao et al. design a prototype-based approach to perform robust traffic classification with novelty detection and achieve good results. However, to our best knowledge, all the existing relevant studies merely use CNN or RNN to extract a single type of feature at flow level and lack a design based on the network flow structure in which we think that distinguishable latent information may be behind.

## 3 Our Approach

### 3.1 Problem Definition

Given a training dataset $D_{tr} = \{(x_k, y_k)\}_{k=1}^{N}$, where $N$ is the total flow number, $x_k \in \mathbb{R}$ is an instance, and $y_k \in Y_{tr} = \{1, 2, ..., c\}$ is the label of $x_k$. In testing stage, there is an open set $D_o = \{(x_k, y_k)\}_{k=1}^{\infty}$, where $y_k \in Y_o = \{1, 2, ..., c, ..., C\}$ with $C > c$. In our approach, we aim at learning a model $M : x_o \to Y_o = \{1, 2, ..., c, unknown\}$, where the instance $x_o$ that flagged as the option $unknown$ is unseen in training phase.

### 3.2 Key Observation

A network flow consists of continuous packets with the same 5-$tuple^1$ in once intercommunication. Inspired by the hierarchical structure of a network flow, we assume that the bytes in a network packet from different positions have a potential correlation and plentiful information. Thus, we deem a byte as a new language symbol, a packet as a sequence, and a flow as a document.

### 3.3 Data Preprocessing

We parse the datasets by the tshark tool. Thereafter, we select the initial $m$ packets in each flow and the original $n$ bytes in each packet starting from the IP layer header instead of the data-link layer header because the latter only has information about the linked devices rather than TCP/UDP. And then, we pad a *zero token*, which is 0, when the flow length is less than $m$ (or the packet length is under $n$). In view of the fact that some new technology, including network address translation, may confuse classifier performance, we mask the IP address and the port number in each packet to get rid of this confusion with four *mask tokens* presenting by 2, 3, 4, and 5. Meanwhile, we insert a *cls token*, which is referring as numeric 1, in front of each packet for aggregating all bytes information in the current packet. After data preprocessing, a flow $x_k \in \mathbb{R}^{m \times n}$ can be regarded as an instance.

---

[1] The 5-*tuple* is $(ip\_src, ip\_dst, port\_src, port\_dst, protocol)$.

## 3.4   Model Design

Based on the key observation, our model SHE-Net is mainly composed of three encoder layers that aim at coherently learning the low-dimensional representation of the input flow $x_k$. The framework is shown in Fig. 1.

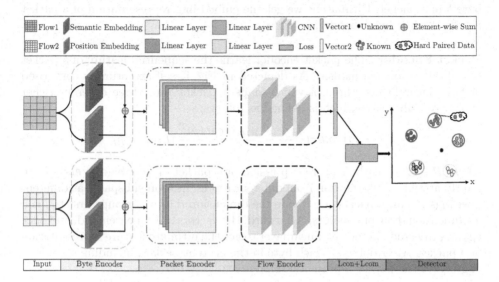

**Fig. 1.** An overview of SHE-Net

**Byte Encoder.** The byte encoder contains two sub-layers including semantic embedding and position embedding. The semantic embedding is to learn the semantic representation of each byte $b$ in a packet. Hence, we have:

$$o_{se} = W onehot(b) \tag{1}$$

where $W$ is the learned transformation matrix during training, onehot is a function that encodes discrete features as a one-hot numeric array, and $o_{se}$ denotes the $d$-dimensional semantic embedding vector of a byte.

The position embedding proposed by [23] is to learn the relative position of the input sequence. As the key observation mentioned in Sect. 3.2, we employ the position embedding to obtain the $d$-dimensional position vector $o_{pe}$ of a byte according to:

$$o_{pe} = \begin{cases} PE_{(pos,2i')} = sin(pos/10000^{2i'/d}) \\ PE_{(pos,2i'+1)} = cos(pos/10000^{2i'/d}) \end{cases} \tag{2}$$

where $2i', 2i'+1 \in [0, d)$ are the even and odd position number of bytes in a packet respectively. Here, we utilize sine and cosine functions of different frequencies with a constant 10000. Each dimension of positional embedding corresponds to a sinusoid, and the wavelengths form a geometric progression in $[2\pi, 10000 \cdot 2\pi]$. For any fixed offset $k$, $PE_{pos+k}$ can be represented as a linear function of $PE_{pos}$, which allows the model to learn the relative position representation effortlessly.

The position embedding has the same dimension $d$ as the semantic embedding so that the two can be element-wise summed for enriching the meaning of every byte $b$ in a packet. Ultimately, we get the embedding representation of a packet $p \in \mathbb{R}^{n \times d}$ after byte encoder.

**Packet Encoder.** The packet encoder learns the sequential feature of a packet by self-attention mechanism. As depicted in Fig. 1, self-attention is comprised of four linear layers, three of which maps a packet $p$ to three different richer expressive sub-space, which is defined as follows:

$$P = softmax(\frac{Linear(p)Linear(p)^T}{\sqrt{d}})Linear(p) \tag{3}$$

where $P \in \mathbb{R}^{d \times n}$ is the feature expression of a packet. To enable the encoder to jointly attend to information from different representation subspaces at different positions within a packet, we divide the dimension $d$ into $h$ groups first to yield $d'$-dimensional output vectors. Afterward, the vectors are concatenated, and once again, projected via the last linear layer to get the ultimate feature representation of a packet. After doing so, each byte in the current packet, including *cls token*, aggregates the packet representation.

**Flow Encoder.** The flow encoder is employed for obtaining the spatial feature of a flow by Eq. 4. More specifically, for each packet $p_t$ ($t \in [1, m]$) getting from the byte encoder, we obtain the packet representation $P_t$ in turn through packet encoder. Next, we take out the *cls* vector $cls_t$ corresponding to the *cls token* from each $P_t$ in a flow, which represents the latent contextual representation of a packet. Thereafter, we concatenate *cls* vectors together in order, so as to form a matrix of a flow. Finally, we employ CNN to encode the flow matrix into representation vectors $x$.

$$M = Concat(cls_1, cls_2, ..., cls_m) \quad (M \in \mathbb{R}^{m \times d})$$
$$x = CNN(M) \quad (x \in \mathbb{R}^2) \tag{4}$$

### 3.5 Loss Function Design

**Contrastive Loss.** To measure the distances of pairwise representation vectors in low dimensional space, we implement the siamese structure [6] with a contrastive loss function and consisting of two identical encoders sharing weight. Its loss function is defined as:

$$L_{con} = \frac{1}{2}YD^2 + \frac{1}{2}(1 - Y)\{max(m - D, 0)\}^2 \quad (m > 0) \tag{5}$$

where $Y$ is the indicator whether pairs $x^i$ and $x^j$ is similar ($Y = 1$) or not ($Y = 0$). $D$ is the Euclidean distance between $x^i$ and $x^j$. $m$ is the margin for dissimilar pairs which contributes to the $L_{con}$ only if their distance is within $m$.

**Complementary Loss.** However, the problems of intra-class diversity and inter-class similarity in attack flows may result in the existence of hard paired samples. Thus, $L_{con}$ suffers from slow convergence and constant fluctuation. To counteract this issue, we, inspired by focal loss, fix the sampling rule in training and design a novel complementary loss function by constructing a complementary feature of the paired data to focus on training the hard examples.

The complementary feature vector $\boldsymbol{v_{com}}$ based on Euclidean distance will be activated by softmax function to get the estimated probability distribution $p$:

$$p = softmax(\boldsymbol{v_{com}}) = softmax([D, max(m - D, 0)]/t) \tag{6}$$

where $t$ ($t \geq 0$) is a temperature normalization factor. After that, we use binary cross entropy loss for $p$:

$$- logPr(Y|p) = -Ylog(p) - (1 - Y)log(1 - p) \tag{7}$$

where $Y$ is the true probability distribution mentioned in Eq. (5). The intuition here is that we want the predicted probability distribution of similar flow vectors to be as close to 1 as possible since $-log(1) = 0$, that is the optimal loss. We want the distribution of the dissimilar examples to be close to 0 since any non-zero values will reduce the value of similar vectors. And then, to take into consideration easy and hard paired samples in training, we design $L_{com}$ in such a way:

$$L_{com} = -(1 - p)^{\gamma}Ylog(p) - p^{\gamma}(1 - Y)log(1 - p) \tag{8}$$

where $(1 - p)^{\gamma}$ or $p^{\gamma}$, with a tweaked parameter $\gamma \geq 0$, is a modulating factor. In the sense that when a pair is misclassified (hard paired sample) and $p$ is near 0, the factor is close to 1 and $L_{com}$ is impervious. In contrast, if $p$ goes to 1, the factor is on the verge of 0 and $L_{com}$ for easy-classified pairs is low-weight that in turn increases the importance of correcting misclassified examples.

We adopt the joint supervision of $L_{con}$ and $L_{com}$ to train our model for discriminative feature learning:

$$L = L_{con} + \lambda L_{com} \tag{9}$$

where $\lambda$ ($\lambda \geq 0$) is a balanced factor between $L_{con}$ and $L_{com}$. Here, the $L_{com}$ term can be considered as a further reinforcement of intra-class compactness and inter-class separateness to save more areas for unknown class in the open world.

### 3.6   Detector Design

After learning the low dimensional representation, we adopt a threshold based nearest prototype matching mechanism for open world traffic classification since the samples in each class approximately follow Gaussian distribution. In addition, to counter that the threshold may be biased in detection performance, we additionally join the SVM simultaneously for classification. We summarise the details in Algorithm 1. Here, twofold need to be noted: First, we get all embedding data $D^{em}$ through SHE-Net. Second, only if the embedding instance vector $\boldsymbol{x}$ falls into the threshold and the predicted label judging by the SVM is equal to its true label can we deem that it belongs to a known class.

**Algorithm 1.** Detector Based on the Prototype and the SVM

---

Input: Data set $D^{em} = \{D^{em}_{tr}, D^{em}_{va}, D^{em}_{te}\}$ and its label set $Y = \{Y_{tr}, Y_{va}, Y_{te}\}$;
    Generalization coefficient $\alpha$

Output: Predicted label set $Y^p$

1: Train a SVM classifier $S$ on $(D^{em}_{tr}, Y_{tr})$ and optimize it on $(D^{em}_{va}, Y_{va})$;
2: Initialize $T \leftarrow \varnothing$ and $Y^p \leftarrow \varnothing$;
3: **for** $i$ in set$(Y_{tr})$ **do**
4:   Select all the instance vectors which belong to class $i$ to form $X_i$ set;
5:   Compute prototype $p_i = \frac{1}{|X_i|} \sum_{x \in X_i} x$ as the centre point of class $i$;
6:   Compute intra-class distance $\mu_i = \frac{1}{|X_i|} \sum_{x \in X_i} \| x - p_i \|_2^2$;
7:   Compute standard deviation distance $\sigma_i = std(\{\| x - p_i \|_2^2, \forall x \in X_i\})$;
8:   Get threshold $t_i = \mu_i + \alpha \sigma_i$ and add $t_i$ to $T$;
9: **end for**
10: **for** $x$ in $D^{em}_{te}$ **do**       % $x$ is an instance in the open world setting
11:   $d_{min} \leftarrow \infty$;
12:   **for** $i$ in set$(Y_{te})$ **do**
13:    **if** $\| x - p_i \|_2^2 < d_{min}$ **then**
14:     $d_{min} \leftarrow \| x - p_i \|_2^2$; $\hat{y} \leftarrow i$; $\hat{t} \leftarrow t_i$;
15:    **end if**
16:   **end for**
17:   **if** $d_{min} < \hat{t} \wedge S(x) = positive$ **then**
18:    $y \leftarrow \hat{y}$;       % $x$ is a known instance
19:   **else**
20:    $y \leftarrow 'unknown'$;     % $x$ is an unknown instance
21:   **end if**
22:   Add $y$ into $Y^p$;
23: **end for**
24: **return** $Y^p$

---

## 4    Evaluation

To prove the effectiveness of our method, we mainly present the experiment process from the horizontal analysis, longitudinal analysis, and sensitivity analysis.

### 4.1    Dataset

We use four datasets from intrusion detection and malware detection. The details of these datasets and the data splitting principles are shown in Table 1.

For intrusion detection, we use public dataset IDS2017 [19] which covering benign and 7 common attack network flows and kept in separated pcap files. After data preprocessing, classes with more than 2000 flows are reserved for known classification and the remains are for unknown detection. To ensure the effectiveness and generalization of our approach, the latest dataset DAPT2020 [17] for Advanced Persistent Threats which captures real-world network behavior spanning all stages of an APT is also used for unknown intrusion traffic detection.

For malware detection, we apply USTC-TFC2016 dataset [25] together with dataset AndMal2017 [14] for the experiment. USTC-TFC2016 dataset contains

**Table 1.** Statistics of traffic datasets

| Intrusion detection traffic datasets | | | | Malware detection traffic datasets | | | |
|---|---|---|---|---|---|---|---|
| For known | | For unknown | | For known | | For unknown | |
| IDS2017 | Flows | DAPT2020 | Flows | TFC2016 | Flows | AndMal2017 | Flows |
| Benign | 100230 | AccountDiscovery | 560 | Cridex | 60050 | Avforandroid | 1000 |
| Ftp-Patator | 3974 | DirectoryBruteforce | 5161 | FTP | 49889 | Dowgin | 1000 |
| Ssh-Patator | 2979 | NetworkScan | 4341 | Weibo | 39806 | Kemoge | 1000 |
| Goldeneye | 7456 | VulnerabilityScan | 1124 | Geodo | 38919 | Shuanet | 1000 |
| Hulk | 100900 | MalwareDownload | 2 | Neris | 36037 | Youmi | 1000 |
| Slowhttptest | 4212 | CSRF | 5 | Warcraft | 7874 | Ransombo | 1000 |
| Slowloris | 3865 | Backdoor | 6 | SMB | 32632 | Simplocker | 1000 |
| DDos | 95116 | CommandInjection | 8 | MySQL | 13975 | Charger | 1000 |
| | | SqlInjection | 55 | | | Wannalocker | 1000 |
| | | AccountBruteforce | 66 | | | Avpass | 1000 |
| Total | 318732 | Total | 11328 | Total | 279182 | Total | 10000 |
| $D_{tr}:D_{va}:D_{te} = 6:2:2$ | | $D_{te}$ | | $D_{tr}:D_{va}:D_{te} = 6:2:2$ | | $D_{te}$ | |

10 classes of malware traffic and 10 classes of benign traffic, we select 8 categories consisting of benign and malware for known detection. In AndMal2017 dataset, we pick out 1000 flows uniformly at random from each one of 10 kinds of malware to verify our approach's ability of unknown malware traffic detection.

### 4.2 Baseline

We compare the performance of SHE-Net with the following relevant methods.

- SEEN [5] discovers unknown traffic by using the CNN network for features extraction and K-Means for unknown clustering on malware traffic dataset.
- DCEMR [29] not only uses CNN network but also imports a distance-based cross entropy (DCE) loss term and a metric regularization (MR) term to enable the robustness of unknown malware traffic detection.
- Flow-WGAN [10] blends with the RNN-based attention structure to get flow feature representations for intrusion detection.

### 4.3 Metric

We concentrate on the True Positive Rate (TPR) and False Positive Rate (FPR) for per class evaluation. TPR means the rate of correctly recognized as a given class, while FPR means the rate of incorrectly identified as another class. Furthermore, we leverage $TPR_{AVE}$, $FPR_{AVE}$ and $FTF$ to access the overall performance [20]. $TPR_{AVE}$ is the ratio between all correctly classified network flows and total network flows. $FPR_{AVE}$ is the ratio between all wrongly classified network flows and total network flows. The definitions are as follows:

$$TPR_{AVE} = \frac{1}{N'} \sum_{i=0}^{C} TPR_i * FlN_i \tag{10}$$

$$FPR_{AVE} = \frac{1}{N'} \sum_{i=0}^{C} FPR_i * FlN_i \qquad (11)$$

$$FTF = \sum_{i=0}^{C} w_i \frac{TPR_i}{1 + FPR_i} \qquad (12)$$

The weight $w_i$ of a category $i$ can be defined as:

$$w_i = \frac{FlN_i}{N'} \qquad (13)$$

where $N'$ is total traffic flow numbers orienting the open world, and $C$ is the number of categories including class *unknown*. $TPR_i$ and $FPR_i$ represent two measures of category $i$. $FlN_i$ is the traffic flow number of category $i$. Our rationale here is that higher $TPR_i$ and lower $FPR_i$ contribute higher $FTF$. Besides, $FTF$ takes the different weights of categories into consideration, which means the classification accuracy of one class can affect the effectiveness of the model to a greater extent if it is given more weight whereas less affected.

### 4.4 Horizontal Analysis

We compare our methods with baselines to prove the effectiveness of SHE-Net (detailed in Table 2 and Table 3) and make an analogy between the results on two types of datasets to bear out the generalization of our approach (see Fig. 2).

**Table 2.** Open world traffic classification results on intrusion detection dataset

| ID | Category | SEEN | | DCEMR | | Flow-WGAN | | SHE-Net | |
|----|----------|------|------|-------|------|-----------|------|---------|------|
| | | $TPR$ | $FPR$ | $TPR$ | $FPR$ | $TPR$ | $FPR$ | $TPR$ | $FPR$ |
| 1 | Benign | 0.9493 | 0.0159 | 0.9129 | 0.0009 | 0.9985 | 0.0209 | 0.9996 | 0.0000 |
| 2 | Ftp-Pat. | 0.9211 | 0.0204 | 0.8781 | 0.0129 | 0.9846 | 0.1128 | 1.0000 | 0.0000 |
| 3 | Ssh-Pat. | 0.9279 | 0.0350 | 0.8923 | 0.0020 | 0.8871 | 0.0004 | 0.9966 | 0.0000 |
| 4 | Goldene. | 0.9403 | 0.0211 | 0.9450 | 0.0015 | 0.9385 | 0.0034 | 0.9973 | 0.0000 |
| 5 | Hulk | 0.9576 | 0.0293 | 0.9362 | 0.0032 | 0.9333 | 0.0006 | 0.9987 | 0.0001 |
| 6 | Slowhtt. | 0.9302 | 0.0249 | 0.9390 | 0.0030 | 0.9155 | 0.0008 | 0.9953 | 0.0000 |
| 7 | Slowlor. | 0.9110 | 0.0225 | 0.9410 | 0.0024 | 0.9561 | 0.0042 | 1.0000 | 0.0000 |
| 8 | DDos | 0.8923 | 0.0024 | 0.9695 | 0.0282 | 0.9622 | 0.0100 | 0.9998 | 0.0000 |
| 9 | Unknown | 0.9641 | 0.0010 | 0.9803 | 0.0286 | – | – | **0.9999** | **0.0007** |
| $AVE$ | | 0.9326 | 0.0192 | 0.9327 | 0.0092 | 0.9470 | 0.0191 | **0.9986** | **0.0001** |

**Table 3.** Open world traffic classification results on malware detection dataset

| ID | Category | SEEN | | DCEMR | | Flow-WGAN | | SHE-Net | |
|---|---|---|---|---|---|---|---|---|---|
| | | *TPR* | *FPR* | *TPR* | *FPR* | *TPR* | *FPR* | *TPR* | *FPR* |
| 1 | Cridex | 0.9037 | 0.0233 | 0.9391 | 0.0023 | 0.9223 | 0.0989 | 1.0000 | 0.0000 |
| 2 | FTP | 0.9210 | 0.0301 | 0.9479 | 0.0092 | 0.9389 | 0.0191 | 1.0000 | 0.0000 |
| 3 | Weibo | 0.9008 | 0.0091 | 0.9334 | 0.0781 | 0.9118 | 0.0150 | 0.9994 | 0.0000 |
| 4 | Geodo | 0.9421 | 0.0105 | 0.9376 | 0.0671 | 0.9410 | 0.0332 | 0.9995 | 0.0000 |
| 5 | Neris | 0.9539 | 0.0292 | 0.9632 | 0.0923 | 0.9070 | 0.0292 | 0.9994 | 0.0000 |
| 6 | Warcraft | 0.9401 | 0.0710 | 0.9747 | 0.0632 | 0.8939 | 0.0105 | 0.9975 | 0.0000 |
| 7 | SMB | 0.9312 | 0.0078 | 0.9708 | 0.0293 | 0.9219 | 0.0282 | 0.9995 | 0.0000 |
| 8 | MySQL | 0.9223 | 0.0098 | 0.9871 | 0.0409 | 0.9373 | 0.0324 | 1.0000 | 0.0002 |
| 9 | Unknown | 0.9522 | 0.0081 | 0.9564 | 0.0301 | – | – | **0.9985** | **0.0004** |
| *AVE* | | 0.9297 | 0.2210 | 0.9567 | 0.0458 | 0.9218 | 0.0333 | **0.9993** | **0.0001** |

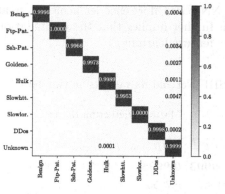

(a) Intrusion detection dataset

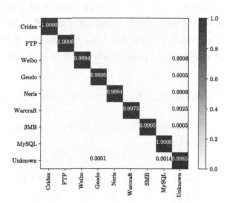

(b) Malware detection dataset

**Fig. 2.** Confusion matrices results on two types of datasets

From the results, we can observe that the *TPR* in SHE-Net is reasonably high. 99.99% of flows are correctly flagged as unseen for intrusion dataset and 99.85% for malware dataset. Moreover, low *FPR* of each known class means existing flows are rarely marked as unseen, reducing the load on many manual checking of alert messages. The other methods have a poor performance, especially for *FPR*. Here, we view that SEEN and DCEMR have tended to focus on extracting spatial features but neglect the sequential feature. Further, DCEMR is better than SEEN since it extra designs a new loss function by taking into account the intra-class and inter-class distance. Flow-WGAN has no result about class *unknown*. It considers the sequential feature but lacks the spatial feature. Our approach has superiority over other methods in consideration of both sequential and spatial features together, which makes SHE-Net robust.

Further investigation revealed that these baselines could not perform well simultaneously on both types of datasets while SHE-Net could, which indicates that our approach has great generalization and robustness.

### 4.5 Longitudinal Analysis

We analyze several properties of the proposed SHE-Net from the perspective of model and loss function.

**Model Property.** As we have mentioned before, the spatial feature can enhance the feature representation and discrimination by enriching the feature diversity since a single feature description cannot accurately represent the flow. To verify it, we implement different variants of the flow encoder. The relevant results are shown in Table 4. Both RNN and self-attention are used for modeling the time series relationship among all packets in a flow, while CNN is designed to automatically and adaptively learn spatial characters of all packets of a flow. No matter which kind of dataset, the CNN-based flow encoder achieves great performance among three variants which further implies that the combination of multiple types of features is critical for feature learning.

**Table 4.** Comparison results between the SHE-Net and its variants on two datasets

| Flow encoder variants | Intrusion detection dataset | | | Malware detection dataset | | |
|---|---|---|---|---|---|---|
| | $TPR_{AVE}$ | $FPR_{AVE}$ | $FTF$ | $TPR_{AVE}$ | $FPR_{AVE}$ | $FTF$ |
| CNN | **0.9994** | **0.0000** | **0.9993** | **0.9995** | **0.0000** | **0.9994** |
| RNN | 0.9829 | 0.0021 | 0.9825 | 0.7253 | 0.0343 | 0.6886 |
| Self Attention | 0.9950 | 0.0006 | 0.9934 | 0.9994 | 0.0001 | 0.9993 |

**Loss Function Property.** In addition to model property, we evaluate the capabilities of $L_{com}$ in hard paired data mining and convergence speed. Here, we embed vectors of all known test data, and the visualization results are shown in Fig. 3. As we can see in Fig. 3(a), for most categories except for DDoS, some of their embedding vectors are far away from their respective class prototype and are relatively discrete. Then, the detector may not find a proper threshold and result in a high false alarm rate. By comparison, the intra-class distances of each class are more compact, and the outliers of each class are significantly reduced in Fig. 3(b). Therefore, with the aid of $L_{com}$, more hard paired examples are reduced, and more space for unknown examples is saved.[2]

---

[2] We only show results about the intrusion dataset in the following experiment since the results can generalize to the other type of dataset.

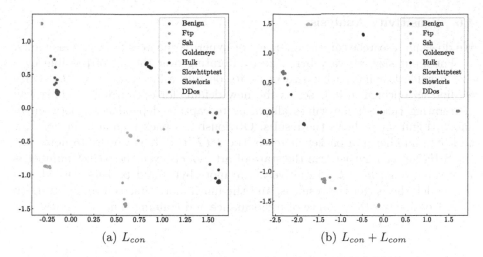

(a) $L_{con}$　　　　　　　　　　(b) $L_{con} + L_{com}$

**Fig. 3.** All test embedding vectors containing 63749 instances of the intrusion dataset

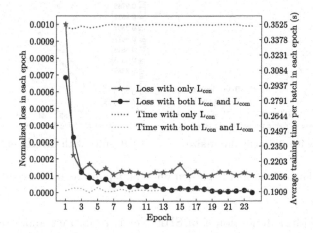

**Fig. 4.** The convergence velocity and time consuming

To further illustrate the advantages of $L_{com}$, we evaluate the convergence speed on the validation dataset and time overhead on the training dataset through the joint supervision of $L_{con}$ with $L_{com}$ or not. Figure 4 shows the result of this experiment. Under the joint supervision of $L_{con}$ and $L_{com}$, the loss curve converges to a lower loss in a shorter time with a smoother trend which reflects that the model inclines to be more robust, while, without $L_{com}$, the loss fluctuates severely and converges relatively hard.

## 4.6   Sensitivity Analysis

We do some researches on the influence of hyper-parameters from various views.

For input size, we vary byte size by iterating over the set of possible byte size $n$ while keeping packet size $m$ as its default value (e.g., $m = 10$). Once we find an optimal $n$, it is set as the new default for optimizing $m$. This way of iterating through the values allows us to capture dependencies between $m$ and $n$. Figure 5(a) shows the results. Obviously, as packet number in per flow and byte number of a packet increase, the $FTF$ rise. It is worthy to note that the SHE-Net can outperform the state-of-art models even the packet number is very small (e.g., $m = 2$). Intuitively, the more bytes and packets are fed into the model, the better the result is, and the much more time consuming to train take. Considering the balance of performance and training time, we choose 128 as $n$ and 10 as $m$.

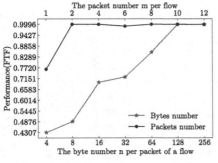 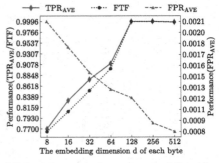

(a) Results with different input size     (b) Results with diverse embedding size

**Fig. 5.** The influence of hyper-parameters: input size and embedding size

The embedding dimension $d$ in SHE-Net is to extract and store the latent information of bytes, and each dimension is an aspect to represent the hidden information. Hence, the dimension of bytes directly affects the performance. To account for this fact, we demonstrate the result in Fig. 5(b). A small value results in a lousy performance due to the weak ability to capture the hidden information, while a large value (e.g., from 128 to 256) performs excellently. An immense value may lead to over-fitting since the model might learn worthless information from the noise data. Therefore, we select 128 as the embedding dimension $d$.

The $\lambda$ is used for balancing the $L_{con}$ and $L_{com}$. $\gamma$ is a modulating factor for concentrating the loss on hard paired examples. We inspect their sensitivity on the validation dataset and amplify the loss change by normalization. The result is shown in Fig. 6(a). As $\lambda$ increase with the same $\gamma$, the amplitude in different loss curve are becoming small (e.g., $\lambda = 4$). Besides, as $\lambda = 4$ and $\gamma$ risen, substantially more weight becomes emphasized the hard examples, and the convergence value decreases sharply (e.g., from 0.2 to 0.01). In fact, with $\lambda = 4, \gamma = 2$ (our default setting), the loss value reduces ideally.

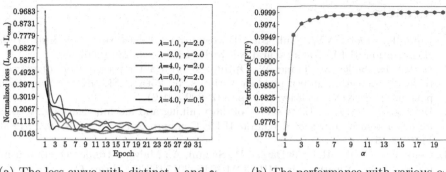

(a) The loss curve with distinct $\lambda$ and $\gamma$     (b) The performance with various $\alpha$

**Fig. 6.** The influence of hyper-parameters: $\lambda$, $\gamma$ and $\alpha$

We optimize the generalization coefficient $\alpha$ with respect to $FTF$ that our approach achieves when recognizing flows. If $\alpha$ is too large, many instances that are not present in the training set will be recognized as seen instances. Otherwise, if $\alpha$ is too small, many known samples will be misclassified as previously unseen samples. We conduct this idea on the validation dataset, and the result is shown in Fig. 6(b). When $\alpha$ is between 6 and 20, the different amplitude of $FTF$ is no more than 0.0001. Note that the higher this parameter, the more conservative we are in flagging flows as originating from novel class. Therefore, it is recommended to set $\alpha$ with values in $[6, 8]$.

## 5   Conclusion

In this paper, we propose SHE-Net, a novel method for open world traffic classification. SHE-Net progressively builds a flow vector by hierarchical encoders that mirroring the hierarchical structure of a flow. Meanwhile, the learned flow vector contains sequential and spatial information, which can enhance the diversity and robustness of flow representation. Moreover, the new loss function $L_{com}$ together with $L_{con}$ jointly forces the model to learn the discriminative feature, which can mine the hard paired examples, increase the intra-class compactness and expand the inter-class separateness. We conduct extensive analysis to evaluate our methods, and the results demonstrate that SHE-Net achieves excellent classification performance in an open world setting and outperforms the state-of-the-art methods.

**Acknowledgement.** This work is supported by the strategic Priority Research Program of Chinese Academy of Sciences, Grant No. XDC02040200.

# References

1. Auld, T., Moore, A.W., Gull, S.F.: Bayesian neural networks for internet traffic classification. IEEE Trans. Neural Networks **18**(1), 223–239 (2007)
2. Bartos, K., Sofka, M., Franc, V.: Optimized invariant representation of network traffic for detecting unseen malware variants. In: 25th {USENIX} Security Symposium ({USENIX} Security 16), pp. 807–822 (2016)
3. Buczak, A.L., Guven, E.: A survey of data mining and machine learning methods for cyber security intrusion detection. IEEE Commun. Surv. Tutorials **18**(2), 1153–1176 (2015)
4. Celik, Z.B., Walls, R.J., McDaniel, P., Swami, A.: Malware traffic detection using tamper resistant features. In: MILCOM 2015–2015 IEEE Military Communications Conference, pp. 330–335. IEEE (2015)
5. Chen, Y., Li, Z., Shi, J., Gou, G., Liu, C., Xiong, G.: Not afraid of the unseen: a siamese network based scheme for unknown traffic discovery. In: 2020 IEEE Symposium on Computers and Communications (ISCC), pp. 1–7. IEEE (2020)
6. Chopra, S., Hadsell, R., LeCun, Y.: Learning a similarity metric discriminatively, with application to face verification. In: 2005 IEEE Computer Society Conference on Computer Vision and Pattern Recognition (CVPR 2005), vol. 1, pp. 539–546. IEEE (2005)
7. Draper-Gil, G., Lashkari, A.H., Mamun, M.S.I., Ghorbani, A.A.: Characterization of encrypted and vpn traffic using time-related. In: Proceedings of the 2nd international conference on information systems security and privacy (ICISSP), pp. 407–414 (2016)
8. Este, A., Gringoli, F., Salgarelli, L.: Support vector machines for tcp traffic classification. Comput. Netw. **53**(14), 2476–2490 (2009)
9. Finsterbusch, M., Richter, C., Rocha, E., Muller, J.A., Hanssgen, K.: A survey of payload-based traffic classification approaches. IEEE Commun. Surv. Tutorials **16**(2), 1135–1156 (2013)
10. Han, L., Sheng, Y., Zeng, X.: A packet-length-adjustable attention model based on bytes embedding using flow-wgan for smart cybersecurity. IEEE Access **7**, 82913–82926 (2019)
11. Khalife, J., Hajjar, A., Diaz-Verdejo, J.: A multilevel taxonomy and requirements for an optimal traffic-classification model. Int. J. Network Manage **24**(2), 101–120 (2014)
12. Kim, J., Kim, J., Thu, H.L.T., Kim, H.: Long short term memory recurrent neural network classifier for intrusion detection. In: 2016 International Conference on Platform Technology and Service (PlatCon), pp. 1–5. IEEE (2016)
13. Kuncheva, L.I., Bezdek, J.C.: Nearest prototype classification: clustering, genetic algorithms, or random search? IEEE Trans. Syst. Man Cybern. Part C (Applications and Reviews) **28**(1), 160–164 (1998)
14. Lashkari, A.H., Kadir, A.F.A., Taheri, L., Ghorbani, A.A.: Toward developing a systematic approach to generate benchmark android malware datasets and classification. In: 2018 International Carnahan Conference on Security Technology (ICCST), pp. 1–7. IEEE (2018)
15. Liu, A., Wang, Y., Li, T.: Sfe-gacn: A novel unknown attack detection under insufficient data via intra categories generation in embedding space. Comput. Secur. **105**, 102262 (2021)
16. Lotfollahi, M., Siavoshani, M.J., Zade, R.S.H., Saberian, M.: Deep packet: a novel approach for encrypted traffic classification using deep learning. Soft Comput. **24**(3), 1999–2012 (2020)

17. Myneni, S., et al.: DAPT 2020 - constructing a benchmark dataset for advanced persistent threats. In: Wang, G., Ciptadi, A., Ahmadzadeh, A. (eds.) MLHat 2020. CCIS, vol. 1271, pp. 138–163. Springer, Cham (2020). https://doi.org/10.1007/978-3-030-59621-7_8

18. Qi, Y., Xu, L., Yang, B., Xue, Y., Li, J.: Packet classification algorithms: from theory to practice. In: IEEE INFOCOM 2009, pp. 648–656. IEEE (2009)

19. Sharafaldin, I., Lashkari, A.H., Ghorbani, A.A.: Toward generating a new intrusion detection dataset and intrusion traffic characterization. ICISSp 1, 108–116 (2018)

20. Shen, M., Wei, M., Zhu, L., Wang, M., Li, F.: Certificate-aware encrypted traffic classification using second-order markov chain. In: 2016 IEEE/ACM 24th International Symposium on Quality of Service (IWQoS), pp. 1–10. IEEE (2016)

21. Shi, H., Li, H., Zhang, D., Cheng, C., Cao, X.: An efficient feature generation approach based on deep learning and feature selection techniques for traffic classification. Comput. Networks 132, 81–98 (2018)

22. Usama, M., et al.: Unsupervised machine learning for networking: techniques, applications and research challenges. IEEE Access 7, 65579–65615 (2019)

23. Vaswani, A., et al.: Attention is all you need. In: Advances in neural information processing systems, pp. 5998–6008 (2017)

24. Wang, W., et al.: Hast-ids: learning hierarchical spatial-temporal features using deep neural networks to improve intrusion detection. IEEE Access 6, 1792–1806 (2017)

25. Wang, W., Zhu, M., Zeng, X., Ye, X., Sheng, Y.: Malware traffic classification using convolutional neural network for representation learning. In: 2017 International Conference on Information Networking (ICOIN), pp. 712–717. IEEE (2017)

26. Yun, X., Wang, Y., Zhang, Y., Zhou, Y.: A semantics-aware approach to the automated network protocol identification. IEEE/ACM Trans. Networking 24(1), 583–595 (2015)

27. Zhang, J., Chen, X., Xiang, Y., Zhou, W., Wu, J.: Robust network traffic classification. IEEE/ACM Trans. Networking 23(4), 1257–1270 (2014)

28. Zhao, L., Cai, L., Yu, A., Xu, Z., Meng, D.: A novel network traffic classification approach via discriminative feature learning. In: Proceedings of the 35th Annual ACM Symposium on Applied Computing, pp. 1026–1033 (2020)

29. Zhao, L., et al.: Prototype-based malware traffic classification with novelty detection. In: ICICS, pp. 3–17 (2019)

# Comprehensive Degree Based Key Node Recognition Method in Complex Networks

Lixia Xie[1], Honghong Sun[1], Hongyu Yang[1,2(✉)], and Liang Zhang[3]

[1] School of Computer Science and Technology, Civil Aviation University of China,
Tianjin 300300, China
[2] School of Safety Science and Engineering, Civil Aviation University of China,
Tianjin 300300, China
[3] School of Information, University of Arizona, Tucson AZ85721, USA

**Abstract.** Aiming at the problem of the insufficient resolution and accuracy of the key node recognition methods in complex networks, a Comprehensive Degree Based Key Node Recognition Method (CDKNR) in complex networks is proposed. Firstly, the K-shell method is adopted to layer the network and obtain the K-shell ($Ks$) value of each node, and the influence of the global structure of the network is measured by the $Ks$ value. Secondly, the concept of Comprehensive Degree (CD) is proposed, and a dynamically adjustable influence coefficient $\mu_i$ is set, and the Comprehensive Degree of each node is obtained by measuring the influence of the local structure of the network through the number of neighboring nodes and sub-neighboring nodes and influence coefficient $\mu_i$. Finally, the importance of nodes is distinguished according to the Comprehensive Degree. Compared with several classical methods and risk assessment method, the experimental results show that the proposed method can effectively identify the key nodes, and has high accuracy and resolution in different complex networks. In addition, the CDKNR can provide a basis for risk assessment of network nodes, important node protection and risk disposal priority ranking of nodes in the network.

**Keywords:** Complex networks · K-shell · Comprehensive Degree · Neighboring nodes · Node importance

## 1 Introduction

In recent years, complex networks have been widely used in the field of network security, identifying key nodes in complex networks is of great significance for understanding the structure and function of the network and maintaining the stable operation of the network [1]. In practical applications, node importance ranking in complex networks can provide a basis for risk assessment of network nodes, protection of important nodes and risk disposal priority ranking of nodes in the network. However, there are the problems of low accuracy and high computational complexity in the existing key node identification methods [2]. Therefore, the research on designing fast and efficient key node identification methods has become an attractive and hot research area.

© Springer Nature Switzerland AG 2021
D. Gao et al. (Eds.): ICICS 2021, LNCS 12918, pp. 348–367, 2021.
https://doi.org/10.1007/978-3-030-86890-1_20

Currently, the key node identification methods for complex networks mainly include Degree Centrality (DC), Betweenness Centrality (BC), Closeness Centrality (CC), K - shell and so on [3]. Degree Centrality is a centrality method based on local characteristics, but Degree Centrality does not consider the influence of node location and surrounding nodes, and its classification effect is not ideal. Closeness Centrality, Betweenness Centrality, and K-shell are methods based on global characteristics. Closeness Centrality and Betweenness Centrality take better account of the connectivity of the network, but the computational complexity is high, so they are not suitable for large networks [4]. K-shell method assigns a *Ks* value to each node to quantify the importance of nodes [5], and this method has good time complexity, but it is difficult to distinguish key nodes in the same layer. To address the shortcomings of the above methods, Liu J G [6] ranked nodes according to the shortest distance between them and the nodes with the largest *Ks* value. Zeng A [7] proposed a Mixed Degree Decomposition method by considering the contributions of both remaining nodes and deleted nodes. Namtirtha A [8] proposed a weighted K-shell degree neighborhood method. Wen T [9] proposed a new method based on the local information dimension of nodes, which mainly referred to the local information around the nodes. Berahmand K [10] proposed a new local ranking method to identify key nodes, which measured the importance of nodes based on important location parameters such as the number of common neighbors, degree, and inverse clustering coefficient. Bae J [11] proposed an extended K-shell, but the resolution of this method was insufficient in some networks. The above methods measure the importance of nodes from different perspectives, but there are some problems: 1) There are various network topologies in reality, and some methods only consider the local structure of the network and ignore the global structure of the network. 2) The methods have high computational complexity. 3) For different networks, some methods require pre-set parameter values each time. 4) Due to the low resolution of some methods, the ability to identify key nodes in some networks is poor.

To solve the above problems and effectively identify the key nodes in complex networks, a Comprehensive Degree Based Key Node Recognition Method (CDKNR) for complex networks is proposed. The contributions of this paper are as follows:

(1) We propose the CDKNR method that considers both the global and local structure characteristics of networks.
(2) Compared with other methods, the CDKNR method does not contain free parameters and reduces the calculation cost.
(3) Experimental results show that CDKNR method has high accuracy and resolution in different complex networks and improves the efficiency of network security protection and emergency disposal in network security applications.

## 2 Related Method and Analysis

### 2.1 Representation of Network Graph

Network graph is represented by $G = (V, E)$, where $V = \{i | i = 1, 2, ..., n\}$ corresponds to node set, $E = \{e_{ij} = e_{ji} = \{i, j\} | i, j \in V\}$ corresponds to edge set, $n$ is the number of nodes and $n = |V|$, $m$ is the number of edges and $m = |E|$, $e_{ij}$ is the edge connecting vertex $i$ and vertex $j$. Moreover, $e_{ij} = e_{ji}$ in undirected graph. The network graph is represented by an adjacency matrix $A = [a_{ij}]_{n \times n}$, if there are edges between nodes $i$ and $j$, then $a_{ij} = 1$, otherwise $a_{ij} = 0$.

## 2.2   K-shell Method

The K-shell method [12] is a coarse-grained decomposition method based on the global structure of the network, which determines the importance of nodes according to their positions in the network. The method is implemented as follows.

(1) Suppose there are no isolated nodes of degree 0 in the network graph $G$. Delete all nodes of degree 1 and their connected edges in the network graph. After the deletion operation is completed, determine whether there are new nodes of degree 1, and if there are such nodes, delete such nodes and their connected edges. Repeat the above process until there are no nodes with degree 1 in the network graph, then all the deleted nodes constitute the 1-shell, and set the nodes' $Ks$ value to 1. For the updated network, the nodes with degree $\geq 2$.
(2) Delete all nodes of degree 2 and their connected edges, and repeat the process until there are no nodes of degree 2 in the network graph, then all the deleted nodes constitute the 2-shell, and set the nodes' $Ks$ value to 2. And so on, until all nodes in the network graph are stratified and assigned $Ks$ values. Among them, the layer of the nodes with the largest $Ks$ value is the core layer of the network, and the nodes in the core layer of the network have the highest influence.

An example of K-shell decomposition for a small network is shown in Fig. 1. The network in Fig. 1 is divided into three layers by the K-shell method, and nodes 1, 2, 3, 4, 5, 6, 9, 14, 16, 17 are assigned the same $Ks$ value, but it is difficult to distinguish their importance. Because of this defect, this paper uses the Comprehensive Degree of nodes in the same layer in the process of K-shell decomposition to further distinguish the importance of different nodes.

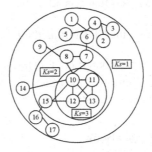

**Fig. 1.**  K-shell decomposition example

## 2.3  θ Method

To solve the defect that the K-shell method is difficult to distinguish the key nodes in the same shell layer, Liu J G [6] proposed the θ method, which sorted the nodes according to the shortest distance between nodes and the network core nodes (the nodes with the maximum $Ks$ value). The closer a node is to the network core nodes, the more important the node is. This method can be expressed as:

$$\theta(i|Ks) = (Ks^{Max} - Ks + 1) \sum_{j \in J} d_{ij}, i \in S_{Ks} \tag{1}$$

Where $Ks^{Max}$ is the maximum value of $Ks$ in the network, $J$ is the core node set with the maximum value of $Ks$, $d_{ij}$ is the shortest distance between node $i$ and node $j$, and $S_{Ks}$ is the set of all nodes, $\theta(i|Ks)$ is the importance value of node $i$.

## 2.4  Mixed Degree Decomposition Method

Zeng A [7] proposed the Mixed Degree Decomposition (MDD) method to overcome the shortcomings of the K-shell method by adding other information of nodes to decompose the network. The MDD method takes into account the influence of the removed nodes and the remaining nodes in the network, and stratifies the network according to the new mixed $Ks$ value. In the MDD method, the impact of the already removed nodes on the designated nodes is measured by the exhausted degree, similarly, the impact of the remaining nodes on the designated nodes is measured by the residual degree. The number of network layers is greatly increased by this kind of improved K-shell with mixed $Ks$ value, however, the parameter λ need to be adjusted for different networks so that the method can achieve the best effect of distinguishing the importance of nodes. In the comparative experiment done in this paper, λ is 0.7. The method is then expressed as:

$$K^m(i) = K^r(i) + \lambda * K^e(i) \tag{2}$$

Where $K^m(i)$ is the mixed $Ks$ value, $K^r(i)$ is the residual degree and $K^e(i)$ is the exhausted degree.

## 2.5  Weighted K-shell Degree Neighborhood Method

Namtirtha et al. [8] proposed a weighted K-shell degree neighborhood method, which uses the degree and the $Ks$ value of the two end nodes of an edge to arrive at the edge weight, and the edge weight between node $i$ and $j$ is expressed as:

$$w_{ij} = (\alpha * K(i) + \mu * Ks(i)) * (\alpha * K(j) + \mu * Ks(j)) \tag{3}$$

where $\alpha$ and $\mu$ are free parameters taking values between 0 and 1, which are varied for the selected network to obtain optimal results. $K(i)$ and $K(j)$ are the degrees of node

$i$ and node $j$ respectively, and $Ks(i)$ and $Ks(j)$ are the $Ks$ values of node $i$ and node $j$ respectively. The weighted K-shell degree neighborhood $ksd^w(i)$ of node $i$ is defined as:

$$ksd^w(i) = \sum_{j \in \Gamma_i} w_{ij} \tag{4}$$

Where $\Gamma_i$ is the nearest neighbor of node $i$ and $w_{ij}$ is the weight of the edge between node $i$ and $j$.

### 2.6  Degree and Clustering Coefficient and Location (DCL) Spreading Method

Berahmand et al. [10] proposed an improved local method, namely DCL method, which measures the importance of nodes based on their significant location parameters such as node degree, neighbor degree, number of common links between a node and its neighbors, and clustering coefficient. The DCL method is expressed as:

$$DCL(i) = K(i) * \left( \frac{1}{CC(i) + 1/K(i)} \right) + \left( \frac{\sum\limits_{j \in \Gamma i} K(j)}{|E(\Gamma_i)| + 1} \right) \tag{5}$$

Where $K(i)$ is the degree of node $i$, $CC(i)$ is the local clustering coefficient of node $i$, $\Sigma K(j)$ and $|E(\Gamma_i)|$ are the sum of the neighbors' degrees of node $i$ and the number of common links between neighbors of node $i$, respectively.

### 2.7  Extended K-shell Method

To improve the shortcomings of the K-shell method and to distinguish the importance of nodes in the same level, Bae J [11] proposed an extended K-shell method, the Ks+ method, which incorporates the node degree and the maximum degree to extend the K-shell, and the method can be expressed as:

$$Ksp(i) = Ks * (K^{Max} + 1) + K \tag{6}$$

Where $K$ is the node degree, $K^{Max}$ is the maximum degree, and $Ksp(i)$ is the extended $Ks$ value.

## 3    Key Node Identification Method

### 3.1  Method Design

To improve the shortcomings of the K-shell method and achieve the goal of distinguishing the importance of nodes, the concept of Comprehensive Degree (CD) is proposed according to the characteristics of the local structure of the network, and the Comprehensive Degree of nodes is defined as:

$$C(i) = K(i) + \mu_i D(i) \tag{7}$$

Where $K(i)$ is the degree of the node, $D(i)$ is the number of sub-neighboring nodes of the node, and $\mu_i$ is the influence coefficient.

Then, based on K-shell and Comprehensive Degree, a Comprehensive Degree Based Key Node Recognition Method (CDKNR) for complex networks is proposed. Among them, the influence of global structure is measured by the K-shell method, and the influence of local structure is measured by the influence degree (Comprehensive Degree) of neighboring and sub-neighboring nodes on the designated node.

The design idea of CDKNR is as follows: Firstly, the K-shell method is used to layer a network in the global scope, and the $Ks$ value of each node is obtained as a measure of the global characteristics; Secondly, the influence of neighboring nodes and sub-neighboring nodes is considered in the local scope, the Comprehensive Degree of each node is calculated and used as a measure of local characteristics. For a node in the network, not only the directly connected neighboring nodes will affect the node, but also the distant nodes will affect it. In order to achieve the goal of both improving the ability to distinguish the importance of nodes and reducing the computational complexity, only the influence of two-step neighborhoods, that is, neighboring nodes and sub-neighboring nodes, is considered in CDKNR. Since the neighboring nodes are closer to the designated node, they have a greater influence on it, while the sub-neighboring nodes are farther away from the designated node, so they have a smaller influence on it. Therefore, the influence coefficient $\mu_i$ is set between 0 and 1 and can be adjusted dynamically to weaken the influence of the sub-neighboring nodes, which makes CDKNR more consistent with the actual situation.

The specific process of the CDKNR method is designed as follows:

Step 1: Applying the K-shell method to stratify the network and obtain the $Ks$ value for each node.

Step 2: Calculate the degree $K(i)$ of all nodes in the network and the total number of nodes $N(i)$ in the two-step neighborhood.

Step 3: Calculate the influence coefficient $\mu_i$ of each node based on the $K(i)$ and $N(i)$ values obtained in step 2.

$$\mu_i = \frac{K(i)}{N(i)} \tag{8}$$

Step 4: Calculate the number of sub-neighboring nodes $D(i)$ based on the $K(i)$ and $N(i)$ values obtained in step 2.

$$D(i) = N(i) - K(i) \tag{9}$$

Step 5: Calculate the Comprehensive Degree $C(i)$ of the nodes based on $\mu_i$, $D(i)$ obtained in steps 3 and 4.

$$C(i) = K(i) + \mu_i D(i) \tag{10}$$

Step 6: Since the $Ks$ value of nodes is the same, it is difficult to judge the importance of nodes in the same level, the importance of nodes in the same level is further distinguished according to the Comprehensive Degree $C(i)$: for nodes with the same $Ks$ value, the nodes with a higher Comprehensive Degree are more important.

From the calculation process of CDKNR method, it can be seen that the method does not contain free parameters and does not need to test parameter values in advance, therefore, the computational complexity can be reduced and the efficiency of the method can be improved.

### 3.2 Example Analysis

Taking an undirected unweighted network with 14 nodes and 20 edges as an example (as shown in Fig. 2). Firstly, CDKNR is used to rank the nodes in Fig. 2, and the ranking effect is compared with other methods, then node 8 is taken as an example to illustrate the calculation process of CDKNR.

(1) Calculate the $Ks$ value of node 8 and get $Ks$ (8) = 2.
(2) Calculate the degree $K(8)$ and the total number of nodes $N(8)$ in the two-step neighborhood. Because the neighboring nodes of node 8 are 5, 6, 7, 9, 10, 11, and the sub-neighboring nodes are 3, 12, 13, 14, so $K(8) = 6, N(8) = 10$.
(3) According to Eq. (8), The influence coefficient $\mu_8$ of node 8 is obtained, $\mu_8 = 6/10 = 0.6$.
(4) According to Eq. (9), the number of sub-neighboring nodes $D(8)$ is calculated, $D(8) = 10–6 = 4$.
(5) According to Eq. (10), the Comprehensive Degree $C(8)$ of node 8 is obtained, $C(8) = 6 + 0.6 * 4 = 8.4$.

Repeat the above process to get the $Ks$ value and Comprehensive Degree of all nodes of the network in Fig. 2. The calculation process of the Comprehensive Degree of network nodes in Fig. 2 is shown in Table 1. To compare with other methods, the node importance values and ranking results obtained by Degree Centrality (K) [1], K-shell method (Ks) [12], extended K-shell method (Ks +) [11], Mixed Degree Decomposition (MDD) method [7], θ method [6], Betweenness Centrality (BC) [9], Closeness Centrality (CC) [3], Weighted K-shell Degree Neighborhood Method (ksd$^w$) [8], Degree and Clustering Coefficient and Location (DCL) Spreading Method [10] and Comprehensive Degree Based Key Node Recognition Method (CDKNR) are respectively presented in Table 2 and Table 3. (In the ksd$^w$ method, $\alpha = 0.1, \mu = 0.9$.) In Table 1, the numbers in the node column represent the nodes, $K(i)$ column is the degree of node $i$, $Ks$ column is the $Ks$ value of node $i$, $N(i)$ column is the number of nodes in the two-step neighborhood, $D(i)$ column is the number of sub-neighboring nodes, $\mu_i$ column is the influence coefficient, and $CD$ column is the Comprehensive Degree.

It can be seen from Table 3 that the resolution of methods such as DC, BC, CC and K-shell are not high. DC only divides the whole network into 6 layers, and the K-shell method only divides the whole network into 2 layers, while we proposed CDKNR method divides the whole network into 11 layers. Therefore, the ranking range is wider and the method resolution is higher compared with other methods.

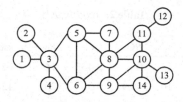

**Fig. 2.** A small network of example analysis

**Table 1.** The Comprehensive Degree calculation process of the network

| Node | $K(i)$ | $Ks$ | $N(i)$ | $D(i)$ | $\mu_i$ | CD |
|------|--------|------|--------|--------|---------|-------|
| 1 | 1 | 1 | 5 | 4 | 0.2 | 1.8 |
| 2 | 1 | 1 | 5 | 4 | 0.2 | 1.8 |
| 3 | 5 | 2 | 8 | 3 | 0.625 | 6.875 |
| 4 | 1 | 1 | 5 | 4 | 0.2 | 1.8 |
| 5 | 4 | 2 | 10 | 6 | 0.4 | 6.4 |
| 6 | 4 | 2 | 11 | 7 | 0.364 | 6.545 |
| 7 | 2 | 2 | 7 | 5 | 0.286 | 3.429 |
| 8 | 6 | 2 | 10 | 4 | 0.6 | 8.4 |
| 9 | 4 | 2 | 9 | 5 | 0.444 | 6.222 |
| 10 | 5 | 2 | 9 | 4 | 0.556 | 7.222 |
| 11 | 3 | 2 | 9 | 6 | 0.333 | 5 |
| 12 | 1 | 1 | 3 | 2 | 0.333 | 1.667 |
| 13 | 1 | 1 | 5 | 4 | 0.2 | 1.8 |
| 14 | 2 | 2 | 6 | 4 | 0.333 | 3.333 |

**Table 2.** Node importance values are obtained by different methods

| Node | $K$ | $Ks$ | MDD | $\theta$ | BC | CC | $Ks+$ | $ksd^w$ | DCL | CDKNR |
|------|-----|------|-----|----------|------|------|-------|---------|------|-------|
| 1 | 1 | 1 | 1 | 52 | 0 | 0.325 | 8 | 5.39 | 6.0 | 1.8 |
| 2 | 1 | 1 | 1 | 52 | 0 | 0.325 | 8 | 5.39 | 6.0 | 1.8 |
| 3 | 5 | 2 | 3.7 | 17 | 33 | 0.464 | 19 | 55.37 | 22.16 | 6.875 |
| 4 | 1 | 1 | 1 | 52 | 0 | 0.325 | 8 | 5.39 | 6.0 | 1.8 |
| 5 | 4 | 2 | 3.7 | 13 | 13.1 | 0.52 | 18 | 67.6 | 9.58 | 6.4 |
| 6 | 4 | 2 | 3.7 | 12 | 20.1 | 0.542 | 18 | 74.8 | 10.08 | 6.545 |

(*continued*)

**Table 2.** (*continued*)

| Node | K | Ks | MDD | θ | BC | CC | Ks + | ksd$^w$ | DCL | CDKNR |
|------|---|----|-----|---|----|----|----|------|-----|-------|
| 7 | 2 | 2 | 2 | 15 | 0 | 0.433 | 16 | 21.56 | 6.33 | 3.429 |
| 8 | 6 | 2 | 4.2 | 10 | 28.9 | 0.565 | 20 | 128.76 | 15.74 | 8.4 |
| 9 | 4 | 2 | 3.7 | 12 | 10.3 | 0.5 | 18 | 67.6 | 9.58 | 6.222 |
| 10 | 5 | 2 | 3.8 | 13 | 16.3 | 0.464 | 19 | 79.38 | 14.0 | 7.222 |
| 11 | 3 | 2 | 2.7 | 15 | 12 | 0.433 | 17 | 36.58 | 10.52 | 5 |
| 12 | 1 | 1 | 1 | 48 | 0 | 0.31 | 8 | 3.41 | 4.0 | 1.667 |
| 13 | 1 | 1 | 1 | 44 | 0 | 0.325 | 8 | 5.39 | 6.0 | 1.8 |
| 14 | 2 | 2 | 2 | 17 | 0 | 0.382 | 16 | 19.58 | 5.83 | 3.333 |

**Table 3.** Node importance rankings are obtained by different methods

| Rank | K | Ks | MDD | θ | BC | CC | Ks + | ksd$^w$ | DCL | CDKNR |
|------|---|----|-----|---|----|----|----|------|-----|-------|
| 1 | 8 | 10, 11, 14, 3, 5, 6, 7, 8, 9 | 8 | 8 | 3 | 8 | 8 | 8 | 3 | 8 |
| 2 | 10, 3 | 1, 12, 13, 2, 4 | 10 | 6, 9 | 8 | 6 | 10, 3 | 10 | 8 | 10 |
| 3 | 5, 6, 9 | | 3, 5, 6, 9 | 10, 5 | 6 | 5 | 5, 6, 9 | 6 | 10 | 3 |
| 4 | 11 | | 11 | 11, 7 | 10 | 9 | 11 | 5, 9 | 11 | 6 |
| 5 | 14, 7 | | 14, 7 | 14, 3 | 5 | 10, 3 | 14, 7 | 3 | 6 | 5 |
| 6 | 1, 2, 4, 12, 13 | | 1, 12, 13, 2, 4 | 13 | 11 | 11, 7 | 1, 12, 13, 2, 4 | 11 | 5, 9 | 9 |
| 7 | | | | 12 | 9 | 14 | | 7 | 7 | 11 |
| 8 | | | | 1, 2, 4 | 1, 12, 13, 14, 2, 4, 7 | 1, 13, 2, 4 | | 14 | 1, 2, 4, 13 | 7 |
| 9 | | | | | | 12 | | 1, 13, 2, 4 | 14 | 14 |
| 10 | | | | | | | | 12 | 12 | 1, 13, 2, 4 |

(*continued*)

**Table 3.** (*continued*)

| Rank | K | Ks | MDD | θ | BC | CC | Ks + | ksd$^w$ | DCL | CDKNR |
|------|---|----|----|---|----|----|------|---------|-----|-------|
| 11 |  |  |  |  |  |  |  |  |  | 12 |
| 12 |  |  |  |  |  |  |  |  |  |  |
| 13 |  |  |  |  |  |  |  |  |  |  |
| 14 |  |  |  |  |  |  |  |  |  |  |

# 4 Experimental Results and Analysis

## 4.1 Experimental Data Set and Experimental Design

To compare and analyze the performance of other methods and CDKNR in terms of accuracy and resolution in six typical complex networks, a comparative experiment of key node identification is carried out on six classical complex networks. The six classical complex networks are as follows: (1) Zachary network; (2) Dolphins network; (3) Jazz network; (4) Blogs network; (5) NetScience network; (6) Powergrid network. The statistical characteristics of the above six networks are shown in Table 4, where $V$ represents the number of network nodes, $E$ represents the number of network edges, $\beta_{th}$ is the epidemic threshold value, and $\beta$ is the infection probability. Both $\alpha$ and $\mu$ are parameters from the literature [8].

**Table 4.** Statistical characteristics of six networks

| Network | V | E | $\beta_{th}$ | $\beta$ | $\alpha$ | $\mu$ |
|---------|-----|-------|--------|--------|--------|--------|
| Zachary | 34 | 78 | 0.129 | 0.15 | 0 | 0.2 |
| Dolphins | 62 | 159 | 0.147 | 0.15 | 0.1 | 0.9 |
| Jazz | 198 | 2742 | 0.026 | 0.05 | 0 | 0.2 |
| Blogs | 1224 | 19025 | 0.012 | 0.1 | 1 | 0.9 |
| NetScience | 1461 | 2742 | 0.144 | 0.15 | 0.1 | 1 |
| Powergrid | 4941 | 6594 | 0.258 | 0.30 | 0 | 0.6 |

In Sect. 4.2, the method performance is evaluated using SIR epidemic model [13], Kendall correlation coefficient [14], and Monotonicity index [15]. In Sect. 4.3, the method performance evaluation is verified in network security applications.

## 4.2 Method Performance Evaluation

In order to evaluate the accuracy and resolution of different methods, ten methods, including DC, BC, CC, K-shell, MDD, θ, Ks +, ksd$^w$, DCL and CDKNR are selected for experiments in six typical networks. Firstly, the SIR model is used to examine the propagation efficiency of single nodes to determine the importance of nodes. Secondly, the Kendall correlation coefficient index is used to compare the ranking results of node importance obtained by different methods with the ranking results of node propagation efficiency of SIR model, and the correlation coefficients are calculated. Finally, the Monotonicity index is used to evaluate the ability of different methods to differentiate the importance of nodes, that is, to check whether there are a large number of nodes with the same importance value in the ranking results of node importance.

**SIR Model.** In the SIR model, each node has three possible states: susceptible (S), infected (I), and recovered (R) states [16]. The goal of this model is to measure the relative importance of nodes during the spread of an epidemic. In the beginning, only the transmission initiation node is in the 'I' state, while the rest of the nodes are in the 'S' state. After that, at each time step, the infected node spreads the infection to its neighboring nodes with infection probability $\beta$. The initially infected node becomes 'R', and the node in 'R' cannot be infected again. The propagation process ends when no new infected node appears, and the number of recovered nodes reflects the influence of the initial node. When the infection probability is large, the epidemic can spread widely in a network, and when the infection probability is small, the epidemic can spread only in a small area. For a disease to spread and become epidemic in a network, the infection probability must be greater than the epidemic threshold, the epidemic threshold $\beta_{th}$ is:

$$\beta_{th} \approx \frac{<k>}{<k^2>} \tag{11}$$

Where $<k>$ is the average degree of the network and $<k^2>$ is the second order average degree of the network.

**Kendall Correlation Coefficient.** Kendall correlation coefficient is expressed as $\tau$, which is used to test the correlation between two sequences. Its value ranges from $-1$ to 1, when $\tau$ is 1, it means that the two sequences are completely consistent; when $\tau$ is $-1$, it means that the two sequences are opposite; when $\tau$ is 0, it means that the two sequences are independent of each other.

The node importance ranking results $x$ obtained from different methods and the node propagation efficiency ranking results $y$ obtained from SIR model are constructed as sequences of the form $(x_1, y_1), (x_2, y_2), \ldots, (x_n, y_n)$, and for any pair of sequences, if $x_i > x_j$ and $y_i > y_j$ or $x_i < x_j$ and $y_i < y_j$, the pair of sequences is said to be consistent; if $x_i > x_j$ but $y_i < y_j$ or $x_i < x_j$ but $y_i > y_j$, the pair of sequences is said to be inconsistent; if $x_i = x_j$ and $y_i = y_j$, the pair of sequences is neither consistent nor inconsistent. $\tau$ is defined as follow:

$$\tau = \frac{2*(C-D)}{N*(N-1)} \tag{12}$$

Where $C$ and $D$ are the number of consistent and inconsistent sequence pairs and $N$ is the network size.

Qiu L Q [5] and Berahmand K [10] through an infection probability $\beta$ to compare the ranking results obtained by different methods and SIR model and then get $\tau$ value. To observe the trend and correlation of the infection probability $\beta$ and the correlation coefficient $\tau$, this paper simulates the propagation process of the SIR model and the infection probability of the model varies from $\beta_{th}$ to $2*\beta_{th}$, and each time increases in steps of $0.1*\beta_{th}$, then $\tau$ is calculated. The probability of infection $\beta$ is:

$$\beta=\beta_{th} + \delta * j \tag{13}$$

Where $\delta$ is the increment of infection probability at each step, $j$ is the number of steps, and $\beta_{th}$ is the epidemic threshold.

**Experimental Results of Methods Accuracy.** In this experiment, small networks (networks with edges less than 10000) are simulated 1000 times and large networks (networks with edges more than 10000) are simulated 100 times. We compare the sequence of node importance ranking results obtained by different methods with that obtained by SIR model simulation, and calculates the corresponding $\tau$ value. The higher the $\tau$ value, the higher the correlation between the two ordered sequences, and the more accurate the node importance ranking results generated by the method.

The ten methods of DC, BC, CC, K-shell, MDD, $\theta$, Ks +, ksd$^w$, DCL and CDKNR are used with the SIR model for Zachary network, Dolphins network, Jazz network, Blogs network, NetScience network, and Powergrid network to calculate the Kendall correlation coefficients $\tau$ with different infection probabilities $\beta$. Then, the average Kendall correlation coefficient $\tau A$ for each network is obtained by averaging the $\tau$ values corresponding to the 10 infection probabilities in the 6 networks (as shown in Table 5). As can be seen from Table 5, among the Dolphins, NetScience, and Powergrid networks, the ranking result of network node importance obtained by CDKNR is closest to the ranking result of node propagation efficiency obtained by the SIR model, and the average Kendall coefficient of CDKNR is higher than other methods. This is the CDKNR not only considers the global characteristics of the nodes ($Ks$ value) but also incorporates the local characteristics of the nodes (Comprehensive Degree), thus improving the accuracy of the ranking results. In the six networks, the infection probability $\beta$ varies from $\beta_{th}$ to $2*\beta_{th}$, increasing in steps of $0.1*\beta_{th}$ each time. The variation trend of $\tau$ is shown in Figs. 3, 4, 5, 6, 7, 8. And the higher $\tau$ value is, the higher accuracy of the method is. From Fig. 3 and Fig. 5, it can be seen that in Zachary and Jazz networks, the correlation between CDKNR and SIR model is slightly lower than that of ksd$^w$ method, but the average $\tau$ value is higher, considering that the accuracy of ksd$^w$ method is very dependent on the selection of parameters, and a large number of repeated experiments are needed to test the parameter taking values for different networks, which increases the computational complexity and lacks some objectivity, while CDKNR method can still achieve good accuracy without testing parameters. In addition, the $\tau$ of CDKNR method in Blogs network is slightly lower than other methods, but the correlation coefficient $\tau$ obtained by CDKNR method increases with the increase of infection probability $\beta$ and the overall trend is upward, in summary CDKNR has high accuracy in different networks.

**Resolution Index.** For the importance ranking of nodes in complex networks, if each node has a different node importance ranking value, it means that the method can clearly distinguish the importance of nodes. The fewer nodes in the same ranking, the higher resolution of the method is. In the experiment, the monotonicity index $M(R)$ is used to

**Table 5.** Average $\tau$ values for 6 networks with different infection probabilities ($\beta_{th}$ to $2\beta_{th}$)

| Network | $\tau A(K)$ | $\tau A(Ks)$ | $\tau A(MDD)$ | $\tau A(\theta)$ | $\tau A(BC)$ | $\tau A(CC)$ | $\tau A(Ks+)$ | $\tau A(ksd^w)$ | $\tau A(DCL)$ | $\tau A(CDKNR)$ |
|---|---|---|---|---|---|---|---|---|---|---|
| Zachary | 0.715 | 0.655 | 0.729 | 0.681 | 0.607 | 0.722 | 0.735 | **0.768** | 0.473 | 0.763 |
| Dolphins | 0.754 | 0.719 | 0.770 | 0.745 | 0.523 | 0.679 | 0.754 | 0.766 | 0.601 | **0.776** |
| Jazz | 0.843 | 0.809 | 0.866 | 0.788 | 0.474 | 0.716 | 0.848 | **0.883** | 0.674 | 0.842 |
| Blogs | 0.811 | **0.822** | 0.814 | 0.787 | 0.654 | 0.725 | 0.819 | 0.808 | 0.246 | 0.815 |
| NetScience | 0.681 | 0.651 | 0.682 | 0.652 | 0.318 | -0.495 | 0.661 | 0.739 | 0.772 | **0.774** |
| Powergrid | 0.431 | 0.398 | 0.452 | 0.349 | 0.294 | 0.417 | 0.446 | 0.482 | 0.441 | **0.506** |

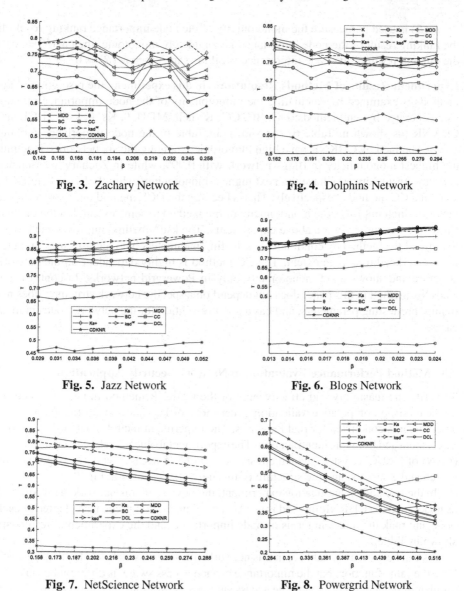

**Fig. 3.** Zachary Network

**Fig. 4.** Dolphins Network

**Fig. 5.** Jazz Network

**Fig. 6.** Blogs Network

**Fig. 7.** NetScience Network

**Fig. 8.** Powergrid Network

evaluate the resolution of different methods. The monotonicity index $M(R)$ is:

$$M(R) = \left[ 1 - \frac{\sum\limits_{r \in R} n_r(n_r - 1)}{n(n-1)} \right]^2 \tag{14}$$

Where $R$ represents the node importance ranking result, $n$ represents the number of nodes in the ranking result, and $n_r$ represents the number of nodes with the same ranking result.

$M(R)$ is used to measure the monotonicity of the node importance ranking result. If the ranking result of the node importance is completely monotonic, then $M(R) = 1$; If the ranking result of the node importance is all the same, then $M(R) = 0$.

**Experiment Results of Methods Resolution.** In this experiment, the resolution of the methods is examined by calculating the monotonicity of the node importance ranking lists generated by ten methods: DC, BC, CC, K-shell, MDD, $\theta$, Ks +, ksd$^w$, DCL and CDKNR (as shown in Table 6). As shown in Table 6, the node importance ranking results obtained by CDKNR have high monotonicity among the six networks, reaching the highest monotonicity in Blogs network with 0.999, while in Zachary, NetScience and Powergrid networks have the next highest monotonicity, slightly lower than DCL, ksd$^w$ and CC methods, respectively. This is because the DCL method incorporates more factors, which improves the monotonicity of the method to some extent, but the method shows poor accuracy in the above experiments. The ksd$^w$ method improves the monotonicity of the method by continuously selecting parameters but increases the computational overhead to a large extent. The CC method is highly dependent on the network structure and shows good monotonicity only in Powergrid networks. In contrast, the CDKNR method in this paper does not depend on a specific network structure, does not require pre-testing parameters, and has a good resolution for the network nodes in most cases.

### 4.3 Method Performance Evaluation in Network Security Applications

To verify the feasibility and effectiveness of the CDKNR method in network security applications, a comparative validation experiment of key node identification and risk assessment is conducted for real networks. The experiment includes network node risk assessment and key node identification. The experimental object is a Local Area Network (LAN) of CAUC's System, which consists of 30 PCs, 5 access switches, 1 aggregation switch and 2 servers, and the structure of this network is shown in Fig. 9.

In the key node identification experiment, the devices in this network are abstracted as nodes, and the CDKNR method is used to calculate the comprehensive degree of each node and rank the nodes in terms of node importance, and the experimental results are shown in Table 7.

As can be seen from Table 7, in this network, the access switches (31, 32, 33, 34, 35) have the same function, but the importance of each access switch is different because the number of devices connected to the access switches is different. The aggregation switch is in the most important position in the whole network, and once the aggregation switch is attacked, it will affect the operation of the whole network.

In the risk assessment section, the risk values of all nodes (devices) in this LAN are calculated according to the risk assessment specification [17]. The data sources are network alarms and fault information from the network management logs for 4 weeks, and network attack detection data from the network security software. In this part of the experiment, based on the assets, alarm and fault information, and attack alarm information, the risk assessment method based on expected loss [18] is used to obtain the threat and risk severity indicators for each node in the network, and the risk value of each node (device) is calculated. The results of the experiments are shown in Table 8.

**Table 6.** Monotonicity of different methods in 6 networks

| Network | M(K) | M(Ks) | M(MDD) | M(θ) | M(BC) | M(CC) | M(Ks +) | M(ksd^w) | M(DCL) | M(CDKNR) |
|---|---|---|---|---|---|---|---|---|---|---|
| Zachary | 0.707 | 0.495 | 0.753 | 0.879 | 0.772 | 0.899 | 0.741 | 0.852 | **0.954** | 0.940 |
| Dolphins | 0.831 | 0.376 | 0.904 | 0.973 | 0.962 | 0.973 | 0.856 | **0.988** | 0.985 | 0.983 |
| Jazz | 0.965 | 0.794 | 0.988 | 0.934 | 0.987 | 0.987 | 0.988 | 0.996 | **0.998** | 0.996 |
| Blogs | 0.932 | 0.905 | 0.944 | 0.996 | 0.944 | 0.997 | 0.944 | 0.861 | 0.998 | **0.999** |
| NetScience | 0.706 | 0.663 | 0.739 | 0.663 | 0.104 | 0.643 | 0.742 | **0.915** | 0.912 | 0.913 |
| Powergrid | 0.592 | 0.245 | 0.692 | 0.960 | 0.831 | **0.999** | 0.660 | 0.751 | 0.940 | 0.952 |

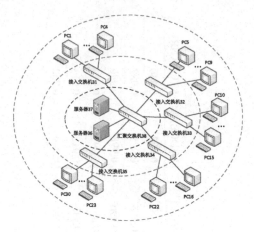

**Fig. 9.** Experimental network structure diagram

**Table 7.** Ranking of network nodes (devices) importance

| Rank | CD | Nodes (Devices) |
|---|---|---|
| 1 | 12.676 | 38 |
| 2 | 12.6 | 35 |
| 3 | 11.429 | 34 |
| 4 | 10.231 | 33 |
| 5 | 9 | 32 |
| 6 | 7.727 | 31 |
| 7 | 1.889 | 23, 24, 25, 26, 27, 28, 29, 30 |
| 8 | 1.875 | 16, 17, 18, 19, 20, 21, 22 |
| 9 | 1.857 | 10, 11, 12, 13, 14, 15, 36, 37 |
| 10 | 1.833 | 5, 6, 7, 8, 9 |
| 11 | 1.8 | 1, 2, 3, 4 |

A comparison of the ranking results of the two experiments mentioned above is shown in Table 9. As can be seen from Table 9, the node importance ranking of this network obtained by the CDKNR method is the same as the TOP10 of the risk value ranking of the network nodes obtained by the risk assessment. This result shows that the node importance ranking results obtained by the CDKNR method can provide a basis for risk assessment of network nodes, important node protection and risk disposal priority ranking of nodes in the network. Based on the node importance ranking, which will improve the efficiency of network security protection and emergency disposal.

**Table 8.** Risk values of network nodes (devices)

| Rank | Risk values | Nodes (Devices) |
| --- | --- | --- |
| 1 | 0.941 | 38 |
| 2 | 0.824 | 35 |
| 3 | 0.733 | 34 |
| 4 | 0.711 | 33 |
| 5 | 0.623 | 32 |
| 6 | 0.542 | 31 |
| 7 | 0.507 | 23, 24, 25, 26, 27, 28, 29, 30 |
| 8 | 0.426 | 16, 17, 18, 19, 20, 21, 22 |
| 9 | 0.389 | 10, 11, 12, 13, 14, 15 |
| 10 | 0.354 | 5, 6, 7, 8, 9, 36, 37 |
| 11 | 0.276 | 1, 2, 3, 4 |

**Table 9.** Network node risk value ranking and network node importance ranking

| | Risk assessment | CDKNR |
| --- | --- | --- |
| Rank | Nodes (Devices) | Nodes (Devices) |
| 1 | 38 | 38 |
| 2 | 35 | 35 |
| 3 | 34 | 34 |
| 4 | 33 | 33 |
| 5 | 32 | 32 |
| 6 | 31 | 31 |
| 7 | 23, 24, 25, 26, 27, 28, 29, 30 | 23, 24, 25, 26, 27, 28, 29, 30 |
| 8 | 16, 17, 18, 19, 20, 21, 22 | 16, 17, 18, 19, 20, 21, 22 |
| 9 | 10, 11, 12, 13, 14, 15 | 10, 11, 12, 13, 14, 15, 36, 37 |
| 10 | 5, 6, 7, 8, 9, 36, 37 | 5, 6, 7, 8, 9 |
| 11 | 1, 2, 3, 4 | 1, 2, 3, 4 |

## 5 Conclusion

In this paper, we propose a new method CDKNR to identify key nodes in complex networks to address the problems of insufficient accuracy and high computational complexity of existing methods. Firstly, the network is decomposed by K-shell method to obtain the $Ks$ value that can measure the location information of nodes in the network. Secondly, the Comprehensive Degree of the designated node is calculated according to the different influence degrees of the neighboring nodes and the sub-neighboring nodes on the designated node. Finally, the importance of nodes is taken into account both the

*Ks* value of nodes and the Comprehensive Degree, that is, for the nodes in the same shell, the nodes with higher Comprehensive Degree are more important. By verifying the effectiveness of the CDKNR method in six typical complex networks and network security applications, experimental results show that, compared with other methods, the method in this paper not only has lower computational complexity, but also has higher accuracy and resolution for different complex networks, which can identify the key nodes in complex networks, improve the efficiency of network security protection and emergency disposal, and provide a theoretical basis for protecting the key nodes and improve the security of the networks in the next step.

**Acknowledgments.** This work was supported by the Civil Aviation Joint Research Fund Project of the National Natural Science Foundation of China under Grant no. U1833107.

# References

1. Yu, E.: A Re-ranking algorithm for identifying influential nodes in complex networks. IEEE Access 8, 211281–211290 (2020)
2. Yu, E.Y.: Identifying critical nodes in complex networks via graph convolutional networks. Knowl.-Based Syst. **198**, 1–8 (2020)
3. Yan, X.L.: Identifying influential spreaders in complex networks based on entropy weight method and gravity law. Chin. Phys. B **29**(4), 582–590 (2020)
4. Ullah, A.: Identification of Influential nodes via effective distance-based centrality mechanism in complex networks. Complexity **2021**(11), 1–6 (2021)
5. Qiu, L.Q.: Ranking influential nodes in complex networks based on local and global structures. Appl. Intell. 1–14 (2021)
6. Liu, J.G.: Ranking the spreading influence in complex networks. Physica A Stat. Mech. Appl. **392**(18), 4154–4159 (2013)
7. Zeng, A.: Ranking spreaders by decomposing complex networks. Phys. Lett. A **377**(14), 1031–1035 (2013)
8. Namtirtha, A.: Weighted kshell degree neighborhood: a new method for identifying the influential spreaders from a variety of complex network connectivity structures. Expert Syst. Appl. **139**, 1–15 (2020)
9. Wen, T.: Identification of influencers in complex networks by local information dimensionality. Inf. Sci. **512**, 549–562 (2020)
10. Berahmand, K.: A new local and multidimensional ranking measure to detect spreaders in social networks. Comput. **101**(11), 1711–1733 (2019)
11. Bae, J.: Identifying and ranking influential spreaders in complex networks by neighborhood coreness. Phys. A: Statal Mech. Appl. **395**, 549–559 (2014)
12. Kitsak, M.: Identification of influential spreaders in complex networks. Nat. Phys. **6**(11), 888–893 (2010)
13. Ibnoulouafi, A.: M-Centrality: identifying key nodes based on global position and local degree variation. J. Stat. Mech. Theory Exp. (7), 1–30 (2018)
14. Maji, G.: Influential spreaders identification in complex networks with potential edge weight based k-shell degree neighborhood method. J. Comput. Sci. **39**, 1–9 (2020)
15. Maji, G.: A systematic survey on influential spreaders identification in complex networks with a focus on K-shell based techniques. Expert Syst. Appl. **161**, 1–18 (2020)
16. Zhang, D.Y.: Identifying and quantifying potential super-spreaders in social networks. Sci. Rep. **9**(1), 1–11 (2019)

17. China National Standardization Administration Committee: Information Security Risk Assessment Specification: GB/T 20984–2007[S]. China Standard Press, Beijing (2007)
18. Peng, J.H.: Utility based security risk measurement model. J. Beijing Univ. Posts Telecommun. **29**(2), 59–61 (2006)

# Improving Convolutional Neural Network-Based Webshell Detection Through Reinforcement Learning

Yalun Wu, Minglu Song, Yike Li, Yunzhe Tian, Endong Tong(✉),
Wenjia Niu(✉), Bowei Jia, Haixiang Huang, Qiong Li, and Jiqiang Liu

Beijing Key Laboratory of Security and Privacy in Intelligent Transportation, Beijing
Jiaotong University, Beijing 100044, China
{wuyalun,18140079,yikeli,tianyunzhe,edtong,niuwj,jiabowei,17281036,
liqiong,jqliu}@bjtu.edu.cn

**Abstract.** Webshell detection is highly important for network security protection. Conventional methods are based on keywords matching, which heavily relies on experiences of domain experts when facing
emerging malicious webshells of various kinds. Recently, machine learning, especially supervised learning, is introduced for webshell detection
and has proved to be a great success. As one of state-of-the-art work,
neural network (NN) is designed to input a large number of features and
enable deep learning. Thus, how to properly combine the advantages
of automatic feature selection and the advantages of expert knowledge-
based way has become a key issue. Considering that special features
to indicate unexpected webshell behaviors for a target business system
are usually simple but effective, in this work, we propose a novel approach for improving webshell detection based on convolutional neural
network (CNN) through reinforcement learning. We utilize the reinforcement learning of asynchronous advantage actor-critic (A3C) for automatic feature selection, aiming to maximize the expected accuracy of
the CNN classifier on a validation dataset by sequentially interacting
with the feature space. Moreover, considering the sparseness of feature
values, we build the CNN classifier with two convolutional layers and a
global pooling. Extensive experiments and analysis have been conducted
to demonstrate the effectiveness of our proposed method.

**Keywords:** Webshell detection · Feature selection · Unexpected
behavior feature · Reinforcement learning · Convolutional neural
network

## 1 Introduction

Webshell, a program that is written in scripting languages, such as ASP, PHP,
JSP, CGI, etc., provides a useful way to communicate with the web server's
operating system (OS) [10,13,20,23]. Unfortunately, the webshell also become

© Springer Nature Switzerland AG 2021
D. Gao et al. (Eds.): ICICS 2021, LNCS 12918, pp. 368–383, 2021.
https://doi.org/10.1007/978-3-030-86890-1_21

one main threat of website security protection. Malicious users can launch web attacks by the normal webshell with malicious functions, generating the so-called webpage backdoor [9]. In such processes, the webshell is installed through vulnerabilities in web applications or weak server security configurations, such as SQL injection [21] and cross-site scripting [11]. Through malicious webshells, malicious attacks for instance, data theft, DDoS attacks [2] , and watering hole attacks can be performed. Actually, in addition to common and easily-used tools of malicious webshells such as WSO, C99, B374K and China Chopper [12], the advanced persistent threat (APT) [7] and criminal groups also use their special and unknown tools to exploit webshells, resulting in a lot of cyberspace damaging incidents. Thus, how to effectively detect malicious webshell to win the longtime game against attacker, is highly important and never out-of-date for network security protection.

For malicious webshell detection, there are different method classifications. From the data capture perspective, they can be divided into two categories: static detection [3] and dynamic detection [18]. The dynamic method requires to set the environment to be a sandbox or a virtual machine, and further utilizes the hook technique for monitoring operating system (OS) process-based operations of file reading and writing. While the static method does not utilize the hook, capture related data as features for further matching, the statistical feature thresholding such as NeoPI, and the association analysis based on the knowledge that malicious webshell generally has lower connections with the existing web files. Due to the difficulty of dynamic detection, the static detection seems more common and is implemented in real systems. From the feature perspective, these detection methods can be classified into the content feature-based and the behavior feature-based. In the content-based method, researchers focus on the content of suspicious files or content in HTTP POST requests. To fully utilize the content feature, the words segmentation for text feature representation is the priority for data preprocessing, and TF-IDF and n-grams are the most frequently used. The behavior-based method aims to analyze the traffic flow in a sequence of packets. Through the traffic flow analysis, unexpected behaviors or even malicious behaviors can be detected.

Recently, machine learning, especially supervised learning, is introduced for webshell detection and has proved to be a great success. A lot of classical models are developed to treat webshell detection as a task of classification, including Naive Bayes (NB), k-Nearest Neighbor (kNN), Support Vector Machine (SVM), Decision Tree (DT). With the development of deep learning, sparse auto-encoder, soft-max regression and convolutional neural network (CNN) are introduced [16, 19, 22, 28]. Although the above models are effective to some extent, when facing real business system, researchers still have to do much challenging work for the feature engineering or the feature selection for neural network (NN). Moreover, many malicious webshells are confused with normal webpage files after confusion and mutation, which requires to extract more features for the collaborative analysis. In other words, such feature space is relatively large to explore. Selecting features closely rely on expert experiences. Moreover, for achieving optimal performance, it is difficult to guarantee that the extracted features comprehensively cover the key distinguishing feature of samples.

Thus, it is highly important to combine the advantages of automatic feature selection and the advantages of expert knowledge-based methods. In this work, we propose a novel approach for improving webshell detection based on convolutional neural network (CNN) through reinforcement learning (RL). To the best of our knowledge, there is hardly any similar work. Especially, we utilize the RL algorithm of asynchronous advantage actor-critic (A3C) for automatic feature selection. In our CNN model, two convolutional layers and a global pooling are built. Through extensive experiments, we find that our method can reduce 32.5% features and can achieve 96% accuracy. Compared with the single CNN without reinforcement learning, our method has an average 2.42% improvement.

The remainder of this paper is organized as follows. Section 2 discusses the related work. In Sect. 3, we propose our A3C-based CNN model. Experiments and the detailed analysis are reported in Sect. 4. Finally, we conclude the paper in Sect. 5.

## 2   Related Work

### 2.1   Content and Behavior Feature Extraction

In feature extraction, there are two kinds of methods: content feature-based and behavior feature-based. Deng et al. [4] proposed the lexical analysis for webshell attacks based on converting the source code syntax into tokens of information. Yang et al. [26] proposed a webshell detection based on the HTTP traffic analysis. Ai et al. [1] proposed a GINI-based method; however, their model has an obvious difference in detection accuracy for different types of files (e.g., .asp and .jsp) even under the same model. Due to the fact that content-based webshell detection is more dependent on syntax, behavior feature is also considered in some works. One direct feature is the connection between existing web files. Wu et al. [25] extracted webshell attack features from raw sequence data in Web logs and proposed a statistical method based on the time interval to identify sessions, using long-term short-term memory model (LSTM) and hidden Markov model (HMM) to learn.

Table 1 lists typical content features, behavior features and corresponding values in our feature extraction (detailed descriptions in Sect. 3). Compared with above related works, our feature extraction integrates both types of features. In particular, we introduce some business-dependent features as indirect features to indicate unexpected behaviors, such as alarm and attack features from business system.

### 2.2   Supervised Learning-Based Webshell Detection

Supervised learning including Naive Bayes (NB), k-Nearest Neighbor (kNN), Support Vector Machine (SVM), Decision Tree (DT) has been proved a success in webshell detection. In recent years, deep learning, is also introduced. Walkowiak et al. [24] proposed a malicious webshell detection method based on word2vec

**Table 1.** Integrated content and behavior features.

| Feature name | Feature value | Feature name | Feature value |
|---|---|---|---|
| Interface | $\{0, 1\}$ | ReqLength | $[0, 1]$ |
| Content_Type | $\{0, 0.5, 1\}$ | D_value | $\{0, 0.5, 1\}$ |
| Accept | $\{0, 1\}$ | Host | $\{0, 1\}$ |
| Alarm | $[0, 1]$ | Method | $\{0, 0.36, 0.88, 1\}$ |
| Attacker | $[0, 1]$ | Code | $\{0, 1\}$ |
| ACL | $\{0, 1\}$ | ResLength | $[0, 1]$ |
| Origin | $\{0, 1\}$ | Structure(i) | $\{0, 1\}$ |
| Referer | $\{0, 1\}$ | Evil_Intentions(i) | $[0, 1]$ |

representation and CNN, which is the first time that CNN has been applied to malicious webshell detection. In their work, word2vec is used to depose HTTP traffic as a fixed-size matrix, and the CNN model is to classify malicious webshells and normal webshells. Zhang et al. [27] proposed access-based webshell traffic detection with deep learning-based of character level functionality. They develop a character-based feature extraction method for continuous content, through obtaining feature vectors as input to CNN and LSTM. The above work shows the effectiveness of CNN.

Compared to fully connected neural networks, the CNN-based webshell detection can reduce computational complexity [17], and focus on local webshell's malicious feature. Although the CNN-based webshell detection works had a good performance, but they use different dataset: Tian et al. [22] simulates the webshell traffic and captures the data by wireshark. Nguyen et al. [15] focuses on php webshell involving Laravel, Wordpress, Joomla, phpMyAdmin, phpPgAdmin, phpbb, and adopts a dataset from github based on a second preprocessing. Jinping et al. [8] adopts a dataset from a security company in Xiamen. While our experiment is based on a real traffic of a bank in China.

In comparison, we redesign the CNN model [6] in two points. First, we build the CNN classifier with two convolutional layers and a global pooling by considering the sparseness of feature values. Such changes belong to a structure improvement. The second one is that we use reinforcement learning for feature selection, connected to CNN input to form a unified training. This is a mechanism-related improvement.

# 3    Our Proposed Method

## 3.1    Framework Overview

As Fig. 1 shows, our method includes four key modules: feature extraction and preparation, A3C-based feature selection, CNN classifier and reward module. The reward module includes reward computing and storage, and reward feedback and parameter updating.

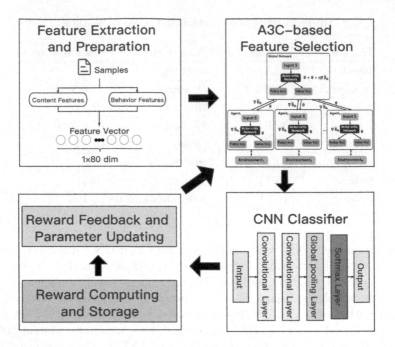

**Fig. 1.** Framework overview

The module of feature extraction and preparation, is responsible for extracting typical features from the traffic flow. We classify features into two categories. One category is the content feature such as source feature, response feature, and others which are captured from specific fields in HTTP request body. For each traffic flow of specific time interval, both content features and behavior features are integrated together to form a feature vector. Thus, for multiple feature vectors, we reorganize them to form a feature matrix by flow time order. A3C-based feature selection module is designed on rebuilding feature matrix through reinforcement learning for automatic feature selection. For feature matrix, we implement a CNN classifier with two convolutional layers and a global pooling, which can achieve the final webshell classification result as reward computation related feedbacks. Since the CNN and A3C network needs a consuming training, the reward computing and storage module is utilized to ensure the stable training. The reward of each classification is computed and stored. Given a specified time setting, we choose an average reward of top-k from agent's replay buffer, and utilize such feedback to update the A3C model parameter. This is an interactive process until the A3C is converged. Finally, our method can find out the optimal feature composition of feature space and improve the CNN-based webshell detection.

## 3.2  Feature Extraction and Preparation

**Feature Vector Extraction:** Our extracted features from traffic flow include types of interface basic feature(4 features), traffic source feature(6 features), traffic request feature(14 features), traffic response feature(16 features) and content feature(40 features). Among them, the basic feature of the interface describes the security feature of the target interface itself. The traffic source feature reflects the legitimacy of the requested source. Traffic request feature and traffic response feature are the features expressed in this HTTP request message. Content feature is the feature of a specific field in the HTTP request body, which is described by the field structure and field content in the body. In total, we have 80 features in the initial feature space to form the feature vector as follows.

$$[Interface, Content\_type, Accept, ACL, Origin, Referer, Alarm,$$
$$Attacker, ReqLength, D\_value, Host, Method, Code, ResLength, \tag{1}$$
$$Structure(1), EvilIntentions(1), ..., Structure(20), EvilIntentions(20)]$$

In the above vector, we choose 14 typical ones and explain them as follows.

**Table 2.** Typical features and explanations.

| Feature name | Feature explanation | Value explanation |
| --- | --- | --- |
| Interface | Describes interface security status | 0: BI API, 1: System API |
| Content_Type | Characterize the type of entity of requester | 0.5: audio/*, 1: text/* 0: application/* |
| Accept | Content type of HTTP | 1: not exists accept field 0: exists accept field |
| Alarm | Confidence-based alarm | Alarm confidence coefficient 1: max, 0: min |
| Attacker | Attacker label | User confidence coefficient 1: max, 0: min |
| ACL | Access control flag | 1: IP legal, 0: IP illegal |
| Origin | Distinguish authority of request | 1:Origin legal, 0:Origin illegal |
| Referer | Determining the url status | 1:Referer legal,0:Referer illegal |
| D_value | Difference of content type | Difference of content type 1: max, 0: min |
| Host | Determining the request whether from host | 1: access website by domain 0: access website by IP |
| Method | Request method | 1: PUT/DELETE 0.36: OPTIONS, 0: GET/POST, 0.88: Others |
| Code | Response code | 1: 2XX/3XX/4XX, 0: Others |
| Structure(i) | Content element in HTTP body | 1: reasonable structure 0: unreasonable structure |
| Evil_Intentions(i) | Corresponding behavior status of structure (i) | Malicious behaviors confidence coefficient, 1: max, 0: min |

**Feature Matrix Generation:**

$$M_{80 \times 80} = \begin{pmatrix} Interface_1 & \cdots & Structure_1(20) & Evil\_Intentions_1(20) \\ Interface_2 & \cdots & Structure_2(20) & Evil\_Intentions_2(20) \\ \vdots & \vdots & \ddots & \vdots \\ Interface_{80} & \cdots & Structure_{80}(20) & Evil\_Intentions_{80}(20) \end{pmatrix} \tag{2}$$

We capture 80 associated traffic flow sequences. For each flow, there are 80 features. Thus, the matrix size is $80 \times 80$ (See Formula 2). In our experiment, a total of 568,936 data streams were processed, the calculation is that: we form a matrix every 5 min, and if there are not enough 80 traffics, we will use mean imputation method to impute up to 80 traffic, we finally obtain 12,876 features matrices.

### 3.3 A3C-Based Feature Selection

**A3C Structure:** A3C utilizes a hierarchical actor-critic framework and enables asynchronous parallel training [14]. Concretely, at the bottom layer, there are multiple agents in parallel to interact with the environment, and each agent has a pair of actor network and critic network, which is also called the local actor-critic network; all agents further connect to a global actor-critic network at an upper layer. For simplicity, we still employ a common two-layer structure to discuss in the paper. Each $agent_i$ copies the global parameters $\boldsymbol{\theta}$ before learning, then $agent_i$ interacts with the $environment_i$ for sampling diverse data. Each $agent_i$ will compute the gradient $\nabla \bar{R}_\theta$ after interacting, then each $agent_i$ will push the gradient to update the parameters of global network $\boldsymbol{\theta} \leftarrow \boldsymbol{\theta} + \eta \nabla \bar{R}_\theta$, where $\eta$ is the learning rate.

In the local actor network, the actor network is responsible for obtaining an approximate state-action value function. While the local critic network is to supervise the learning of the local actor network with the state value function, the global actor-critic network does not train itself but manages cumulated updates and then cooperates with local actor-critic networks for improving training efficiency. A3C is to learn a policy $\pi$, whose input is the observation of agents(also called actors in A3C) represented as a state $a$, and its output is a vector of action probability for any state. Then, the probability to choose $a_t$ at $t$ time, can use the policy $\pi(a_t|s_t) = p(a_t|s_t)$.

To learn a good policy $\pi_\theta$, it is natural to update the $\boldsymbol{\theta}$ to maximize the expected total reward of all sampled trajectories, namely $\bar{R}_\theta = \sum_\tau R(\tau) p_{\theta(\tau)}$, thus a gradient ascend can be used $\boldsymbol{\theta} \leftarrow \boldsymbol{\theta} + \eta \nabla \bar{R}_\theta$. In A3C, the advantage function is defined as: $r_t^n + V_\pi(s_{t+1})^n - V_\pi(s_t^n)$, and the advantage function-based gradient is $\nabla \bar{R}_\theta \approx \frac{1}{N} \sum_{n=1}^N \sum_{t=1}^{T_n} (r_t^n + V_\pi(s_{t+1})^n - V_\pi(s_t^n)) \nabla log p_\theta(a_t^n|s_t^n)$.

**Action Space and Reward:** The action space in the proposed framework can be defined as:

$$a = \{a \in N \cap a \leq |F|\} \tag{3}$$

where $N$ represents all the alternative original features to be selected and a termination state to be selected. $|F|$ represent the index of the action which selects the final terminal action. When $a < |F|$, action a means to set the element with index $a$ in the environment state to 1. When $a = |F|$, action $a$ means that the agent stops the selection procedure. The reward of A3C is the detection accuracy of classifier.

## 3.4    CNN Model Construction with Two Convolutional Layer

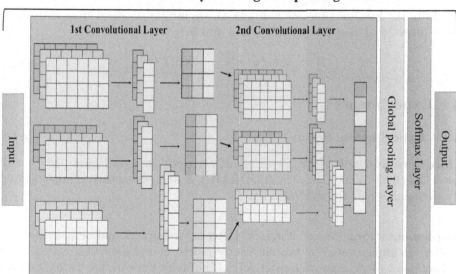

**Fig. 2.** CNN model construction

As depicted in Fig. 2, the network contains four layers. The first two are convolutional and the remaining are global pooling layer and softmax layer. The first convolutional layer filters the $80 \times 80 \times 1$ input feature with 256 kernels of different sizes $(3 \times 80 \times 1, 4 \times 80 \times 1, 5 \times 80 \times 1)$ with a stride of 1 pixels (this is the distance between the receptive field centers of neighboring neurons in a feature map). We stitch the output of the first convolutional layer into a matrix as the second convolutional layer input. In the second Convolutional layer, each feature map filters the input with 256 kernels of different sizes$(3 \times 256 \times 1, 4 \times 256 \times 1, 5 \times 256 \times 1)$, then gets a $225 \times 1 \times 256$ feature map by stitching the output feature map. Global pooling layer summarizes the outputs of neighboring groups of neurons in the same kernel map and we use max pooling and get a $256 \times 1 \times 1$.

The softmax layer takes a vector $z$ of K real numbers as input, and normalizes it into a probability distribution consisting of K probabilities proportional to the exponentials of the input numbers. That is, prior to applying the softmax, some vector components could be negative, or greater than one; and might not sum to 1; but after applying the softmax, each component will be in the interval (0,1), and the components will add up to 1, so that they can be interpreted as probabilities. Furthermore, the larger input components will correspond to larger probabilities. In this work, we employ the standard softmax function $\sigma :$ $\mathbb{R}^K \to [0,1]^K$ as follows:

$$\sigma(\mathbf{z})_i = \frac{e^{z_i}}{\sum_{j=1}^{K} e^{z_j}} \quad \text{for } i = 1, \ldots, K \text{ and } \mathbf{z} = (z_1, \ldots, z_K) \in \mathbb{R}^K \quad (4)$$

As to the loss function (see Formula 4), we use the classical cross entropy to get a measure of dissimilarity between $y$ and $\hat{y}$.

$$H = -y \log \hat{y} - (1 - y) \log(1 - \hat{y}) \quad (5)$$

## 4  Experiments

### 4.1  Experimental Setup

**Hardware and Software:** Experiments are performed on Ubuntu 16.04 with Intel Gold 6240 @2.60 GHz, 32 GB DDR4 RAM, and our method is implemented by python 3.8 and TensorFlow 2.4.1.

**Implementation:** We use 3 agents. Both actor-learners and actor-critic controller are implemented by two differently paired neural networks( called *Actor* network and *Critic* network respectively). The *Actor* network has $2 \times 200 \times 4$ neurons from the input layer to output layer, and the *Critic* network has $2 \times 100 \times 1$ neurons.

We also set the parameters as follows: the learning rate $LR$ is set to be 0.001, the regularization parameter $\delta = 0.001$, the value function accumulate parameter $\gamma = 0.9$ and the bias $b = 0$. We initialize the convolution kernels of *Actor* and *Critic* neural network with *random_normal_initializer(mean* $= 0.0, stddev =$ 0.1), where *mean* refers to the average value and *stddev* refers to the standard deviation.

**Dataset:** Table 3 shows the source of the data traffic and the relevant information. The dataset uses 4 types of traffic totaling 568,936 pieces, occupying 5.26 TB of storage space. The traffic used in the experiment is divided into two parts, one part comes from the real business traffic of a banking system (including real attack traffic), and the other part comes from the artificially generated attack traffic by the bank researchers through tools, manuals, etc.

**Table 3.** Traffic source information.

| Number | System sevel | Data amount | Number of positive sample | Number of negative samplc |
|--------|-------------|-------------|---------------------------|---------------------------|
| A1 - A2 | Business systems | 93,221 | 2,857 | 90,364 |
| B1 - B3 | Promotion system | 86,894 | 5,240 | 81,654 |
| C1 - C3 | Internal support system | 178,908 | 32,252 | 146,656 |
| D1 | Test system | 95,424 | 53,045 | 42,379 |
| D2 | Honeypot system | 114,489 | 81,287 | 33,202 |

Among them, the real business traffic part is the monthly traffic of a certain bank. During the busy time period of the banking business, a fixed-size time window is randomly selected for traffic acquisition (the time window size is set to 1 h). One time window is selected every day, and 30 days of traffic data are accumulated as the real business traffic. The part of the artificially generated attack traffic is obtained by professionals using AWVS, Weblogic batch utilization tools, Metspoit tools and other tools to attack the system.

We extracted the traffic feature by full traffic equipment, then we formed a total of 568,936 data traffic into 12,876 traffic feature matrices, of which 4,190 were positive samples (webshell attack samples), and 8,686 were negative samples (non-webshell attack samples), involving 2,746 users (IP addresses). There were 1,125 users involved in webshell attacks and 1,621 normal users.

**Table 4.** Evaluation metrics.

| Metrics name | Definition |
|--------------|------------|
| Hit rate(HR) | TP/(TP+FP) |
| False positive rate (FPR) | TP/(TP+FN) |
| ACC | (TN+TP)/(TN+FN+TP+FP) |
| AUC | Area under the ROC curve |

In order to reduce the effect of test set and training set on the result, we use the $K$-fold cross-validation (CV) method [5]. The whole data will be divided into $k$ parts. Each data will take turns as a test set and a validation set and do not repeat each time one of them as a test set, with other $k - 1$ parts to do the training set, and then calculate the model on the test set.

**Evaluation Metrics:** Webshell detection is a two-classification problem. In this paper, hit rate (HR), false positive rate (FPR), ACC and AUC are selected as evaluation metrics (see Table 4).

## 4.2    Experimental Results

**Detection Evaluation:** In the experiment, we used the grid search to traverse our proposed model. Table 5 shows the results of the grid search (fill mode V-Valid, S-Same; activation function R-RELU, T-Tanh). From the ACC, HR, AUC and other aspects, the number of 5 hyperparameter group performance is the best (ACC is 96%, HR reaches 96%, AUC is 96.7%, while FPR can be as low as 6%).

**Table 5.** Grid search-based parameter influence. (Bold:best.)

| Number | Padding | Activation function | Kernel num | Kernel size | Learning rate | ACC | AUC | HR | FPR |
|--------|---------|---------------------|------------|-------------|---------------|------|------|------|------|
| 1 | V | R | 256 | 3,4,5 | 1e-3 | 94.6% | 94.1% | 94% | 6.2% |
| 2 | V | R | 256 | 3,4,5 | 1e-2 | 94.8% | 94.3% | 95% | 6.1% |
| 3 | V | R | 256 | 4,5,6 | 1e-3 | 92.9% | 92.7% | 92% | 6.3% |
| 4 | V | R | 256 | 4,5,6 | 1e-2 | 91.9% | 91.1% | 92% | 14.1% |
| 5 | S | R | 256 | 3,4,5 | 1e-3 | **96.0%** | **96.7%** | **96%** | 6.0% |
| 6 | S | R | 256 | 3,4,5 | 1e-2 | 94.0% | 94.0% | 94% | 4.2% |
| 7 | S | R | 256 | 4,5,6 | 1e-3 | 94.5% | 94.0% | 94% | 6.2% |
| 8 | S | R | 256 | 4,5,6 | 1e-2 | 93.2% | 91.8% | 95% | 15.5% |
| 9 | V | T | 256 | 3,4,5 | 1e-3 | 94.2% | 93.8% | 93% | 6.3% |
| 10 | V | T | 256 | 3,4,5 | 1e-2 | 95.2% | 94.8% | 95% | **4.1%** |
| 11 | V | T | 256 | 4,5,6 | 1e-3 | 95.1% | 94.5% | 95% | 6.1% |
| 12 | V | T | 256 | 4,5,6 | 1e-2 | 95.6% | 94.6% | 95% | 6.1% |
| 13 | S | T | 256 | 3,4,5 | 1e-3 | 94.5% | 94.0% | 94% | 6.1% |
| 14 | S | T | 256 | 3,4,5 | 1e-2 | 94.0% | 93.6% | 95% | 10.0% |
| 15 | S | T | 256 | 4,5,6 | 1e-3 | 93.0% | 91.7% | 95% | 17.3% |
| 16 | S | T | 256 | 4,5,6 | 1e-2 | 93.2% | 91.8% | 95% | 15.5% |

**Robustness Analysis:** In order to test the adversarial effect of the model presented in this paper, we have conducted the cross-language testing. We select three different types of language honeypots as a breakthrough in file upload vulnerability, three different types of language honeypots to attack, to achieve Trojan upload. In this experiment, in order to facilitate the processing of traffic, we intercept a portion of the continuous attack traffic in each type of attack as a detection sample, and if the model could detect abnormal traffic, it was considered a hit.

As Table 6 shows, when model presented in this paper are in the face of different Trojan horses [25], which Big Trojan has a large size and comprehensive functions for command execution, database operations, and other malicious intentions. Small Trojan is small and easy to hide but generally only has an upload function. We can see that the test results are stable. In addition, due to the fact that the same Trojans differ greatly in text feature between different languages, content-based traffic detection devices usually enable different detection loads to detect the same piece of text multiple times, and the behavior-based detection model proposed in this paper can detect the traffic of multiple slice

**Table 6.** Detection analysis under different Trojan type.

| Number | Trojan type | Language | Number of Webshell traffic | Our method HR |
|---|---|---|---|---|
| 1 | Small Trojan | Java | 800 | 95.5% |
| 2 | Big Trojan | Java | 800 | 97.3% |
| 3 | One word Trojan | Java | 800 | 92.5% |
| 4 | Photo Trojan | Java | 80 | 77.0% |
| 5 | Small Trojan | PHP | 400 | 97.3% |
| 6 | Big Trojan | PHP | 400 | 97.3% |
| 7 | One word Trojan | PHP | 400 | 92.5% |
| 8 | Photo Trojan | PHP | 80 | 77.0% |
| 9 | Small Trojan | ASP.NET | 240 | 95.3% |
| 10 | Big Trojan | ASP.NET | 240 | 97.3% |
| 11 | One word Trojan | ASP.NET | 240 | 92.9% |
| 12 | Photo Trojan | ASP.NET | 80 | 77.0% |

windows at the same time; thus, the processing efficiency is significantly higher than the content-based detection mechanism.

**Table 7.** Comparison between our method and commercial detection device.

| Detection model | Trojan file type | Language | Small Trojan HR | Big Trojan HR | One word Trojan HR | Photo Trojan HR |
|---|---|---|---|---|---|---|
| Our method | Original Trojan file | Java | 95.5% | 97.3% | 92.5% | 77% |
| | Confusing Trojan files | Java | 98.9% | 100% | 95.8% | 81% |
| A commercial APT | Original Trojan file | Java | 100% | 100% | 100% | 83.3% |
| | Confusing Trojan files | Java | 88% | 56% | 92% | 0% |
| A commercial WAF | Original Trojan file | Java | 95.9% | 97.5% | 100% | 0% |
| | Confusing Trojan files | Java | 0% | 0% | 0% | 0% |

In order to test the detection effect of the model presented in this paper on different forms of webshell, we construct different forms of webshell based on four common java Trojan types. Communication encryption uses AES256-bit fixed passwords, code confusion is confused using open source confusion tools, code encoding is encoded using Moss encoding, and the experimental results are shown in Table 7. We can see that, the detection method of this paper is still effective even when coding, encryption and other bypass mechanisms appear. Our method experimental results are worse than the commercial APT on the original Trojan File, but better than them to some extent when there are obfuscated samples. Such commercial APT analysis system is well developed based on a large number of expert rules for real applications. While in the scenario

**Table 8.** Comparison between single-layer CNN and two-layer CNN.

| Number | ACC of Single-layer CNN | ACC of Two-layer CNN | AUC of Single-layer CNN | AUC of Two-layer CNN |
|--------|-------------------------|----------------------|-------------------------|----------------------|
| A0 | 57% | 60% | 27% | 42% |
| A1 | 92% | 92% | 86% | 95% |
| A2 | 92% | 93% | 90% | 94% |
| A3 | 55% | 63% | 29% | 66% |
| A4 | 91% | 93% | 82% | 95% |
| A5 | 91% | 96% | 83% | 94% |
| A6 | 62% | 61% | 47% | 64% |
| A7 | 91% | 93% | 85% | 95% |

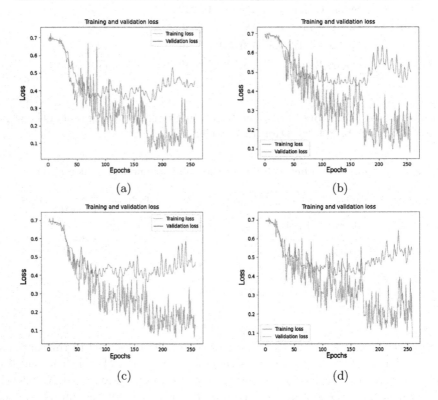

(a)　　　　　　　　　　　　　　(b)

(c)　　　　　　　　　　　　　　(d)

**Fig. 3.** Loss analysis under different hyperparameters when disable RL. (a) Kernel_size = $\{3, 4, 5\}$, Learning Rate = 1e-3; (b) Kernel_size = $\{4, 5, 6\}$, Learning Rate = 1e-3; (c) Kernel_size = $\{3, 4, 5\}$, Learning Rate = 1e-2; (d) Kernel_size = $\{4, 5, 6\}$, Learning Rate = 1e-2

with obfuscated samples, the static rules and corresponding rule-based methods are not always effective.

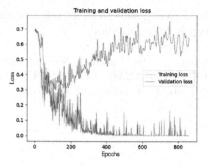 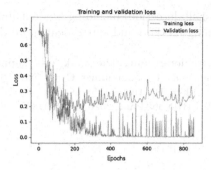

**Fig. 4.** Loss analysis comparison between our method without RL (a) and with RL (b).

**Parameters and Efficiency Analysis:** To compare the differences between the two-layer convolution global pooling CNN model and the single-layer convolution maximum pooling CNN model, we trained the two models to configure the same hyperparametrics. Table 8 shows the comparison. We find that the two-layer convolution global pooling CNN model is better than the single-layer convolution maximum pooling model.

From the Fig. 3 we can see that, without RL, even given some optimized hyperparameters with grid search, the performance dramatically decreases, even drop average 40% loss on validation set in the fourth subfigure. Moreover, the RL helps to reduce features in our experiment from 80 features down to 54 features, with optimal ACC 96%. The filtered features include "x-cache", "general referrer policy", "cache-control" etc.

From the Fig. 4 we can see that, the loss of our method without RL dramatically increases on validation dataset after 200 epochs. As a comparison, in our method with RL on the validation set, the loss can be reduced to about 25%.

## 5    Conclusion

In this work, we propose a novel approach for improving the convolutional neural network-based webshell detection through Reinforcement Learning. To the best of our knowledge, no similar work has focused on. Especially, we utilize the RL algorithm of A3C for automatic feature selection. In our CNN model, two convolutional layers and a global pooling are built. Through extensive experiment, we find that our method can reduce 32.5% features and can achieve 96% accuracy. Compared with single CNN without reinforcement learning, our method has an average 2.42% improvement.

This work serves as a first step to further investigate the A3C-based CNN for webshell detection. It is expected to inspire a series of follow-up studies for webshell detection, including but not limited to experiment more reinforcement learning method and experiment more deep learning method.

**Acknowledgment.** The work was supported by the National Natural Science Foundation of China under Grant Nos. 61972025, 61802389, 61672092, U1811264, and 61966009, the National Key R&D Program of China under Grant Nos. 2020YFB1005604 and 2020YFB2103802.

# References

1. Ai, Z., Luktarhan, N., Zhao, Y., Tang, C.: Ws-lsmr: malicious webshell detection algorithm based on ensemble learning. IEEE Access **8**, 75785–75797 (2020)
2. Ben-Porat, U., Bremler-Barr, A., Levy, H.: Vulnerability of network mechanisms to sophisticated ddos attacks. IEEE Trans. Comput. **62**(5), 1031–1043 (2012)
3. Bergeron, J., Debbabi, M., Desharnais, J., Erhioui, M.M., Lavoie, Y., Tawbi, N., et al.: Static detection of malicious code in executable programs. Int. J. Req. Eng. **2001**(184–189), 79 (2001)
4. Deng, L.Y., Lee, D.L., Chen, Y.H., Yann, L.X.: Lexical analysis for the webshell attacks. In: 2016 International Symposium on Computer, Consumer and Control (IS3C), pp. 579–582. IEEE (2016)
5. Fushiki, T.: Estimation of prediction error by using k-fold cross-validation. Stat. Comput. **21**(2), 137–146 (2011)
6. Gong, L., Ji, R.: What does a textcnn learn? arXiv preprint arXiv:1801.06287 (2018)
7. Haq, T., Zhai, J., Pidathala, V.K.: Advanced persistent threat (apt) detection center (Apr 18 2017), uS Patent 9,628,507
8. Jinping, L., Zhi, T., Jian, M., Zhiling, G., Jiemin, Z.: Mixed-models method based on machine learning in detecting webshell attack. In: Proceedings of the 2020 International Conference on Computers, Information Processing and Advanced Education, pp. 251–259 (2020)
9. Kang, W., Zhong, S., Chen, K., Lai, J., Xu, G.: RF-AdaCost: webshell detection method that combines statistical features and opcode. In: Xu, G., Liang, K., Su, C. (eds.) FCS 2020. CCIS, vol. 1286, pp. 667–682. Springer, Singapore (2020). https://doi.org/10.1007/978-981-15-9739-8_49
10. Kim, J., Yoo, D.H., Jang, H., Jeong, K.: Webshark 1.0: A benchmark collection for malicious web shell detection. JIPS **11**(2), 229–238 (2015)
11. Le, V.-G., Nguyen, H.-T., Lu, D.-N., Nguyen, N.-H.: A solution for automatically malicious web shell and web application vulnerability detection. In: Nguyen, N.-T., Manolopoulos, Y., Iliadis, L., Trawiński, B. (eds.) ICCCI 2016. LNCS (LNAI), vol. 9875, pp. 367–378. Springer, Cham (2016). https://doi.org/10.1007/978-3-319-45243-2_34
12. Matsuda, W., Fujimoto, M., Mitsunaga, T.: Real-time detection system against malicious tools by monitoring dll on client computers. In: 2019 IEEE Conference on Application, Information and Network Security (AINS), pp. 36–41. IEEE (2019)
13. Mingkun, X., Xi, C., Yan, H.: Design of software to search asp web shell. Procedia Eng. **29**, 123–127 (2012)
14. Mnih, V., et al.: Asynchronous methods for deep reinforcement learning. In: International conference on machine learning, pp. 1928–1937. PMLR (2016)

15. Nguyen, N.H., Le, V.H., Phung, V.O., Du, P.H.: Toward a deep learning approach for detecting php webshell. In: Proceedings of the Tenth International Symposium on Information and Communication Technology, pp. 514–521 (2019)

16. Qi, L., Kong, R., Lu, Y., Zhuang, H.: An end-to-end detection method for webshell with deep learning. In: 2018 Eighth International Conference on Instrumentation & Measurement, Computer, Communication and Control (IMCCC), pp. 660–665. IEEE (2018)

17. Qin, X., Peng, S., Yang, X., Yao, Y.D.: Deep learning based channel code recognition using textcnn. In: 2019 IEEE International Symposium on Dynamic Spectrum Access Networks (DySPAN), pp. 1–5. IEEE (2019)

18. Salois, M., Charpentier, R.: Dynamic detection of malicious code in cots software. Technical Report, DEFENCE RESEARCH ESTABLISHMENT VALCARTIER (QUEBEC) (2000)

19. Sun, X., Ma, X., Ni, Z., Bian, L.: A new lSTM network model combining TextCNN. In: Cheng, L., Leung, A.C.S., Ozawa, S. (eds.) ICONIP 2018. LNCS, vol. 11301, pp. 416–424. Springer, Cham (2018). https://doi.org/10.1007/978-3-030-04167-0_38

20. Sun, X., Lu, X., Dai, H.: A matrix decomposition based webshell detection method. In: Proceedings of the 2017 International Conference on Cryptography, Security and Privacy, pp. 66–70 (2017)

21. Šuteva, N., Mileva, A., Loleski, M.: Computer forensic analisys of some web attacks. In: World Congress on Internet Security (WorldCIS-2014), pp. 42–47. IEEE (2014)

22. Tian, Y., Wang, J., Zhou, Z., Zhou, S.: Cnn-webshell: malicious web shell detection with convolutional neural network. In: Proceedings of the 2017 VI International Conference on Network, Communication and Computing, pp. 75–79 (2017)

23. Tianmin, G., Jiemin, Z., Jian, M.: Research on webshell detection method based on machine learning. In: 2019 3rd International Conference on Electronic Information Technology and Computer Engineering (EITCE), pp. 1391–1394. IEEE (2019)

24. Walkowiak, T., Datko, S., Maciejewski, H.: Bag-of-words, bag-of-topics and word-to-vec based subject classification of text documents in polish - a comparative study. In: Zamojski, W., Mazurkiewicz, J., Sugier, J., Walkowiak, T., Kacprzyk, J. (eds.) DepCoS-RELCOMEX 2018. AISC, vol. 761, pp. 526–535. Springer, Cham (2019). https://doi.org/10.1007/978-3-319-91446-6_49

25. Wu, Y., Sun, Y., Huang, C., Jia, P., Liu, L.: Session-based webshell detection using machine learning in web logs. Secur. Commun. Netw. **2019**, 11 p. (2019). Article ID 3093809. https://doi.org/10.1155/2019/3093809

26. Yang, W., Sun, B., Cui, B.: A webshell detection technology based on HTTP traffic analysis. In: Barolli, L., Xhafa, F., Javaid, N., Enokido, T. (eds.) IMIS 2018. AISC, vol. 773, pp. 336–342. Springer, Cham (2019). https://doi.org/10.1007/978-3-319-93554-6_31

27. Zhang, H., et al.: Webshell traffic detection with character-level features based on deep learning. IEEE Access **6**, 75268–75277 (2018)

28. Zhongzheng, X., Luktarhan, N.: Webshell detection with byte-level features based on deep learning. J. Intell. Fuzzy Syst. (Preprint) **40**(1), 1585–1596 (2021)

# Exploring the Security Issues of Trusted CA Certificate Management

Yanduo Fu[1,2,3], Qiongxiao Wang[1,2]([✉]), Jingqiang Lin[4,5], Aozhuo Sun[1,2], and Linli Lu[1,3]

[1] State Key Laboratory of Information Security, Institute of Information Engineering, CAS,Beijing 100089, China
`wangqiongxiao@iie.ac.cn`
[2] School of Cyber Security, University of Chinese Academy of Sciences, Beijing 100089, China
[3] Data Assurance and Communication Security Research Center, CAS, Beijing, China
[4] School of Cyber Security, University of Science and Technology of China, Hefei 230027, Anhui, China
[5] Beijing Institute, University of Science and Technology of China, Beijing, China

**Abstract.** Public Key Infrastructure (PKI) is widely used in security protocols, and the root certification authority (CA) plays a role as the trust anchor of PKI. However, as researches show, not all root CAs are trustworthy and malicious CAs might issue fraudulent certificates, which can cause Man-in-the-Middle attacks and eavesdropping attacks. Besides, massive CAs and CA certificates make it hard for users to manage the CA certificates by themselves. Though PKI applications generally provide the implementation of trusted CA certificate management (called CA manager in this paper) to store, manage, and verify CA certificates, security incidents still exist, and a malicious CA certificate can damage the entire security. This work explores the security issues of CA managers for three popular operating systems and eight applications installed on them. We make a systematic analysis of the CA managers, such as the modification of the certificate trust list, the source of trust, and the security check of the CA certificates, and propose the functionalities that a CA manager should have. Our work shows that all CA managers we analyzed have security issues, e.g., silent addition of CA certificates, inefficient validation on CA certificates, which will result in insecure CA certificates being falsely trusted. We also make some suggestions on the security enhancement for CA managers.

**Keywords:** Certification authority · Public key infrastructure · CA certificate management

## 1 Introduction

Public key infrastructure (PKI) plays a critical role in secure networking, offering security functionalities such as confidentiality, data integrity, and authen-

© Springer Nature Switzerland AG 2021
D. Gao et al. (Eds.): ICICS 2021, LNCS 12918, pp. 384–401, 2021.
https://doi.org/10.1007/978-3-030-86890-1_22

tication. Well-known applications based on PKI include but are not limited to TLS, HTTPS, code signing, eSign documents, S/MIME, and OpenID Connect. Research [15] shows that as of March 2020, more than 60% of the top one million websites in *Alexa* have used HTTPS, and the number of PKI deployments is still growing.

The security of a PKI system is based on a secure and trustworthy root certification authority (CA). Authoritative organizations such as WebTrust [30] will audit public CAs to ensure their legitimacy and security. However, not all CAs (including public CAs and unaudited self-built CAs) are trustworthy. Recent researches and events have shown that due to malicious attacks or misbehaviors, CAs might issue fraudulent certificates [10,12], which may cause Man-in-the-Middle (MITM) attacks or phishing attacks to the users. The common way to realize a PKI system is that users need to choose which CA can be trusted. A user can trust and accept an end-entity certificate based on the trust of the root CA. While if a root CA is not trusted by the verifier, any certificate issued by this CA cannot be trusted.

Since there are a large number of CA companies and the quantity of root *CA certificates* is much larger, users even with professional knowledge cannot identify the trustiness of each CA certificate by themselves [6,7,9]. According to statistics from *Censys* [21], there have been more than 68 million self-signed CA certificates, and 88% are unexpired. Many applications, such as Windows, macOS, Firefox, and Acrobat, have integrated with *trusted CA certificate management* to help the users with certificate verification and storage. Some of them also offer preset Certificate Trust List (CTL), which usually contains the globally trusted root CA certificates and some platform/application specified root CA certificates. When a root CA certificate is added to the *local CTL*, all the certificates issued by it can be accepted.

Previous researches have disclosed the security incidents caused by the improperly *preset CTL*. *Lenovo* shipped some laptops with a pre-installed traffic scanning software called *Superfish* [3] in 2014, which installed a CA certificate and then actually injected advertisements to even encrypted web pages through a MITM attack. The same private key across laptops made things worse as a third-party entity could interpret or modify encrypted traffic without triggering any security warnings. Besides the preset CTL, if a trusted CA certificate can be added to local CTL arbitrarily, the security problem mentioned above still exists [1,5,8,32].

In this paper, we focus on the implementation of trusted CA certificate management integrated with operating systems (OSs) and applications to explore the security issues of trusted CA certificate management. For simplicity, we refer to the implementation of trusted CA certificate management as a *"CA manager,"* which focuses on the local management of CA certificates such as the modification of the CTL, the storage of local certificate files, and the security check of CA certificates. Our work is accomplished by exploring the CA managers on three OSs (i.e., Windows, macOS, and Linux) and eight applications (i.e., seven Web Browsers and Adobe Acrobat) installed on these OSs, which are carefully chosen according to the market share and security features. We find that all the CA managers we studied have some security issues. For example, we find that there is no requirement for explicit participation of the user (e.g., inputting a password) while importing a CA certificate to the local CTL of some CA managers (e.g., Windows), which means malicious CA certificates can be installed silently and the attackers can even accomplish the addition by replacing the CTL files with a file including malicious CA certificates. Besides, some CA managers have a weak verification on the certificate security and key usage (or certificate purpose) of the CA certificate, and some CA managers (e.g., Firefox) do not verify the complete certificate chain when trusting an intermediate CA certificate.

The main contributions are the followings:

1) We investigate the most popular desktop OSs and PKI applications (e.g., web browsers) for studying the implementation and the use of CA managers. We propose the functionalities a CA manager should have.
2) We conduct a comprehensive test on different CA managers from the perspective of the source of trust, modification to the local CTL, control of the certificate purpose, and security check of CA certificates. Several security issues are disclosed and described in this paper.
3) Based on the security issues we found, we make some suggestions to enhance the security of the CA managers.

The rest of this paper is organized as follows. Related work is introduced in Sect. 2. In Sect. 3 we explore the CA managers in the wild, including the source of trust and the functionalities. Section 4 specifies our method for analyzing the CA managers and reports the security issues we found. We give our suggestions in Sect. 5 and offer the conclusions of the paper in Sect. 6.

## 2   Related Work

In recent years, CA managers have been analyzed extensively. Some researches have exposed many security issues of the CA managers, and several organizations also develop a *root store* (or *root program*), which contains the preset CTL. In the meantime, there have been some schemes proposed for specific situations.

**Schemes for CA Manager.** Many well-known organizations (such as Microsoft [25], Mozilla [27], Apple [16], and Adobe [14]) maintain their CTLs and root stores for global users. A CA intending to be included in these lists must comply with the baseline requirements of the CA/Browser Forum [19,20]. Besides, it should be audited by European Telecommunications Standards Institute (ETSI) [22] or WebTrust [30] to ensure its security and legality. These CA certificates will be installed by default in the OSs and applications and are considered secure and trustworthy enough to be used to provide users with security services.

Besides, due to the arbitrary addition of CA certificates and the lack of security knowledge of users, some researchers have introduced some new technologies for improvements based on the current CA managers. Li et al. [11] propose a locally-centralized CA certificate management solution named *vCertGuard* in private cloud environments combined with desktop virtualization technology characteristics that realize centralized CA management by a professional administrator based on the granularity of trust. CA-TMS [4] proposes a CA reputation evaluation system based on a computational trust model, which determines the number of trusted CA certificates and decision-making rules by learning the browser's certificate-related parameters and the users' behavior habits.

**Terrible Situations of CA Manager.** Some researches have revealed the terrible situations of the current CA managers. Perl et al. [13] surveyed the local CTL of eleven OSs or applications, finding that 34% CAs in the root store were not used to sign HTTPS server certificates and can be removed to reduce the attack interface without triggering any security warnings in browsers. Vallina-Rodriguez et al. [32] analyzed the CTLs of thousands of Android devices installed by hardware vendors, mobile operators, and Android OS, finding that the Android CAs have no distinction between trust levels and no restrictions on the purpose of the certificate and the rooted applications might install malicious CA certificates without any barriers. Besides, it was found that some manufacturers [2,3] would pre-install some insecure CA certificates, which caused severe security problems. Malicious CA certificates can also be imported by audio drivers [31], antivirus and parental-control software [5], and programming [1].

These works have made a huge contribution to the security of CA managers. However, there is no study analyzing the specific implementation of current CA managers, and we focus on this part, exploring the security issues of CA managers.

## 3   CA Manager in the Wild

This section specifies the studied target objects and then elaborates their actual situation in the wild from two perspectives: source of trust and functionality of the CA manager.

### 3.1   Target Object

Our research is based on the most mainstream desktop OSs, including Windows, Linux, and macOS. The specific version is Windows 10 Professional, Ubuntu 16.04 with Linux kernel 4.15.0, and macOS High Sierra 10.13.6.

*OS.* The OS generally maintains a *system-level* local CTL shared by all accounts in the same machine, and each account maintains a *user-level* local CTL, which will not affect other users in this machine. We treat them as two different CTLs for analysis in the following. Note that two system-level local CTL exists in macOS. One CTL stores the system-level CA certificates which can be modified at any time by users, and the other CTL stores the original CA certificates maintained by Apple, which is referred to as *"root-level" CTL* in this paper. For Ubuntu, it stores the system-level CTL in the system directory (i.e., */usr/share/ca-certificates/*), and the CTL, which is stored in the *.pki* folder under the user directory, is considered as user-level CTL in this paper.

*Application.* We investigate several PKI applications containing seven desktop browsers and one PDF reader named Adobe Acrobat DC (Acrobat). The seven browsers consist of the top six desktop browsers (i.e., Chrome, Safari, Edge, Firefox, Opera, and IE), which are ranked by market share in *StatCounter* [29] during the last 12 months, and the Tor browser (Tor), which is known for its open-source, anonymity and privacy technologies. Table 1 shows the specific version of applications. It is worth noting that two different installation methods in Ubuntu need to be considered: (*a*) the *apt*, which gets the applications from the Ubuntu repositories; and (*b*) the *snap*, which is cross-platform, dependency-free, and more secure than *apt*.

We also conduct experiments on Windows 7 Ultimate and CentOS 8.3 with Linux kernel 4.18.0. Windows 7 has similar experiment phenomena with Windows 10. CentOS is different from Ubuntu in the addition and storage location of the system-level CTL. CentOS stores the CTLs in */usr/share/pki/ca-trust-source* and */etc./pki/ca-trust/source*. Many browsers including Firefox will trust the certificates in the two folders, while the Tor browser and the applications installed by *snap* maintain their own CTLs. Overall, CentOS is similar to Ubuntu in terms of addition, storage security, and verification of certificate security. The following mainly introduces Windows 10, macOS 10.13.6, and Ubuntu 16.04.

**Table 1.** The version number of applications studied

| OS<br>App | Windows | macOS | Linux | |
| --- | --- | --- | --- | --- |
| | | | Via *snap* | Via *apt* |
| Chrome | 87.0.4280.88 | 87.0.4280.88 | ⊘ | 87.0.4280.88 |
| Safari | ⊘ | 13.1.2 | ⊘ | ⊘ |
| Edge | 87.0.664.55 | 87.0.664.60 | ⊘ | 89.0.723.0 |
| Firefox | 83 | 83 | 83 | 83 |
| Opera | 76.0.4017.123 | 76.0.4017.123 | 76.0.4017.123 | 76.0.4017.123 |
| IE | 11.630.19041.0 | ⊘ | ⊘ | ⊘ |
| Tor | 10.0.15 | 10.0.15 | ⊘ | 10.0.15 |
| Acrobat | 2019.021.20061 | 2019.021.20058 | 2020.013.20064 | ⊘ |

⊘: It indicates that the OS does not support the installation of the application.

**Challenge.** (*i*) There is no unified standard and specification for the CA manager. Different OS platforms have various policies and methods for managing CTLs, which makes it necessary to employ several unique processing and observation methods. For example, Windows OS manages the local CTL using Group Policy, and we can access and modify the CA certificates in the CTL through the user interface; while in Ubuntu, each CA certificate is stored as a separate file, and we can only modify the CTL by moving the file. (*ii*) The internal implementation logic of the applications' local CA managers is different and opaque. In order to explore the vulnerabilities of the CA manager, it is necessary to design numerous black box testings. For example, to study the security check of CA certificate and the use of CA certificate purpose, we create a website and a document, generate many CA certificates with insecure fields and special certificate purpose, and conduct various tests on each browser and Acrobat separately. (*iii*) Different CA managers may have various forms of unpredictable phenomena for the same test case. Therefore, many parts of the experiment require manual intervention, making it difficult to automate. For example, importing a CA certificate through the user interface to the user-level CTL requires no password but shows the information about the certificate on Windows, while the system-level CTL requires the administrator's permission without a password and shows no notification about the addition. We have to perform experiments manually on each surveyed object and record the results.

## 3.2   Exploring the Source of Trust

In a PKI system, OSs and applications generally configure a preset CTL as their source of trust. The preset CTLs of our target objects mainly originate from the four mainstream platforms, namely Microsoft, Mozilla, Apple, and Adobe. Among them, Microsoft and Apple maintain the local CTL by themselves, Mozilla root store is a part of the Network Security Services (NSS) cryptographic library [28], and Adobe manages the Adobe Approved Trust List (AATL) by itself and regards European Union Trusted Lists (EUTL) as a third-party source. The number of CA certificates in each CTL is shown in Table 2, which is obtained on April 2021.

**Table 2.** The widely acknowledged CTLs

| Platform | Quantity | CTL |
|---|---|---|
| Microsoft | 417 | Microsoft Included CA Certificate List [24] |
| Mozilla | 142 | Mozilla Included CA Certificate List [26] |
| Apple | 217 | Apple Included CA Certificate [17] |
| Adobe | 246 | AATL [14] & EUTL [23] |

To figure out the local default source of each OS and application, we carry out a systematic exploration of the CA managers. Then, we classify the target objects' management mode of their local CTL into two categories: *Global Level (global-level)* and *Application Level.* Table 3 shows the source of trust and the category of each manager.

**Table 3.** The source of trust and the category of each CA manager

| OS | CA manager | | Source of trust | | | | Category |
|---|---|---|---|---|---|---|---|
| | | | Microsoft | Apple | Mozilla | Adobe | |
| Windows | System-level/User-level | | √ | – | – | – | ⋆ |
| | Chrome | | √ | – | – | – | ∓ |
| | Edge | | √ | – | – | – | ∓ |
| | Opera | | √ | – | – | – | ∓ |
| | IE | | √ | – | – | – | ∓ |
| | Firefox | | – | – | √ | – | ◇ |
| | Tor | | – | – | √ | – | ◇ |
| | Acrobat | | – | – | – | √ | ◇ |
| macOS | System-level/User-level | | – | √ | – | – | ⋆ |
| | Chrome | | – | √ | – | – | ∓ |
| | Edge | | – | √ | – | – | ∓ |
| | Safari | | – | √ | – | – | ∓ |
| | Opera | | – | √ | – | – | ∓ |
| | Firefox | | – | – | √ | – | ◇ |
| | Tor | | – | – | √ | – | ◇ |
| | Acrobat | | – | – | – | √ | ◇ |
| Linux | System-level/User-level | | – | – | √ | – | ⋆ |
| | Install via *snap* | Firefox | – | – | √ | – | ◇ |
| | | Opera | – | – | √ | – | ◇ |
| | | Acrobat | – | – | – | √ | ◇ |
| | Install via *apt* | Chrome | – | – | √ | – | ∓ |
| | | Firefox | – | – | √ | – | ◇ |
| | | Opera | – | – | √ | – | ∓ |
| | | Edge | – | – | √ | – | ∓ |
| | | Tor | – | – | √ | – | ◇ |

⋆: Global level.   ◇: Application level.   ∓: Use global-level CA manager from OS.

**Global Level.** At the Global Level, CA managers maintain a global-level CTL, including user-level CTL and system-level CTL, and can be modified by any authorized entities. The modification to the global-level CTL will affect all the entities trusting it. Specifically, OSs maintain a local CTL on the computer as a system-level CTL that can be trusted directly by some applications (see Table 3 for details). Besides, some browsers installed via *apt* package in Ubuntu jointly trust a user-level CTL from Mozilla located in the *.pki* folder under the user directory. Note that the applications that trust the global-level CTL actually do not have an independent CA manager, and they use the third-party CA managers directly, which are provided by the user-level CTL of the OS by default, according to our observation.

**Application Level.** At the Application Level, each CA manager maintains its own CTL, which can only be used and modified by itself. The preset CTL can be customized or originate from mainstream platforms. For instance, every browser installed by *snap* in Ubuntu will occupy a separate folder to store the CTL deriving from Mozilla. Tor, Firefox, and Acrobat manage their CTLs on their own, and they can decide whether to put extra trust in the OS's CTL or not by *preferences*. But the extra trust in Ubuntu is not valid according to our experiments.

### 3.3    Exploring the Functionalities of CA Manager

Once the CA manager obtains the preset CTL from the source of trust, the CTL can be modified by entities (e.g., users, system, application, etc.) for some usage, as shown in Fig. 1. According to the requirements of the audit agency, such as WebTrust and ETSI, baseline requirements of CA/Browser Forum and our operation, and our observation of each CA manager, we conclude that the CA manager should possess the functionalities including but not limited to the followings.

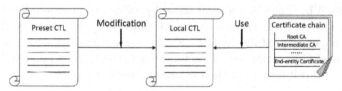

**Fig. 1.** Overview of the CA manager

**Storage Protection.** CA manager should store the local CTL and the corresponding trust relationship of these CA certificates in the form of a file locally. The storage of the local CTL should be sufficiently secure, and it should be protected by a security mechanism (such as digital signature and authority management) and not be modified and moved arbitrarily. Moreover, even the file is replaced, the trust relationship of the maliciously replaced CA certificate will not change with the replacement.

**Modifying CTL.** The CTL can be modified through the user interface and the command line, and users can customize their local CTL. The allowed operations include adding, deleting, and blocking the CA certificate. In particular, the adding operation should require the explicit participation of the user for security, such as inputting the user's password.

**Restricting CA Certificate Usage.** The usage of CA certificates should be restricted. By default, an added CA certificate cannot be used for any purpose, and users can specify the required CA certificate purpose to be available, which cannot exceed the purposes specified by the key usage and extended key usage (EKU).

**Security Audit.** Whenever a CA certificate is added or used, it should be checked for security, such as validity period, revocation status, certificate purpose, key size, supported cipher suites, and hash algorithms used. For insecure CA certificates, the CA managers should display a security warning on the user interface.

**Isolation.** Different accounts (or applications) on the same computer should maintain their local CTLs without affecting each other. However, the OSs maintain a system-level CTL shared by all accounts, which is contrary to isolation. In this case, the system-level CTL should not be modified arbitrarily.

**Updating CTL.** The CA manager should provide the functionality of updating the local CTL from the source of trust automatically or manually. In this way, the CA manager can obtain the latest CTL timely when the source of trust adds or blocks some CA certificates. This functionality can exist independently or as a part of an overall update of the OS or the applications.

## 4    Security Analysis on CA Manager

In this section, we conduct a comprehensive security analysis on the CA managers with a systematic and customized experiment method. Our work for the local CA managers mainly concerns the modifications of the local CTL, security checks of CA certificate, restrictions and inspections of CA certificate purposes, and some problems are found during the processes. We elaborate on the disclosed problems and make a detailed analysis separately in the end.

**For simplicity**, we hide the applications that do not have a CA manager in the following tables. Their behaviors are the same as the user-level CA manager.

### 4.1    Methodology

Given the diversity of experimental scenarios, our methodology is mainly categorized into two parts, *the black box testing* and *manual alteration of CTL*.

***Black Box Testing.*** It is applied for three purposes, including security checks of CA certificates, inspections of CA certificate purpose, and the verification policy of the certificate chain. To prepare for our experiments, we utilize *OpenSSL 1.1.1* to generate various self-signed root CA certificates with different fields, and create the corresponding three-tier certificate chains. Besides, we build a website using *Apache* and sign some PDF documents to check the CA managers of browsers and Acrobat.

Among these self-signed root CA certificates with security risks which are used for checking the security of CA certificate, some employ weak hash algorithms such as MD5 and SHA-1, or weak key pairs such as RSA-512, RSA-1024, and ECC-192, and others have disparate certificate purposes (especially the EKU) for the inspections of the key usage. Moreover, expired CAs are also considered. For the sake of eliminating interference factors from other irrelevant aspects, the other certificates in those three-tier certificate chains adopt identical and secure algorithms, specifically, RSA-4096 and SHA-256.

***Manual Alteration of CTL.*** This method places emphasis on the operations of importing new CA certificates to the existing CTLs and replacing the *CTL file* with another file. To achieve our goals, we import our root CA certificates which are secure enough to the local CTL of each CA manager using the user interface or the command line and make them trusted. Meanwhile, we record the required authorities and prompt messages in the process. Besides, we try to replace the CTL file with another file that has the same format and can be identified by the CA manager, observe and record the behaviors of the local CA managers.

## 4.2   Silent Addition of CA Certificate

Generally, importing a new CA certificate to the local CTL requires the user's knowledge or/and involvement, such as system prompt message or/and password authentication. However, an attacker can maliciously revise the CTL without the user's awareness, such as bypassing the prompt message and adding a CA certificate or directly replacing unprotected files through a malicious program when asking users to install software or drivers [1,5,8]. In our experiment, we adopt the method of manually altering the CTL and find there are several cases demonstrating that a local CTL can be modified silently.

***No Password Required for Addition.*** When adding a new CA certificate to the local CTL and making it trusted, we figure out that the CA managers of Windows, all browsers except Tor, and Acrobat do not require password authentication. Worse still, the system prompt messages can also be bypassed by programs, which has already been disclosed in related research [1].

These issues make it easy for attackers to add malicious or insecure CA certificates to the local CTL arbitrarily in silence. Once a malicious CA is trusted, all certificates issued by it will be used with complete trust, which may result in severe results, such as launching man-in-the-middle attacks, monitoring the behavior of the user's computer, injecting advertisements into the user's computer, or legally installing malicious software. Besides, when we install Alipay's security controls, we also find that the application installed several CA certificates in our Windows without any prompt of certificate information and adding quantity, which may leave the user in a monitored state.

**Table 4.** Silent addition of CA certificate

| OS | CA manager | Vulnerability¶ | | |
|---|---|---|---|---|
| | | Authority | Password | Storage |
| Windows | User-level | • | • | ⊙ |
| | System-level | ○ | • | ⊙ |
| | Firefox | • | • | ⊗ |
| | Tor | * | * | ⊙ |
| | Acrobat | • | • | ⊗ |
| macOS | User-level | ○ | ○ | ⊙ |
| | System-level | ○ | ○ | ⊙ |
| | Root-level | * | * | ⊙ |
| | Firefox | • | • | ⊗ |
| | Tor | * | * | ⊙ |
| | Acrobat | • | • | ⊗ |
| Linux | Ca-certificates (system-level) | ○ | ○ | ⊙ |
| | .*pki* (user-level) | • | • | ⊗ |
| | Firefox (via *snap/apt*) | • | • | ⊗ |
| | Opera (via *snap*) | • | • | ⊗ |
| | Tor (via *apt*) | * | * | ⊙ |
| | Acrobat (via *snap*) | • | • | ⊗ |

¶: *Authority* means whether the administrator's authority is required; *Password* means the administrator's password is required; *Storage* means whether the CTL file is protected.
○ means this feature is required, while • means this feature is not required, and * means this CA manager does not support the functionality of addition.
⊙ means the trust relationship or the file itself cannot be replaced, while ⊗ means this trust relationship can be changed by replacing the file.

***Malicious Replacement of CTL File.*** Since the CTL file generally stores many CA certificates and the corresponding trust relationship, it is significantly important to maintain its security. Nevertheless, our experiment reveals that Acrobat and browsers that trust raw CTL of Mozilla, excluding Tor, can replace the existing CTL file with another file coming from a disparate machine, and it only requires the current account's authorities and shows no security warning. Meanwhile, the trust relationship in the new CTL file is taken too, which means the trust relationship in the old CTL file is replaced. Consequently, attackers can use this physical method (which can also be achieved by programming) to add many malicious CA certificates to the local CTL without the user's knowledge, which may pose a more severe threat than adding CA certificates randomly.

Besides, implementations of some current CA managers may cause the consequences of these issues to be more serious, which are the followings:

*a*) **Security Issues of Shared CTL.** There are local CTLs shared by several applications on each OS. A malicious CA certificate added to the global-level CTL will be trusted by these applications, which increases the influence scope of risk. Besides, due to the dependence of trust, the applications may not perform any certificate verification but directly trust the global-level CTL, which shows no support for **isolation**.

*b*) **Possible Failure in Deleting and Blocking the CA certificates.** The CA certificates in the CA managers of Windows and Acrobat can restore automatically with the update after deletion. Since every CA manager can update its preset CTL, it is difficult for us to delete these CA certificates thoroughly in some CA managers. There are also some managers that do not support the blacklist. For example, Tor does not allow users to make any modifications to the CA manager for security. The system-level CTL of Ubuntu is allowed to cancel the trust of the CA certificates, but this certificate can be added again and trusted. The user may want to cancel the trust of malicious CA certificates, but the automatic recovery mechanism and ineffective blocking make it difficult.

We try to add secure CA certificates and replace files in various OSs and applications, and experiment results are shown in Table 4, including the required authorities during addition and the protection of the CTL file. In summary, we can spot that Acrobat, most browsers, and Windows OS lack a good security mechanism for modifications to CTL. The management methods of Ubuntu and macOS are worth learning.

## 4.3   Non-strict Security Check of CA Certificate

As required by CA/Browser Forum [19], MD5, SHA-1, RSA with a key size fewer than 2048 bits, and ECC with a key size fewer than 256 bits are viewed as insecure or not recommended. CA managers are expected to check these fields. To observe the security check of each CA manager, we employ the black box testing and use a three-tier certificate chain to do the experiments. We import the insecure root CA certificates to the local CTL and attempt to visit the website using browsers or verify the document signatures using Acrobat, whose end-entity certificate is issued by one of the insecure CAs. We record the behaviors of different CA managers, including warning messages on the user interface and application verification results. Besides, as mentioned in Sect. 4.1, the intermediate CA certificates and end-entity certificates are secure and valid.

To determine the certificate verification policies of these CA managers, only the intermediate CA certificates are added to the CTL and trusted, and we check whether the CA manager verifies the complete certificate chain or not.

For global-level CTLs, we observe the behaviors through the applications trusting their CTL, such as Safari in macOS. The system-level CTL of Ubuntu stores the certificates and trust status but does not provide verification functions. Many command lines such as *wget* and *curl* will trust and use the certificates, and the verification policy mainly depends on the cryptographic libraries such as OpenSSL and NSS. We only display the results of *wget* with OpenSSL. Tor and the root-level CTL of macOS are not considered in this section.

***False Trust in CA Certificate with Insecure Fields.*** During our experiment, it is indicated that only the CA managers of the macOS consider MD5 to be insecure, and ECC-192 is regarded as secure except for the browsers that trust Mozilla's CTL. In addition, SHA-1 and RSA-1024 are thought as secure in all CA managers. Acrobat can verify the signature of any end-entity certificates issued by insecure CA, as long as the CA certificate is imported and trusted. All of these situations should not happen and need to be warned by the CA managers. But only the CA managers of Windows and macOS show the security warnings for the insecure CA certificates in the user interface. Such CA certificates have great potential to be exploited by attackers. For instance, if a CA certificate uses MD5 as the hash algorithm, the attacker already has the ability to forge it with the same hash value. And if a CA certificate encrypts a message with a public key whose length is no longer considered secure, it will be under the risk that the information can be cracked in a limited time. Any of these can pose a great threat to the security of applications, OSs, and users. We reported to Mozilla the issue of not checking the MD5 algorithm in a trusted CA certificate and got a response. They stated that if the certificate was trusted by the user, the security of the digest algorithm would not be verified.

**Table 5.** Trust status of every CA manager for insecure root CA certificate*

| OS | CA manager | SHA-1 | | MD5 | | RSA-512 | | RSA-1024 | | ECC-192 | | Expired | |
|---|---|---|---|---|---|---|---|---|---|---|---|---|---|
| | | Inter | Root | Inter | Root | Inter | Root | Inter | Root | Inter | Root | Inter | Root |
| Windows | User-level[1,2] | ⊕ | △ | ⊕ | △ | ⊕ | ⊕³ | ⊕ | △ | ⊕ | △ | ⊕ | ⊕³ |
| | System-level[1,2] | ⊕ | △ | ⊕ | △ | ⊕ | ⊕³ | ⊕ | △ | ⊕ | △ | ⊕ | ⊕³ |
| | Firefox | △ | △ | △ | △ | ⊕ | ⊕ | △ | △ | △ | ⊕ | △ | △ |
| | Acrobat | △ | △ | △ | △ | △ | △ | △ | △ | △ | △ | △ | △ |
| macOS | User-level[1,2] | ⊕ | △ | ⊕³ | ⊕³ | ⊕ | ⊕³ | ⊕ | △ | ⊕ | △ | ⊕ | ⊕³ |
| | System-level[1,2] | ⊕ | △ | ⊕³ | ⊕³ | ⊕ | ⊕³ | ⊕ | △ | ⊕ | △ | ⊕ | ⊕³ |
| | Firefox | △ | △ | △ | △ | ⊕ | ⊕ | △ | △ | △ | ⊕ | △ | ⊕ |
| | Acrobat | △ | △ | △ | △ | △ | △ | △ | △ | △ | △ | △ | △ |
| Linux | Ca-certificates (system-level)[2] | ⊕ | △ | ⊕ | △ | ⊕ | ⊕ | ⊕ | △ | ⊕ | △ | ⊕ | ⊕ |
| | .pki (user-level) | △ | △ | △ | △ | ⊕ | ⊕ | △ | △ | △ | ⊕ | △ | ⊕ |
| | Firefox via snap/apt | △ | △ | △ | △ | ⊕ | ⊕ | △ | △ | △ | ⊕ | △ | ⊕ |
| | Opera via snap | △ | △ | △ | △ | ⊕ | ⊕ | △ | △ | △ | ⊕ | △ | △ |
| | Acrobat via snap | △ | △ | △ | △ | △ | △ | △ | △ | △ | △ | △ | △ |

△: (Intermediate) CA certificate is trusted.    ⊕: (Intermediate) CA certificate is not trusted.
*: **Inter** indicates that only the intermediate CA certificate with RSA-1096 and SHA-256 is added to the CTL and set as trusted. **Root** indicates that the root CA certificate is added to the CTL and form a complete trusted certificate chain.
1: This CA manager has a user interface to show the security warning.
2: This CA manager verifies the complete certificate chain.
3: The user interface shows this root CA certificate is insecure.

***Incomplete Verification of Certificate Chain.*** Due to the different policies of certification path validation between CA managers, some CA managers will not verify the complete certificate chain and stop verifying the rest certificates of the chain when they encounter a trusted CA certificate (not necessarily a root CA certificate). With only the intermediate CA certificates added into the CTL, we retry the above tests and find that CA managers of Acrobat, Firefox, and other applications (except Tor) that trust Mozilla's CTL in Ubuntu do not verify the complete certificate chain. For example, when we trust an intermediate CA certificate issued by a root CA using ECC-192 or an expired CA (the intermediate CA certificate is issued during the validity of the root CA and is secure enough), the website can still be accessed successfully in Firefox. Since the incomplete verification of certificate chains can cause some insecure CA certificates to bypass the verification, the risk of the non-strict security check is much larger.

The results are displayed in Table 5. It shows the behaviors of various CA managers for different insecure fields in the certificate. In particular, CA managers with user interfaces also show us the trust status of the CA manager. The results of different verification policies are also displayed. In general, we can see that many managers make no warnings about the insecure fields. Insecure CA certificates with weak algorithms weaken the security and may bring undetectable attacks to users, and the risks may be larger if the certificate chain is incompletely validated.

### 4.4   Potential Abuse of CA Certificate Purpose

The CA certificate usually contains several fields for specifying the certificate purpose, such as *Key Usage* and *Extended Key Usage*, and the certificate should be used for the purpose for which it is intended. CA managers may have their unique methods to process the certificate purpose. For example, as EKU is not required, a CA manager (e.g., Windows) may consider that a certificate without the EKU field has all the certificate purposes, which include the ones to manage the OS and can be abused to tamper with or attack the OS. Furthermore, CA managers (e.g., Firefox) can also provide some commonly used certificate purpose options on its user interface for authorized users to choose manually, such as verifying websites and emails. We will describe in detail later. These purposes may not exist in the certificate purpose field, resulting in inconsistent certificate purposes. To explore the verification of the certificate purpose, we add the self-created CA certificate to each CA manager and observe their behaviors.

*Loose Verification on CA Certificate Purpose.* Our findings manifest that different OSs vary greatly in verification on certificate purposes of CAs. The certificate purpose selected by the user on the user interface may be inconsistent with the actual key usage of the CA certificate. For macOS, it behaves as if it does not verify the key usage and EKU of the CA certificate after we import the self-created CA with an EKU *Timestamp* only to its CTL. If the key usage of *SSL* is selected in the user interface, we can still visit the website successfully whose end-entity certificate is issued by the CA normally. This issue can lead to the abuse of CA certificates, which may cause some malicious CA certificates to issue fraudulent certificates and have a great impact on the local users. We reported that macOS had the issue of inconsistent certificate purposes to apple and had not received a response yet.

*No Restriction on CA Certificate Purpose.* For Windows, when the EKU field of a CA certificate is empty, any purposes in the EKU list are selected and allowed, including some system's functionalities such as *Windows Update* and *Microsoft Trust List Signing*. In our work, we create a CTL using a CA certificate with the corresponding EKU and successfully add it to the local CTL in Windows. Then all CAs in this CTL are directly trusted. Besides, we find that no additional prompt or authentication is required for importing a CA certificate with such a special purpose. Attackers can inject such a CA certificate and a CTL signed by it through programs, which can add a set of CA certificates to the user's computer at one time. The malicious CA certificates with such high-privileged certificate purposes can always pass the verification, which can cause a significant impact on users.

## 5   Suggestions for Secure CA Manager

Based on the above problems of the CA managers, we put forward the following suggestions for building a more secure CA manager.

**User Participation during Modification.** When modifying the local CTL, especially when a new CA certificate is imported, the user's participation is necessary, and the need for a password is a recommended way, which can reduce the risk of being tampered with by malicious entities. For example, the system-level CTL and user-level CTL of macOS and system-level CTL of Ubuntu stored in */usr/share/ca-certificates/* need the administrator's password for addition, which is considered secure. Furthermore, it is recommended that displaying an explicit prompt when new CA certificates are imported to the CTL, which can tell the users about the target local CTL and the specific information and the quantity of the CA certificates. Therefore, users can be aware of what has happened and take action to ensure its security.

Meanwhile, deleting or blocking the insecure CA certificates should also be allowed and supported by CA managers so that users can cancel the trust of some certificates permanently. For example, when a user discovers a vulnerability caused by a CA certificate and that the certificate also exists in his computer, he should be able to delete or block the CA certificate.

**Enhanced Security of Local Storage.** There should be a certain security mechanism in the storage of the local CTL so that the corresponding files should not be moved, copied, or deleted at will. The non-replicability of the trust relationship is a sign of a secure file, too. The even better proposal is that the file could be signed by the current user or OS to ensure its integrity and security. Additionally, we also advise that each application or OS can maintain the CTL by itself instead of sharing it with others since the isolation of different entities can greatly reduce the possibility of being attacked.

Here are some pretty good cases in our research. Each CA certificate in the system-level CTL of Ubuntu is stored in a separate file, and the file replacement has the same requirement as adding, which demands the password of the administrator. Tor stores its CTL in a dynamic link library file, which can not be modified at all. Besides, the root-level CTL file in macOS is protected by System Integrity Protection (SIP) [18]. As for Windows, the CTL file is occupied since the user logins in, and we can't perform any operations on the file.

**Strict Security Check on CA Certificate.** Though the verification policy of some CA managers is based on the trust of the user, users may not be able to determine whether to trust a CA certificate, and a strict security check on the CA certificate when importing it or periodically is recommended. There are some fields requiring special attention, such as the validity period, the security of the cryptographic algorithm, and the certificate purpose/key usage. Besides, it is worth noting that those applications that directly trust the OS's list should also strictly check the security of the list when using it.

More importantly, it is necessary to verify the complete certificate chain for applications. In detail, the validity, the hash algorithm, and key size of all certificates in the certificate chain should be verified. Furthermore, we suggest that if the security strength of the upper-level CA certificate in a certificate chain is not stronger than the lower-level CA certificate, then the latter should not be considered secure. Last but not least, updating the CTL from the trusted source timely can mitigate the risk of trusting in CAs which have been deleted or blocked.

**Restrictions on Certificate Purpose.** We recommend that CA certificates in the CTL provide users with no certificate purpose by default and users can turn on the certificate purposes by selecting them from common certificate purposes such as SSL and S/MIME. Besides, the CA manager should verify whether the selected key usage is consistent with the purpose declared in the certificate. For the key usage related to some system functionalities, additional authentication or completely disabling is recommended.

# 6   Conclusion

This work has analyzed and reported the security issues of the trusted CA certificate management in current OSs and PKI applications. We explored three OSs and eight applications installed on each OS, focused on the source of trust, the functionalities of the CA managers, the modification to the local CTL, control of the certificate purpose and security check of the CA certificate, and found several security issues which may bring troubles and risks to users. Furthermore, we propose some suggestions for these security problems.

**Acknowledgment.** We thank all the reviewers and our shepherd for their helpful feedback and advice. This work was partially supported by the National Cyber Security Key Research and Development Program of China (No. 2018YFB0804600).

# References

1. Alsaid, A., Mitchell, C.J.: Installing fake root keys in a PC. In: Chadwick, D., Zhao, G. (eds.) EuroPKI 2005. LNCS, vol. 3545, pp. 227–239. Springer, Heidelberg (2005). https://doi.org/10.1007/11533733_16
2. Sloppy Security Software Exposes Dell Laptop to Hackers — Laptop Mag. https://www.laptopmag.com/articles/dell-certificate-security-flaw
3. Superfish - Wikipedia. https://en.wikipedia.org/wiki/Superfish
4. Braun, J., Volk, F., Classen, J., Buchmann, J., Mühlhäuser, M.: CA trust management for the web PKI. J. Comput. Secur. **22**(6), 913–959 (2014)
5. de Carnavalet, X.D.C., Mannan, M.: Killed by proxy: analyzing client-end tls interception software. In: Network and Distributed System Security Symposium (2016)
6. Chung, T., et al.: Measuring and applying invalid ssl certificates: the silent majority. In: IMC (2016)
7. Durumeric, Z., Kasten, J., Bailey, M., Halderman, J.A.: Analysis of the https certificate ecosystem. In: IMC (2013)
8. Durumeric, Z., et al.: The security impact of https interception. In: NDSS (2017)
9. Krombholz, K., Mayer, W., Schmiedecker, M., Weippl, E.: " I have no idea what i'm doing"-on the usability of deploying {HTTPS}. In: 26th {USENIX} Security Symposium ({USENIX} Security 17) (2017)
10. Li, B., et al.: Certificate transparency in the wild: exploring the reliability of monitors. In: Proceedings of the 2019 ACM SIGSAC Conference on Computer and Communications Security (2019)
11. Li, B., Lin, J., Wang, Q., Wang, Z., Jing, J.: Locally-centralized certificate validation and its application in desktop virtualization systems. IEEE Trans. Inf. Forensics Secur. **16**, 1380–1395 (2020)
12. Li, B., Wang, W., Meng, L., Lin, J., Liu, X., Wang, C.: Elaphurus: ensemble defense against fraudulent certificates in TLS. In: Liu, Z., Yung, M. (eds.) Inscrypt 2019. LNCS, vol. 12020, pp. 246–259. Springer, Cham (2020). https://doi.org/10.1007/978-3-030-42921-8_14
13. Perl, H., Fahl, S., Smith, M.: You won't be needing these any more: on removing unused certificates from trust stores. In: Christin, N., Safavi-Naini, R. (eds.) FC 2014. LNCS, vol. 8437, pp. 307–315. Springer, Heidelberg (2014). https://doi.org/10.1007/978-3-662-45472-5_20
14. Adobe Approved Trust List members. https://helpx.adobe.com/acrobat/kb/approved-trust-list1.html, Accessed 30 Apr 2021
15. Top 1 Million Analysis - March 2020. https://scotthelme.co.uk/top-1-million-analysis-march-2020/
16. Apple Root Certificate Program. https://www.apple.com/certificateauthority/ca_program.html
17. List of available trusted root certificates in macOS High Sierra. https://support.apple.com/en-us/HT208127, Accessed 11 May 2021
18. About System Integrity Protection on your Mac. https://support.apple.com/en-us/HT204899
19. Certificate Contents for Baseline SSL - CAB Forum. https://cabforum.org/baseline-requirements-certificate-contents/
20. CA/Browser Forum - CAB Forum. https://cabforum.org/
21. Censys. https://censys.io/certificates?q=, Accessed 30 Apr 2021
22. ETSI - Welcome to the World of Standards!. https://www.etsi.org/

23. European Union Trusted Lists. https://helpx.adobe.com/document-cloud/kb/european-union-trust-lists.html, Accessed 30 Apr 2021

24. MICROSOFT Included CA Certificate List. https://ccadb-public.secure.force.com/microsoft/IncludedCACertificateReportForMSFT, Accessed 30 Apr 2021

25. Program Requirements - Microsoft Trusted Root Program. https://docs.microsoft.com/en-us/security/trusted-root/program-requirements

26. MOZILLA Included CA Certificate List. https://ccadb-public.secure.force.com/mozilla/IncludedCACertificateReport, Accessed 30 Apr 2021

27. Mozilla Root Store Policy-Mozilla. https://www.mozilla.org/en-US/about/governance/policies/security-group/certs/policy/, Accessed 1 May 2021

28. Why Does Mozilla Maintain Our Own Root Certificate Store? - Mozilla Security Blog. https://blog.mozilla.org/security/2019/02/14/why-does-mozilla-maintain-our-own-root-certificate-store/

29. StatCounter Global Stats - Browser, OS, Search Engine including Mobile Usage Share. https://gs.statcounter.com/, Accessed 30 Apr 2021

30. Principles and criteria and practitioner guidance. https://www.cpacanada.ca/en/business-and-accounting-resources/audit-and-assurance/overview-of-webtrust-services/principles-and-criteria

31. Root CA Certificate: When you shouldn't trust a trusted root certificate — Malwarebytes Labs. https://blog.malwarebytes.com/security-world/technology/2017/11/when-you-shouldnt-trust-a-trusted-root-certificate/

32. Vallina-Rodriguez, N., Amann, J., Kreibich, C., Weaver, N., Paxson, V.: A tangled mass: the android root certificate stores. In: Proceedings of the 10th ACM International on Conference on emerging Networking Experiments and Technologies (2014)

# Effective Anomaly Detection Model Training with only Unlabeled Data by Weakly Supervised Learning Techniques

Wenzhuo Yang[✉] and Kwok-Yan Lam[✉]

School of Computer Science and Engineering, Nanyang Technological University, Singapore, Republic of Singapore
wenzhuo001@e.ntu.edu.sg, kwokyan.lam@ntu.edu.sg

**Abstract.** Intrusion detection systems (IDS) play an important role in security monitoring to identify anomalous or suspicious activities. Traditional IDS could be signature-based (or rule-based) or anomaly-based (or analytics-based). With the objectives of detecting zero-day attacks, analytics-based IDS have attracted great interest of the cybersecurity community. Furthermore, machine learning (ML) techniques have been extensively explored for advancing analytics-based IDS. Many ML techniques have been studied to improve the efficiency of intrusion detection and some have shown good performance. However, traditional supervised learning algorithms need strong supervision information, fully correctly labeled (FCL) data, to train an accurate model. Whereas, with the rapid development of network and communication technologies, the volume of network traffic and system logs has increased drastically in recent years, especially with the introduction of Next Generation Broadband Network (NGBN) and 5G networks. This caused huge pressure on analytics-based IDS because, for ML to train predictive models, security-relevant data need to be labeled manually, hence leading to practical barriers to achieving effective IDS. In order to avoid being overly dependent on strong supervision information, weakly supervised learning techniques, which utilize incomplete, inexact, or possibly inaccurate labels, have been studied by cybersecurity researchers in that such weak supervision information are easier and cheaper to obtain than FCL data. This research aims to explore the feasibility of weakly supervised learning techniques in IDS tasks so as to reduce the reliance on a massive amount of strong supervision information, which will only continue to grow tremendously in the big data society. We also investigated the detection stability of the proposed scheme when inaccurate weak supervision information is provided. In this article, we propose an IDS model training scheme that is based on a weakly supervised learning algorithm, which requires only unlabeled data. Experiments have been performed on three publicly available IDS evaluation datasets. The results showed that the proposed scheme performs well and is even better than some supervised learning-based IDS (SL-IDS) models. Experimental results also indicated that the weakly supervised learning based IDS model is robust and can be applied in real world situations. Besides, we examined detection performance of

D. Gao et al. (Eds.): ICICS 2021, LNCS 12918, pp. 402–425, 2021.
https://doi.org/10.1007/978-3-030-86890-1_23

the proposed method when it faces class-imbalanced training data and the experiment results show that it performs better than the compared methods.

**Keywords:** Intrusion detection · Network traffic · Weakly supervised learning · Class prior

# 1 Introduction

Intrusion detection system (IDS) is one of the most important measures in any holistic security architecture. Cybersecurity puts a lot of emphasis on protection, detection and reactive measures such as data protection, identification and trust management, authorization and accountability enforcement, behavior analytics for anomaly detection, malware analysis, digital forensic of computer and network nodes, privacy protection and anonymity of individuals. As one of the most significant cybersecurity detection tools, IDS can identify anomalous or suspicious activities in network traffic and system logs [10,19]. Anomalous or suspicious activities may include but are not limited to command and control behind the distributed denial of service (DDoS), port scanning attacks, unauthorized access, and remote control. Sensitive information, like credentials and private data of users, is in danger of being leaked or modified by these attacks. As a consequence, it becomes a crucial task to design an IDS with high detection efficiency and accuracy to enhance the integrity, confidentiality, and availability of the communication information and secure the privacy of user data in cyberspace [1,32].

Traditional IDS could be signature-based (or rule-based) or anomaly-based (or analytics-based). Anomaly-based (or analytics-based) IDS has received great interest from the cybersecurity community as it has the potential to detect novel intrusion, like zero-day attacks. Many well-developed machine learning (ML) and deep learning (DL) techniques, like Support Vector Machine (SVM), Logistic Regression (LR), Decision Tree (DT) C4.5, Recurrent Neural Networks (RNN), Deep Belief Network (DBN), and Convolutional Neural Network (CNN) have been applied to help improve the detection efficiency and performance of analytics-based IDS to better deal with the novel attacks. These are supervised learning (SL) algorithms which traditionally require a fully correctly labeled (FCL) dataset to train an effective model. FCL data can also be called strong supervision information. For most SL algorithms, especially deep learning approaches, the more FCL data is provided, the more accurate the detection result can be obtained. However, it is time and cost consuming to continuously get enough strong supervision information (FCL data). Semi-supervised learning-based IDS (SSL-IDS) and unsupervised learning-based IDS (UL-IDS) may detect attacks with partially labeled or unlabeled data but they usually are accompanied by high computational complexity or a high false-positive rate. Clustering-based unsupervised learning methods face another problem, which is the difficulty of interpreting different clusters as specific meaningful categories.

With the rapid development of network and communication technology, the frequency of network connections among different people and organizations continues to increase, especially with the widespread adoption of Next Generation Broadband Network (NGBN) and 5G networks. Network traffic data is enormous in quantity and keeps increasing every year. According to Cisco's annual report (2018–2023) [8], there are more and more people involved in network communication. From 2018 to 2019, global attacks between 100 Gbps and 400 Gbps increased by 776% year over year. The total number of DDoS attacks will increase from 7.9 million to 15.4 million from 2018 to 2023. Faced with the huge amount of traffic data in the big data society, the current annotation capabilities for labeling intrusion data are far from sufficient. It causes difficulty for cybersecurity personnel to obtain enough FCL data to train an accurate analytics-based intrusion detection model by traditional supervised learning techniques. Therefore, techniques that can construct an effective intrusion detection model by avoiding overly dependent on FCL data are needed.

As it is challenging to prepare FCL data, research works focused on weakly supervised learning have been extensively studied in recent years. Weakly supervised learning is a special branch of machine learning involving various studies that attempt to build predictive models by learning with weak supervision information on a supervised learning setting [37,56]. Weak supervision information refers to high-level or noisy labels, including incomplete, inexact, and inaccurate supervision information that is different from strong supervision information (fully correct labels).

In reality, weak supervision information is much easier to obtain at a lower cost than strong supervision information, especially in special circumstances when the data providers need to maintain their data privacy. The weak supervision information can be obtained directly based on the domain knowledge/experience or estimated from mixed data by some statistical methods. For estimated weak supervision information, one problem is that there may be a slight bias between the estimated parameters and the true values. Hence, it is important to check the stability of the weakly supervised learning techniques when biased weak supervision parameters are provided.

Therefore, we expect that weakly supervised learning in cybersecurity has promising potential in developing of analytics-based intrusion detection techniques. In this paper, we focus on analytics-based intrusion detection and the term IDS refers to anomaly detection (anomaly detection in this paper is different with its traditional meaning). In fact, weakly supervised learning techniques have been successfully applied in a wide range of application domains, including automatic image annotation [55], web mining [27], text classification [21], and ecoinformatics [23].

Motivated by the aforementioned, we propose a method to construct prediction models for IDS with only unlabeled data by a weakly supervised learning technique. Unlike the traditional supervised learning method, we only need the class priors of two training datasets, instead of knowing the specific ground-truth labels of every sample, to train an intrusion detection model. Class prior is a kind of weak supervision information that preserves the proportion of positive samples in a given dataset. The class priors are easier and cheaper to obtain or construct than the accurate annotation in intrusion detection tasks. Our purpose is to construct an effective and robust intrusion detection system. Our contributions in this paper can be summarized as follows:

- First, for constructing a cost-effective IDS, we explore the feasibility of applying weakly supervised learning techniques on intrusion detection tasks. We propose a scheme to generate weak supervision labels and train the intrusion detection models by a weakly supervised learning technique.
- Second, for evaluating the detection accuracy, we utilize three publicly available IDS datasets, NSL-KDD [47], UNSW-NB15 [30], and CIC-IDS2017 [42] to compare the proposed weakly supervised learning (unlabeled training data) trained IDS model with two baseline supervised learning (labeled training data) trained models.
- Third, considering the problem in weakly supervised learning that the estimated class prior may exhibit a little bias from real industrial scenarios, we investigate their influence on model performance by giving different parameter settings. We provide 11 groups of biased class priors to train the weakly supervised learning based IDS and compare their detection rates with the true class priors based IDS to check the stability of the proposed method.
- Fourth, considering that it is common to handle class-imbalanced datasets in real intrusion detection tasks, we checked the practicality of the proposed method by examining its performance and time overhead when detecting three types of class-imbalanced malicious traffic data. Five baseline methods (including two SL-IDSs, one SSL-IDS, and two UL-IDSs) are compared with the proposed method based on the CIC-IDS2017 dataset.

The rest of this paper is organized as follows: Sect. 2 introduces the relevant terminology and technical background. Section 3 demonstrates the technical details and the whole scheme of the proposed method. Experiment settings and results analysis are shown in Sect. 4. We have a discussion in Sect. 5, and Sect. 6 then reviews the related works. Finally, Sect. 7 makes a conclusion and summarizes the future work.

## 2    Background

In this section, we mainly introduce the technology background for our work and explain the terminologies in weakly supervised learning.

### 2.1    Weakly Supervised Learning

Machine learning has received increasing attention in recent years. It has achieved great success in many fields, including computer vision, text/speech recognition, recommendation system, marketing forecasting, safety warning, etc. Many well-developed supervised learning (SL) algorithms and traditional SL algorithms need a fully correctly labeled dataset to learn an accurate and effective model. For most SL algorithms, the more FCL data are provided, the more accurate the predicted result can be obtained. However, it is time-consuming and expense-consuming to collect enough amount of FCL data in reality [39]. In the area of intrusion detection, it is more difficult to label the traffic data accurately since there is enormous network traffic every day but a limited number of security experts. Fortunately, it could be easier to collect *weak supervision* information.

There are mainly three types of weak supervision [56], including incomplete supervision, inexact supervision, and inaccurate supervision information. Figure 1 gives a simple illustration to show the difference among traditional supervised learning and the three typical types of weakly supervised learning (with different types of weak supervision labels). We assume that there are only positive and negative samples. Small rectangles represent feature vectors. Red and blue denote positive (P) and negative (N) labels, respectively. Label with a question mark means that the label may be inaccurate. Blank circles without "P" or "N" denote unlabeled samples.

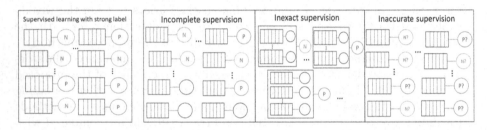

**Fig. 1.** An illustration for the traditional supervised learning and three typical weakly supervised learning. Small rectangles denote feature vectors. Red and blue denote positive and negative labels. Question marks mean that the label maybe inaccurate. Blank circle represents unlabeled data. (Color figure online)

In supervised learning, the ground-truth labels of all the training samples are provided. In incomplete supervision, only a subset of the training data is labeled. Usually, most of the remaining data is unlabeled. In inexact supervision, a fine-grained label is not given for each sample. Some higher-level, less precise information like heuristic rules, expected label distributions will be provided [37]. The data samples are grouped into different data bags. One labeling rule is that if there is one positive sample, the bag can be labeled as positive. If all the samples are negative, the bag can be labeled as negative. The purpose of the task in this case is usually to predict the category of unlabeled data bags. In inaccurate supervision, the data is labeled but may be incorrect. Sometimes, noisy information is given. Sometimes, we may know the probability that one sample belongs to positive or negative. This situation is common in real life when the annotator is not an expert (e.g., crowdsourcing), the sample is difficult to identify, or the classifier is not strong enough.

Weakly supervised learning aims to train a predictive model [37,38,56] with only weak supervision information (incomplete, inaccurate, inexact, noisy information or existing related knowledge). Such kind of weak supervision information is much easier and cheaper to collect than strong supervision information. Using weak supervision information can relieve the pressure of lacking FCL data in some new research areas that require the application of machine learning techniques. In practice, the weak supervision information of a dataset may not be given directly sometimes, while the weak supervision information for mixed data can be constructed or estimated. Therefore, some estimated weak supervision parameters may be slightly different from the true values. As a result, when we utilize weakly supervised learning techniques to train a model, we should consider its stability of predictive performance. The stability means that the intrusion detection system can keep high detection accuracy even biased weak supervision parameters are provided.

## 2.2   Unlabeled-Unlabeled Learning

In our work, we propose to construct intrusion detection models by unlabeled-unlabeled (UU) learning [25], which is an emerging weakly supervised binary classification framework that has not received much attention. The advantage of UU learning lies in that it only needs two unlabeled training datasets with different class priors to train a classification model. Class prior is weak supervision information representing the ratio of the number of samples belonging to a certain class to the number of all samples. Bekker and Davis summarized that there are mainly three ways to get the class prior [5]. The first method is to directly get class prior from the data providers or obtain class prior based on the domain knowledge/experience. The second approach is deducing it by a small fully labeled data set [9]. The third way is estimating it from positive and unlabeled (PU) data. Elkan and Noto first had the idea of estimating positive label frequency when training a binary classifier by positive and unlabeled (PU) learning under the condition that the labeled positive samples are selected completely at random [11]. Some research works applied kernel density estimator to estimate class prior [6,17,36]. The authors of [35] proposed a class prior estimation approach that does not need the assumption that selecting the observed labels completely at random. Zeiberg et al. provided a simple and fast class proportions estimation algorithm using a distance curve generated by repeated sampling and nearest neighbor calculation [54]. These research works provide detailed methods for estimating class priors, which makes weakly supervised learning techniques more practical in real-world systems. Our work mainly focuses on exploring the feasibility of applying weakly supervised learning techniques to alleviate the burden of continuously acquiring large amounts of FCL network traffic data in today's data-driven environment for intrusion detection tasks.

Unlike supervised learning, UU learning can successfully train a classifier by using only two unlabeled training sets instead of fully correctly labeled data. Practically, it is much easier to collect or estimate the class prior than knowing the specific label of every sample in a dataset. As mentioned before, collecting the correct label for each network traffic record is extremely difficult because of the huge amount of data and the limited number of security professionals. Semi-supervised and unsupervised learning techniques can also avoid relying too much on FCL data, while their predictive performance is not as good as supervised learning techniques and usually have high computational complexity. Compared with some clustering-based unsupervised learning approaches, UU learning can overcome the problem of translating the clusters into meaningful categories by *empirical risk minimization* [48], which theoretically provides a performance guarantee of the learned model. Therefore, we aim to explore the applicability, stability, and practicality of UU learning based-IDS models to mitigate the problem of that lacking FCL data in intrusion detection tasks.

## 3    Our Scheme and Methodology

### 3.1    Overview of the Scheme

An overview of the proposed intrusion detection scheme is shown in Fig. 2. In this paper, we do binary classification to study the performance of applying weakly supervised learning techniques on intrusion detection tasks. First, we describe the three selected commonly used IDS evaluation datasets. Then we introduce the data preprocessing steps. We do numeralization and normalization for all the collected data. We randomly split the dataset into training and testing sets. Then, we generate weak supervision labels for the three datasets based on the selected unlabeled-unlabeled learning algorithm. After these steps, we use the same processed training data as input to train intrusion detection models by different machine learning techniques. The processed testing data is utilized to evaluate the intrusion detection performance of the different models.

### 3.2    Dataset Description

**NSL-KDD.** There are several publicly available datasets for evaluating intrusion detection systems, one of the most popular datasets is KDD Cup 99 [29]. MIT Lincoln Labs prepared this dataset for an intrusion detection competition task in 1998. It is a compressed binary Transmission Control Protocol (TCP) dump dataset collected from a simulating Air Force local-area network (LAN) [46]. The raw data is processed into approximately 7 million connection records. Each connection record consists of a sequence of TCP packets collected from a fixed period of time and is labeled as normal or a specific attack. However, this dataset has some obvious defects. It contains a large number of redundant records, and the sample distribution is biased in terms of quantity and difficulty level. Hence, NSL-KDD, an improved version of KDD Cup 99 is generated [47]. NSL-KDD dataset removes the redundant records of KDD Cup 99 and reorganizes the distribution of different kinds of network traffic records to make the number of samples in training and testing sets more reasonable. This dataset

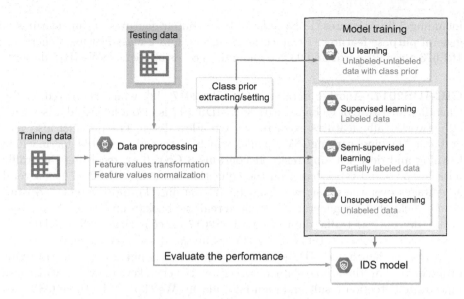

**Fig. 2.** An illustration of the proposed scheme for constructing an intrusion detection model.

contains 148,517 samples in total. Each sample owns 41 features and 1 label. The features are classified into three types: basic features of user TCP connections, content features, and traffic features that are computed by a two-second time window [29].

Many researchers use NSL-KDD as the benchmark dataset for research related to intrusion detection problems. Hence, we also choose this dataset to evaluate the performance of our proposed intrusion detection model. Though NSL-KDD solves most problems in KDD Cup 99, it still suffers the problem that it is too old to represent the modern attack traffic. It is more meaningful and convinced to use the recent IDS evaluation dataset. Because the recent datasets can represent the most up-to-date common attacks. Therefore, in our work, we also choose another two IDS datasets created in recent years.

**UNSW-NB15.** One is UNSW-NB15 [30], which is another commonly used IDS dataset provided in 2015. Cyber Range Lab of the Australian Centre for Cyber Security (ACCS)[1] created this dataset by using IXIA PerfectStorm tool[2]. The attack types in UNSW-NB15 are more recent than NSL-KDD, which can better represent the modern network traffic attack types. There are 2,540,044 samples in total in UNSW-NB15 dataset and they are distributed into four CSV files. The generator also constructs a training set and a testing set that contains 175,341 and 82,332 samples, respectively. In UNSW-NB15, each sample has 49 features,

---

[1] https://www.unsw.adfa.edu.au/unsw-canberra-cyber/.

[2] http://www.ixiacom.com/products/perfectstorm.

including 5 flow features, 13 basic features, 8 content features, 9 time features, 5 general purpose features, 7 connection features, and 2 class features. There are 164,673 normal records and 93,000 anomaly records in the UNSW-NB15 dataset.

**CIC-IDS2017.** Another dataset is CIC-IDS2017 [42], which is created by the Canadian Institute of Cybersecurity. CIC-IDS2017 is also a labeled dataset with 80 network traffic features extracted by CICFlowMeter. The provider also generates a MachineLearningCSV.zip file, which is specially created for machine learning and deep learning tasks. It contains 8 CSV files, recording the traffic data features and types based on the collection date. CIC-IDS2017 has several advantages over the other two datasets. It contains the most recent and common attacks that are generated in a more realistic background. The amount and attack diversity of traffic data in CIC-IDS2017 exceeds that of NSL-KDD and UNSW-NB15, with a total of 3,119,345 instances. It satisfies the eleven criteria for a reliable benchmark IDS evaluation dataset that put forward in [14]. One problem is that the whole data distribution is very skewed, which will cause inaccurate detection result and high false alarm. We choose the three CSV files collected on Friday to be our target dataset for CIC-IDS2017. Because we found that the total traffic data distribution on Friday is not at high-class imbalance for normal and anomaly samples.

In summary, we use the three datasets to evaluate the performance of the proposed IDS model and other baseline models in our work.

### 3.3 Data Preprocessing

As mentioned, we use the three publicly available datasets, NSL-KDD, UNSW-NB15, and CIC-IDS2017 to evaluate the performance of the proposed intrusion detection model. There are different types of features (symbolic and numeric) in both NSL-KDD and UNSW-NB15 datasets. For these two datasets, we need to convert the values of symbolic features to numbers. For CIC-IDS2017 dataset, we use the MachineLearningCSV.zip file. The feature values in CIC-IDS2017 are all numbers, which is more convenient and suitable for machine learning task. But there exist some infinite values and redundant information in CIC-IDS2017 dataset which need to be processed. All the three datasets face the problem that the values of many features differ greatly. We need to do normalization for the three datasets accordingly. Besides, there is still a problem of data imbalance. We will solve these problems in the data preprocessing and model training steps. We first explain the data preprocessing works for the three datasets before sending the training sets to the intrusion detection models as input.

**Preparation Works.** The CSV files in CIC-IDS2017 are separated. We merge the CSV files into one CSV file first for the convenience of later process steps. There are two columns named "Fwd Header Length" and "Fwd Header Length.1" contain the same information. As a result, we remove the redundant column "Fwd Header Length.1".

**Feature Values Transformation.** The values of some features in NSL-KDD and UNSW-NB15 are symbolic or text values. For example, the values for feature "service" are "http", "telnet", "ftp", "smtp", etc. The values for feature "proto-col_type" are "tcp", "udp", and "icmp". Only numerical vectors or matrix can be fed into machine learning classifiers. Hence, we need to transfer the symbolic/text features into numeric values. One hot encoding is used to convert the text format values to numeric values. As mentioned, the feature values in CIC-IDS2017 CSV files are all numbers. But some of the values are infinite numbers, which can not be processed by the system. So we need to convert the infinite numbers to a specific finite number. The raw datasets of the three evaluation IDS datasets are labeled datasets. The values of the "label" attribute are "NOR-MAL/BENIGN" and "ANOMALY/ATTACK". We need to convert the former and later items to "−1" (negative) and "+1" (positive) respectively.

**Feature Values Normalization.** There are different types of features in the three datasets. The values of different features have different scales. The values of some features (same_srv_rate, tcprtt, ACK Flag Count, etc.) are below 10. But the values of some features (duration, sload, stcpb, Subflow Bwd Bytes, etc.) range from 0 to 1,000 or even above 10,000. Therefore, we need to standardize the values of the features in the datasets. The feature values are normalized based on Eq. (1). The new normalized feature value is denoted by $x_{new}$. The minimum and maximum values of the normalized feature are represented as $x_{min}$ and $x_{max}$ respectively.

$$x_{new} = \frac{x - x_{min}}{x_{max} - x_{min}} \qquad (1)$$

### 3.4 Model Training

**Training Objective.** Let $\mathcal{X}$ denotes the feature space of the dataset. As we do binary classification in this work, the binary label space is $\dagger = \{+1, -1\}$. The *underlying joint density* is represented by $p(x, y)$. The positive and negative *class-conditional densities* are expressed by $p_+(x) = p(x|y = +1)$ and $p_-(x) = p(x|y = -1)$ [16]. $\pi_+ = p(y = +1)$ is the *class-prior probability* of the raw dataset.

Let $f$ and $\ell$ be the decision function and loss function. The purpose of the classification task is to minimize the following risk:

$$R(f) = \mathbb{E}_{(x,y)\sim p(x,y)}[\ell(yf(x))] = \pi_+\mathbb{E}_{p_+(x)}[\ell(f(x))] + \pi_-\mathbb{E}_{p_-(x)}[\ell(-f(x))], \quad (2)$$

where $\pi_- = 1 - \pi_+$ is the negative class prior, $\mathbb{E}_{p_+(x)}$ and $\mathbb{E}_{p_-(x)}$ are the expectations of positive and negative samples over $p_+(x)$ and $p_-(x)$. If the positive and negative samples are balanced, the value of $\pi_+$ is 0.5.

In UU learning, there are two unlabeled training sets $\{x_i\}_{i=1}^n$ and $\{x'_j\}_{j=1}^{n'}$ with different class priors $\theta$ and $\theta'$ ($\theta \neq \theta'$), which are drawn from the following marginal distributions:

$$p_{tr}(x) = \theta p_+(x) + (1 - \theta)p_-(x), \quad p'_{tr}(x) = \theta' p_+(x) + (1 - \theta')p_-(x) \quad (3)$$

The specific labels of each sample in the two training sets $\{x_i\}_{i=1}^{n} \sim p_{tr}(x)$ and $\{x_j'\}_{j=1}^{n'} \sim p_{tr}'(x)$ are unknown. Given the above notations, it was shown [25] that we can successfully learn an effective binary classifier by minimizing the following objective function:

$$\widehat{R}_{uu}(f) = \frac{1}{n} \sum_{i=1}^{n} \left( \frac{(1-\theta')\pi_+}{\theta - \theta'} \ell(f(x_i)) - \frac{\theta'(1-\pi_+)}{\theta - \theta'} \ell(-f(x_i)) \right)$$
$$+ \frac{1}{n'} \sum_{j=1}^{n'} \left( -\frac{(1-\theta)\pi_+}{\theta - \theta'} \ell(f(x_j')) - \frac{\theta(1-\pi_+)}{\theta - \theta'} \ell(-f(x_j')) \right) \quad (4)$$

This is because Eq. (4) is an *unbiased estimator* of the classification risk $R(f)$, which means the expectation of the empirical risk $\widehat{R}_{uu}(f)$ is equivalent to $R(f)$, i.e., $\mathbb{E}[\widehat{R}_{uu}(f)] = R(f)$. Equation (4) is a tool that is used to measure the classification risk of weakly supervised learning methods of the proposed system based on the concept of an unbiased estimator of classification risk. Interested readers can refer to the proof of Theorem 4 in Lu et al. [25] for more details about the derivation process.

**Class Prior Settings.** As mentioned above, it is assumed that the ground-truth label of each sample is unknown in UU learning. But the weak supervision information, i.e., class priors of the training and testing sets, is assumed to be known. Therefore, we need to set the class priors of the processed data. After data preprocessing, we get 71,463 normal records and 77,054 anomaly records for the NSL-KDD dataset, 164,673 normal records and 93,000 anomaly records for the UNSW-NB15 dataset, 592,347 normal records and 285,549 anomaly records for the CIC-IDS2017 Friday dataset.

As explained, there are two training sets in UU learning. Our paper constructs different groups of class priors for the two training sets to verify several different problems. Here, we take NSL-KDD as an example to explain the class priors setting process. First, we randomly select 55,000 anomaly and normal samples from the training set of the original NSL-KDD dataset, respectively. Then we construct two new sub-training data sets by the 110,000 samples as shown in Fig. 3. Here we set the positive class priors $\theta$ and $\theta'$ as 0.3 and 0.7 in the two new sub-training sets. The proportion of anomaly and normal samples is 3:7 in set A. The distribution of anomaly and normal samples is 7:3 in set B. In the experiment, the positive class priors $\theta$ and $\theta'$ in the two unlabeled training data sets can be any values between 0 and 1 but not equal to each other. According to the settings in UU learning, the specific label of each traffic record is removed from the two sub-training sets when training the intrusion detection model. All the class priors setting processes in this paper are followed the above method for all evaluation datasets.

Since the testing set is randomly selected from the original dataset, the positive class prior of the testing set is equal to the class prior of the original dataset $\pi_+$. Therefore, the testing set class priors of NSL-KDD, UNSW-NB15, and CIC-IDS2017 are set as 0.5 and 0.6, and 0.67, respectively.

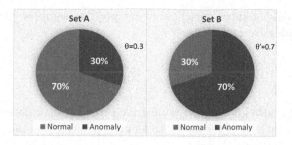

**Fig. 3.** An illustration of generating two unlabeled training data sets with class priors based on UU learning.

The constructed two unlabeled training sets are used to train the proposed intrusion detection model by the UU learning algorithm. To ensure a fair comparison with other methods, we use the same processed training and testing sets when training the IDS models by other methods. We combine the processed two sub-training sets into one set and provide the ground-truth label of every samples when comparing with supervised learning based intrusion detection models.

## 4 Experiments

### 4.1 Evaluation Metrics

We use positive to denote anomaly sample and use negative to represent normal sample. Positive sample is labeled as "1" and negative sample is annotated as "−1". The confusion matrix is shown in Table 1. True Positive (TP) and True Negative (TN) mean the number of correctly identified anomaly samples and normal samples. False Negative (FN) refers that anomaly samples are wrongly predicted as normal samples. False Positive (FP) indicates the normal samples that are incorrectly classified as anomaly samples.

**Table 1.** This is the confusion matrix.

|              |         | Predicted class      |                      |
|--------------|---------|----------------------|----------------------|
|              |         | Anomaly              | Normal               |
| Actual class | Anomaly | True positive (TP)   | False negative (FN)  |
|              | Normal  | False positive (FP)  | True negative (TN)   |

In our work, we use the following indicators to evaluate the performance of different network intrusion detection models:

**Precision.** Precision represents how many positive samples out of all positive predictive samples are really positive.

$$Precision = \frac{TP}{TP + FP} \tag{5}$$

**Recall.** Recall is also known as true positive rate (TPR) or detection rate (DR). It reflects the sensitivity of the model to identify positive (anomaly) samples from all real positive samples.

$$Recall = \frac{TP}{TP + FN} \tag{6}$$

**F Score (Fs).** F score is used to evaluate a binary classification model. It is the harmonic mean of recall and precision, which can give a more comprehensive assessment for the evaluated network intrusion detection model.

$$Fs = 2 * \frac{Precision * Recall}{Precision + Recall} = \frac{2 * TP}{2 * TP + FP + FN} \tag{7}$$

The real values of the above metrics are between 0 and 1. The larger the value of the indicators, the better the model is.

## 4.2  Experiment Results

We have implemented three experiments in our work to explore the applicability, stability, and practicality of the UU learning based-IDS (UUL-IDS).

**Experiment 1.** In the first experiment, we use the same selected training data to train different intrusion detection models and use the same testing data to evaluate their performance.

We compare the UU learning method (i.e., directly minimizing Eq. (4)) with the following methods:

- **UUrelu**, which denotes the method that minimizes Eq. (4) with the *ReLU function*, for alleviating overfitting when training with complex models [26];
- **UUabs**, which denotes the method that minimizes Eq. (4) with the *absolute value function*, for alleviating overfitting when training with complex models [26];
- **SL100**, which is a supervised learning method trained by 100% correctly labeled data;
- **Biased**, which uses the supervised learning method to treat the unlabeled dataset with larger class prior as positive data and the other unlabeled dataset with smaller class prior as negative data to train a model. This baseline method is the same as the "Binary-Biased" in [12].

Because UUrelu and UUabs are both UU learning based techniques, we also apply the two methods to validate their applicability on intrusion detection tasks. All UU learning, UUrelu, and UUabs trained intrusion detection models are fed by unlabeled data with class priors. We use "UUL" to represent the UU learning trained IDS models in the later experiment. In this work, a linear decision model $f(x) = \omega^T x + b$ and a hinge loss $\ell_{hinge}(y) = max[0, 1 - f(x)y]$ are used for the above five methods (including UUL). We can also use other decision models and loss functions. Adam optimization algorithm is used and the learning rate and weight decay are set as 0.001 and 0.0005, respectively. After preprocessing, we got 110,000/32,000, 155,000/77,500, 380,000/190,000 training/testing samples for NSL-KDD, UNSW-NB15, and CIC-IDS2017 datasets respectively. The class priors $\theta/\theta'$ are set as 0.7/0.3, 0.9/0.3, and 0.8/0.2 for the NSL-KDD, UNSW-NB15, and CIC-IDS2017 datasets. Precision, recall, and F score are used to evaluate the detection performance of the above models. Each model is trained by 100 epochs and we calculate the evaluation metrics' average values of the last 10 epochs. We trained every model several times to get the average performance and time cost values.

**Table 2.** The experiment results of the five intrusion detection models based on the three selected evaluation datasets and the average time overhead for training and testing a model once.

| Model | NSL-KDD | | | | UNSW-NB15 | | | | CIC-IDS2017 | | | |
|---|---|---|---|---|---|---|---|---|---|---|---|---|
| | Precision | Recall | F score | Time | Precision | Recall | F score | Time | Precision | Recall | F score | Time |
| SL100 | 0.9537 | 0.9534 | 0.9534 | 1.93 s | 0.8917 | 0.8795 | 0.8756 | 2.71 s | 0.9344 | 0.9274 | 0.9271 | 6.59 s |
| Biased | 0.9214 | 0.9207 | 0.9206 | 1.86 s | 0.8185 | 0.8196 | 0.8186 | 2.75 s | 0.9047 | 0.8870 | 0.8857 | 6.60 s |
| UUrelu | 0.9500 | 0.9500 | 0.9500 | 1.92 s | 0.8822 | 0.8824 | 0.8822 | 2.82 s | 0.9268 | 0.9175 | 0.9171 | 6.91 s |
| UUabs | 0.9502 | 0.9501 | 0.9502 | 1.93 s | 0.8820 | 0.8823 | 0.8820 | 2.76 s | 0.9265 | 0.9170 | 0.9165 | 6.79 s |
| UUL | 0.9519 | 0.9517 | 0.9517 | 1.97 s | 0.8918 | 0.8798 | 0.8758 | 2.76 s | 0.9258 | 0.9163 | 0.9159 | 6.86 s |

This experiment mainly wants to check whether the UU learning technique is applicable to intrusion detection tasks. Hence, we compare UUL-IDS with SL-IDS to check its detection performance. Table 2 shows the experiment results and the time overhead of the five intrusion detection models based on NSL-KDD, UNSW-NB15, and CIC-IDS2017 datasets. In general, it is reasonable to expect that SL100 should perform best as its training data are fully correctly labeled. The SL100 shows the best detection performance for NSL-KDD and CIC-IDS2017 datasets. But we can find that UUL performs a little better than SL100 for UNSW-NB15 due to the imbalanced characteristics of this dataset. SL-IDS cannot always perform well when handling imbalanced data. The proposed UUL-IDS can perform well and is stable when facing imbalanced data, which is another advantage over SL-IDS. We implemented Experiment 3 to validate this. The proposed method can almost reach the SL100 performance and perform much better than another baseline method (Biased) only with the class priors. The time cost of UUL training and testing a model is also reasonable, which shows that the proposed method is effective and suitable for handling big data challenges in real-world intrusion detection tasks. Further time overhead comparison with semi-supervised learning and unsupervised learning can be seen in Experiment 3. As explained before, UUrelu and UUabs are modified versions of UU learning which overcome the overfitting problem. The number of traffic samples in UNSW-NB15 and CIC-IDS2017 are more than the data samples in NSL-KDD dataset. Therefore, the overfitting problem is easier to arise when applying the UU learning trained IDS model on UNSW-NB15 and CIC-IDS2017 datasets. As shown in Table 2, the detection results of UUrelu and UUabs based IDS models are a little better than the UU learning based IDS models.

We can see from Fig. 4(a), Fig. 4(b), and Fig. 4(c) more intuitively that the UU learning trained IDS models (blue lines) can perform as well as the fully correctly labeled data trained supervised learning based IDS models (green lines) on the NSL-KDD and UNSW-NB15 datasets. The detection rates of UU learning trained IDS models have obvious advantage over another baseline method (Biased) trained IDS models for these two IDS evaluation datasets. For the CIC-IDS2017 dataset, the detection rate of UUL trained IDS is a little bit lower than the SL100 trained IDS. But it still obviously outperforms the Biased method trained IDS on CIC-IDS2017 dataset.

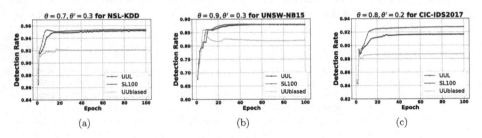

**Fig. 4.** The comparison of detection rate between UUL and two supervised learning based baseline methods on NSL-KDD, UNSW-NB15, and CIC-IDS2017 datasets. (Color figure online)

**Experiment 2.** In the second experiment, we evaluate the stability of the proposed intrusion detection model. As mentioned in Sect. 2, we can not always obtain the class priors directly in practice. The constructed or estimated class priors may deviate a little from the real values. It is necessary to check the stability of UU learning based IDS in a more realistic scene. We use two real numbers $\beta$ and $\beta'$ that are around 1 to be the biased coefficients. The biased class priors $\alpha$ and $\alpha'$ can be denoted as $\alpha = \beta\theta$ and $\alpha' = \beta'\theta'$ respectively. We use 12 groups of different biased coefficients to generate different biased class priors to train the intrusion detection model by UU learning algorithm to check the models' stability. Other settings are the same as the first experiment. Experiment 2 is also based on the three selected IDS evaluation datasets.

**Table 3.** Experiment results (detection rate %) of checking the stability of the proposed method when giving inaccurate class priors based on NSL-KDD, UNSW-NB15, and CIC-IDS2017 datasets.

| Dataset | $\theta, \theta'$ | $\beta = \beta' = 0.8$ | $\beta = \beta' = 0.9$ | $\beta = \beta' = 1$ | $\beta = \beta' = 1.05$ | $\beta = \beta' = 1.1$ | $\beta = \beta' = 1.11$ |
|---|---|---|---|---|---|---|---|
| NSL-KDD | 0.7, 0.3 | 89.16 | 91.98 | 95.17 | 95.36 | 95.17 | 95.04 |
| UNSW-NB15 | 0.9, 0.3 | 85.15 | 87.02 | 87.98 | 86.85 | 86.56 | 86.58 |
| CIC-IDS2017 | 0.8, 0.2 | 92.04 | 92.61 | 91.63 | 91.18 | 90.81 | 90.74 |
|  | $\theta, \theta'$ | $\beta, \beta' = 0.8, 1.11$ | $\beta, \beta' = 0.9, 1.05$ | $\beta, \beta' = 0.95, 1$ | $\beta, \beta' = 1, 0.95$ | $\beta, \beta' = 1.05, 0.9$ | $\beta, \beta' = 1.11, 0.8$ |
| NSL-KDD | 0.7, 0.3 | 95.67 | 95.48 | 93.24 | 95.17 | 94.40 | 92.36 |
| UNSW-NB15 | 0.9, 0.3 | 84.50 | 87.31 | 87.94 | 87.25 | 86.00 | 85.58 |
| CIC-IDS2017 | 0.8, 0.2 | 92.26 | 92.61 | 92.40 | 91.57 | 91.05 | 89.83 |

Table 3 shows the experiment results for experiment 2. We use detection rate to compare the performance of different class priors based IDS models. We notice that when $\beta = \beta' = 1$, the detection rate is the true value, which is higher than most detection rates of the other groups. The value of $\alpha$ and $\alpha'$ should be between 0 and 1, so we set the largest biased coefficient as 1.11. When we set $\beta$ and $\beta'$ as 0.8, there generates the most biased $\alpha$ and $\alpha'$. Even under the worst situation, UU learning based intrusion detection models still achieve desirable detection rates for all the three evaluation datasets.

Another interesting point that needs to be mentioned is that some detection rates under biased class priors are a little higher than the real detection rate for all the three selected evaluation datasets. This is because the proposed model can perform better when the values of class priors for the two training sets are away from 0.5. This means if the distribution of anomaly and normal samples is more skewed, the trained model will perform better. This is a practical ability for IDS, because it is easy to encounter scenarios that deal with imbalanced datasets in real life. Therefore, we implemented the third experiment to further examine the detection performance of the proposed method when detecting class-imbalanced malicious traffic data.

**Experiment 3.** In this step, we mainly check the detection ability of the proposed method when dealing with an imbalanced dataset. We compare the proposed UU learning based IDS model with another five IDS models. Except for SL100 and Biased, we utilize one semi-supervised learning technique (SelfTraining) and two unsupervised learning techniques (OneClassSVM and Kmeans) from scikit-learn [33] to train the different intrusion detection models. Semi-supervised learning uses partially labeled data and unsupervised learning techniques use unlabeled data to train a predictive model, which is fairer to compare with our proposed Unlabeled-Unlabeled learning-based IDS model than the supervised learning methods. In experiment 3, we also use precision, recall, and F score to evaluate the detection performance of the six IDS models. We know that the benefit of semi-supervised and unsupervised learning techniques is that they do not need much FCL data to train a model. But some of the techniques have high computational complexity. Therefore, we record the training and testing time for every method in experiment 3 for convenience to compare their detection efficiency.

CIC-IDS2017 Friday dataset is selected to evaluate the detection performance of different models in experiment 3. After preprocessing, the CIC-IDS2017 Friday dataset consists of 592,347 benign traffic samples, 125,670 DDoS traffic samples, 156,012 Portscan traffic samples, and 3,867 Botnet traffic samples. Compared with the number of benign samples, the quantities of these three types of malicious samples are obviously small. According to the distribution of different traffic samples in CIC-IDS2017 Friday dataset, the class priors of DDoS, Portscan, and Botnet can be set as 15%, 18%, and 0.5% respectively. We construct the training sets and testing sets for the three types of class-imbalanced malicious network traffic data based on the setting of UU learning. The proportion of labeled training data, unlabeled data, and testing data for SelfTraining are set as 10%, 70%, and 20%, respectively. All the six IDS models are trained and evaluated by the same training and testing data. Other settings are the same as experiment 1.

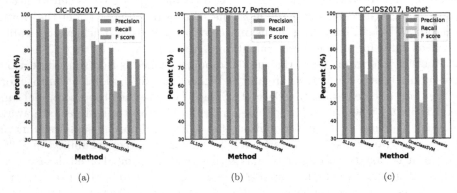

**Fig. 5.** Detection performance comparison among UUL and other five compared methods (SL100, Biased, SelfTraining, OneClassSVM, Kmeans) for three types of class-imbalanced malicious traffic data (DDoS, Portscan, Botnet) based on CIC-IDS2017 Friday dataset.

Figure 5 shows the experiment results when detecting the three class-imbalanced malicious traffic data by the six machine learning techniques based IDS models. UU learning based IDS models show the best detection performance for all the three types of class-imbalanced malicious traffic data. It can achieve 99.00% precision, 99.50% recall, and 99.25% F score even for the most biased attack type. As there are only 3,867 Botnet samples, the IDS model tends to predict a sample as "−1" instead of "1" when detecting Botnet traffic. Hence, the false negative values will be large and the false positive values will be small. This will easily cause a high precision and low recall. As shown in Fig. 5(c), almost all the selected methods reach a high detection precision. But the values of recall are not desirable for all the approaches except UU learning and SelfTraining based IDS models. However, the time cost of SelfTraining is very high, which needs more than 15 h to train an IDS model based on the CIC-IDS2017 Friday dataset. The two unsupervised learning techniques based IDS models, OneClassSVM and Kmeans, can train the models by unlabeled data. But their detection performance is not good when detecting class-imbalanced data. Besides, OneClassSVM based IDS is also time-consuming, which costs more than 16 h to train an IDS model with the training set generated from CIC-IDS2017 Friday dataset. The proposed UU learning based IDS is much more efficient than SelfTraining and OneClassSVM approaches, which only costs no more than 10 s to train an IDS model. It is a vital advantage for the proposed IDS that can detect the imbalanced malicious traffic data efficiently with a good detection performance.

## 5   Discussion

We have implemented three experiments to explore the applicability, stability, and practicality of the proposed weakly supervised learning (UU learning) based intrusion detection model. Experiment results show that the proposed intrusion detection model performs as well as the SL-IDS model in accurately detecting anomaly traffic. It also has the same benefit as SSL-IDS and UL-IDS that does not demand much FCL data. The UUL-IDS has another advantage that it can perform better even than the SL-IDS when dealing with imbalanced data. Besides, UU learning-based IDS is efficient which can train a detection model at a fast speed. Our experiment results demonstrate the applicability, stability, and practicality of the proposed UUL-IDS model trained by weak supervision information (class prior).

There are mainly two reasons that motivate us to try to utilize weak supervision information rather than strong supervision information for training an intrusion detection model. First, the class priors (weak supervision information) can be obtained easier than specific labels in some special circumstances where the data providers want to maintain data privacy. Second, constructing or estimating weak supervision information saves time and workforce than obtaining strong supervision information, especially in the domain of cybersecurity. This gives an insight that weakly supervised learning technology may be a suitable method to mitigate the problem that lacking FCL data in intrusion detection tasks, and it is worthy of further research. One of our previous works focuses on proactive cyber threat analysis by keeping up with the emerging threat intelligence from open sources [52]. This paper focuses on threat detection from insider network traffic records. We hope to combine the external threat intelligence analysis with insider intrusion detection works together to enhance the next-generation security operation centers in the future.

## 6   Related Work

Effectively monitoring the abnormal network traffic is an important requirement for IDS [15]. To mitigate the burden of continuously get enough fully correctly labeled network traffic data in the big data society, designing an efficient intrusion detection model with high economic efficiency and accuracy has become an important task for a long time. Many machine learning [18,22,32,40,44] and deep learning techniques [13,24,45,49,53] are applied to improve the intrusion detection efficiency. Some supervised learning based methods have achieved good performance on intrusion detection tasks. However, the supervised learning techniques need fully correctly labeled data to train the classifiers. The more labeled data involved, the more accurate the detection model performs for the deep learning-based method. As explained before, it is not practical to get much FCL data when considering the limited annotation capacity, high time and expense consumption in the real situation. Therefore, some semi-supervised and unsupervised learning techniques are studied for constructing IDS models.

Some research works utilize partially labeled data to train the intrusion detection models by semi-supervised learning techniques to reduce the dependency on FCL data. As one of the earliest and the most typical semi-supervised learning algorithms, self-training trains a classifier by a small portion of labeled data and a large portion of unlabeled data. The unlabeled samples are given pseudo labels by the labeled data trained model. Then the predicted unlabeled samples are added into the labeled data to retrain a classifier. Self-training method is utilized to overcome the difficulty of getting a large amount of FCL data to train an IDS model in these works [41,50,51]. Rana et al. proposed a fuzzy-based semi-supervised learning method for constructing an IDS model [4]. Their training steps are based on the idea of self-training. Other semi-supervised learning approaches like co-training are also motivated by self-training. Mao et al. applied a co-training semi-supervised method to design a multi-view intrusion detection model [28]. Most of the above research works only use KDD Cup 99 or NSL-KDD dataset to evaluate the performance of the proposed models. It would be better to use modern datasets to validate their methods. Another problem for these semi-supervised learning techniques is that the model will be affected by the incorrect predicted unlabeled samples.

Unsupervised learning techniques can train a model by unlabeled datasets, which also alleviates the burden of requiring a large amount of FCL data. As a typical and simple clustering-based unsupervised learning method, Kmeans algorithm has been applied by many researchers to construct an intrusion detection model [2,31,34]. Though the Kmeans based IDS model is easy and quick to train, it is sensitive to initialization. The predicted result can not always be transferred into a meaningful category. Except Kmeans, there are also many other unsupervised learning based techniques. The authors in [20] use an unsupervised learning approach that applying density-based and grid-based clustering algorithms to do the anomaly detection works. Casas et al. utilize a multi-clustering-based method to construct a knowledge independent intrusion detection model [7]. They combine the Sub-Space Clustering, Density-based Clustering, and multiple Evidence Accumulation algorithms to make an outliers detection. Some unsupervised deep learning techniques, like auto-encoder, are used to assign the label to unlabeled data in the intrusion detection tasks [3,43]. These methods can solve the problem that a limited amount of annotated data is available in the cybersecurity domain. However, it is difficult to interpret the different clusters into specific, meaningful categories for clustering-based unsupervised learning methods. In addition, the unsupervised learning-based IDS models can not perform as well as the supervised learning-based IDS and tend to be computationally complex.

Our proposed UU learning-based IDS can train a model by two unlabeled training sets with class priors. Class priors are weak supervision information which is easier and faster to get or estimate at a low cost than the strong supervision information in reality. The proposed method solves the problem of continuously getting a large amount of FCL data. Besides, it can achieve good detection performance, high detection efficiency, and keep robust when biased class priors are provided or facing class-imbalanced datasets. Furthermore, we choose three IDS evaluation datasets to examine the detection performance of the different machine learning-based IDS models in our work, which is more convinced.

## 7    Conclusion and Future Work

This paper proposed a novel direction for efficient and robust intrusion detection that mitigates the burden of being overly dependent on continuously obtaining massive FCL data to construct an effective anomaly detection (or analytics-based intrusion detection) model. The proposed scheme applies weak supervision information to train an intrusion detection model by weakly supervised learning techniques in this paper. Experiment results indicated that our proposed IDS model can achieve good detection performance as FCL data trained supervised learning-based IDS model and even much better than another baseline model that also trained under the supervised learning settings.

Weak supervision information, like class prior, is easier and cheaper to obtain than strong supervision information in real industrial scenarios. Considering the fact that it may not always be possible to obtain the true class priors directly, we examined the stability of the proposed IDS model when different pairs of biased class priors are provided. The experiment results showed that even under the worst biased coefficients, the detection rates are just 6.01%, 3.48%, and 1.8% less than the real detection rates based on NSL-KDD, UNSW-NB15, and CIC-IDS2017 datasets, respectively. This implies that our intrusion detection model is stable to overcome the negative influence from biased weak supervision information. Additionally, we examined the detection ability for class-imbalanced traffic data of the proposed method. It keeps good detection performance for all the three types of selected class-imbalanced traffic data and outperforms other baseline methods. The results also confirm that the proposed method is much efficient than two out of the five compared algorithms.

This work mainly focuses on exploring the applicability, stability, and practicality of applying weakly supervised learning techniques on intrusion detection tasks. As future work, we expect to design a multi-class classification IDS by weakly supervised learning algorithms or combine some attack behavior analysis techniques to enhance the proposed IDS model. Besides, we plan to select more intrusion detection datasets, especially the dataset that contains emerging traffic attacks, to validate the performance of the proposed model in more complicated scenarios.

**Acknowledgments.** This research is supported by the Cyber Security Agency of Singapore (CSA), under its repertoire of initiatives leveraging on research institutes and think-tanks to contribute to the international community "towards a secure and trusted IoT ecosystem".

# References

1. Ahmad, Z., Khan, A.S., Shiang, C.W., Abdullah, J., Ahmad, F.: Network intrusion detection system: a systematic study of machine learning and deep learning approaches. Trans. Emerg. Telecommun. Technol. **32**(1), e4150 (2021)
2. Al-Yaseen, W.L., Othman, Z.A., Nazri, M.Z.A.: Multi-level hybrid support vector machine and extreme learning machine based on modified k-means for intrusion detection system. Expert Syst. Appl. **67**, 296–303 (2017)
3. Alom, M.Z., Taha, T.M.: Network intrusion detection for cyber security using unsupervised deep learning approaches. In: 2017 IEEE National Aerospace and Electronics Conference (NAECON), pp. 63–69. IEEE (2017)
4. Ashfaq, R.A.R., Wang, X.Z., Huang, J.Z., Abbas, H., He, Y.L.: Fuzziness based semi-supervised learning approach for intrusion detection system. Inf. Sci. **378**, 484–497 (2017)
5. Bekker, J., Davis, J.: Estimating the class prior in positive and unlabeled data through decision tree induction. In: Thirty-Second AAAI Conference on Artificial Intelligence (2018)
6. Blanchard, G., Lee, G., Scott, C.: Semi-supervised novelty detection. J. Mach. Learn. Res. **11**, 2973–3009 (2010)
7. Casas, P., Mazel, J., Owezarski, P.: Knowledge-independent traffic monitoring: unsupervised detection of network attacks. IEEE Network **26**(1), 13–21 (2012)
8. Cisco, F.: Cisco annual internet report (2018–2023). White Paper. https://www.cisco.com/c/en/us/solutions/collateral/executive-perspectives/annual-internet-report/white-paper-c11-741490.html (2020)
9. De Comité, F., Denis, F., Gilleron, R., Letouzey, F.: Positive and unlabeled examples help learning. In: Watanabe, O., Yokomori, T. (eds.) ALT 1999. LNCS (LNAI), vol. 1720, pp. 219–230. Springer, Heidelberg (1999). https://doi.org/10.1007/3-540-46769-6_18
10. Debar, H., Dacier, M., Wespi, A.: Towards a taxonomy of intrusion-detection systems. Comput. Netw. **31**(8), 805–822 (1999)
11. Elkan, C., Noto, K.: Learning classifiers from only positive and unlabeled data. In: Proceedings of the 14th ACM SIGKDD International Conference on Knowledge Discovery and Data Mining, pp. 213–220 (2008)
12. Feng, L., et al.: Pointwise binary classification with pairwise confidence comparisons. In: International Conference on Machine Learning, pp. 3252–3262. PMLR (2021)
13. Gao, N., Gao, L., Gao, Q., Wang, H.: An intrusion detection model based on deep belief networks. In: 2014 Second International Conference on Advanced Cloud and Big Data, pp. 247–252. IEEE (2014)

14. Gharib, A., Sharafaldin, I., Lashkari, A.H., Ghorbani, A.A.: An evaluation framework for intrusion detection dataset. In: 2016 International Conference on Information Science and Security (ICISS), pp. 1–6. IEEE (2016)

15. Guo, Z., Lam, K.-Y., Chung, S.-L., Gu, M., Sun, J.-G.: Efficient presentation of multivariate audit data for intrusion detection of web-based internet services. In: Zhou, J., Yung, M., Han, Y. (eds.) ACNS 2003. LNCS, vol. 2846, pp. 63–75. Springer, Heidelberg (2003). https://doi.org/10.1007/978-3-540-45203-4_5

16. Hou, M., Chaib-Draa, B., Li, C., Zhao, Q.: Generative adversarial positive-unlabelled learning. arXiv preprint arXiv:1711.08054 (2017)

17. Jain, S., White, M., Radivojac, P.: Estimating the class prior and posterior from noisy positives and unlabeled data. Adv. Neural. Inf. Process. Syst. **29**, 2693–2701 (2016)

18. Kuang, F., Xu, W., Zhang, S.: A novel hybrid KPCA and SVM with GA model for intrusion detection. Appl. Soft Comput. **18**, 178–184 (2014)

19. Lam, K.Y., Hui, L., Chung, S.L.: Multivariate data analysis software for enhancing system security. J. Syst. Softw. **31**(3), 267–275 (1995)

20. Leung, K., Leckie, C.: Unsupervised anomaly detection in network intrusion detection using clusters. In: Proceedings of the Twenty-Eighth Australasian Conference on Computer Science, vol. 38, pp. 333–342 (2005)

21. Li, X., Bing, L.: Learning to classify texts using positive and unlabeled data. In: International Joint Conference on Artificial Intelligence (2003)

22. Li, Y., Guo, L.: An active learning based TCM-KNN algorithm for supervised network intrusion detection. Comput. Secur. **26**(7–8), 459–467 (2007)

23. Liu, L.P., Dietterich, T.G.: A conditional multinomial mixture model for superset label learning. In: NeurIPS. pp. 548–556 (2012)

24. Lopez-Martin, M., Carro, B., Sanchez-Esguevillas, A.: Application of deep reinforcement learning to intrusion detection for supervised problems. Expert Syst. Appl. **141**, 112963 (2020)

25. Lu, N., Niu, G., Menon, A.K., Sugiyama, M.: On the minimal supervision for training any binary classifier from only unlabeled data. arXiv preprint arXiv:1808.10585 (2018)

26. Lu, N., Zhang, T., Niu, G., Sugiyama, M.: Mitigating overfitting in supervised classification from two unlabeled datasets: a consistent risk correction approach. In: International Conference on Artificial Intelligence and Statistics, pp. 1115–1125. PMLR (2020)

27. Luo, J., Orabona, F.: Learning from candidate labeling sets. In: NeurIPS, pp. 1504–1512 (2010)

28. Mao, C.H., Lee, H.M., Parikh, D., Chen, T., Huang, S.Y.: Semi-supervised co-training and active learning based approach for multi-view intrusion detection. In: Proceedings of the 2009 ACM Symposium on Applied Computing, pp. 2042–2048 (2009)

29. MIT, L.L.: KDD Cup 1999 Data. http://kdd.ics.uci.edu/databases/kddcup99/kddcup99.htmll. Accessed 20 Jan 2021

30. Moustafa, N., Slay, J.: UNSW-NB15: a comprehensive data set for network intrusion detection systems (UNSW-NB15 network data set). In: 2015 Military Communications and Information Systems Conference (MilCIS), pp. 1–6. IEEE (2015)

31. Muda, Z., Yassin, W., Sulaiman, M., Udzir, N.: Intrusion detection based on k-means clustering and Naïve Bayes classification. In: 2011 7th International Conference on Information Technology in Asia, pp. 1–6. IEEE (2011)

32. Mukkamala, S., Janoski, G., Sung, A.: Intrusion detection using neural networks and support vector machines. In: Proceedings of the 2002 International Joint Conference on Neural Networks. IJCNN 2002 (Cat. No. 02CH37290), vol. 2, pp. 1702–1707. IEEE (2002)

33. Pedregosa, F., et al.: Scikit-learn: machine learning in Python. J. Mach. Learn. Res. 12, 2825–2830 (2011)

34. Peng, K., Leung, V.C., Huang, Q.: Clustering approach based on mini batch Kmeans for intrusion detection system over big data. IEEE Access 6, 11897–11906 (2018)

35. Perini, L., Vercruyssen, V., Davis, J.: Class prior estimation in active positive and unlabeled learning. In: Proceedings of the 29th International Joint Conference on Artificial Intelligence and the 17th Pacific Rim International Conference on Artificial Intelligence (IJCAI-PRICAI 2020), pp. 2915–2921. IJCAI-PRICAI (2020)

36. Ramaswamy, H., Scott, C., Tewari, A.: Mixture proportion estimation via Kernel embeddings of distributions. In: International Conference on Machine Learning, pp. 2052–2060. PMLR (2016)

37. Ratner, A., Bach, S., Varma, P., Ré, C.: Weak supervision: the new programming paradigm for machine learning. Hazy Research. Available via https://dawn.cs.stanford.edu//2017/07/16/weak-supervision/. Accessed 5 Sept 2019

38. Ratner, A.J., De Sa, C.M., Wu, S., Selsam, D., Ré, C.: Data programming: creating large training sets, quickly. Adv. Neural. Inf. Process. Syst. 29, 3567–3575 (2016)

39. Roh, Y., Heo, G., Whang, S.E.: A survey on data collection for machine learning: a big data-AI integration perspective. IEEE Trans. Knowl. Data Eng. 33(4), 1328–1347 (2021)

40. Ryan, J., Lin, M.J., Miikkulainen, R.: Intrusion detection with neural networks. In: Advances in Neural Information Processing Systems, pp. 943–949 (1998)

41. Shao, G., Chen, X., Zeng, X., Wang, L.: Labeling malicious communication samples based on semi-supervised deep neural network. China Commun. 16(11), 183–200 (2019)

42. Sharafaldin, I., Lashkari, A.H., Ghorbani, A.A.: Toward generating a new intrusion detection dataset and intrusion traffic characterization. In: ICISSp, pp. 108–116 (2018)

43. Shone, N., Ngoc, T.N., Phai, V.D., Shi, Q.: A deep learning approach to network intrusion detection. IEEE Trans. Emerg. Topics Comput. Intell. 2(1), 41–50 (2018)

44. Sindhu, S.S.S., Geetha, S., Kannan, A.: Decision tree based light weight intrusion detection using a wrapper approach. Expert Syst. Appl. 39(1), 129–141 (2012)

45. Singla, A., Bertino, E., Verma, D.: Preparing network intrusion detection deep learning models with minimal data using adversarial domain adaptation. In: Proceedings of the 15th ACM Asia Conference on Computer and Communications Security, pp. 127–140 (2020)

46. Stolfo, S.J., Fan, W., Lee, W., Prodromidis, A., Chan, P.K.: Cost-based modeling for fraud and intrusion detection: results from the jam project. In: Proceedings DARPA Information Survivability Conference and Exposition. DISCEX 2000, vol. 2, pp. 130–144. IEEE (2000)

47. Tavallaee, M., Bagheri, E., Lu, W., Ghorbani, A.A.: A detailed analysis of the KDD cup 99 data set. In: 2009 IEEE Symposium on Computational Intelligence for Security and Defense Applications, pp. 1–6. IEEE (2009)

48. Vapnik, V.: Principles of risk minimization for learning theory. In: Advances in Neural Information Processing Systems, pp. 831–838 (1992)

49. Vinayakumar, R., Soman, K., Poornachandran, P.: A comparative analysis of deep learning approaches for network intrusion detection systems (N-IDSs): deep learning for N-IDSs. Int. J. Digital Crime Forensics (IJDCF) **11**(3), 65–89 (2019)

50. Wagh, S.K., Kolhe, S.R.: Effective intrusion detection system using semi-supervised learning. In: 2014 International Conference on Data Mining and Intelligent Computing (ICDMIC), pp. 1–5. IEEE (2014)

51. Wurzenberger, M., Skopik, F., Landauer, M., Greitbauer, P., Fiedler, R., Kastner, W.: Incremental clustering for semi-supervised anomaly detection applied on log data. In: Proceedings of the 12th International Conference on Availability, Reliability and Security, pp. 1–6 (2017)

52. Yang, W., Lam, K.-Y.: Automated cyber threat intelligence reports classification for early warning of cyber attacks in next generation SOC. In: Zhou, J., Luo, X., Shen, Q., Xu, Z. (eds.) ICICS 2019. LNCS, vol. 11999, pp. 145–164. Springer, Cham (2020). https://doi.org/10.1007/978-3-030-41579-2_9

53. Yin, C., Zhu, Y., Fei, J., He, X.: A deep learning approach for intrusion detection using recurrent neural networks. IEEE Access **5**, 21954–21961 (2017)

54. Zeiberg, D., Jain, S., Radivojac, P.: Fast nonparametric estimation of class proportions in the positive-unlabeled classification setting. In: Proceedings of the AAAI Conference on Artificial Intelligence, vol. 34, pp. 6729–6736 (2020)

55. Zeng, Z.N., et al.: Learning by associating ambiguously labeled images. In: CVPR, pp. 708–715 (2013)

56. Zhou, Z.H.: A brief introduction to weakly supervised learning. Natl. Sci. Rev. **5**(1), 44–53 (2018)

# Data-Driven Cybersecurity

# CySecAlert: An Alert Generation System for Cyber Security Events Using Open Source Intelligence Data

Thea Riebe[1]([⊠]) [iD], Tristan Wirth[2] [iD], Markus Bayer[1] [iD], Philipp Kühn[1] [iD],
Marc-André Kaufhold[1] [iD], Volker Knauthe[2] [iD], Stefan Guthe[2] [iD],
and Christian Reuter[1] [iD]

[1] Science and Technology for Peace and Security (PEASEC),
Department of Computer Science, Technical University of Darmstadt,
Darmstadt, Germany
riebe@peasec.tu-darmstadt.de
[2] Interactive Graphics Systems Group, Technical University of Darmstadt,
Darmstadt, Germany
tristan.wirth@gris.informatik.tu-darmstadt.de

**Abstract.** Receiving relevant information on possible cyber threats, attacks, and data breaches in a timely manner is crucial for early response. The social media platform Twitter hosts an active cyber security community. Their activities are often monitored manually by security experts, such as Computer Emergency Response Teams (CERTs). We thus propose a Twitter-based alert generation system that issues alerts to a system operator as soon as new relevant cyber security related topics emerge. Thereby, our system allows us to monitor user accounts with significantly less workload. Our system applies a supervised classifier, based on active learning, that detects tweets containing relevant information. The results indicate that uncertainty sampling can reduce the amount of manual relevance classification effort and enhance the classifier performance substantially compared to random sampling. Our approach reduces the number of accounts and tweets that are needed for the classifier training, thus making the tool easily and rapidly adaptable to the specific context while also supporting data minimization for Open Source Intelligence (OSINT). Relevant tweets are clustered by a greedy stream clustering algorithm in order to identify significant events. The proposed system is able to work near real-time within the required 15-min time frameand detects up to 93.8% of relevant events with a false alert rate of 14.81%.

**Keywords:** Cyber security event detection · Twitter · Active learning · CERT

## 1 Introduction

Social Media has become a viable source for cyber security incident prevention and response, helping to gain situational awareness for Computer Emergency

© Springer Nature Switzerland AG 2021
D. Gao et al. (Eds.): ICICS 2021, LNCS 12918, pp. 429–446, 2021.
https://doi.org/10.1007/978-3-030-86890-1_24

Response Teams (CERTs). Therefore, the trend towards processing Social Media data in real-time to support emergency management [1] continues to grow. Husák et al. [2] show how Cyber Situational Awareness (CSA) is an adaptation of situational awareness to the cyber domain and supports operators to make strategic decisions. To perform such informed, situational decision-making, CERTs have to gain CSA by gathering and processing threat data from different closed and open sources [3]. These include Open Source Intelligence (OSINT), which uses any publicly available open source to accumulate relevant intelligence [4]. Especially the micro-blogging service Twitter has proven itself as a valuable source of OSINT due to its popularity among the cyber security community [5], as well as its available content and metadata for analysis [6]. Alves et al. [7] have shown that there is a small but impactful subset of vulnerabilities being discussed on Twitter before they are included into a vulnerability database. Increasingly big amounts of data make the use of more complex models possible. While concentrating on volume might be the best variable for some use cases, focusing on near real-time and data minimizing [8] approaches have been neglected in the recent state of research. Therefore, this paper seeks to answer the following main research question: **(RQ) How can relevant cyber security related events be detected automatically in near real-time based on Twitter data?**

By answering this research question the proposed paper aims to make the following contributions **(C)**: The first contribution **(C1)** deducts the concept and presents the implementation of an automated near real-time alert generation system for cyber security events based on Twitter data (Sect. 2). The second contribution **(C2)** covers the evaluation of the *CySecAlert* system that assists CERTs with the detection of cyber security events in order to improve CSA by automatically generating alerts on the basis of Twitter data (Sect. 3). The near real-time capability is achieved by labelling and clustering the Twitter stream within the required 15-min time frame [9]. The third contribution **(C3)** provides a comparison of existing tools based on the systematic of Atafeh and Khreich [10] that are suitable to detect relevant cyber security related events based on Twitter data (Sect. 4). Lastly, the results are summed up (Sect. 5). To enable further improvement of our work, we will make the source code and the labelled Twitter dataset available.[1]

## 2   Concept

This section presents the concept of *CySecAlert*, including the data source and architecture (Sect. 2.1), data preprocessing (Sect. 2.2), and training of the relevance classifier (Sect. 2.3) which serve as input to detect novel cyber security events (Sect. 2.4). It concludes with a concise description of the concept's implementation (Sect. 2.5).

---

[1] https://github.com/PEASEC/CySecAlert.

## 2.1  Data Source and Architecture

Twitter offers a multitude of advantages over other Social Media platforms. Firstly, Twitter is frequently used for the early discussion and disclosure of software vulnerabilities [7]. Secondly, Twitter accommodates a broad variety of participants, that are involved in the discourse evolving around cyber security topics. Since most important cyber security news feeds (e.g., NVD, ExploitDB, CVE) are present on the platform, Twitter serves as a cyber security news feed aggregate [11] and is used by both individuals and organisations [12]. In addition, tweets can be processed fast and easily [11], due to their limited length.

Hasan et al. [13] propose a general framework for *Event Detection* systems. We added a relevance classifier to the architecture that filters out irrelevant tweets. By classifying relevance per tweet, the individual relevance of each tweet was determined before the clustering process, reducing the number of tweets at an early stage. This extension was necessary because our tweet retrieval method is account-based, leveraging preexisting lists of cyber security experts' Twitter accounts (see Appendix A).

## 2.2  Preprocessing and Representation

In a preprocessing step, we standardized the tweet representation by converting their content to a lower case and removing any textual part that is unlikely to contain relevant information, i.e., stop words, URLs, and Social Media specific terms and constructs (e.g. "tweet", "retweet", user name mentions) as well as non-alphanumerical characters. Then the text was tokenized and stemmed.

We applied a clustering-based approach to *Event Detection*. Therefore, a representation of individual tweets was necessary. To address this issue we adopted the setting of Kaufhold et al. [14], where a *Bag-of-Words* approach was applied. Clustering and classification were performed online. Therefore, the Inverse Document Frequency (IDF) regularization term would have had to be updated after every iteration, undermining the benefits of online techniques. In the context of crisis informatics, it has been suggested that the regularization via IDF does not necessarily yield a relevant benefit on classification performance [14]. Therefore, we omitted IDF regularization and represented tweets by Term Frequency (TF) vectorization only.

## 2.3  Relevance Classifier

To filter relevant tweets, we used an active learning approach [15], which has been found to reduce the amount of labelled data that is required to reach a certain accuracy level [16,17]. We employed *uncertainty sampling* in order to obtain beneficial tweet samples for labeling. Therefore, we examined the suggestion of Kaufhold et al. [14] regarding rapid relevance classification. Lewis and Catlett [18] point out that it is reasonable to label the post which the current classifier instance is least confident about. Thus, the *Relevance Classification* is performed by application of *pool-based sampling* with the *least confidence* metric. *Pool-based sampling* refers to an algorithm class that picks an optimal data point out of the set of non-labelled data points utilizing a metric that refers to the data's information content [16]. We applied the *least confidence* metric that regarded a data point as the most optimal labeling sample if the classifier was least confident about its classification [16]. Therefore, the datum with a prediction confidentiality closest to the decision boundary was selected.

*Uncertainty sampling* requires retraining of the classifier after every labeling process [18], which is not done in online learning. Kaufhold et al. [14] have shown, that this improvement in training time comes at the price of classifier accuracy, which can be addressed by using a fast online learning algorithm for the selection of data to be labelled, while batchwise creating a more sophisticated offline classifier with the same labelled data in parallel [18]. The combination of an incremental k Nearest Neighbor (kNN) classifier for *uncertainty sampling* and Random Forest (RF) is suggested to perform well on datasets in crisis informatics [14]. The Evaluation shows that this is true for the domain of cyber security as well (Sect. 3.2). Despite the increase of deep learning algorithms in this field, the utilization of classical machine learning algorithms suits best for this use case as the retraining can be performed automatically without the need for long training phases and specific training optimizations for every batch.

### 2.4    Detecting Events and Generating Alerts

*Clustering* based event detection approaches utilize vectorized representations of Social Media posts. In this scenario, every cluster represented a candidate event. We applied a simple greedy clustering algorithm that utilizes similarity metrics of new Social Media posts to old ones by considering them part of a new cluster if they exceeded a certain similarity threshold and otherwise adding them to the most similar preexisting cluster [19]. We performed the clustering based on nearest-neighbor search and used cosine similarity to the nearest cluster's centroids.

Alves et al. [11] propose a more sophisticated method that applies regular offline k-means clustering to improve the cluster quality. However, we chose not to do so as we put a special emphasis on near real-time applicability on our system. Furthermore, we justify the choice of relatively simple event detection techniques by the fact that the active learning approach for relevance classification in the cyber security *event detection* domain constitutes the core novelty of our contribution.

To *obtain significant events*, candidate events are filtered by their significance. Depending on the costs of alert processing and underlying costs regarding false alerts, it is reasonable to allow a system operator to configure the system's alert generation sensibility. *CySecAlert* supports the prediction of candidate events based on (1) overall post count associated with the event, (2) count of experts covering the event, and (3) the number of retweets.

The significance of candidate events based on the system operator's configuration was evaluated when a new tweet was added to the respective cluster. If the cluster met the significance criteria and no alert had been issued based on the candidate event before, an alert was issued to the system operator. In order to assure the application's near-real-time capabilities tweets older than a certain time threshold (14 days by default) were removed from their respective cluster.

To *summarize events*, research suggests that textual clusters can be represented by display of their respective centroid [20, 21]. We chose this event representation because it is cost-efficient and maintains the feeling of handling original Twitter data. We additionally allowed the display of the entirety of posts associated with an event to allow a system operator to further examine the event.

## 2.5 Implementation

*CySecAlert* was implemented in Java 11 and utilized a MongoDB database because of its high performance in handling textual documents. Figure 1 serves as an overview of the implementation's architecture.

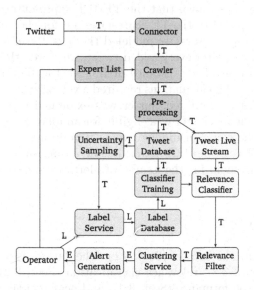

**Fig. 1.** Architecture of proposed Information and Communication Technology (ICT) illustrating the information flow for [T]weets, [L]abels and [E]vents. The ICT is divided into Tweet Retrieval (blue), Relevance Classifier Training (green) and Real-Time Event Detection (yellow). (Color figure online)

The Crawler module requested the most recent tweets of a list of trusted Twitter users in a regular manner. For this purpose, it used the Connector module. This functionality was implemented using Twitter4J[2]. To train a relevance

---

[2] Twitter4J Version 4.0.7 (twitter4j.org/en/index.html on 14.08.2020).

classifier, it is necessary to manually label a set of tweets. The proposed application offers the use of active learners to reduce labeling effort. We evaluated an active batch RF, an active Naive Bayes, and an active kNN classifier. We used the classifier implementation of Weka[3]. A Relevance Classifier was trained based on the labelled data. The tweets to be labelled depended on the chosen sampling method. We chose an RF because its performance is well-proven in the context of Twitter Analysis, which was verified by qualitative evaluation. Our implementation utilized the Weka (See Footnote 3) implementation of an RF in its default configuration. The Relevance Classifier was used to filter out irrelevant tweets.

Then relevant tweets that covered the same topics were clustered to candidate events. This allowed an estimation of how much coverage a topic has on Twitter and helped to avoid alerts being used twice for the same topic. Therefore, we employed a greedy streaming clustering algorithm, which assigned each new tweet to the cluster with the most similar centroid according to the cosine similarity. If this similarity was smaller than a certain operator-defined threshold (*Similarity Threshold*) the tweet was designed to a new cluster.

A pre-evaluation has shown that the TF-IDF representation yielded performance benefits compared to the TF representation for the clustering task. Due to the sparsity of these vectors, we modeled them as *HashMaps*. Since classical IDF had to be updated after every added tweet, we stored the tweets in TF vectorized form and a centralized instance of IDF vector. The IDF regularization was applied on-demand if calculations required a vectorized representation. After every tweet insertion, the altered cluster was examined regarding its qualifications for an alert. Such a cluster was eligible for an alert if no alert had yet been issued for it and the count of unique tweets it contained exceeds a predefined threshold (*Alert Tweet Count Threshold*). The cosine similarity threshold and the tweet count threshold for the issuing of alerts were passed during program initialization.

# 3 Evaluation

This section presents the dataset (Sect. 3.1). The dataset is used to evaluate the active learning (Sect. 3.2), relevance classification (Sect. 3.2), alert generation (Sect. 3.3), system performance (Sect. 3.4), and near real-time capability (Sect. 3.5) of *CySecAlert*.

## 3.1 Dataset

We gathered 350,061 English tweets (151,861 tweets excl. retweets) published by 170 Twitter accounts of leading cyber security experts in the time period between 1st January 2019 and 31st July 2020. The list of accounts was derived based on a set of blog entries that provide lists of leading cyber security experts on Twitter (see Appendix A, Table 4).

---

[3] Weka v3.8.4(https://www.cs.waikato.ac.nz/ml/weka/ on 14.08.2020).

**Table 1.** Class distribution over tweets of ground truth datasets.

|          | $S1$            | $S2$            |
|----------|-----------------|-----------------|
| From     | 01/12/2019      | 01/05/2020      |
| To       | 31/12/2019      | 14/05/2020      |
| Irrelevant | 5,801 (88.9%) | 5,780 (85.25%)  |
| Relevant | 724 (11.10%)    | 1000 (14.25%)   |
| Total    | 6,525           | 6,780           |
| $\kappa$ | 0.9318          | 0,9377          |

In Relevance Classification, it is common to apply a binary classification into *relevant* and *irrelevant* tweets [11,22,23]. The class definitions of *relevant* and *not relevant* we applied are illustrated in a codebook (see Appendix B, Table 5) after Mayring [24].

Based on the dataset and the proposed annotation scheme, we created an annotated ground truth dataset consisting of two subsets ($S1$, $S2$) covering different time frames. The Datasets $S1$ and $S2$ were annotated by an additional researcher to estimate the inter-rater reliability of the coding scheme as shown in the codebook (Appendix B). Our ground truth shows a high level of inter-rater reliability ($\kappa > 0.90$) measured by Cohen's kappa ($\kappa$). We used $S2$ for evaluation purposes. The class distributions of these datasets are illustrated in Table 1.

### 3.2   Relevance Classification

**Sampling Method.** We evaluated the influence of active learning and the selection of a sampling method and sampling classifier on the performance of a relevance classifier in order to choose a high-performing classifier. Therefore, we used the preprocessed and stemmed ground truth datasets $S1$ and $S2$. In this evaluation, a scenario was simulated where no labelled data is available initially. A virtual expert incrementally labelled tweets that were chosen by different sampling methods. The labels were taken from the respective ground-truth dataset. We examined a Naive Bayes classifier, a kNN classifier with $k = 50$ and an RF classifier. As *uncertainty sampling* technique we applied *least confidence* measure in a *pool-based sampling* scenario were examined.

While Naive Bayes and kNN can be implemented in an incremental manner and thus allow to add single tweets without retraining, the RF classifier did not offer this property. For this reason, kNN and Naive Bayes were updated after every new labelled tweet and the next uncertainty sampling step was performed on the updated classifier. In contrast, the RF classifier sampled a set of most uncertain tweets (rather than one) which were labelled as batches before being added to the training set. Thereafter, the classifier was retrained on the updated dataset.

An evaluation of the experiment (see Appendix C, Fig. 3) showed, that the active version of the Naive Bayes classifier performed worst, representing nearly random classification behaviour. However, the kNN classifier was able to train a model whose AUC measure plateaus around roughly 0.75 for both datasets. This finding is similar to the results of Kaufhold et al. [14]. In contrast to them, we also considered active learning with an RF classifier. In our evaluation setting, it performed best with an AUC in the range of 0.9. Therefore, we choose a RF classifier for our system.

**Classification Model.** In this subsection, we analyse whether the use of a different active learning algorithm-based sampling method is useful for an RF relevance classifier. We compare (1) kNN and (2) batchwise RF uncertainty sampling with (3) random sampling and (4) batchwise Random-RF-Hybrid Sampling. This hybrid approach picks 50% of tweets per batch by RF-based uncertainty sampling and 50% tweets at random. By determining a threshold of Random Trees, which is needed to classify an instance as positive, a classifier is instantiated from the learned RF. In the context of this contribution, we chose the $F_1$ metric for evaluation purposes, as it is suitable for imbalanced datasets.

We evaluated the performance of the RF instances based on the $F_1$ measure of the classifier instance with the highest $F_1$ measure for every 100 labelled tweets. The evaluation was conducted by leaving out 1,000 tweets and using them as a test set. In order to mitigate performance issues, the uncertainty sampling was performed on a randomly chosen subsample of size 200 (500 for active batch RF), which changed in every iteration, rather than on the complete data pool. The results of this evaluation are illustrated in Fig. 2.

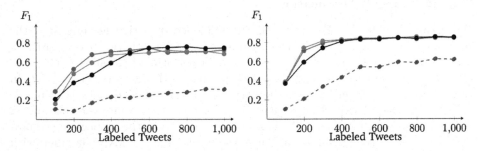

**Fig. 2.** Performance comparison of RF Classifier trained on dataset $S1$ (left) and $S2$ (right) with uncertainty sampling by different classifiers: Random (blue), RF Classifier (red), 50% RF and 50% Random (brown) and by kNN classifier with $k = 50$ (black). Average over 5 Executions using a 1,000 tweet holdout set measured in $F_1$. (Color figure online)

The experimental results show that every examined type of uncertainty sampling leads to classifier out-performance compared to random sampling. For every experiment, the classifier instance that used a randomly sampled dataset was not able to achieve the performance of uncertainty sampled classifier with 300 or more labelled tweets, even if it was trained based on 1,000 randomly sampled tweets. Furthermore, the results indicate that there are no significant performance differences between the tested uncertainty sampling classifiers.

Due to the fact that there are no substantial classification quality implications, we opted for the kNN based uncertainty sampling because it can be executed in an online manner. Additionally, the results suggest that the overall classification performance suffered for datasets with higher class imbalance. Nevertheless, the results indicate that after around 600 labelled tweets the classifier achieved its best classification quality and therefore did not show significant improvements for a bigger training dataset. This constituted a reduction of manual tweet annotation of up to 90% compared to a randomly sampled approach, which makes it necessary to label the whole dataset (roughly 6,000 tweets each).

### 3.3  Alert Generation

In this section, we jointly evaluate the clustering algorithm and the alert generation process. Therefore, we executed the combination of these modules using different parameters for *Similarity Threshold* and *Alert Tweet Count Threshold*. Even though there are multiple configurations for alert generation thresholds, the evaluation was performed based on the relevant tweet count per cluster metric only. Thereby, we received a list of clusters that represent a list of relevant events and their associated tweets. By comparing this list to the ground truth dataset (Sect. 3.1), the quality of the alert generation process could be estimated.

Therefore, clusters that were found by the clustering algorithm and flagged as alerts are classified as *topic related*, *mixed* or *duplicate*. A cluster was regarded as *topic* related if more than half of its tweets belong to the same topic of the ground truth topic list. If a *topic related* cluster that discussed this topic had been found before, the cluster was marked as *duplicate*. If there was no major topic in the cluster, it was defined as *mixed*. *Topic related* clusters were marked as positive, while *mixed* and *duplicate* clusters were marked as negative. Combining this information we derived a calculation for precision and recalled measures as follows:

$$Precision = \frac{\#truepositives}{\#truepositives + \#falsepositives} = \frac{\#topicrelated}{\#clusters} \quad (1)$$

$$Recall = \frac{\#truepositives}{\#truepositives + \#falsenegatives} = \frac{\#topicrelated}{\#topics} \quad (2)$$

In order to decouple the evaluation of clustering and alert generation from the performance of the relevance classifier, we tested the clustering-based alert generation algorithm on the set of *relevant* and *potentially relevant* tweets from our ground truth datasets $S1$ and $S2$. We used TF-IDF as tweet vectorization in order to avoid the formation of big clusters based on frequently used common words. The results show that an increase in the value of the used similarity threshold (in the observed range) decreases the recall (see Appendix D). Intuitively, this can be explained by the creation of more clusters due to similarity failing the threshold. Therefore, clusters are smaller on average and stay under the alert generation threshold, which leads to suppression of alert generation for relevant topics. In contrast, the influence of similarity threshold on cluster precision (which is the invert of the wrongful alert quote) is lower. This is the reason why operators should be advised to prefer lower values for the Cosine Similarity Threshold. Even though this configuration increases the wrongful alert rate, it increases the recall. Nevertheless, if the similarity threshold is chosen too low, this does not hold. For example, a similarity threshold of 0 led to every tweet being part of one giant cluster. This led to a low recall as well. The alert generation instance with the best performance regarding the F1 score resulted in a precision of 96.08% and a recall of 96.23%.

Our experiment shows that the value of the *Cosine Similarity Threshold* leading to an optimal F1-measure depends on the *Alert Tweet Count Threshold*. Furthermore, the results indicate that minor changes in *Alert Tweet Count Threshold* have no significant effect on the Alert Generation System's performance. Comparing the best performing configurations for every examined *Alert Tweet Count Threshold* (similarity threshold of 0.3 for 3, similarity threshold of 0.25 for 5) shows that the performance differences are lower than 5%. Therefore, the system operator is advised to choose the *Alert Tweet Count Threshold* based on an alert frequency, that s/he is willing to process.

### 3.4  System Performance

This section examines the performance of the overall system combining Uncertainty Sampling, Relevance Classification, and Alert Generation. The evaluation is conducted based on the datasets $S1$ and $S2$. After data preprocessing, an RF classifier was trained based on 600 tweets that were chosen by Uncertainty Sampling using a kNN classifier. Every tweet in the dataset that the resulting classifier deemed relevant was passed to the Alert Generation System which is configured according to the findings in Sect. 3.3: *Alert Tweet Count Threshold* $= 5$, *Cosine Similarity Threshold* $= 0.25$. The evaluation of the clusters was performed analogous to the procedure in Sect. 3.3 with *irrelevant* clusters as additional cluster class. A cluster was thereby considered *irrelevant* if it contained at least 50% tweets that are labelled as irrelevant. The experimental results (Table 2) suggest that the system is capable of detecting 90% of the events occurring in the ground truth data while 15% of reported alerts were not part of the ground truth data (false alert rate).

### 3.5  (Near-)Real-Time Capability

The run-time tests were performed on a computer with an *AMD Phenom II X6* CPU and 12 GB DDR3 RAM running Windows 10. We divided the alert generation system into two stages and measured their execution time separately: (TU1) the Relevance Classifier and (TU2) combining the clustering process with the alert generation process. We conducted the experiments using dataset $S1$. Since individual tweet frequency is highly volatile, we conducted our simulation

**Table 2.** Combined performance of relevance classifier, clustering algorithm and alert generation for datasets $S1$ and $S2$.

| Dataset | $S1$ | $S2$ |
|---|---|---|
| Precision | 95% | 85.19% |
| Recall | 90.48% | 93.88% |
| $F_1$ | 92.68% | 89.32% |

assuming the following worst-case scenario: Every user sends twice his/her average daily tweet count in the same one our frame: 2.5 Tweets per user per 15 min time-frame.

Sabottke et al. [9] suggest that the cyber security community on Twitter consists of about 32,000 accounts. Assuming that the system is used to issue alerts based on the tweets of 25% of these accounts, 20,000 have to be processed in a 15-min time frame in order to allow near real-time execution. Our experiments show that the execution of (TU1) takes 17.5 s for 20,000 Tweets. Based on the class distribution, we determined in Sect. 3.1, $\approx$2,000 of these tweets are going to be labelled as positive. Assuming that tweets that are older than 14 days are discarded, the clusters of the clustering service contain about 112,000 tweets at any time in this scenario. Extrapolation of the experiment on the execution time for the proposed clustering algorithm suggests that the clustering of 500 tweets takes about 210 s in this case. That corresponds to around 840 s (or 14 min) for the given 2,000 tweets. Adding the execution times of (TU1) and (TU2) up shows that an execution in the given 15-min time frame is possible. An execution in a timely manner for more accounts or accounts that are more active is possible using a more powerful machine.

# 4   Related Work and Discussion

To use Twitter as an OSINT source for CERTs, we conducted a comparative analysis of existing tools and approaches which are suitable to complete this task (Sect. 4.1). Based on our contributions (Sect. 4.2), we identified limitations and potentials for future work (Sect. 4.3).

## 4.1   Cyber Security Event and Hot Topic Detection

Previous work has examined the possibilities of Twitter as an information source for cyber security event detection (overview in Table 3). As the techniques for event detection using Twitter differ, Atafeh and Khreich [10] offer a systematic approach that allows a comparison based on the of the necessary parts. Most previous work [12,21–23,25] examines the detection of generic cyber security threats. The majority of these publications [12,21,23] employs some kind of clustering algorithm on a Term Frequency-Inverse Document Frequency (TF-IDF) representation of single tweets compared by the cosine similarity distance.

Even though the publications' core approach is related, they differ in details concerning the preprocessing of tweets and usage of the detected clusters. On closer inspection, most methodologies use human-generated input that serves as a filter for user-generated content and automatically expands these filters configuration by utilizing Twitter data [26]. These filters are either represented by lists of relevant keywords [26] or a set of credible experts [27]. To our knowledge, the scientific literature has not discussed the advantages and disadvantages of either approach extensively. This is especially true for the performance of machine learning algorithms on the respective databases. While a keyword-based retrieval approach is less prone to miss relevant tweets regarding a certain objective, it may attract a lot of tweets that contain a relevant keyword in a different semantic. Account-based approaches reduce the number of tweets that have to be processed and therefore reduce performance requirements for the underlying hardware. However, these accounts have to be known beforehand.

**Table 3.** An overview of event detection techniques with application to the cyber security domain, categorized by Retrieval Method (RM, [A]ccount-based or [K]eyword-based (* is filtering)), Detection Method (DM, [S]upervised or [U]nsupervised), as well as Pivot Technique (PT, [D]ocument- or [F]eature-based) and Detection Technique (DT) and Model, based on Atefeh and Khreich [10].

| Work | RM | | DM | | PT | | Application | DT | Model |
|---|---|---|---|---|---|---|---|---|---|
| | A | K | S | U | D | F | | | |
| [11] | ✓ | * | ✓ | | ✓ | | Summarization | CluStream, SVM, NN | TF-IDF |
| [23] | | ✓ | | ✓ | | ✓ | Threats | DBSCAN | TF-IDF |
| [21] | | ✓ | | ✓ | ✓ | | Novel malware | Counting, K-Means | #, TF-IDF |
| [22] | ✓ | * | ✓ | | ✓ | | Threats | NER by NN | Word Emb. |
| [28] | ✓ | * | ✓ | | ✓ | | Threats | NER by MTL | Word Emb. |
| [29] | | ✓ | ✓ | | ✓ | | Threat events | MTL | Word Emb. |
| [30] | | ✓ | ✓ | | ✓ | | Cur. incidents | Prob. learning | TF |
| [31] | | ✓ | | ✓ | | ✓ | Attacks | Clustering | Exp. queries |
| [27] | ✓ | | | ✓ | ✓ | | Topics | Clustering | TF, Corr. |
| [26] | | ✓ | | ✓ | ✓ | | Classification | Clustering | TF-IDF |
| [32] | | ✓ | | ✓ | | ✓ | IT-Sec. alerts | Rule-based reason. | Graph(VKG) |
| [20] | | ✓ | ✓ | | ✓ | | IT-Sec. events | Expect. Reg. | Diff. feat. |
| [25] | ✓ | | | ✓ | ✓ | | Ident. Attacks | Term Filtering | TF |
| [33] | | ✓ | ✓ | | ✓ | | Threat indicators | CNN-GRU | Random Emb. |
| [12] | | ✓ | | ✓ | | ✓ | 0-day exploits | K-Means | Documents |
| CySecAlert | | | | | | | | | |
| | ✓ | ✓ | ✓ | | ✓ | | IT-Sec Events | Rel. Filter, Clustering | TF-IDF |

## 4.2 Contributions

For the **CySecAlert concept (C1)**, we opted for an account-based retrieval approach, that retrieves tweets based on a list of credible cyber security experts' accounts. Active learning using uncertainty sampling has shown to be beneficial for training supervised classifiers with limited data in other domains [14,16,17,34]. Literature of crisis informatics in combination with our evaluation suggests that an incremental kNN classifier outperforms a Naive Bayes classifier and an active batch sampling version of an RF classifier if they are used as uncertainty sampling classifier for a batch RF classifier. Therefore, they allow high-quality classifiers with a smaller training set. This is valuable for the privacy by design principle of data minimization [8]. This means that fewer accounts and tweets are needed. In detail, our **evaluations (C2)** show that a training set containing only 600 tweets gathered by Uncertainty Sampling (10% of ground truth database) is suited to build a sufficient classifier. A classifier based on a training set consisting of 1,000 randomly sampled tweets is outperformed by a set of 200 uncertainty sampled tweets. The evaluation shows that *CySecAlert* scores a maximal $F_1$ measure of 92.68% (Precision: 95%, Recall: 90.48%) (Sect. 3.4). In **comparison to other approaches (C3)**, this exceeds the performance of Bose et al. [23] with an $F_1$ measure of 78.26% (Precision: 81.82%, Recall: 75%) and is comparable to the results of Dionísio et al. [28] with an $F_1$ measure of 95.1%, who have examined a related task. Although these papers are most comparable as they conduct similar experiments, a direct comparison of the evaluation results is nevertheless impractical because they refer to datasets of different time periods gathered from different sets of accounts. Regarding the real-time capability to our knowledge, only Le Sceller et al. [26] included a simple evaluation in their experiments. We extend the research in this direction as we perform a more in-depth analysis also incorporating the usage behavior. The near real-time of the system is not only supported by its capability to analyse the real-time Twitter stream [21,25,26], it also performs almost as fast as the SONAR system [26] (17.5 s for 20,000 tweets compared to 12 s).

### 4.3 Limitations and Future Work

As the *CySecAlert* system is designed to support CERTs, further improvements and evaluations as part of larger-scale incident monitoring are planned, such as the deployment on other social media platforms and longitudinal testing with larger datasets. The tests will include further studies regarding the security of the system against hacked or fake accounts as well as the risk of model poisoning. Further, controlled experiments will be conducted to exclude the impact of the dataset. Additionally, in recent times more sophisticated clustering algorithms have been proposed. For instance, Alves et al. [11] extends a greedy clustering approach by offline re-clustering if the cluster affiliation of a new tweet is unclear. This approach may be suited to avoid *duplicate* clusters in our clustering algorithm but may have a negative impact on the real-time properties. Furthermore, re-clustering, in general, interferes with the used online event selection process by changing cluster affiliation of past tweets. Future work should examine streaming clustering algorithms that are suited to enhance the proposed system's overall performance without strongly influencing the capability of processing tweets of many users in a timely manner and the need for re-clustering.

Following the proposed system by Kaufhold et al. [14], we used the bag-of-word approach to represent text. However, recent contributions suggest that *Word Embeddings* can have relevant performance advantages over a multitude of other textual representation methods, including the bag of word approach applied in this contribution [35]. Future research should examine if the application of Word Embeddings is suited to further improve the proposed alert generation system's performance without the negative influence of the system's timing constraints. Furthermore, NNs in general and in the domain of cyber security related event detection enjoy increasing popularity and show high performance in relevance classification tasks [22]. While the current state of the system with its real-time, low-resource, and robust applicability is only suited for classical machine learning algorithms, future work should examine the influence of different uncertainty sampling classifiers on the performance of NNs as relevance classifiers.

## 5    Conclusion

This work proposes a framework for timely detection of novel and relevant cyber security related events based on data from the social media platform Twitter (*CySecAlert*). *CySecAlert* is capable of collecting tweets based on a list of trusted user accounts, filtering them by relevance, dividing them into clusters by topic similarity, and issuing alerts if one such topic surpasses a predefined significance threshold. The system further aims to support data minimization for OSINT by focussing on a network of expert accounts. Further, it is easy for an expert community, such as CERTs, to adopt as well as quick to train with little labelling and runs in near real-time. Our study based on manually labelled ground truth data shows that the amount of labelled data to train a classifier can be substantially reduced by the application of uncertainty sampling for training set generation in contrast to random sampling. The proposed classifier achieves a precision of 87.18% and a recall of 84.12%, while the cluster-based alert generation subsystem achieves a false alert rate of 3.77% and detects 96.08% of relevant events in the ground truth dataset. An evaluation of the overall system shows that it is able to detect up to 93.88% of relevant events in a ground truth dataset with a false alert rate of 14.81%.

**Acknoledgements.** This work was supported by the German Federal Ministry for Education and Research (BMBF) in the projects CYWARN (13N15407) and KontiKat (13N14351), as well as by the BMBF and the Hessian Ministry of Higher Education, Research, Science and the Arts within their joint support of the National Research Center for Applied Cybersecurity ATHENE. We would like to thank the anonymous reviewers for their valuable and constructive comments.

## Appendix A    Dataset

Table 4 provides the websites and blogs we used to retrieve 170 accounts of the leading cyber security experts on Twitter, from which we gathered the dataset of 350,061 English tweets (see Sect. 3.1).

**Table 4.** Sources for cyber security experts on Twitter

| List of security expert sources |
| --- |
| The top 25 infosec leaders to follow on Twitter[a] |
| Top 15 security experts to follow on Twitter in 2018[b] |
| Best cyber security Twitter profiles to follow 2018[c] |
| 100 security experts you could follow on Twitter[d] |
| 10 cybersecurity Twitter profiles to watch[e] |
| 21 cyber security Twitter accounts you should be following[f] |

[a] techbeacon.com/security/top-25-infosec-leaders-follow-twitter, accessed 2021-07-08
[b] resources.whitesourcesoftware.com/blog-whitesource/top-15-security-experts-to-foll-ow-on-twitter-in-2018, accessed 08.07.2021
[c] cyberdb.co/best-cyber-security-twitter-profiles-follow-2018, accessed 08.07.2021
[d] bridewellconsulting.com/100-security-experts-follow-twitter, accessed 08.07.2021
[e] darkreading.com/vulnerabilities—threats/10-cybersecurity-twitter-profiles-to-watch/d/d-id/1325031, accessed 08.07.2021
[f] sentinelone.com/blog/21-cybersecurity-twitter-accounts-you-should-follow/, accessed 08.07.2021

# Appendix B     Codebook

In Table 5 the codebook [24] for the annotation of tweets is presented, which is applied to the coding of the dataset (see Sect. 3.1). Table 5 gives an overview of the codes' definitions.

**Table 5.** Codebook for tweet relevance classification.

| Code | Definition | Example |
| --- | --- | --- |
| Relevant (2) | Information on existence, properties, assessment, real-world application or warning of (1) vulnerabilities in software, (2) vulnerabilities in hardware, (3) malware, or (4) attack vectors, that are (a) currently in use, (b) may be (ab-used) or (c) in theory | "Zeppelin, a new #ransomware variant of Vega family, is targeting #technology and health companies across Europe, the US and Canada."[a], "Frankfurt City IT Network Taken Offline to Stop #Emotet #Botnet Infection"[b], "Citrix Vulnerability Puts 80K Companies at Risk"[c] |
| Irrelevant (1) | None of the above | |

[a] Twitter (twitter.com/unix_root/status/1204813126371295238)
[b] Twitter (twitter.com/neirajones/status/1208817022295068672)
[c] Twitter (twitter.com/InfosecurityMag/status/1209175732695523330)

## Appendix C    Classifier Comparison

Figure 3 depicts the results of active classifier comparison. Experiment details are discussed in Sect. 3.2.

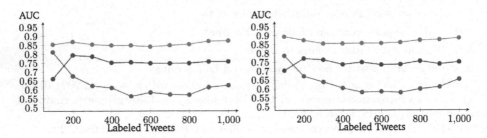

**Fig. 3.** Performance comparison of Naive Bayes (red), kNN with $k = 50$ (blue) and Random Forest (brown) classifier with uncertainty sampling based on their respective model on dataset $S1$ (left) and $S2$ (right). Average over 5 executions using Cross-Validation. (Color figure online)

## Appendix D    Alert Generation by Similarity Threshold

Table 6 depicts how recall and alert generation is impacted by the similarity threshold of the greedy clustering (see Sect. 3.3).

**Table 6.** Performance measures of greedy clustering-based generated alerts for different similarity thresholds and for alert count thresholds 3 and 5 for the datasets $S1$ and $S2$, respectively.

| Alert count thresh. | 3 ($S1$) | | | | 5 ($S2$) | | |
|---|---|---|---|---|---|---|---|
| Similarity-thresh. | 0.25 | 0.3 | 0.4 | 0.5 | 0.2 | 0.25 | 0.3 |
| Precision | 81.54% | **96.08%** | 90.63% | 94.11% | 75% | **95.24%** | 86.67% |
| Recall | **100%** | 96.23% | 60.41% | 30.18% | **100%** | 95.24% | 61.9% |
| $F_1$ | 89.83% | **96.15%** | 72.5% | 45.7% | 86% | **95.24%** | 72.22% |

## References

1. Reuter, C., Kaufhold, M.A.: Fifteen years of social media in emergencies: a retrospective review and future directions for crisis informatics. J. Contingencies Crisis Manage. **26**(1), 41–57 (2018)
2. Husák, M., Jirsík, T., Yang, S.J.: SoK: contemporary issues and challenges to enable cyber situational awareness for network security. In: Proceedings of the 15th International Conference on Availability, Reliability and Security. ARES 2020. Association for Computing Machinery, New York, NY, USA (2020)

3. Yang, W., Lam, K.Y.: Automated cyber threat intelligence reports classification for early warning of cyber attacks in next generation SOC. In: International Conference on Information and Communication Systems (ICICS), pp. 145–164 (2020)

4. Mittal, S., Das, P.K., Mulwad, V., Joshi, A., Finin, T.: CyberTwitter: using Twitter to generate alerts for cybersecurity threats and vulnerabilities. In: 2016 IEEE/ACM International Conference on Advances in Social Networks Analysis and Mining (ASONAM), pp. 860–867. IEEE (2016)

5. Behzadan, V., Aguirre, C., Bose, A., Hsu, W.: Corpus and deep learning classifier for collection of cyber threat indicators in Twitter stream. In: 2018 IEEE International Conference on Big Data (Big Data), pp. 5002–5007. IEEE (2018)

6. Tundis, A., Ruppert, S., Mühlhäuser, M.: On the automated assessment of open-source cyber threat intelligence sources. In: Krzhizhanovskaya, V.V., et al. (eds.) ICCS 2020. LNCS, vol. 12138, pp. 453–467. Springer, Cham (2020). https://doi.org/10.1007/978-3-030-50417-5_34

7. Alves, F., Andongabo, A., Gashi, I., Ferreira, P.M., Bessani, A.: Follow the blue bird: a study on threat data published on Twitter. In: Chen, L., Li, N., Liang, K., Schneider, S. (eds.) ESORICS 2020. LNCS, vol. 12308, pp. 217–236. Springer, Cham (2020). https://doi.org/10.1007/978-3-030-58951-6_11

8. Koops, B.J., Hoepman, J.H., Leenes, R.: Open-source intelligence and privacy by design. Comput. Law Secur. Rev. **29**(6), 676–688 (2013)

9. Sabottke, C., Suciu, O., Dumitras, T.: Vulnerability disclosure in the age of social media: exploiting Twitter for predicting real-world exploits. In: 24th USENIX Security Symposium USENIX Security 15, pp. 1041–1056 (2015)

10. Atefeh, F., Khreich, W.: A survey of techniques for event detection in Twitter. Comput. Intell. **31**(1), 132–164 (2015)

11. Alves, F., Bettini, A., Ferreira, P.M., Bessani, A.: Processing tweets for cybersecurity threat awareness. arXiv preprint arXiv:1904.02072 (2019)

12. Trabelsi, S., et al.: Mining social networks for software vulnerabilities monitoring. In: 2015 7th International Conference on New Technologies, Mobility and Security (NTMS), pp. 1–7. IEEE (2015)

13. Hasan, M., Orgun, M.A., Schwitter, R.: A survey on real-time event detection from the Twitter data stream. J. Inf. Sci. **44**(4), 443–463 (2018)

14. Kaufhold, M.A., Bayer, M., Reuter, C.: Rapid relevance classification of social media posts in disasters and emergencies: A system and evaluation featuring active, incremental and online learning. Inf. Process. Manage. **57**(1), 102132 (2020)

15. Habdank, M., Rodehutskors, N., Koch, R.: Relevancy assessment of tweets using supervised learning techniques: mining emergency related tweets for automated relevancy classification. In: 2017 4th International Conference on Information and Communication Technologies for Disaster Management (ICT-DM), pp. 1–8. IEEE (2017)

16. Settles, B.: Active learning literature survey. University of Wisconsin (2010)

17. Imran, M., Mitra, P., Srivastava, J.: Enabling rapid classification of social media communications during crises. Int. J. Inf. Syst. Crisis Response Manage. (IJIS-CRAM) **8**(3), 1–17 (2016)

18. Lewis, D.D., Catlett, J.: Heterogeneous uncertainty sampling for supervised learning. In: Machine Learning Proceedings 1994, pp. 148–156. Elsevier (1994)

19. Allan, J., Lavrenko, V., Jin, H.: First story detection in TDT is hard. In: Proceedings of the Ninth International Conference on Information and Knowledge Management, pp. 374–381 (2000)

20. Ritter, A., Wright, E., Casey, W., Mitchell, T.: Weakly supervised extraction of computer security events from Twitter. In: Proceedings of the 24th International Conference on World Wide Web, pp. 896–905 (2015)
21. Concone, F., De Paola, A., Re, G.L., Morana, M.: Twitter analysis for real-time malware discovery. In: 2017 AEIT International Annual Conference, pp. 1–6. IEEE (2017)
22. Dionisio, N., Alves, F., Ferreira, P.M., Bessani, A.: Cyberthreat detection from twitter using deep neural networks. In: 2019 International Joint Conference on Neural Networks (IJCNN), pp. 1–8. IEEE (2019)
23. Bose, A., Behzadan, V., Aguirre, C., Hsu, W.H.: A novel approach for detection and ranking of trendy and emerging cyber threat events in Twitter streams. In: Proceedings of the 2019 IEEE/ACM International Conference on Advances in Social Networks Analysis and Mining, pp. 871–878 (2019)
24. Mayring, P.: Qualitative content analysis. Companion Qual. Res. 1(2004), 159–176 (2004)
25. Sapienza, A., Ernala, S.K., Bessi, A., Lerman, K., Ferrara, E.: Discover: mining online chatter for emerging cyber threats. In: Companion Proceedings of the The Web Conference 2018, pp. 983–990 (2018)
26. Le Sceller, Q., Karbab, E.B., Debbabi, M., Iqbal, F.: Sonar: automatic detection of cyber security events over the Twitter stream. In: Proceedings of the 12th International Conference on Availability, Reliability and Security (ARES), pp. 1–11 (2017)
27. Lee, K.C., Hsieh, C.H., Wei, L.J., Mao, C.H., Dai, J.H., Kuang, Y.T.: Sec-buzzer: cyber security emerging topic mining with open threat intelligence retrieval and timeline event annotation. Soft. Comput. 21(11), 2883–2896 (2017)
28. Dionísio, N., Alves, F., Ferreira, P.M., Bessani, A.: Towards end-to-end cyberthreat detection from twitter using multi-task learning. In: 2020 International Joint Conference on Neural Networks (IJCNN), pp. 1–8. IEEE (2020)
29. Fang, Y., Gao, J., Liu, Z., Huang, C.: Detecting cyber threat event from twitter using IDCNN and BiLSTM. Appl. Sci. 10(17), 5922 (2020)
30. Ji, T., Zhang, X., Self, N., Fu, K., Lu, C.T., Ramakrishnan, N.: Feature driven learning framework for cybersecurity event detection. In: Proceedings of the 2019 IEEE/ACM International Conference on Advances in Social Networks Analysis and Mining, pp. 196–203 (2019)
31. Khandpur, R.P., Ji, T., Jan, S., Wang, G., Lu, C.T., Ramakrishnan, N.: Crowdsourcing cybersecurity: Cyber attack detection using social media. In: Proceedings of the 2017 ACM on Conference on Information and Knowledge Management, pp. 1049–1057 (2017)
32. Mittal, S., Joshi, A., Finin, T.: Cyber-all-intel: an AI for security related threat intelligence. arXiv preprint arXiv:1905.02895 (2019)
33. Simran, K., Balakrishna, P., Vinayakumar, R., Soman, K.P.: Deep learning approach for enhanced cyber threat indicators in Twitter stream. In: Thampi, S.M., Martinez Perez, G., Ko, R., Rawat, D.B. (eds.) SSCC 2019. CCIS, vol. 1208, pp. 135–145. Springer, Singapore (2020). https://doi.org/10.1007/978-981-15-4825-3_11
34. Bernard, J., Zeppelzauer, M., Lehmann, M., Müller, M., Sedlmair, M.: Towards user-centered active learning algorithms. In: Computer Graphics Forum, vol. 37, pp. 121–132. Wiley Online Library (2018)
35. Mikolov, T., Chen, K., Corrado, G., Dean, J.: Efficient estimation of word representations in vector space. arXiv preprint arXiv:1301.3781 (2013)

# CyberRel: Joint Entity and Relation Extraction for Cybersecurity Concepts

Yongyan Guo, Zhengyu Liu, Cheng Huang$^{(\boxtimes)}$, Jiayong Liu, Wangyuan Jing, Ziwang Wang, and Yanghao Wang

School of Cyber Science and Engineering, Sichuan University, Chengdu, China
`codesec@scu.edu.cn`

**Abstract.** Cyber threats are becoming increasingly sophisticated, while new attack techniques are emerging, causing serious harm to businesses and even countries. Therefore, how to analyze attack incidents and trace the attack groups behind them becomes extremely important. Threat intelligence provides a new technical solution for attack traceability by constructing Cybersecurity Knowledge Graph (CKG). The CKG cannot be constructed without a large number of entity-relation triples, and the existing entity and relation extraction for cybersecurity concepts uses the traditional pipeline model that suffers from error propagation and ignores the connection between the two subtasks. To solve the above problem, we propose CyberRel, a joint entity and relation extraction model for cybersecurity concepts. We model the joint extraction problem as a multiple sequence labeling problem, generating separate label sequences for different relations containing information about the involved entities and the subject and object of that relation. CyberRel introduces the latest pretrained model BERT to generate word vectors, then uses BiGRU neural network and the attention mechanism to extract features, and finally decodes them by BiGRU combined with CRF. Experimental results on Open Source Intelligence (OSINT) data show that the F1 value of CyberRel is 80.98%, which is better than the previous pipeline model.

**Keywords:** Relation extraction · Joint model · Threat intelligence · Knowledge graph

## 1 Introduction

Nowadays, the damage and impact caused by malicious behavior in cyberspace such as hacker attacks, frauds, and rumors have become more serious. Therefore, how to effectively and accurately detect cyber attacks as early as possible, analyze attack incidents, and trace the source of attackers and groups has become a severe problem for enterprises and countries.

The concept of Cyber Threat Intelligence (CTI) was developed supplying new theoretic support for cyber-attack source tracing, making it possible to trace the source of a wide range of attacks. Therefore, many researchers extract and analyze different threat intelligence to generate the Cybersecurity Knowledge Graph

© Springer Nature Switzerland AG 2021
D. Gao et al. (Eds.): ICICS 2021, LNCS 12918, pp. 447–463, 2021.
https://doi.org/10.1007/978-3-030-86890-1_25

(CKG). The CKG has the characteristic of strong timeliness and high accuracy, which can timely and easily detect, respond and defend against specific targets providing a new measure for attack source tracking, and can even effectively deal with sophisticated cyberattacks (e.g., zero-day attacks, advanced persistent threat).

The key step in constructing CKG is cyber threat intelligence information extraction, which involves subtasks such as entity recognition, relation extraction, and event extraction. Currently, many research groups have conducted research on the automated construction and analysis of CKG [3–9]. In terms of CTI information extraction, previous studies are dedicated to extracting cybersecurity concepts [10–12] and entities [13–15] from unstructured data.

However, the construction of CKG is inseparable from a large number of cybersecurity entity-relation triples. The CKG consists of a number of nodes and edges, where the nodes represent entities and the edges represent the relations between entities. Because that information comes from a large scale of unstructured data through various sources like system logs, vulnerability databases, cybersecurity reports, hacker forums, and social media, it has the characteristics of multisource, heterogeneous, polysemy, and highly dependent on domain knowledge. Therefore, relation extraction of cybersecurity is still a great challenge. Existing researches on cybersecurity relation extraction [16,17] uses the traditional pipeline model, named entity recognition first and then relation extraction, which leads to error propagation and losses sight of the relevance between entity recognition and relation extraction.

To solve the above problem, we propose CyberRel, a joint entity and relation extraction model for cybersecurity concepts, which extracts both cybersecurity entities and relations and generates the semantic triples. Specifically, we use a tagging scheme to convert the joint extraction problem into a multiple sequence labeling problem by generating separate label sequences for different relations containing information about the related entities and the subject and object of that relation. CyberRel applies the pre-trained model, BERT, to generate word vectors. After extracting semantic features by BiGRU, the model assigns higher weights to relation-related words in the sentences by an attention mechanism. Finally, BiGRU combined with CRF is used to decode and construct cybersecurity triples.

In summary, the main contribution of this paper are as follows:

- We propose a joint entity and relation extraction model for cybersecurity concepts, named CyberRel. The model employs deep learning techniques to extract entities and relations in sentences simultaneously, avoiding the error propagation of traditional pipeline models.
- We model the joint extraction problem as a multiple sequence labeling problem by generating separate label sequences for different relations. Each label sequence contains information about the related entities and the subject and object of that relation. This method can effectively solve the entity overlapping problem commonly found in the cybersecurity corpus.

- To validate the effectiveness of CyberRel, we collected and manually labeled OSINT data including vulnerability databases, security bulletins, and APT reports. The experimental results show that CyberRel outperforms the traditional pipeline model with an F1 value of 80.98%.

The rest of the paper is organized as follows: Sect. 2 discusses related work, and Sect. 3 presents the details of the joint entity and relation extraction model for cybersecurity concepts (CyberRel) which we proposed in this paper. Section 4 provides the experiments and analysis related to this work. Section 5 summarizes conclusion and proposes future works.

## 2   Related Work

In this section, we first review the methods for automated construction and analysis of CKG. Secondly, since the pivotal step of CKG construction is threat intelligence information extraction, we review the work related to CTI extraction including entity recognition, relation extraction, and event extraction subtasks. Finally, we present the related research on relation extraction.

### 2.1   Cybersecurity Knowledge Graph

The Knowledge Graph (KG) was originally proposed by Google. It is a knowledge base that integrates information from multiple sources, links real-world entities or concepts, and provides search services through semantic retrieval. In the field of cybersecurity, correlating and fusing threat intelligence data from different sources to generate the CKG can provide new technical means for situational awareness and attack traceability.

Building a CKG first requires abstracting a myriad of concepts and complex relations in the cybersecurity domain into a semantic network. Iannacone et al. [1] proposed STUCCO, an ontology for building CKGs, integrating 13 different formats of cybersecurity data sources. Building on this foundation, Syed et al. [2] proposed a Unified Cybersecurity Ontology (UCO). The UCO ontology provides a general understanding of the cybersecurity domain and, in addition to mapping to STIX, UCO extends several related cybersecurity standards, vocabularies, and ontologies such as CVE, CCE, CVSS, CAPEC, CYBOX, KillChain, and STUCCO.

In the area of automated construction and analysis of CKG, researchers have also proposed several ideas and approaches in recent years [3–9]. Jia et al. [3] introduced a cybersecurity knowledge base and deduction rules based on a quintuple model. Gao et al. [4] proposed EFFHUNTER, a system that facilitates threat hunting in computer systems using OSINT. The system uses an unsupervised, lightweight, and accurate NLP pipeline to extract structured threat behaviors from unstructured OSINT text. Piplai et al. [5] described a system that extracts information from After Action Reports (AARs) and represents the extracted information in a CKG. Zhao et al. [6] demonstrated a threat intelligence framework (HINTI). HINTI first recognizes IOCs and models the interdependent relations between IOCs using heterogeneous information networks

(HINs), and then proposes a threat intelligence computing framework based on graph convolutional networks to explore complex security knowledge. Although these approaches have made initial attempts and achieved good results in CKG construction, further research is needed in the key steps of knowledge graph construction.

## 2.2  Threat Intelligence Information Extraction

The construction of a knowledge graph can be divided into three steps, including information extraction, knowledge fusion, and knowledge reasoning. Among them, information extraction plays a decisive role in the quality of the generated knowledge graph. Information extraction for threat intelligence is divided into several subtasks, including entity recognition [10–15], relation extraction [16,17] and event extraction [18].

In terms of cybersecurity entity and concept recognition, Mittal et al. [10] proposed a framework for extracting threat intelligence from Twitter, Cyber-Twitter, which automates the extraction of security vulnerability concepts. Liao et al. [11] introduced iACE for automatically extracting IOCs and their context in the sentences of technical articles. Zhu et al. [12] designed Chainsmith, an IOC extraction system that collects IOCs from security articles and classifies them according to the stages of the Kill Chain. Ghazi et al. [13] used natural language processing to extract threat sources from unstructured web threat information sources and provided comprehensive threat reports in the STIX standard.

Due to the lack of a well-labeled corpus for training, relatively few studies have been conducted on cybersecurity relation extraction and event extraction compared to entity recognition. Pingle et al. [16] proposed RelExt, a deep learning-based cybersecurity relation extraction method for constructing CKGs. The model uses a pipeline approach, first identifying entities in the text by an entity recognizer then classifying the relations by a deep learning model. Jones et al. [17] implemented a semi-supervised cybersecurity relation extraction method based on a bootstrapping algorithm to extract relations. Satyapanich et al. [18] proposed CASIE, a security event extraction system that uses deep neural networks and can incorporate rich linguistic features and word embeddings for extracting security events related to cyber-attacks and vulnerabilities.

## 2.3  Relation Extraction

As a subtask of information extraction, relation extraction has a long research history. The main approaches to relation extraction can be broadly divided into three categories, including early rule-based approaches [19,20]; traditional machine learning-based approaches [21,22]; and deep learning-based approaches [23–27]. In recent years, the latest research results in the field of relation extraction have focused on deep learning models [28–31]. The advantage of deep learning methods is that they do not require manual extraction of features nor a large amount of domain knowledge.

Currently, there are two main approaches to relation extraction based on deep learning: the pipeline approach and the joint approach. The pipeline approach performs relation classification after extracting all the entities. Zeng et al. [23] first applied CNN to relation extraction to automatically extract lexical and sentence-level features. Wei et al. [24] proposed a novel cascaded binary annotation framework (CASREL) that models relations as functions that map subjects to objects in a sentence, which naturally handles the overlapping triple problems. Although these methods achieve promising results, the pipeline architectures suffer from the problem of error propagation. In addition, neglecting the relationship between the two tasks of entity recognition and relation extraction for training can also affect the effectiveness of relation extraction. Therefore, to construct the bridge between the two subtasks, building a joint model that extracts entities together with relations simultaneously has attracted much attention. Miwa et al. [25] proposed a joint relation extraction model based on shared parameters, which captures both word sequences and dependency tree substructure information for end-to-end relation extraction via LSTM. Bekoulis et al. [26] propose a joint model that uses a CRF layer to model the entity recognition task and the relation extraction task as a multi-headed selection problem. Zheng et al. [27] proposed a new tagging scheme that can convert the joint extraction task to a sequence labeling problem. Yuan et al. [30] proposed a relation-based attention network (RSAN) to jointly extract entities and relations using a relation-aware attention mechanism.

In the construction of CKG, a lot of research has been conducted on the extraction of cybersecurity entities and concepts, while research on cybersecurity relation extraction is still in its infancy. Existing approaches use traditional pipeline methods, which leads to error propagation and loses sight of the relevance between entity recognition and relation extraction. Different from these above works, this paper proposes a joint entity and relation extraction model for cybersecurity concepts, which extracts entities and relations simultaneously, effectively avoiding the shortcomings of the traditional pipeline model.

## 3    Methodology

In this section, we introduce CyberRel, a joint entity and relation extraction model for cybersecurity concepts. We briefly outline the overall strategy here before discussing details in the following subsections. The overall architecture of CyberRel is shown in Fig. 1. CyberRel takes threat intelligence data collected from multiple sources as raw input. Then the data undergoes a pre-processing process including data cleaning, sentence segmentation, and tokenization to obtain the training corpus, which will be fed into the joint extraction model subsequently (see Sect. 3.1 for details). We adopt the cybersecurity entities and relations defined in the UCO 2.0 [2] ontology and model the joint entity and relation extraction problem as a multiple sequence labeling problem by generating a sequence of labels for each relation through a specific tagging schema (see Sect. 3.2 for details). Each relation label sequence contains information about the entities involved and the subject and object of the relation. Our proposed multiple sequence labeling model is structured into an embedding layer, an encoding layer, an attention layer, and a decoding layer (see Sect. 3.3 for details). Finally, CyberRel constructs cybersecurity triples based on the label sequences predicted by the model, and these triples will eventually be used to construct CKGs.

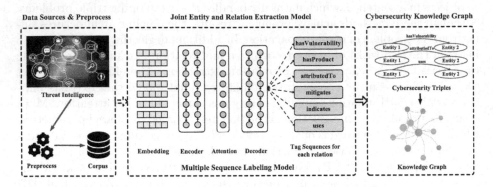

**Fig. 1.** CyberRel architecture.

### 3.1   Data Preprocess

CyberRel can extract cybersecurity triples from massive amounts of heterogeneous threat intelligence data. Threat intelligence data can be sourced from vulnerability databases, security bulletins, APT reports, security or technology blogs, hacking forums. This data is usually stored in rich text documents such as PDF, HTML/XML, JSON, and other formats. First, we use various text parsing tools (e.g. HTMLParser, PDFLib) to extract the raw text from these documents. But the extracted raw text is not well-formatted. Therefore, we devised some data pre-processing procedures as follows.

The first step in preprocessing is data cleaning, where we remove non-ASCII characters from the text and whitespace characters at the beginning and end of each sentence. It is worth noting that in some threat intelligence data, special types of entities are often rewritten to prevent readers from clicking on them by mistake. For example, the IP address "136.244.119.85" is rewritten as "136. 244.119[.]85"; the URL "http://www.test.com" is rewritten to "http://www.test[.]com"; the email address "hacker@test.com" is rewritten as "hacker[at]test.com". We revert this rewritten form to its original form.

The next step in preprocessing is special entity substitution. In the field of cybersecurity, some entities are very different in form from the normal natural language, such as IP, MAC, Hash, URL, Email, domain name, file name, and file path. We build regular expressions to match these entities from text and replace them with natural language strings in the form of "sub type", where "type" is the type of the special entity. For example, we would replace the IP address "136.244.119.85" with "sub ip".

The last step in the preprocessing process is text segmentation, which is the process of converting text into sequences. We use the NLTK library for sentence segmentation and WordPiece for word tokenization.

## 3.2 Tagging Scheme

In this section, we will introduce the tagging schema for the joint entity and relation extraction. The entities and relations applied by CyberRel are derived from UCO 2.0 [2], which provides a general understanding of the cybersecurity domain.

- The main entity types in UCO 2.0 include: Indicator, Threat Actor, Attack Pattern, Malware, Tool, Campaign, Course of Action, Vulnerability.
- The main relation types in UCO 2.0 include: hasProduct, hasVulnerability, uses, attributedTo, mitigates, indicates.

In the field of relation extraction, there has been related work [27,30,31] on the joint entity and relation extraction through the construction of a specific tagging schema. For cybersecurity concepts, the extracted relation usually suffers from the entity overlapping problem that different types of relations sharing the same entities, so the tagging scheme has to overcome this issue. CyberRel generates a sequence of labels for each relation in UCO 2.0. In each tag sequence, we use the typical "BIO" signs to locate the entities in the sentence, where "B" represents the starting part of the entity, "I" represents the middle part of the entity, and "O" is the non-entity part. At the same time, we also label the entity as subject or object in the relation, with "1" representing the subject in the triple and "2" representing the object in the triple.

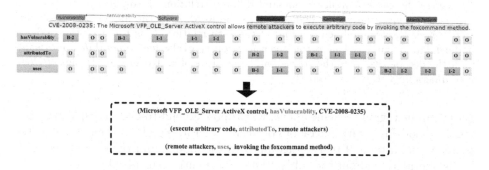

**Fig. 2.** An example for tagging scheme.

Figure 2 shows an example of our tagging scheme. The first label sequence describes the "hasVulnerablity" relation, where "Microsoft VFP_OLE _Server ActiveX control" is an entity of type "Software", as the subject of the "hasVulnerablity" relation; "CVE-2008-0235" is an entity of type "Vulnerability", as the object of the "hasVulnerablity" relation. Through the label sequence, we can generate triple ("Microsoft VFP_OLE_Server ActiveX control", "hasVulnerability", "CVE-2008-0235"). Likewise, other label sequences can be used to generate triples of corresponding relations. If a relation does not exist in a sentence, the label sequence for that relation will be all "O". Besides, as we can see, the "*attributedTo*" and "*uses*" relations have the over-lapped entity "*remote attackers*", and they can be extracted without conflict based on the separate label sequences.

### 3.3   Multiple Sequence Labeling Model

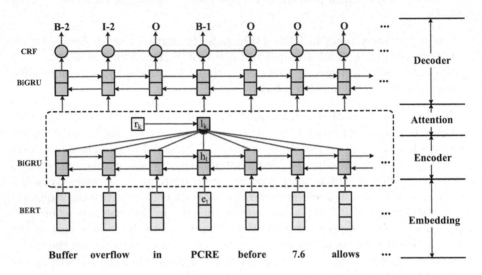

**Fig. 3.** The multiple sequence labeling model for joint entity and relation extraction. It receives the same sentence input and different relation $r_k$ to extract all triples in the sentence. $e_t$ is the BERT embedding of the word, $h_t$ is the hidden vector of time step $t$, $r_k$ is the trainable embedding of the $k$-th relation, $l_k$ is the attention weights under relation type $r_k$. Under the given relation $r_k$ (Take *hasVulnerability* for example), the decoder extracts the corresponding entities of $r_k$ to generate triples (*PCRE, hasVulnerability, Buffer overflow*).

Based on the tagging scheme above, we propose an end-to-end multiple sequence labeling model to jointly extract cybersecurity entities and relations. We take the sentence and a type of relation as input to the model, and the output sequence holds information about the subject and object entities involved in that relation. Thus, for a sentence, when we traverse all the relation types, the model generates a label sequence for each type of relation, resulting in a joint extraction of entities and relations. Figure 3 gives an overall structure of the model, which is divided into four parts. The embedding layer generates a word vector $e_t$ for each word $x_t$ in sentence $X$. In the encoding layer, the embedding sentence is fed into the bi-directional Gated Recurrent Units (BiGRU) neural network to generate a hidden state representation $h_t$. Then we apply the attention mechanism to assign different weights to the context words under different relations and constructs a relation-specific sentence representation $l_k$. Finally, in the decoding layer, we use another BiGRU neural network and joined it with CRF for decoding to obtain the label sequence and extract corresponding entities under the specific relation.

**Embedding.** Given a sentence as a sequence of tokens, the word embedding layer is responsible to map each token to a word vector. In this paper, we propose to use a pre-trained model to generate word vectors. The pre-trained word embedding model converts words in natural language into dense vectors, and semantically similar words will have similar vector representations. The latest pre-trained model BERT [35] can solve the problem of polysemy, generating different word vectors for the same word according to the context, which can better express the semantic features of the words. This situation often occurs in the cybersecurity corpus. For a piece of software, when describing the vulnerabilities that exist in that software, this entity should then be recognized as a *"Software"* type, and the triple (*"Software"*, *"hasVulnerability"*, *"Vulnerability"*) can be extracted. In another context, the software may be exploited as a tool by an attacker, at which point the entity should be recognized as a *"Tool"* type, and the triple (*"Threat Actor"*, *"uses"*, *"Tool"*) can be extracted. So, we use the BERT model to generate word embedding vectors in the embedding layer. For the input sentence $X = \{x_1, x_2, x_3, ..., x_n\}$, where $x_t$ is the $t$-th word in the sentence. After the computation of the BERT pre-trained model, the word embedding vector $E = \{e_1, e_2, e_3, ..., e_n\}$ of the sentence is generated, where $e_t$ is the word vector of the $t$-th word in the sentence.

**Encoder.** Compared with the traditional recurrent neural network (RNN), GRU consists of an update gate and a reset gate, which can alleviate the gradient disappearance or explosion problem that occurs during training. The GRU hidden state $h_t$ is generated by the previous hidden state $h_{t-1}$ and the input $e_t$ of the current state together. The GRU only calculates the correlation between time step $t$ and the previous time step. However, in the cybersecurity corpus, entities may constitute relations with the entities before or after. So, for the word vectors generated by the embedding layer, we further extract the semantic features of the sentences $H = \{h_1, h_2, h_3, ..., h_n\}$ using BiGRU and then concatenate the forward and backward GRU hidden states as the contextual word representation. The transformations are as follows:

$$h_t = \left[\overrightarrow{GRU}(e_t), \overleftarrow{GRU}(e_t)\right] \tag{1}$$

**Attention Mechanism.** In the cybersecurity corpus, a sentence usually contains many entities and complex relations. As shown in Fig. 2, the sentence contains five different entities (*"Vulnerability"*, *"Software"*, *"Threat Actor"*, *"Campaign"*, *"Attack Pattern"*) and three different relations (*"hasVulnerability"*, *"attributedTo"*, *"uses"*). Therefore, it is necessary to assign different weights to the words in a sentence according to different types of relations. For example, for the "*hasVulnerability*" relation, the words in the sentence indicating a software name or identify a specific vulnerability should be paid higher attention. Thus, we have referred to the relation-based attention mechanism proposed by Yuan et al. [30]. The attention mechanism can assign different weights to the words in a sentence under each relation, and the attention score can be calculated as follows:

$$h_g = avg\{h_1, h_2, h_3, \ldots, h_n\} \tag{2}$$

$$e_{tk} = v^T \tanh(W_r r_k + W_g h_g + W_h h_t) \tag{3}$$

$$a_{tk} = \frac{\exp(e_{tk})}{\sum_{j=1}^{n} \exp(e_{jk})} \tag{4}$$

where $h_g$ indicates the global representation of the sentence, $r_k$ is the embedding of the $k$-th relation. $v$, $W_r$, $W_g$, and $W_h$ are all trainable parameters. The attention score generated reflects the importance of the sentence's words in the context as well as relational expression in the current relation. The sentence representation $l_k$ under the $r_k$ relation is generated by the weighted sum of the sentence words, which is calculated as shown in Eq. 5. The attention layer combines the generated $l_k$ and the sentence representations output by the encoding layer as input to the decoding layer, as shown in Eq. 6.

$$l_k = \sum_{t=1}^{n} a_{tk} h_t \tag{5}$$

$$h_t^k = h_t \oplus l_k \tag{6}$$

**Decoder.** The decoding layer generates the label sequences of the sentences under the $r_k$ relation and returns the relational triples through the tagging scheme described in Sect. 3.2. We first used another BiGRU to produce sentence representations $H^o = \{h_1^o, h_2^o, h_3^o, \ldots, h_n^o\}$ and generate sequence scores $Z = \{z_1, z_2, z_3, \ldots, z_n\}$ using features from the encoding and attention layers. The calculation process is as follows, where W is the parameter:

$$h_t^o = \left[\overrightarrow{GRU}(h_t^k), \overleftarrow{GRU}(h_t^k)\right] \tag{7}$$

$$z_t = W h_t^o \tag{8}$$

Next, the sequence is decoded by the CRF layer, which is able to obtain constrained rules from the training data, to ensure that the predicted cybersecurity entity labels are valid. The decoding process is shown as follows:

$$score(Z, y) = \sum_{t=0}^{n} A_{y_t, y_{t+1}} + \sum_{t=1}^{n} Z_{t, y_t} \tag{9}$$

$$p(y \mid Z) = \frac{\exp(score(Z, y))}{\sum_{y' \in Y_Z} \exp(score(Z, y'))} \tag{10}$$

$$y^* = \arg \max_{y \in Y_Z} score(Z, y) \tag{11}$$

where A is the transition matrix between labels, $score(Z, y)$ is the position score, and $p(y \mid Z)$ is the normalized probability function. Finally, the label sequence $y^*$ is generated.

## 4    Experiments

### 4.1    Datasets

The datasets used in this paper are collected from publicly available OSINT data, including the CVE vulnerability database, security bulletins, and Advanced Persistent Threats (APT) reports. To train the CyberRel model, we invited five graduate students majoring in cybersecurity to annotate the dataset, using the BRAT annotation platform [34]. In total, we annotated 13,262 sentences containing 75,990 triples.

- **CVE vulnerability database:** CVE is the Common Vulnerabilities and Exposures, a list of various computer security vulnerabilities that have been publicly disclosed. The CVE Automation Working Group is piloting the use of git to share information about public vulnerabilities [32].
- **Security bulletins:** Many vendors (e.g. Microsoft, Adobe, Oracle, Vmware) regularly publish security bulletins that are intended to disclose security vulnerabilities in their software, describe remedies, and provide applicable updates for the affected software.
- **APT reports:** APT reports are publicly available papers and blogs related to malicious activities and associated with APT organizations or toolsets [33].

### 4.2    Evaluation Metrics

We use standard Precision (P), Recall (R), and F1-score to measure the performance of CyberRel. A triple is considered to be correctly extracted if and only if its relation type and both entities are correctly matched.

### 4.3    Experimental Settings

To evaluate the effectiveness of CyberRel, we design a set of experiments. Since the previous work used the traditional pipeline model, we compare CyberRel (joint model) with the previous work [16] (pipeline model) in the main experiment. As CyberRel is built with the word embedding model and neural networks, we designed two comparison experiments to analyze the effects of different word embedding models and different neural networks on the performance of Cyber-Rel.

We use StratifiedKFold to create train/test splits and set k = 5. The size of the BERT embedding vector is 768 dimensions. The size of the BiGRU hidden layer and relational embedding vector are both set to 300. We choose RMSprop as our model optimizer, the learning rate is 0.0001, and the batch size is 64. We use the dropout mechanism to avoid overfitting with a rate of 0.5.

## 4.4   Experimental Result

**Main Results.** The CyberRel proposed in this paper is a joint entity and relation extraction model, so we compare it with the existing pipeline approach, RelExt [16]. From Table 1, we can see that CyberRel outperforms RelExt, significantly improving precision (83.00%), recall (79.09%) and F1-score (80.98%). This indicates that the joint model extracts both entities and relations, which avoids the error propagation between the two subtasks of the pipeline model, and effectively improves the performance of entity-relation triples extraction.

**Table 1.** Main results of the compared models.

| Models | Precision | Recall | F1-score |
|---|---|---|---|
| RelExt [16] | 57.48% | 63.90% | 60.52% |
| **CyberRel** | **83.00%** | **79.09%** | **80.98%** |

**Effect of Different Word Embeddings.** As the word vectors generated by the word embedding model serve as the input to the downstream model, the quality of the word vectors has an important impact on the model performance. In this section, we experiment with two representative word embedding models, BERT [35] and Word2Vec [36], where the BERT model is the "cased_L-12_H-768_A-12" version, and the Word2Vec model is trained by Youngja et al. [36] through cybersecurity corpus. It can be seen from Table 2 that using BERT for word embedding has a certain improvement compared to Word2Vec. The F1 value is improved by 13.10% when GRU is used and by 14.03% when LSTM is used. This is attributed to the fact that BERT can generate different word vectors for the same word depending on the context thus making better use of the contextual information of the text, while Word2Vec can only generate a fixed word vector representation for each word.

**Table 2.** Results for different word embeddings and different neural networks in Cyber-Rel.

| Methods | | Precision | Recall | F1-score |
|---|---|---|---|---|
| Embedding | Neural Network | | | |
| Word2Vec | LSTM | 70.84% | 62.55% | 66.40% |
| | GRU | 73.91% | 62.81% | 67.88% |
| **BERT** | LSTM | 82.40% | 78.63% | 80.43% |
| | **GRU** | **83.00%** | **79.09%** | **80.98%** |

**Effect of Different Neural Networks.** Since a neural network is used in our model for the sequence labeling task, we investigated the effect of different neural networks on the model performance, specifically, we experimented with the performance of LSTM and GRU neural networks, respectively. As shown in Table 2, when using the same word embedding model, such as BERT, GRU (Precision: 83.00%, Recall: 79.09%, F1-score: 80.98%) performs slightly better than LSTM (Precision: 82.40%, Recall: 78.63%, F1-score: 80.43%). The experimental results show that GRU is more suitable for the problem of the joint entity and relation extraction for cybersecurity concepts. So we take BiGRU neural network in CyberRel.

### 4.5  Case Study

In this section, we illustrate the advantages of the joint model over the pipeline model by two examples, as shown in Appendix Table 3. In both examples, our proposed joint model predicts all the triples in the sentences correctly.

For Case 1, although the pipeline model correctly predicts all the *"Software"* entities in the entity recognition task, in the relation prediction between the three *"Software"* entities, the model predicts the three entities in two-by-two combinations and comes up with the wrong relations (*"mitigates"*). This indicates that the pipeline model does not take into account the connection between entity recognition and relation extraction tasks, while the joint model is able to predict the two *"hasProduct"* triples between the three *"Software"* entities well.

For Case 2, the pipeline model only recognizes the *"patches/Course-of-Action"* entity but not the *"CVE-2008-3138/Vulnerability"* entity, resulting in a null input to the relation extraction model that fails to predict the relation between them. This indicates that the pipeline model has the defect of error propagation, implying that if an entity is not predicted or is incorrectly predicted, it will affect the subsequent relation extraction task.

## 5  Conclusion

In this paper, we propose CyberRel, a joint entity and relation extraction model for cybersecurity concepts, which can extract both entities and relations in the cybersecurity corpus. Specifically, we use an tagging scheme to convert the joint extraction problem into a multiple sequence labeling problem by generating separate label sequences for different relations containing information about the related entities and the subject and object of that relation. In addition, CyberRel employs BERT model, BiGRU neural network, and attention mechanism to extract the features of sentences and generate label sequences under different relations. In the experimental part, our results on OSINT data demonstrate that CyberRel achieves better results compared to the traditional pipeline approach. To further improve the quality of CKG generation, our future research work will focus on document-level relation extraction and cybersecurity entity disambiguation.

**Acknowledgment.** This research is funded by the National Natural Science Foundation of China (No. 61902265), Sichuan Science and Technology Program (No. 2020YFG0047, No. 2020YFG0374).

# A    Appendix

**Table 3.** The examples of the triples to the given sentences extracted by joint model and pipeline model.

| #Case 1 | |
|---|---|
| Raw text | CVE-2008-3138: The (1) PANA and (2) KISMET dissectors in Wireshark (formerly Ethereal) 0.99.3 through 1.0.0 allow remote attackers to cause a denial of service (application stop) via unknown vectors. |
| Joint model | (PANA/Software, hasVulnerability, CVE-2008-3138/Vulnerability) |
| | (KISMET/Software, hasVulnerability, CVE-2008-3138/Vulnerability) |
| | (Wireshark/Software, hasProduct, PANA/Software) |
| | (Wireshark/Software, hasProduct, KISMET/Software) |
| | (remote attackers/Threat-Actor, uses, unknown vectors/Attack-Pattern) |
| | (denial of service/Campaign, attributedTo, remote attacker/Threat-Actor) |
| | (denial of service/Campaign, attributedTo, unknown vectors/Attack-Pattern) |
| Pipeline model | (PANA/Software, mitigates, Wireshark/Software) |
| | (KISMET/Software, mitigates, Wireshark/Software) |
| | (PANA/Software, mitigates, KISMET/Software) |
| | (KISMET/Software, mitigates, PANA/Software) |
| | (Wireshark/Software, mitigates, PANA/Software) |
| | (Wireshark/Software, mitigates, KISMET/Software) |
| | (remote attackers/Threat-Actor, uses, unknown vectors/Attack-Pattern) |
| | (denial of service/Campaign, attributedTo, remote attackers/Threat-Actor) |
| | (denial of service/Campaign, attributedTo, unknown vectors/Attack-Pattern) |
| #Case2 | |
| Raw text | To remediate CVE-2020-3956 apply the patches listed in the 'Fixed Version' column of the 'Response Matrix' found below. |
| Joint model | (patches/Course-of-Action, mitigates, CVE-2008-3138/Vulnerability) |
| Pipeline model | Only "patches/Course-of-Action" found |

# References

1. Iannacone, M., et al.: Developing an ontology for cyber security knowledge graphs. In: Proceedings of the 10th Annual Cyber and Information Security Research Conference, pp. 1–4 (2015)
2. Syed, Z., Padia, A., Mathews, M.L., Finin, T., Joshi, A., et al.: UCO: a unified cybersecurity ontology. In: Proceedings of the AAAI Workshop on Artificial Intelligence for Cyber Security (2016)
3. Jia, Y., Qi, Y., Shang, H., Jiang, R., Li, A.: A practical approach to constructing a knowledge graph for cybersecurity. Engineering 4(1), 53–60 (2018)
4. Gao, P., et al.: Enabling efficient cyber threat hunting with cyber threat intelligence. arXiv preprint arXiv:2010.13637 (2020)
5. Piplai, A., Mittal, S., Joshi, A., Finin, T., Holt, J., Zak, R.: Creating cybersecurity knowledge graphs from malware after action reports. IEEE Access 8, 211:691–211:703 (2020)
6. Zhao, J., Yan, Q., Liu, X., Li, B., Zuo, G.: Cyber threat intelligence modeling based on heterogeneous graph convolutional network. In: 23rd International Symposium on Research in Attacks, Intrusions and Defenses ({RAID} 2020), pp. 241–256 (2020)
7. Husari, G., Al-Shaer, E., Ahmed, M., Chu, B., Niu, X.: TTPDrill: automatic and accurate extraction of threat actions from unstructured text of CTI sources. In: Proceedings of the 33rd Annual Computer Security Applications Conference, pp. 103–115 (2017)
8. Piplai, A., Mittal, S., Abdelsalam, M., Gupta, M., Joshi, A., Finin, T.: Knowledge enrichment by fusing representations for malware threat intelligence and behavior. In: 2020 IEEE International Conference on Intelligence and Security Informatics (ISI). IEEE, pp. 1–6 (2020)
9. Milajerdi, S.M., Gjomemo, R., Eshete, B., Sekar, R., Venkatakrishnan, V.: Holmes: real-time apt detection through correlation of suspicious information flows. In: IEEE Symposium on Security and Privacy (SP), vol. 2019, pp. 1137–1152. IEEE (2019)
10. Mittal, S., Das, P.K., Mulwad, V., Joshi, A., Finin, T.: CyberTwitter: using Twitter to generate alerts for cybersecurity threats and vulnerabilities. In: 2016 IEEE/ACM International Conference on Advances in Social Networks Analysis and Mining (ASONAM), pp. 860–867. IEEE (2016)
11. Liao, X., Yuan, K., Wang, X., Li, Z., Xing, L., Beyah, R.: Acing the IOC game: toward automatic discovery and analysis of open-source cyber threat intelligence. In: Proceedings of the 2016 ACM SIGSAC Conference on Computer and Communications Security, pp. 755–766 (2016)
12. Zhu, Z., Dumitras, T.: ChainSmith: automatically learning the semantics of malicious campaigns by mining threat intelligence reports. In: IEEE European Symposium on Security and Privacy (EuroS&P), vol. 2018, pp. 458–472. IEEE (2018)
13. Ghazi, Y., Anwar, Z., Mumtaz, R., Saleem, S., Tahir, A.: A supervised machine learning based approach for automatically extracting high-level threat intelligence from unstructured sources. In: 2018 International Conference on Frontiers of Information Technology (FIT), pp. 129–134. IEEE (2018)
14. Zhao, J., Yan, Q., Li, J., Shao, M., He, Z., Li, B.: TIMiner: automatically extracting and analyzing categorized cyber threat intelligence from social data. Comput. Secur. 95, 101867 (2020)

15. Husari, G., Niu, X., Chu, B., Al-Shaer, E.: Using entropy and mutual information to extract threat actions from cyber threat intelligence. In: 2018 IEEE International Conference on Intelligence and Security Informatics (ISI), pp. 1–6. IEEE (2018)
16. Pingle, A., Piplai, A., Mittal, S., Joshi, A., Holt, J., Zak, R.: ReLExt: relation extraction using deep learning approaches for cybersecurity knowledge graph improvement. In: Proceedings of the 2019 IEEE/ACM International Conference on Advances in Social Networks Analysis and Mining, pp. 879–886 (2019)
17. Jones, C.L., Bridges, R.A., Huffer, K.M., Goodall, J.R.: Towards a relation extraction framework for cyber-security concepts. In: Proceedings of the 10th Annual Cyber and Information Security Research Conference, pp. 1–4 (2015)
18. Satyapanich, T., Ferraro, F., Finin, T.: CASIE: extracting cybersecurity event information from text. In: Proceedings of the AAAI Conference on Artificial Intelligence, vol. 34, no. 05, pp. 8749–8757 (2020)
19. Iria, J.: T-rex: a flexible relation extraction framework. In: Proceedings of the 8th Annual Colloquium for the UK Special Interest Group for Computational Linguistics (CLUK 2005), vol. 6, p. 9. Citeseer (2005)
20. McDonald, R., Pereira, F., Kulick, S., Winters, S., Jin, Y., White, P.: Simple algorithms for complex relation extraction with applications to biomedical IE. In: Proceedings of the 43rd Annual Meeting of the Association for Computational Linguistics (ACL 2005), pp. 491–498 (2005)
21. Jiang, J., Zhai, C.: A systematic exploration of the feature space for relation extraction. In: Human Language Technologies: The Conference of the North American Chapter of the Association for Computational Linguistics. Proceedings of the Main Conference, vol. 2007, pp. 113–120 (2007)
22. Culotta, A., Sorensen, J.: Dependency tree kernels for relation extraction. In: Proceedings of the 42nd Annual Meeting of the Association for Computational Linguistics (ACL-04), pp. 423–429 (2004)
23. Zeng, D., Liu, K., Lai, S., Zhou, G., Zhao, J.: Relation classification via convolutional deep neural network. In: Proceedings of COLING 2014, the 25th International Conference on Computational Linguistics: Technical Papers, pp. 2335–2344 (2014)
24. Wei, Z., Su, J., Wang, Y., Tian, Y., Chang, Y.: A novel cascade binary tagging framework for relational triple extraction. In: Proceedings of the 58th Annual Meeting of the Association for Computational Linguistics, pp. 1476–1488 (2020)
25. Miwa, M., Bansal, M.: End-to-end relation extraction using LSTMs on sequences and tree structures. In: Proceedings of the 54th Annual Meeting of the Association for Computational Linguistics (Volume 1: Long Papers), pp. 1105–1116 (2016)
26. Bekoulis, G., Deleu, J., Demeester, T., Develder, C.: Joint entity recognition and relation extraction as a multi-head selection problem. Expert Syst. Appl. **114**, 34–45 (2018)
27. Zheng, S., Wang, F., Bao, H., Hao, Y., Zhou, P., Xu, B.: Joint extraction of entities and relations based on a novel tagging scheme. In: Proceedings of the 55th Annual Meeting of the Association for Computational Linguistics (Volume 1: Long Papers), pp. 1227–1236 (2017)
28. Sun, C., et al.: Extracting entities and relations with joint minimum risk training. In: Proceedings of the 2018 Conference on Empirical Methods in Natural Language Processing, pp. 2256–2265 (2018)
29. Fu, T.-J., Li, P.-H., Ma, W.-Y.: GraphRel: modeling text as relational graphs for joint entity and relation extraction. In: Proceedings of the 57th Annual Meeting of the Association for Computational Linguistics, pp. 1409–1418 (2019)

30. Yuan, Y., Zhou, X., Pan, S., Zhu, Q., Song, Z., Guo, L.: A relation-specific attention network for joint entity and relation extraction. In: International Joint Conference on Artificial Intelligence 2020. Association for the Advancement of Artificial Intelligence (AAAI), pp. 4054–4060 (2020)
31. Dai, D., Xiao, X., Lyu, Y., Dou, S., She, Q., Wang, H.: Joint extraction of entities and overlapping relations using position-attentive sequence labeling. In: Proceedings of the AAAI Conference on Artificial Intelligence, vol. 33, no. 01, pp. 6300–6308 (2019)
32. MITRE: Cvelist project (2021). https://github.com/CVEProject/cvelist
33. CyberMonitor: Apt & cybercriminals campaign collection (2021). https://github.com/CyberMonitor/APT_CyberCriminal_Campagin_Collections
34. Stenetorp, P., Pyysalo, S., Topić, G., Ohta, T., Ananiadou, S., Tsujii, J.: Brat: a web-based tool for NLP-assisted text annotation. In: Proceedings of the Demonstrations at the 13th Conference of the European Chapter of the Association for Computational Linguistics, pp. 102–107 (2012)
35. Devlin, J., Chang, M.-W., Lee, K., Toutanova, K.: Bert: pre-training of deep bidirectional transformers for language understanding. arXiv preprint arXiv:1810.04805 (2018)
36. Youngja, P.: Cybersecurity embeddings. https://ebiquity.umbc.edu/resource/html/id/379/Cybersecurity-embeddings (2018)

# Microblog User Location Inference Based on POI and Query Likelihood Model

Yimin Liu[1,2], Xiangyang Luo[1,2(✉)], and Han Li[3]

[1] State Key Laboratory of Mathematical Engineering and Advanced Computing, Zhengzhou 450001, Henan, China
luoxy_ieu@sina.com
[2] Key Laboratory of Cyberspace Situation Awareness of Henan Province, Zhengzhou 450001, Henan, China
[3] 360 Future Security Labs, Beijing 10089, China

**Abstract.** Location inference of microblog users is of great significance for disaster monitoring, public opinion tracing and tracking, and extensive location-based services. However due to the noisy content of microblog text and the ambiguity of geographic location, it is quite difficult to infer user location based only on user-generated text. This paper proposes a microblog user location inference algorithm based on POI and query likelihood model, named PaQL. First, the POI (Point of Interest) model of each region is constructed based on the electronic map. Then, from the word segmentation results of the user's blog texts, the POIs with stronger location orientation are extracted as user features. Next, the inverse region frequency of POIs is calculated, based on which the correlation between users and the candidate regions is calculated based on the query likelihood model. Finally, the candidate region with the highest correlation is considered as the user's inferred location. The location inference experiment is conducted on the provincial-level data set (3,862k blogs of 154k users) and the city-level data set (3,086k blogs of 103k users) of Sina Weibo platform. The results show that: Compared with three existing typical algorithms, GP-FLIW, GP-LIWTF and WC-EFS, which are only based on user text, the precision of provincial-level inference is improved by 7.80%, 4.99% and 1.41%, respectively, and the city-level inference precision is improved by 10.67%, 8.38% and 3.72%, respectively. Moreover, the proposed algorithm also outperforms the existing methods in terms of recall and $F_1$.

**Keywords:** User location inference · Query likelihood model · Inverse region frequency · Likelihood probability · POI

## 1 Introduction

With the development of mobile Internet technology and the popularity of intelligent mobile terminals, more and more people are accustomed to obtaining information and sharing anecdotes around them anytime and anywhere, which further promotes the application and development of location-based social media platforms. As a high-quality

D. Gao et al. (Eds.): ICICS 2021, LNCS 12918, pp. 464–480, 2021.
https://doi.org/10.1007/978-3-030-86890-1_26

information resource in social media big data [1], user geolocation can connect the virtual network world and the real society. It can help government departments to carry out election prediction [2], event monitoring [3], disease transmission control [4], etc., help businesses to accurately place advertisements [5] and assist security personnel to analyze location-based attack technology [6]. It is of great importance to study the location inference technology of social media users.

However, the real location information contained in social network data is extremely sparse, making location-based social media applications lack sufficient support data [7, 8]. Social media platforms such as Twitter, Facebook, Foursquare, and Sina Weibo provide users with ways to share their locations. Taking Sina Weibo as an example, users need to fill in their locations when registering an account, as shown in Fig. 1(a); users can choose to add geo-tags when publishing blogs to associate them with geographical coordinates, as shown in Fig. 1(b); users can also mention geolocations in their blogs, such as tourist attractions they have visited, restaurants they want to try, as shown in Fig. 1(c). However, due to the increasing awareness of privacy protection, most social media users are reluctant to disclose their geolocation information. For example, Chen et al. [7] found that only 26% of Twitter users filled in city-level locations in their profiles, and only 0.42% of tweets were geo-tagged with GPS. Ryoo et al. [8] reported that only 34% of Twitter users filled in meaningful location information in their profiles, and less than 1% of tweets added GPS location tags.

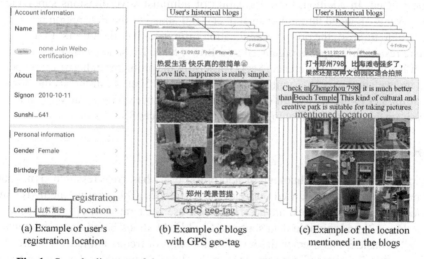

(a) Example of user's registration location

(b) Example of blogs with GPS geo-tag

(c) Example of the location mentioned in the blogs

**Fig. 1.** Sample diagram of three types of user location information in Sina Weibo

In this paper, the goal is to find a way to accurately infer a user's geographic location from the user's published blogs. This avoids the need for user's private information, IP addresses or other sensitive data. However, it is a difficult task to effectively locate the microblog users based solely on the blog text [7]. Because of the randomness of people's writing in the daily use of microblogs, the blog data becomes noisy due to abbreviations, spelling errors, etc. In addition, the location nouns mentioned in the blog also have geographic ambiguities; for example, "Yuntai Mountain" can point to 19 locations in China. These factors make it difficult to infer the user's location accurately.

In response to the above issues, this paper proposes a microblog user location inference algorithm based on POI and query likelihood model, named PaQL. In PaQL, the geographic location inference problem of microblog users is transformed into a document retrieval problem. The candidate region is regarded as a document, and the POI list mentioned by users is regarded as a query. PaQL calculates the correlation between the user and each region based on the query likelihood model, and selects the region with the highest correlation as the user's inferred location. The experimental results show that PaQL is superior to the existing three typical algorithms based only on user-generated text in terms of precision, recall and $F_1$. The main contributions of this paper are as follows:

(1) A microblog user location inference algorithm based on POI and query likelihood model is proposed. The algorithm refers to IDF (inverse document frequency) and query likelihood models in the information retrieval field, and proposes IRF (inverse region frequency) of POI and the query likelihood probability of POI generated by candidate regions. This helps to extract more location-oriented words from user blogs, thus improving the precision of location inference.

(2) A nationwide POI-region map is constructed. Based on the electronic map API, the POI information of all provinces, cities, districts/counties is crawled. After machine processing and manual screening, a relatively accurate and complete POI-region map is constructed. In addition, the information of administrative divisions subdivided to street level is obtained through manual query and added to the POI-region map, which is helpful to improve the precision of location inferences based on users' POI.

The rest of this paper is organized as follows: Sect. 2 discusses the related work of microblog user location inference; Sect. 3 describes the problem studied in this paper and gives an explanation of the symbols used in the paper; Sect. 4 introduces the proposed algorithm and its main steps in detail; Sect. 5 conducts experimental verification and analysis on the effect of microblog user location inference; Sect. 6 summarizes the paper.

## 2   Related Work

The location inference technology of microblog users has great research significance, and many scholars have carried out relevant researches. The existing methods can be divided into three categories: methods based on user-generated text [9, 10], social networks [11, 12] and multiple data sources [13, 14]. Methods based on social networks generally consider that users with social relationships or frequent interactions are closer in geographic location, so as to infer the users' location based on their social friends' location [15]. Methods based on multiple data sources integrate multiple data sources such as geographic information in the text, user's social network information, context information (timestamp, topic, etc.), and user's other social platform information [13].

The emphasis of this paper is to infer the user's location when only the blog texts are available. The method based on user-generated text generally uses language regionalism to analyze the relationship between words in the text and locations [16].

Hecht et al. [17] first used a polynomial naive Bayes model based on word frequency to infer users' location. Ryoo et al. [8] used a generative probability model to construct the geographic distribution of words, and calculated weighted centroids based on user's mentioned words as inferred locations, and used data binning technology to reduce computational costs. Hosseini et al. [18] proposed a spatial phrase detection method. Inspired by TwiNER [19], the geo-tagged tweets of target regions were segmented into phrases based on NER, and calculate the probability that the phrase links to the candidate region based on the Twitter Search API. Zola et al. [1] used Google Trends to query the words in the text, converted the returned multiple cities into two-dimensional polygons, then extracted multiple points and used Gaussian mixture model and DBSCAN for clustering, and took the centroid of the largest class as the user's location.

Han et al. [20] first formally proposed the concept of Location Indicative Words (LIW), that is, the word associated with a specific location, and proposed a LIW selection method based on information gain ratio (IGR) and maximum entropy. Han et al. [21] compared a variety of LIW selection methods on the basis of Ref. [20], and the results showed that the method based on IGR performed best in user location inference. Chi et al. [22] used LIW and textual features (including city/country names, topic tags, and mentioned user names) to construct a feature set, then trained a polynomial naive Bayes classifier to infer users' location. Tian et al. [23] improved Han's method, after selecting LIW based on IGR, clustering words based on the distance of word vectors, then using the package feature selection algorithm to select the best cluster subset as the location indicative words.

Above methods can locate users only based on generating text, and can effectively solve the problem that the methods based on social networks cannot locate isolated users. However, the key of these methods lies in the extracted statistical features and location indicative words, and there may still exist many common words without location directivity in the extracted words, which has a negative influence on the accuracy of user location inference. Therefore, this paper only retains the POI mentioned in the blog as user's features, and each POI has an explicit location. Besides, this paper also adds the national administrative division information subdivided into streets to the POI-region map, so that the reserved POI words have stronger location-oriented, thereby improving the accuracy of user location inference.

## 3   Problem Formulation

To ease the understanding of the proposed algorithm, the symbols and concepts used in this paper are defined, and the problem formulation is also presented here.

*User location.* For each user $u \in U$, the user's inferred location is denoted as $L_u$, which refers to the user's home location, that is, the location where the user lives in.

*User text.* For each user $u \in U$, the user text is denoted as $\mathbf{T}_u = \{t_1, t_2, \ldots\}$, which refers to the user's historical blog set, and $t_i \in \mathbf{T}_u$ represents a single historical blog text.

*Candidate region.* Denote the set of candidate regions as $L$. The candidate regions are determined by the location inference granularity, which can be provinces, cities, or locations with specified granularity, also can be a set of a certain number of regions.

*Point of Interest (POI).* Denote the set of POIs as $\mathbf{P}$, and each POI entry as $poi \in \mathbf{P}$. POI refers to a geographic location entity with known location information, such as restaurants, movie theaters, and scenic spots.

*POI-region map.* POI-region map includes a mapping $M_{poi \to l}$ from each POI to a geographic region, and a mapping $M_{l \to poi}$ from each geographic region to a POI set.

*Inverse Region Frequency (IRF).* Defining the inverse region frequency as the distinguishing ability of a certain $poi \in \mathbf{P}$ to different geographic regions.

*User location inference problem.* For each user $u \in U$, given a set of historical blog $\mathbf{T}_u$ posted by user $u$, and a set of candidate regions $\mathbf{L}$, calculate the probability $P(L|\mathbf{T}_u)$ that $u$ is located in each candidate region $L \in \mathbf{L}$. Select the candidate region with the highest probability as the user's inferred location $L_u$.

The symbols used in this paper are described in Table 1.

**Table 1.** List of notations.

| Notations | Description |
|---|---|
| $L_u$ | The inferred location of user $u$ |
| $\mathbf{T}_u$ | A collection of historical blog posted by user $u$ |
| $L, \mathbf{L}$ | Candidate region, The set of candidate regions |
| $poi, \mathbf{P}$ | Point of interest, The set of point of interest |
| $M_{poi \to l}$ | A mapping from POI to a geographic location |
| $M_{l \to poi}$ | A mapping from geographic location to POIs |
| $W_u$ | The set of words after text segmentation |
| $W_u'$ | The set of words after excluding stop words |
| $\mathbf{P}_u$ | POIs mentioned by user $u$ |

## 4   The Proposed PaQL Algorithm

Aiming at the low accuracy of existing methods based only on user-generated text, this paper proposes a microblog user location inference algorithm based on POI and query likelihood model (named PaQL). This algorithm considers user location inference as a document retrieval problem, candidate regions as pseudo documents, the POI list mentioned by users as queries, and uses a variant of query likelihood model to infer the most likely location of users. The algorithm framework is shown in Fig. 2.

The main steps of the algorithm are as follows.

Step1: Build the POI library of the candidate region. POI information of 19 categories in candidate regions is obtained based on electronic map API, and POI-region map is constructed.

Step2: Data preprocessing. Preprocessing includes text merging, data cleaning, text segmentation, and stop words elimination.

Step3: Obtain the mentioned POI vector of users. According to the POI library, the BOW (Bag of Word) model is used to generate the mentioned POI vector from the word list after the user's word segmentation.

Step4: Train the classifier. The process of training classifier is to calculate the probabilities needed in the query likelihood model, including the inverse region frequency of POI and the conditional probability that POI is mentioned in each candidate region.

Step5: Infer user location. The trained classifier is used to infer the user location; that is, the correlation between the target user's mentioned POI vector and each candidate region is calculated based on the query likelihood model, and the region with the highest correlation is regarded as the inferred location.

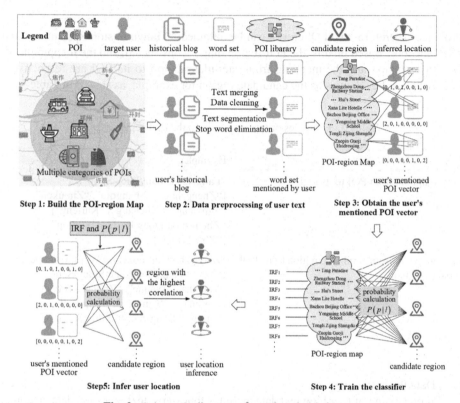

**Fig. 2.** Schematic diagram of user location inference

Step 1–3 is the POI vector construction module. Step 4–5 is the user location inference module. These two modules are described in detail in Sect. 4.1 and Sect. 4.2, respectively.

## 4.1 The Construction of Mentioned POI Vector Based on BOW

(1) POI Library Construction for Candidate Regions

Firstly, POI information in all candidate regions is crawled based on electronic map API, including 19 categories. The POI category and the content contained in the POI information are shown in Table 2.

More POI entries can be obtained from the POI name and POI address. For example, "Manyu Fusion Restaurant (Xintian 360)" is a POI located in Zhengzhou, Henan

**Table 2.** POI category and POI information content.

|  | Content |
|---|---|
| Category | Food, hotel, shopping, service life, beauty, tourist sites, leisure, exercise, education and training, cultural media, health, auto service, transportation, financial, real estate, company, government, gateway, natural features |
| Information | Name, address, province, city, district/county, street, longitude, latitude, type |

Province. Therefore, two POI entries can be obtained: "Manyu Fusion Restaurant" and "Xintian 360". Based on this, the POI-region map is constructed, and the mapping from POI to its region and the mapping from candidate region to its POI are respectively established. Take all cities as the candidate region for example, as shown in Table 3.

**Table 3.** Examples of two mappings in the POI-region map.

|  | Examples |
|---|---|
| The mapping from POI to its region | "Fangzhongshan Spicy Soup": ["Zhengzhou", "Shangqiu", "Zhoukou", "Xinxiang", "Xuchang", "Kaifeng"] "Zhengzhou Dong Railway Station": ["Zhengzhou"] |
| The mapping from candidate region to its POI contained in it | "Zhengzhou": ["Foxconn Wojin Commercial Plaza", "Fangzhongshan Spicy Soup ", "Yanhuang Avenue", "Zhongji Yuman Building", "Zhengzhou Dong Railway Station", "Songshan Building",...] |

(2) Data Preprocessing

Preprocessing of user blog text data includes four steps: text merging, data cleaning, word segmentation and stop word elimination, as shown in Fig. 3.

- Text merge. In order to facilitate the subsequent data processing, the user's historical texts $T_u = \{t_1, t_2, \ldots\}$ are first concatenated before and after, and merged into one text.
- Data cleaning. Remove the emoticons, URLs, English words, etc., in the merged text, and only retain Chinese characters and numbers.
- Text segmentation. The existing Chinese word segmentation tools are used for text segmentation, and all the POIs in the POI library are added to the word library of the word segmenter. In the segmentation results, only the words with part of speech of n./ns./nt./nz./f./an./x./s. are retained, and words with other parts of speech are deleted to obtain the word set, denoted as $W_u$. An example is shown in Table 4.

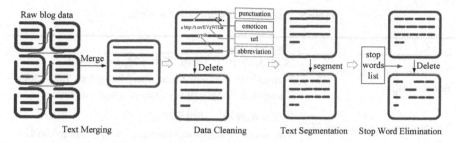

**Fig. 3.** Schematic diagram of the preprocessing process of blog text data

**Table 4.** Examples of text word segmentation.

| | The text content |
|---|---|
| The original text | 1) 终于喝到家门口的方中山胡辣汤啦. 比在郑州东站的好喝多了, 果然还是这种街头小巷里的地道。2) 今天一整天都在望着天空发呆, 思考理想能否顺利实现(ˇ_ˇ) <br><br> Translation: 1) I finally had the spicy soup of Fangzhongshan near my home. It is much better than that of Zhengzhou Dong Railway Station. Sure enough, it is more delicious in streets and lanes. 2) Today, I spent the whole day staring at the sky, thinking about whether my ideal can come true (ˇ_ˇ) |
| Data cleaning | 终于喝到家门口的方中山胡辣汤啦比在郑州东站的好喝多了果然还是这种街头小巷里的地道今天一整天都在望着天空发呆思考理想能否顺利实现 |
| Text segmentation | 终于(d.)/喝(vg.)/到(v.)/家门口(s.)/的(uj.)/方中山胡辣汤(x.)/啦(y.)/比(p.)/在(p.)/郑州东站(x.)/的(uj.)/好喝(v.)/多(m.)/了(ul.)/果然(d.)/还是(c.)/这种(r.)/街头(s.)/小巷(n.)/里(f.)/的(uj.)/地道(n.)/今天(t.)/一整天(m.)/都(d.)/在(p.)/望(v.)/着(uz.)/天空(n.)/发呆(v.)/思考(v.)/理想(n.)/能否(v.)/顺利实现(i.) |
| Screening of part of speech | 家门口/方中山胡辣汤/郑州东站/街头/小巷/地道/天空/理想 <br><br> Translation: near my home/ Fangzhongshan Spicy Soup / Zhengzhou Dong Railway Station/ street/ alley/ authentic/ sky/ idea |

- Stop word elimination. The stop words are eliminated based on the stop word library, and the word set of the target user's blogs $W_u'$ is obtained.

(3) Getting the Mentioned POI Vector

All the words in $W_u'$ are matched with the POI library to extract the mentioned POI list. Then the mentioned POI vector is obtained based on the BOW model. Taking the POI library in Table 3 and the results in Table 4 as examples, an example of mentioned POI vector is shown in Table 5.

**Table 5.** Example of mentioned POI vector.

| | The text content |
|---|---|
| The word set with stop words removed $W'_u$ | 家门口, 方中山胡辣汤, 郑州东站, 小巷, 地道<br><br>Translation: near my home/ Fangzhongshan Spicy Soup / Zhengzhou Dong Railway Station/ alley/ authentic |
| The target user' mentioned POI list | 方中山胡辣汤, 郑州东站<br><br>Translation: Fangzhongshan Spicy Soup, Zhengzhou Dong Railway Station |
| The target user' mentioned POI vector $\mathbf{P}_u$ | [0, 1, 0, 0, 1, 0, …] |

## 4.2 User Location Inference Based on Query Likelihood Model

After obtaining the POI library of each candidate region and the POI vector mentioned by the user, this paper trains the classifier and infers the user location based on the query likelihood model [24], as shown in Fig. 4. The classifier training is mainly to calculate the inverse region frequency of POI and the query likelihood probability between the user's blogs and the candidate region.

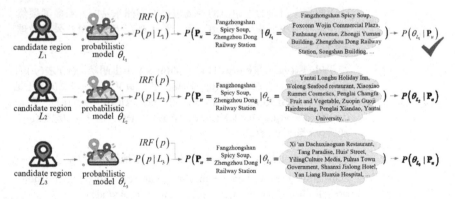

**Fig. 4.** Schematic diagram of user location inference

(1)  Inverse Region Frequency

Different POIs have different capabilities to reflect geolocation. For example, the scenic spot "Tang Paradise" is only associated with a place in "Yanta District, Xi'an City, Shaanxi Province", while the "Manyu fusion restaurant" is associated with 8 places in Henan Province, Beijing and Shaanxi Province. In order to reflect the ability of different POIs to distinguish different regions, this paper proposes IRF (Inverse Region

Frequency) by referring to IDF (Inverse Document Frequency) [25]. The IRF of each POI entry $poi \in \mathbf{P}$ is calculated by the following formula.

$$IRF(poi) = log \frac{|R|}{|M_{poi \rightarrow l}(poi)|} \tag{1}$$

where, $R$ is all candidate regions, $M_{poi \rightarrow l}$ is the mapping from POI to the candidate region, and $M_{poi \rightarrow l}(poi)$ is the regions containing the POI.

(2) Query Likelihood Probability

The set of all POIs is denoted as $\mathbf{P} = \{poi_1, poi_2, \ldots, poi_{|\mathbf{P}|}\}$. The user's historical blog and candidate region can be represented by POI sequence. $\mathbf{P}_u = p_1 p_2 p_3 \cdots p_m, p_i \in \mathbf{P}$ represents the POI sequence mentioned in the user's blogs; $L = l_1 l_2 l_3 \ldots l_n, l_j \in \mathbf{P}$ represents the POI sequence contained in the candidate region $L$. Model each candidate region $L$ on the POI library $\mathbf{P}$, to get the probability model $\theta_L$. Then the probability of generating the user's mentioned POI vector under $\theta_L$ is $P(\mathbf{P}_u | \theta_L)$, which is called query likelihood probability.

The multinomial model is used to calculate the query likelihood probability $P(\mathbf{P}_u | \theta_L)$, abbreviated as $P(\mathbf{P}_u | L)$:

$$P(\mathbf{P}_u | L) = \prod_{i=1}^{m} P(p_i | L) = \prod_{p \in \mathbf{P}_u} P(p | L)^{TF(p, \mathbf{P}_u)} \tag{2}$$

where, $TF(p, \mathbf{P}_u)$ represents the word frequency of $p \in \mathbf{P}_u$ in the user's blogs.

Use the maximum likelihood estimation method to estimate $P(p|L)$ in Formula (2):

$$P(p|L) = P_{ML}(p|L) = \frac{TF(p, L)}{\sum_{poi \in L} TF(poi, L)} \tag{3}$$

where, $TF(poi, L)$ is the number of times that $poi$ appears in the candidate region.

Due to the data sparsity, a certain POI in the mentioned POI list is probably not in the POI library of the candidate region. In this case, $TF(p, L) = 0$; that is, the user cannot be in the region, which is clearly wrong. In order to solve the zero probability problem, the Linear Interpolation Smoothing method is used to improve the model:

$$P(p|L) = (1 - \lambda) \frac{TF(p, L)}{\sum_{poi \in L} TF(poi, L)} + \lambda \frac{TF(p, \mathbf{L})}{\sum_{poi \in \mathbf{L}} TF(poi, \mathbf{L})}, \lambda \in [0, 1] \tag{4}$$

Refer to the TF-IDF model, we can obtain Formula (5) by combining the query likelihood probability and the IRF:

$$P(p|L) = (1 - \lambda) \frac{TF(p, L)}{\sum_{poi \in L} TF(poi, L)} IRF(p) + \lambda \frac{TF(p, \mathbf{L})}{\sum_{poi \in \mathbf{L}} TF(poi, \mathbf{L})} IRF(p), \lambda \in [0, 1] \tag{5}$$

$P(p|L)$ is expressed as the general form shown below:

$$P(p|L) = \begin{cases} P_{Seen}(p|L), \text{ if } p \in L \\ \alpha_L P(p|L), \text{ if } p \notin L \end{cases} \tag{6}$$

(3)  User Location Inference

To infer the user's location, we need to calculate the posterior probability $P(L|\mathbf{P}_u)$, that is, given the user's blogs, to analyze which candidate region the user is most likely to be in. Because of $P(L|\mathbf{P}_u) \propto P(\mathbf{P}_u|L)P(L)$, and $P(L)$ is usually assumed uniform distribution, so the candidate regions can be sorted according to $P(\mathbf{P}_u|L)$ to infer the user's location.

Taking the logarithm of both sides of Eq. (2), we can get:

$$logP(\mathbf{P}_u|L) = \sum_{p \in \mathbf{P}_u} TF(p, \mathbf{P}_u)logP(p|L) \tag{7}$$

Substitute Formula (6) into Formula (7):

$$logP(\mathbf{P}_u|L) = \sum_{p \in L} TF(p, \mathbf{P}_u)log\frac{P_{Seen}(p|L)}{\alpha_L P(p|\mathbf{L})} + |\mathbf{P}_u|\alpha_L + \sum_p TF(p, \mathbf{P}_u)logP(p|\mathbf{L}) \tag{8}$$

Since the last term in the above equation has nothing to do with candidate region L and has no influence on the sorting result, Formula (8) can be equivalent to:

$$logP(\mathbf{P}_u|L) = \sum_{p \in L} TF(p, \mathbf{P}_u)log\frac{P_{Seen}(p|L)}{\alpha_L P(p|\mathbf{L})} + |\mathbf{P}_u|\alpha_L \tag{9}$$

Finally, the candidate region with the highest probability obtained by Formula (10) is taken as the user's inferred location:

$$L_u = argmax_{L \in \mathbf{L}} logP(\mathbf{P}_u|L) \tag{10}$$

## 5   Experimental Results and Analysis

In order to verify the feasibility and effectiveness of the proposed algorithm, this section carried out provincial-level and city-level location inference experiments based on Sina Weibo, and compared the results with three existing typical algorithms: GP-FLIW [21] proposed by Han et al., GP-LIWTF [22] proposed by Chi et al., and WC-EFS [23] proposed by Tian et al.

### 5.1   Experimental Data Set

(1)  POI-Region Map

This paper crawled nationwide POI information based on Baidu Map API, with a total of 741,240 results. In order to expand the POI library, more POI name-region mapping relationships are extracted from the original POI information using the method described in Sect. 4.1.

In addition, in order to supplement more and high-quality POI location information, the nationwide administrative division information is obtained by manual query, including provinces, cities, districts/counties, towns, streets, and some scenic spots. It is refined and converted into POI-region information, totaling 49,222 items. After combining the data from the above two sources and remove duplicates, the statistics are shown in Table 6.

Considering the update timeliness of administrative division information on the Internet and the Baidu Map POI information, this section has manually adjusted the alignment of both data.

Table 6.  Statistics of POI-region database.

| Items | Data |
|---|---|
| The number of POI categories | 19 |
| The number of provinces covered | 34 |
| The number of cities covered | 392 |
| The number of Districts/counties covered | 3,066 |
| The number of POI-region information | 1,203,543 |

(2)   Data of Users and Blogs

This section uses Sina Weibo as an example to conduct experiments. The Sina Weibo data comes from Ref. [23], including two parts: a provincial-level data set and a city-level data set. The statistical results are shown in Table 7. The registered location in the user profile is used as the ground truth of experiments.

Table 7.  Statistics of the experimental data set.

| Items | Provincial-level | City-level |
|---|---|---|
| The number of provinces and cities covered | 34 | 197 |
| The number of user data | 154,478 | 102,735 |
| The number of blog data | 3,862,117 | 3,085,972 |

The proposed algorithm only needs the POI information when training the classifier, but does not need users' blog data. However, the three comparison algorithms all need to use blog data to train the model, so this section randomly selects 80% of all users in each location as the training set, and the remaining 20% as the test set.

## 5.2   Experimental Setup

(1)   Parameter Setting

The experimental parameter settings are shown in Table 8, and the parameters are selected through experiments with the optimum results.

**Table 8.** Experimental parameter settings.

| Parameter | Description | Value |
|---|---|---|
| $\lambda_1$ | Smoothing parameters of provincial location inference | 0.35 |
| $\lambda_2$ | Smoothing parameters of city-level location inference | 0.15 |
| $I$ | The threshold of information gain ratio in comparison algorithms | 0.2 |

(2)  Evaluation Metrics

Considering that when users post fewer blogs or the blogs do not contain location information, the location inference algorithm cannot locate them [15]. So we use precision, recall and $F_1$, which are commonly used in previous location inference researches [7, 26], to evaluate the algorithm's effectiveness. Given a target user set $\mathbf{U}$, where each user is denoted as $u$, the user's real location is $L_s$, and the inferred location is $L_s^*$. The definition and calculation formula of the evaluation metrics are as follows.

- Precision: among all users who got the inferred location, the proportion of the correct inferred results.

$$Precision = \frac{\left|\{u \in \mathbf{U} : L_s^* = L_s\}\right|}{\left|\{u \in \mathbf{U} : L_s^* \neq NULL\}\right|} \tag{11}$$

- Recall: among all users, the proportion of the correct inferred results.

$$Recall = \frac{\left|\{u \in \mathbf{U} : L_s^* = L_s\}\right|}{|\mathbf{U}|} \tag{12}$$

- $F_1$: the harmonic average of precision and recall.

$$F_1 = \frac{2 \times Precision \times Recall}{Precision + Recall} \tag{13}$$

### 5.3  Comparative Experimental Results of Location Inference

In this section, the user's provincial-level and city-level location inference experiments are conducted on Sina Weibo. The experimental results of the proposed algorithm are compared with three existing typical algorithms, GP-FLIW [21], GP-LIWTF [22] and WC-EFS [23] on the same data set. The results are shown in Fig. 5 and Table 9.

The bolded part in the table is the highest value in the comparison results. It can be seen from Table 9 and Fig. 5 that the algorithm proposed outperforms the three existing baseline algorithms on both the provincial-level and city-level data sets. The precision of provincial-level location inference is 7.80%, 4.99%, and 1.41% higher than the baseline algorithms, respectively, and the precision of city-level is 10.67%, 8.38%, and 3.72% higher than the three algorithms, respectively.

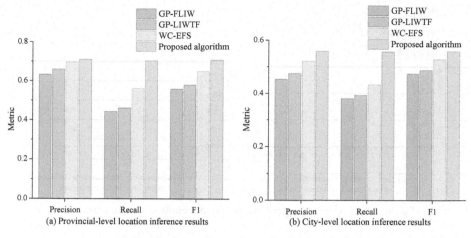

**Fig. 5.** Comparison of experimental results

**Table 9.** Comparison results between the proposed algorithm and the baseline algorithms.

| Algorithm | Provincial-level location inference | | | City-level location inference | | |
|---|---|---|---|---|---|---|
| | Precision | Recall | $F_1$ | Precision | Recall | $F_1$ |
| GP-FLIW [22] | 0.6315 | 0.4429 | 0.5580 | 0.4523 | 0.3798 | 0.4724 |
| GP-LIWTF [23] | 0.6596 | 0.4607 | 0.5778 | 0.4752 | 0.3926 | 0.4872 |
| WC-EFS [24] | 0.6954 | 0.5607 | 0.64815 | 0.5218 | 0.4324 | 0.5282 |
| Proposed algorithm | **0.7095** | **0.7027** | **0.7061** | **0.5590** | **0.5573** | **0.5582** |

The inference precision improvement is mainly because the features selected by the proposed algorithm are more location-oriented than baseline algorithms. The baseline algorithms use information gain ratio, word clustering and other methods to extract location indicative words from all contents posted by users; thus, there are still common words without location-oriented in the word screening results, which will affect the accuracy of location inference. However, the proposed PaQL algorithm only retains the POIs mentioned by the user as features. Each POI crawled by electric map has a corresponding geographic location, and the national administrative division information subdivided into streets is also added to the POI library, so that the user features obtained by PaQL have a stronger location-oriented.

## 5.4 The Influence of the Smoothing Parameter on Location Inference

This section analyzes the influence of the smoothing parameter on the location inference results of the proposed algorithm. Different smoothing parameters are used to carry out experiments at provincial and city levels, and the results are shown in Fig. 6.

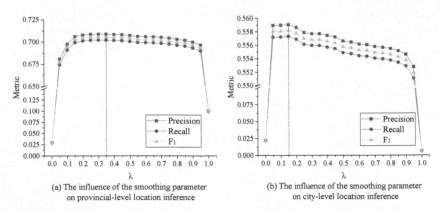

(a) The influence of the smoothing parameter
on provincial-level location inference

(b) The influence of the smoothing parameter
on city-level location inference

**Fig. 6.** Effect of smoothing parameters on the results of location inference

Figure 6 shows the results of the provincial-level and city-level location inference when the smoothing parameter changes from 0 to 1 in a span of 0.05. It can be seen from the figure that the smoothing process will greatly improve the precision of user location inference. According to the experimental results, it is more reasonable to set the smoothing parameter between 0.15 and 0.35.

## 6    Conclusion

Aiming to improve the precision of user location inference based only on user-generated text, this paper proposes a microblog user location inference algorithm based on POI and query likelihood model. This algorithm only retains the POI mentioned in the blog as user features, then calculates the correlation between the candidate region and the user based on the query likelihood model, which improves the accuracy of user location inference. This paper builds a POI-region map for each province and city in China based on electronic map API and manual processing, which improves the precision of location inference, and avoids the computational and time overhead of training a language model for each candidate region. Experiments conducted on the provincial-level and city-level data set of Sina Weibo show that: compared with three existing typical algorithms, the proposed algorithm has better performance in the precision, recall and $F_1$ of location inference. However, this algorithm cannot infer the location of users without blogs. In future work, a hybrid method combining user-generated text and social relationship networks will be studied to infer microblog users' location.

**Acknowledgments.** This work was supported by the National Natural Science Foundation of China (No. U1804263, U1736214), the Zhongyuan Science and Technology Innovation Leading Talent Project (No. 214200510019).

# References

1. Zola, P., Ragno, C., Cortez, P.: A google trends spatial clustering approach for a worldwide twitter user geolocation. Information Processing & Management **57**(6), 102312 (2020).
2. Tumasjan, A., Sprenger, T.O., Sandner, P.G., Welpe I.M.: Predicting elections with twitter: what 140 characters reveal about political sentiment. In: 4th International Conference on Weblogs and Social Media, pp. 178–185. AAAI, Washington DC, USA (2010).
3. Carmela, C., Agostino, F., Clara, P.: Bursty event detection in twitter streams. ACM Trans. Knowl. Discov. Data **13**(4), 1–28 (2019)
4. Lan, L., Malbasa, V., Vucetic, S.: Spatial scan for disease mapping on a mobile population. In: 28th AAAI conference on Artificial Intelligence, pp. 431–437. AAAI, Québec, Canada (2014).
5. Heba, A., John, K., Gireeja, R., Horvitz, E.: To buy or not to buy: computing value of spatiotemporal information. ACM Transactions on Spatial Algorithms and Systems **5**(40), 1–25 (2019)
6. Halimi, A., Ayday, E.: Profile Matching Across Online Social Networks. In: Meng, W., Gollmann, D., Jensen, C.D., Zhou, J. (eds.) ICICS 2020. LNCS, vol. 12282, pp. 54–70. Springer, Cham (2020). https://doi.org/10.1007/978-3-030-61078-4_4
7. Chen, Z.Y., James, C., Kyumin, L.: You are where you tweet: a content-based approach to geolocating twitter users. In: 19th ACM International Conference on Information and Knowledge Management, pp. 759–768. ACM, Toronto, Canada (2010).
8. Ryoo, K.M., Moon, S.S.: Inferring twitter user locations with 10km accuracy. In: 23rd International Conference on World Wide Web, pp. 643–648. ACM, Seoul, Korea (2014).
9. Rahimi, A., Vu, D., Cohn, T., Baldwin, T.: Exploiting text and network context for geolocation of social media users. In: 2015 Conference of the North American Chapter of the Association for Computational Linguistics: Human Language Technologies, pp. 1362–1367. NAACL, Denver, USA (2015).
10. Huang, B., Carley, K.M.: A hierarchical location prediction neural network for twitter user geolocation. In: 2019 Conference on Empirical Methods in Natural Language Processing and the 9th International Joint Conference on Natural Language Processing, pp. 4731–4741. ACL, Hong Kong, China (2019).
11. Longbo, K., Zhi, L., Yan, H.: Spot: Locating social media users based on social network context. Proceedings of the VLDB Endowment **7**(13), 1681–1684 (2014)
12. Tian, H.C., Zhang, M., Luo, X.Y., Liu, F.L., Qiao, Y.Q.: Twitter user location inference based on representation learning and label propagation. In: The Web Conference 2020, pp. 2648–2654. ACM, Taipei, Taiwan (2020).
13. Miura, Y., Taniguchi, M., Taniguchi, T., Ohkuma, T.: Unifying text, metadata, and user network representations with a neural network for geolocation prediction. In: 55th Annual Meeting of the Association for Computational Linguistics, pp. 1260–1272. ACL, Vancouver, Canada (2017).
14. Rahimi, A., Cohn, T., Baldwin, T.: Semi-supervised user geolocation via graph convolutional networks. In: 56th Annual Meeting of the Association for Computational Linguistics, pp. 2009–2019. ACL, Melbourne, Australia (2018).
15. Ryan, C., David, J., David, A.: Geotagging one hundred million twitter accounts with total variation minimization. In: 2014 IEEE International Conference on Big Data, pp. 393–401. IEEE, Washington DC, USA (2014).
16. Xin, Z., Han, J., Sun, A.: A survey of location prediction on twitter. IEEE Trans. Knowl. Data Eng. **30**(9), 1–20 (2017)
17. Hecht, B., Hong, L.C., Suh, B.W., Chi, E.H.: Tweets from Justin Bieber's heart: the dynamics of the location field in user profiles. In: 2011 CHI Conference on Human Factors in Computing Systems, pp. 237–246. ACM, Vancouver, Canada (2011).

18. Hosseini, S., Unankard, S., Zhou, X., Sadiq, S.: Location Oriented Phrase Detection in Microblogs. In: Bhowmick, S.S., Dyreson, C.E., Jensen, C.S., Lee, M.L., Muliantara, A., Thalheim, B. (eds.) DASFAA 2014. LNCS, vol. 8421, pp. 495–509. Springer, Cham (2014). https://doi.org/10.1007/978-3-319-05810-8_33

19. Li, C., Weng, J.S., He, Q., Yao, Y.X., Datta, A., Sun, A., et al.: Twiner: Named entity recognition in targeted twitter stream. In: 35th International ACM SIGIR Conference on Research and Development in Information Retrieval, pp. 721–730. ACM, Portland, USA (2012).

20. Han, B., Cook, P., Baldwin, T.: Geolocation prediction in social media data by finding location indicative words. In: 24th International Conference on Computational Linguistics, pp. 1045–1062. ACM, Mumbai, India (2012).

21. Han, B., Cook, P., Baldwin, T.: Text-based twitter user geolocation prediction. Journal of Artificial Intelligence Research **49**(1), 451–500 (2014)

22. Chi, L.H., Lim, K.H., Alam, N., Butler, C.: Geolocation prediction in twitter using location indicative words and textual features. In: 2nd Workshop on Noisy User-Generated Text, pp. 227–234. COLING, Osaka, Japan (2016).

23. Tian, H.C.: Research on social network user location prediction technology, University of Information Engineering (2019).

24. Jay, M., Croft, Bruce, W.: A language modeling approach to information retrieval. In: 21st Annual International ACM SIGIR Conference on Research and Development in Information Retrieval, pp. 275–281. ACM, Melbourne, Australia (1998).

25. Salton, G., Buckley, C.: Term-weighting approaches in automatic text retrieval. Inf. Process. Manage. **24**(5), 513–523 (1988)

26. Huang, B., Carley, K.M.: On Predicting Geolocation of Tweets Using Convolutional Neural Networks. In: Lee, D., Lin, Y.-R., Osgood, N., Thomson, R. (eds.) SBP-BRiMS 2017. LNCS, vol. 10354, pp. 281–291. Springer, Cham (2017). https://doi.org/10.1007/978-3-319-60240-0_34

# Author Index

Printed in the United States
by Baker & Taylor Publisher Services